Linear Model Theory

Dale L. Zimmerman

Linear Model Theory

Exercises and Solutions

Springer

Dale L. Zimmerman
Department of Statistics and Actuarial
Science
University of Iowa
Iowa City, IA, USA

ISBN 978-3-030-52076-2 ISBN 978-3-030-52074-8 (eBook)
https://doi.org/10.1007/978-3-030-52074-8

Mathematics Subject Classification: 62J05, 62J10, 62F03, 62F10, 62F25

This Springer imprint is published by the registered company Springer Nature Switzerland AG.
The registered company address is: Gewerbestrasse 11, 6330 Cham, Switzerland

In loving gratitude to my late father, Dean Zimmerman, and my mother, Wendy Zimmerman

Contents

1 A Brief Introduction

This book contains 296 solved exercises on the theory of linear models. The exercises are taken from the author's graduate-level textbook, *Linear Model Theory: With Examples and Exercises*, which was published by Springer in 2020. The exercises themselves have been restated, when necessary and feasible, to make them as comprehensible as possible independently of the textbook, but the solutions refer liberally to theorems and other results therein. They are arranged in chapters, the numbers and titles of which are identical to those of the chapters in the textbook that have exercises.

Some of the exercises and solutions are short, while others have multiple parts and are quite lengthy. Some are proofs of theorems presented but not proved in the aforementioned textbook, but most are specializations of said theorems and other general results to specific linear models. In this respect they are quite similar to the textbook's examples. A few of the exercises require the use of a computer, but none involve the analysis of actual data.

The author is not aware of any other published set of solved exercises for a graduate-level course on the theory of linear models. It is hoped that students and instructors alike, possibly even those not using *Linear Model Theory: With Examples and Exercises* for their course, will find these exercises and solutions useful.

D. L. Zimmerman, *Linear Model Theory*,
https://doi.org/10.1007/978-3-030-52074-8_1

Selected Matrix Algebra Topics and Results 2

This chapter presents exercises on selected matrix algebra topics and results and provides solutions to those exercises.

▶ **Exercise 1** Let V_n represent a vector space, and let S_n represent a subspace of V_n. Prove that the orthogonal complement of S_n (relative to V_n) is a subspace of V_n.

Solution By definition, the orthogonal complement of S_n (relative to V_n) is $S_n^{\perp} = \{\mathbf{v} \in V_n : \mathbf{w}^T\mathbf{v} = 0 \text{ for all } \mathbf{w} \in S_n\}$. Obviously $\mathbf{0} \in S_n^{\perp}$, so $S_n^{\perp}$ is nonempty. For any $\mathbf{v}_1, \mathbf{v}_2 \in S_n^{\perp}$, $\mathbf{w}^T(\mathbf{v}_1 + \mathbf{v}_2) = \mathbf{w}^T\mathbf{v}_1 + \mathbf{w}^T\mathbf{v}_2 = 0$ for all $\mathbf{w} \in S_n$, so $\mathbf{v}_1 + \mathbf{v}_2 \in S_n^{\perp}$. Also, for any $\mathbf{v} \in S_n^{\perp}$, $\mathbf{w}^T(c\mathbf{v}) = c\mathbf{w}^T\mathbf{v} = 0$ for any $c \in \mathbb{R}$ and all $\mathbf{w} \in S_n$, so $c\mathbf{v} \in S_n^{\perp}$. Therefore, $S_n^{\perp}$ is a vector space, hence a subspace of V_n.

▶ **Exercise 2** Let $\mathbf{A}$ represent an $n \times m$ matrix. Prove that $\mathcal{N}(\mathbf{A})$ is a subspace of $\mathbb{R}^m$.

Solution By definition, $\mathcal{N}(\mathbf{A}) = \{\mathbf{v} : \mathbf{A}\mathbf{v} = \mathbf{0}\} \subseteq \mathbb{R}^m$. Obviously $\mathbf{0} \in \mathcal{N}(\mathbf{A})$, so $\mathcal{N}(\mathbf{A})$ is nonempty. For any $\mathbf{v}_1, \mathbf{v}_2 \in \mathcal{N}(\mathbf{A})$, $\mathbf{A}(\mathbf{v}_1 + \mathbf{v}_2) = \mathbf{A}\mathbf{v}_1 + \mathbf{A}\mathbf{v}_2 = \mathbf{0}$, so $\mathbf{v}_1 + \mathbf{v}_2 \in \mathcal{N}(\mathbf{A})$. Also, for any $\mathbf{v} \in \mathcal{N}(\mathbf{A})$, $\mathbf{A}(c\mathbf{v}) = c\mathbf{A}\mathbf{v} = \mathbf{0}$ for any $c \in \mathbb{R}$, so $c\mathbf{v} \in \mathcal{N}(\mathbf{A})$. Therefore, $\mathcal{N}(\mathbf{A})$ is a vector space, hence a subspace of $\mathbb{R}^m$.

▶ **Exercise 3** Prove Theorem 2.8.2: For any matrix $\mathbf{A}$, $\text{rank}(\mathbf{A}^T) = \text{rank}(\mathbf{A})$.

Solution $\text{rank}(\mathbf{A}^T) = \dim\{\mathcal{R}(\mathbf{A}^T)\} = \dim\{\mathcal{C}(\mathbf{A})\} = \text{rank}(\mathbf{A})$.

▶ **Exercise 4** Prove Theorem 2.8.3: For any matrix $\mathbf{A}$ and any nonzero scalar c, $\text{rank}(c\mathbf{A}) = \text{rank}(\mathbf{A})$.

Solution $\text{rank}(c\mathbf{A}) = \dim\{\mathcal{R}(c\mathbf{A})\} = \dim\{\mathcal{R}(\mathbf{A})\} = \text{rank}(\mathbf{A})$.

D. L. Zimmerman, *Linear Model Theory*,
https://doi.org/10.1007/978-3-030-52074-8_2

▶ **Exercise 5** Prove Theorem 2.15.1: If $\mathbf{A}$ is a nonnegative definite matrix, then all of its diagonal elements are nonnegative; if $\mathbf{A}$ is positive definite, then all of its diagonal elements are positive.

Solution Denote the diagonal elements of the $n \times n$ matrix $\mathbf{A}$ by $a_{11}, \ldots, a_{nn}$. If $\mathbf{A}$ is nonnegative definite, then $\mathbf{x}^T\mathbf{A}\mathbf{x} \geq 0$ for all $\mathbf{x}$. Choosing $\mathbf{x} = \mathbf{u}_i$ $(i = 1, \ldots, n)$ yields $a_{ii} \geq 0$ $(i = 1, \ldots, n)$. Alternatively, if $\mathbf{A}$ is positive definite, then $\mathbf{x}^T\mathbf{A}\mathbf{x} > 0$ for all $\mathbf{x} \neq \mathbf{0}$. Again choosing $\mathbf{x} = \mathbf{u}_i$ $(i = 1, \ldots, n)$ yields $a_{ii} > 0$ $(i = 1, \ldots, n)$.

▶ **Exercise 6** Prove Theorem 2.15.2: Let c represent a positive scalar. If $\mathbf{A}$ is a positive definite matrix, then so is $c\mathbf{A}$; if $\mathbf{A}$ is positive semidefinite, then so is $c\mathbf{A}$.

Solution If $\mathbf{A}$ is positive definite, then $\mathbf{x}^T(c\mathbf{A})\mathbf{x} = c\mathbf{x}^T\mathbf{A}\mathbf{x} > 0$ for all $\mathbf{x} \neq \mathbf{0}$ [and trivially $\mathbf{0}^T(c\mathbf{A})\mathbf{0} = 0$], implying that $c\mathbf{A}$ is positive definite. Alternatively, if $\mathbf{A}$ is positive semidefinite, then $\mathbf{x}^T(c\mathbf{A})\mathbf{x} = c\mathbf{x}^T\mathbf{A}\mathbf{x} \geq 0$ for all $\mathbf{x}$ and $\mathbf{x}^T(c\mathbf{A})\mathbf{x} = c\mathbf{x}^T\mathbf{A}\mathbf{x} = 0$ for some $\mathbf{x} \neq \mathbf{0}$, implying that $c\mathbf{A}$ is positive semidefinite.

▶ **Exercise 7** Prove Theorem 2.15.3: Let $\mathbf{A}$ and $\mathbf{B}$ represent $n \times n$ matrices. If $\mathbf{A}$ and $\mathbf{B}$ are both nonnegative definite, then so is $\mathbf{A}+\mathbf{B}$; if either one is positive definite and the other is nonnegative definite, then $\mathbf{A}+\mathbf{B}$ is positive definite.

Solution If $\mathbf{A}$ and $\mathbf{B}$ are both nonnegative definite, then for any $\mathbf{x} \in \mathbb{R}^n$, $\mathbf{x}^T(\mathbf{A}+\mathbf{B})\mathbf{x} = \mathbf{x}^T\mathbf{A}\mathbf{x}+\mathbf{x}^T\mathbf{B}\mathbf{x} \geq 0$, implying that $\mathbf{A}+\mathbf{B}$ is nonnegative definite. Alternatively, suppose without loss of generality that $\mathbf{A}$ is positive definite and $\mathbf{B}$ is nonnegative definite. Then for any nonnull $\mathbf{x} \in \mathbb{R}^n$, $\mathbf{x}^T(\mathbf{A}+\mathbf{B})\mathbf{x} = \mathbf{x}^T\mathbf{A}\mathbf{x} + \mathbf{x}^T\mathbf{B}\mathbf{x} \geq \mathbf{x}^T\mathbf{A}\mathbf{x} > 0$, and $\mathbf{0}^T(\mathbf{A}+\mathbf{B})\mathbf{0} = 0$. Thus, $\mathbf{0}$ is the only value of $\mathbf{x}$ for which $\mathbf{x}^T(\mathbf{A}+\mathbf{B})\mathbf{x}$ equals 0, implying that $\mathbf{A}+\mathbf{B}$ is positive definite.

▶ **Exercise 8** Prove Theorem 2.15.4: The block diagonal matrix $\begin{pmatrix} \mathbf{A} & \mathbf{0} \\ \mathbf{0} & \mathbf{B} \end{pmatrix}$ is nonnegative definite if and only if both $\mathbf{A}$ and $\mathbf{B}$ are nonnegative definite and is positive definite if and only if both $\mathbf{A}$ and $\mathbf{B}$ are positive definite.

Solution First observe that for any vector $\mathbf{x}$ with dimension equal to the number of rows (or columns) of $\begin{pmatrix} \mathbf{A} & \mathbf{0} \\ \mathbf{0} & \mathbf{B} \end{pmatrix}$, and which is partitioned as $\begin{pmatrix} \mathbf{x}_1 \\ \mathbf{x}_2 \end{pmatrix}$ where the dimension of $\mathbf{x}_1$ is equal to the number of rows (or columns) of $\mathbf{A}$, we have

$$\mathbf{x}^T \begin{pmatrix} \mathbf{A} & \mathbf{0} \\ \mathbf{0} & \mathbf{B} \end{pmatrix} \mathbf{x} = \left(\mathbf{x}_1^T, \mathbf{x}_2^T\right) \begin{pmatrix} \mathbf{A} & \mathbf{0} \\ \mathbf{0} & \mathbf{B} \end{pmatrix} \begin{pmatrix} \mathbf{x}_1 \\ \mathbf{x}_2 \end{pmatrix} = \mathbf{x}_1^T\mathbf{A}\mathbf{x}_1 + \mathbf{x}_2^T\mathbf{B}\mathbf{x}_2.$$

Now, if $\mathbf{A}$ and $\mathbf{B}$ are nonnegative definite, then $\mathbf{x}_1^T\mathbf{A}\mathbf{x}_1 \geq 0$ for all $\mathbf{x}_1$ and $\mathbf{x}_2^T\mathbf{B}\mathbf{x}_2 \geq 0$ for all $\mathbf{x}_2$, implying (by the result displayed above) that $\mathbf{x}^T \begin{pmatrix} \mathbf{A} & \mathbf{0} \\ \mathbf{0} & \mathbf{B} \end{pmatrix} \mathbf{x} \geq 0$ for all

$\mathbf{x}$, i.e., that $\begin{pmatrix} \mathbf{A} & \mathbf{0} \\ \mathbf{0} & \mathbf{B} \end{pmatrix}$ is nonnegative definite. Conversely, if $\begin{pmatrix} \mathbf{A} & \mathbf{0} \\ \mathbf{0} & \mathbf{B} \end{pmatrix}$ is nonnegative definite, then

$\left(\mathbf{x}_1^T, \mathbf{0}^T\right) \begin{pmatrix} \mathbf{A} & \mathbf{0} \\ \mathbf{0} & \mathbf{B} \end{pmatrix} \begin{pmatrix} \mathbf{x}_1 \\ \mathbf{0} \end{pmatrix} \geq 0$ for all $\mathbf{x}_1$ and $\left(\mathbf{0}^T, \mathbf{x}_2^T\right) \begin{pmatrix} \mathbf{A} & \mathbf{0} \\ \mathbf{0} & \mathbf{B} \end{pmatrix} \begin{pmatrix} \mathbf{0} \\ \mathbf{x}_2 \end{pmatrix} \geq 0$ for all $\mathbf{x}_2$, i.e., $\mathbf{x}_1^T \mathbf{A} \mathbf{x}_1 \geq 0$ for all $\mathbf{x}_1$ and $\mathbf{x}_2^T \mathbf{B} \mathbf{x}_2 \geq 0$ for all $\mathbf{x}_2$. Thus $\mathbf{A}$ and $\mathbf{B}$ are nonnegative definite.

On the other hand, if $\mathbf{A}$ and $\mathbf{B}$ are positive definite, then $\mathbf{x}_1^T \mathbf{A} \mathbf{x}_1 > 0$ for all $\mathbf{x}_1 \neq \mathbf{0}$ and $\mathbf{x}_2^T \mathbf{B} \mathbf{x}_2 > 0$ for all $\mathbf{x}_2 \neq \mathbf{0}$, implying (by the result displayed above and by the trivial results $\mathbf{0}^T \mathbf{A} \mathbf{0} = \mathbf{0}^T \mathbf{B} \mathbf{0} = 0$) that $\mathbf{x}^T \begin{pmatrix} \mathbf{A} & \mathbf{0} \\ \mathbf{0} & \mathbf{B} \end{pmatrix} \mathbf{x} > 0$ for all $\mathbf{x} \neq \mathbf{0}$, i.e., that $\begin{pmatrix} \mathbf{A} & \mathbf{0} \\ \mathbf{0} & \mathbf{B} \end{pmatrix}$ is positive definite. Conversely, if $\begin{pmatrix} \mathbf{A} & \mathbf{0} \\ \mathbf{0} & \mathbf{B} \end{pmatrix}$ is positive definite, then

$\left(\mathbf{x}_1^T, \mathbf{0}^T\right) \begin{pmatrix} \mathbf{A} & \mathbf{0} \\ \mathbf{0} & \mathbf{B} \end{pmatrix} \begin{pmatrix} \mathbf{x}_1 \\ \mathbf{0} \end{pmatrix} > 0$ for all $\mathbf{x}_1 \neq \mathbf{0}$ and $\left(\mathbf{0}^T, \mathbf{x}_2^T\right) \begin{pmatrix} \mathbf{A} & \mathbf{0} \\ \mathbf{0} & \mathbf{B} \end{pmatrix} \begin{pmatrix} \mathbf{0} \\ \mathbf{x}_2 \end{pmatrix} > 0$ for all $\mathbf{x}_2 \neq \mathbf{0}$, i.e., $\mathbf{x}_1^T \mathbf{A} \mathbf{x}_1 > 0$ for all $\mathbf{x}_1 \neq \mathbf{0}$ and $\mathbf{x}_2^T \mathbf{B} \mathbf{x}_2 > 0$ for all $\mathbf{x}_2 \neq \mathbf{0}$. Thus $\mathbf{A}$ and $\mathbf{B}$ are positive definite.

▶ **Exercise 9** Prove Corollary 2.15.12.1: If $\mathbf{A}$ is an $n \times n$ nonnull nonnegative definite matrix and $\mathbf{C}$ is any matrix having n columns, then $\mathbf{C}\mathbf{A}\mathbf{C}^T$ is nonnegative definite.

Solution By Theorem 2.15.12, a matrix $\mathbf{B}$ exists such that $\mathbf{A} = \mathbf{B}\mathbf{B}^T$. Thus $\mathbf{C}\mathbf{A}\mathbf{C}^T = \mathbf{C}\mathbf{B}\mathbf{B}^T\mathbf{C}^T = (\mathbf{B}^T\mathbf{C}^T)^T(\mathbf{B}^T\mathbf{C}^T)$, which is nonnegative definite by Theorem 2.15.9.

▶ **Exercise 10** Prove Theorem 2.15.13: Let $\mathbf{A}$ and $\mathbf{B}$ represent nonnegative definite matrices. Then $\mathrm{tr}(\mathbf{AB}) \geq 0$, with equality if and only if $\mathbf{AB} = \mathbf{0}$.

Solution By Theorem 2.15.12, symmetric matrices $\mathbf{A}^{\frac{1}{2}}$ and $\mathbf{B}^{\frac{1}{2}}$ exist such that $\mathbf{A}^{\frac{1}{2}}\mathbf{A}^{\frac{1}{2}} = \mathbf{A}$ and $\mathbf{B}^{\frac{1}{2}}\mathbf{B}^{\frac{1}{2}} = \mathbf{B}$. Then by Theorem 2.10.3,

$$\begin{aligned} \mathrm{tr}(\mathbf{AB}) &= \mathrm{tr}(\mathbf{A}^{\frac{1}{2}}\mathbf{A}^{\frac{1}{2}}\mathbf{B}^{\frac{1}{2}}\mathbf{B}^{\frac{1}{2}}) \\ &= \mathrm{tr}(\mathbf{B}^{\frac{1}{2}}\mathbf{A}^{\frac{1}{2}}\mathbf{A}^{\frac{1}{2}}\mathbf{B}^{\frac{1}{2}}) \\ &= \mathrm{tr}[(\mathbf{A}^{\frac{1}{2}}\mathbf{B}^{\frac{1}{2}})^T(\mathbf{A}^{\frac{1}{2}}\mathbf{B}^{\frac{1}{2}})]. \end{aligned}$$

Therefore, by Theorem 2.10.4, $\mathrm{tr}(\mathbf{AB}) \geq 0$, with equality if and only if $\mathbf{A}^{\frac{1}{2}}\mathbf{B}^{\frac{1}{2}} = \mathbf{0}$. If $\mathbf{A}^{\frac{1}{2}}\mathbf{B}^{\frac{1}{2}} = \mathbf{0}$, then pre-multiplication by $\mathbf{A}^{\frac{1}{2}}$ and post-multiplication by $\mathbf{B}^{\frac{1}{2}}$, respectively, of this matrix equation yield $\mathbf{AB} = \mathbf{0}$. Conversely, if $\mathbf{AB} = \mathbf{0}$, then immediately $\mathrm{tr}(\mathbf{AB}) = 0$.

▶ **Exercise 11** Prove Theorem 2.18.1: If $\mathbf{A}_1, \mathbf{A}_2, \ldots, \mathbf{A}_k$ are nonsingular, then so is $\oplus_{i=1}^k \mathbf{A}_i$, and its inverse is $\oplus_{i=1}^k \mathbf{A}_i^{-1}$.

Solution

$$\left(\oplus_{i=1}^k \mathbf{A}_i\right)\left(\oplus_{i=1}^k \mathbf{A}_i^{-1}\right) = \begin{pmatrix} \mathbf{A}_1 & \mathbf{0} & \cdots & \mathbf{0} \\ \mathbf{0} & \mathbf{A}_2 & \cdots & \mathbf{0} \\ \vdots & \vdots & \ddots & \vdots \\ \mathbf{0} & \mathbf{0} & \cdots & \mathbf{A}_k \end{pmatrix} \begin{pmatrix} \mathbf{A}_1^{-1} & \mathbf{0} & \cdots & \mathbf{0} \\ \mathbf{0} & \mathbf{A}_2^{-1} & \cdots & \mathbf{0} \\ \vdots & \vdots & \ddots & \vdots \\ \mathbf{0} & \mathbf{0} & \cdots & \mathbf{A}_k^{-1} \end{pmatrix}$$

$$= \begin{pmatrix} \mathbf{I} & \mathbf{0} & \cdots & \mathbf{0} \\ \mathbf{0} & \mathbf{I} & \cdots & \mathbf{0} \\ \vdots & \vdots & \ddots & \vdots \\ \mathbf{0} & \mathbf{0} & \cdots & \mathbf{I} \end{pmatrix} = \mathbf{I}.$$

▶ **Exercise 12** Prove Theorem 2.19.1: Let $\mathbf{x}$ represent an n-vector of variables.

(a) If $f(\mathbf{x}) = \mathbf{a}^T\mathbf{x}$ where $\mathbf{a}$ is an n-vector of constants, then $\nabla f = \frac{\partial(\mathbf{a}^T\mathbf{x})}{\partial \mathbf{x}} = \mathbf{a}$.
(b) If $f(\mathbf{x}) = \mathbf{x}^T\mathbf{A}\mathbf{x}$ where $\mathbf{A}$ is an $n \times n$ symmetric matrix of constants, then $\nabla f = \frac{\partial(\mathbf{x}^T\mathbf{A}\mathbf{x})}{\partial \mathbf{x}} = 2\mathbf{A}\mathbf{x}$.

Solution

(a)

$$\nabla f = \frac{\partial(\mathbf{a}^T\mathbf{x})}{\partial \mathbf{x}} = \begin{pmatrix} \frac{\partial(\sum_{i=1}^n a_i x_i)}{\partial x_1} \\ \vdots \\ \frac{\partial(\sum_{i=1}^n a_i x_i)}{\partial x_n} \end{pmatrix} = \begin{pmatrix} a_1 \\ \vdots \\ a_n \end{pmatrix} = \mathbf{a}.$$

(b)

$$\nabla f = \frac{\partial(\mathbf{x}^T\mathbf{A}\mathbf{x})}{\partial \mathbf{x}} = \begin{pmatrix} \frac{\partial(\sum_{i=1}^n \sum_{j=1}^n a_{ij} x_i x_j)}{\partial x_1} \\ \vdots \\ \frac{\partial(\sum_{i=1}^n \sum_{j=1}^n a_{ij} x_i x_j)}{\partial x_n} \end{pmatrix} = \begin{pmatrix} 2a_{11}x_1 + \sum_{j=2}^n a_{1j}x_j + \sum_{i=2}^n a_{i1}x_i \\ \vdots \\ 2a_{nn}x_n + \sum_{j=2}^n a_{nj}x_j + \sum_{i=2}^n a_{in}x_i \end{pmatrix}$$

$$= \begin{pmatrix} 2\sum_{j=1}^n a_{1j}x_j \\ \vdots \\ 2\sum_{j=1}^n a_{nj}x_j \end{pmatrix} = 2\mathbf{A}\mathbf{x}.$$

Generalized Inverses and Solutions to Systems of Linear Equations 3

This chapter presents exercises on generalized inverses and solutions to systems of linear equations, and provides solutions to those exercises.

▶ **Exercise 1** Find a generalized inverse of each of the following matrices.

(a) $c\mathbf{A}$, where $c \neq 0$ and $\mathbf{A}$ is an arbitrary matrix.
(b) $\mathbf{ab}^T$, where $\mathbf{a}$ is a nonnull m-vector and $\mathbf{b}$ is a nonnull n-vector.
(c) $\mathbf{K}$, where $\mathbf{K}$ is nonnull and idempotent.
(d) $\mathbf{D} = \text{diag}(d_1, \ldots, d_n)$ where $d_1, \ldots, d_n$ are arbitrary real numbers.
(e) $\mathbf{C}$, where $\mathbf{C}$ is a square matrix whose elements are equal to zero except possibly on the cross-diagonal stretching from the lower left element to the upper right element.
(f) $\mathbf{PAQ}$, where $\mathbf{A}$ is any nonnull $m \times n$ matrix and $\mathbf{P}$ and $\mathbf{Q}$ are $m \times m$ and $n \times n$ orthogonal matrices.
(g) $\mathbf{P}$, where $\mathbf{P}$ is an $m \times n$ matrix such that $\mathbf{P}^T\mathbf{P} = \mathbf{I}_n$.
(h) $\mathbf{J}_{m \times n}$, and characterize the collection of *all* generalized inverses of $\mathbf{J}_{m \times n}$.
(i) $a\mathbf{I}_n + b\mathbf{J}_n$, where a and b are nonzero scalars. (Hint: Consider the nonsingular and singular cases separately, and use Corollary 2.9.7.1 for the former.)
(j) $\mathbf{A} + \mathbf{bd}^T$, where $\mathbf{A}$ is nonsingular and $\mathbf{A} + \mathbf{bd}^T$ is singular. (Hint: The singularity of $\mathbf{A} + \mathbf{bd}^T$ implies that $\mathbf{d}^T\mathbf{A}^{-1}\mathbf{b} = -1$ by Corollary 2.9.7.1.)
(k) $\mathbf{B}$, where $\mathbf{B} = \oplus_{i=1}^k \mathbf{B}_i$ and $\mathbf{B}_1, \ldots, \mathbf{B}_k$ are arbitrary matrices.
(l) $\mathbf{A} \otimes \mathbf{B}$, where $\mathbf{A}$ and $\mathbf{B}$ are arbitrary matrices.

Solution

(a) Let $\mathbf{G}$ represent a generalized inverse of $\mathbf{A}$. Then $(1/c)\mathbf{G}$ is a generalized inverse of $c\mathbf{A}$ because $c\mathbf{A}[(1/c)\mathbf{G}]c\mathbf{A} = c\mathbf{AGA} = c\mathbf{A}$.

D. L. Zimmerman, *Linear Model Theory*,
https://doi.org/10.1007/978-3-030-52074-8_3

(b) $\frac{1}{\|\mathbf{a}\|^2\|\mathbf{b}\|^2}\mathbf{b}\mathbf{a}^T$ is a generalized inverse of $\mathbf{a}\mathbf{b}^T$ because

$$\mathbf{a}\mathbf{b}^T\left(\frac{1}{\|\mathbf{a}\|^2\|\mathbf{b}\|^2}\mathbf{b}\mathbf{a}^T\right)\mathbf{a}\mathbf{b}^T = \frac{1}{\|\mathbf{a}\|^2\|\mathbf{b}\|^2}\mathbf{a}\|\mathbf{b}\|^2\|\mathbf{a}\|^2\mathbf{b}^T = \mathbf{a}\mathbf{b}^T.$$

(c) $\mathbf{K}$ is a generalized inverse of itself because $\mathbf{KKK} = (\mathbf{KK})\mathbf{K} = \mathbf{KK} = \mathbf{K}$. Another generalized inverse of $\mathbf{K}$ is $\mathbf{I}$ because $\mathbf{KIK} = \mathbf{KK} = \mathbf{K}$.

(d) Let $\mathbf{C} = \text{diag}(c_1, \ldots, c_n)$ where $c_k = \begin{cases} 1/d_k, & \text{if } d_k \neq 0 \\ 0, & \text{if } d_k = 0 \end{cases}$ for $k = 1, \ldots, n$. Then $\mathbf{C}$ is a generalized inverse of $\mathbf{D}$ because

$$\begin{aligned}\mathbf{DCD} &= \text{diag}(d_1, \ldots, d_n)\text{diag}(c_1, \ldots, c_n)\text{diag}(d_1, \ldots, d_n) \\ &= \text{diag}(d_1c_1d_1, \ldots, d_nc_nd_n) = \mathbf{D}.\end{aligned}$$

(e) Let

$$\mathbf{C} = \begin{pmatrix} & & c_1 \\ & \iddots & \\ c_n & & \end{pmatrix} \quad \text{and} \quad \mathbf{D} = \begin{pmatrix} & & d_n \\ & \iddots & \\ d_1 & & \end{pmatrix},$$

where $d_k = \begin{cases} 1/c_k, & \text{if } c_k \neq 0 \\ 0, & \text{if } c_k = 0 \end{cases}$ for $k = 1, \ldots, n$. Then $\mathbf{D}$ is a generalized inverse of $\mathbf{C}$ because

$$\begin{aligned}\mathbf{CDC} &= \begin{pmatrix} & & c_1 \\ & \iddots & \\ c_n & & \end{pmatrix}\begin{pmatrix} & & d_n \\ & \iddots & \\ d_1 & & \end{pmatrix}\begin{pmatrix} & & c_1 \\ & \iddots & \\ c_n & & \end{pmatrix} \\ &= \begin{pmatrix} c_1d_1 & & \\ & \ddots & \\ & & c_nd_n \end{pmatrix}\begin{pmatrix} & & c_1 \\ & \iddots & \\ c_n & & \end{pmatrix} \\ &= \begin{pmatrix} & & c_1d_1c_1 \\ & \iddots & \\ c_nd_nc_n & & \end{pmatrix} \\ &= \begin{pmatrix} & & c_1 \\ & \iddots & \\ c_n & & \end{pmatrix}.\end{aligned}$$

(f) Let $\mathbf{G}$ represent a generalized inverse of $\mathbf{A}$. Then $\mathbf{Q}^T\mathbf{G}\mathbf{P}^T$ is a generalized inverse of $\mathbf{PAQ}$ because $\mathbf{PAQ}(\mathbf{Q}^T\mathbf{G}\mathbf{P}^T)\mathbf{PAQ} = \mathbf{PAIGIAQ} = \mathbf{PAGAQ} = \mathbf{PAQ}$.
(g) $\mathbf{P}^T$ is a generalized inverse of $\mathbf{P}$ because $\mathbf{P}\mathbf{P}^T\mathbf{P} = \mathbf{P}(\mathbf{P}^T\mathbf{P}) = \mathbf{P}\mathbf{I}_n = \mathbf{P}$.
(h) $(1/mn)\mathbf{J}_{n\times m}$ is a generalized inverse of $\mathbf{J}_{m\times n}$ because

$$\mathbf{J}_{m\times n}[(1/mn)\mathbf{J}_{n\times m}]\mathbf{J}_{m\times n} = (1/mn)\mathbf{1}_m\mathbf{1}_n^T\mathbf{1}_n\mathbf{1}_m^T\mathbf{1}_m\mathbf{1}_n^T = (1/mn)\mathbf{1}_m(nm)\mathbf{1}_n^T = \mathbf{J}_{m\times n}.$$

To characterize the entire collection of generalized inverses of $\mathbf{J}_{m\times n}$, let $\mathbf{G} = (g_{ij})$ represent an arbitrary member of the collection. Then $\mathbf{J}_{m\times n}\mathbf{G}\mathbf{J}_{m\times n} = \mathbf{J}_{m\times n}$, i.e., $\mathbf{1}_m\mathbf{1}_n^T\mathbf{G}\mathbf{1}_m\mathbf{1}_n^T = \mathbf{1}_m\mathbf{1}_n^T$, i.e., $\left(\sum_{i=1}^n\sum_{j=1}^m g_{ij}\right)\mathbf{1}_m\mathbf{1}_n^T = \mathbf{1}_m\mathbf{1}_n^T$, implying that $\sum_{i=1}^n\sum_{j=1}^m g_{ij} = 1$. Thus the collection of all generalized inverses of $\mathbf{J}_{m\times n}$ consists of all $n\times m$ matrices whose elements sum to one.
(i) $a\mathbf{I} + b\mathbf{J}_n = a\mathbf{I} + (b\mathbf{1}_n)\mathbf{1}_n^T$, which by Corollary 2.9.7.1 is nonsingular if $\mathbf{1}_n^T(a\mathbf{I})^{-1}(b\mathbf{1}_n) \neq -1$, i.e., if $b \neq -a/n$. In that case the corollary yields

$$\begin{aligned}(a\mathbf{I}+b\mathbf{J}_n)^{-1} &= a^{-1}\mathbf{I} - [1+\mathbf{1}_n^T(a^{-1}\mathbf{I})(b\mathbf{1}_n)]^{-1}(a^{-1}\mathbf{I})(b\mathbf{1}_n)\mathbf{1}_n^T(a^{-1}\mathbf{I})\\ &= \frac{1}{a}\mathbf{I} - \frac{1}{1+bn/a}\cdot\frac{b}{a^2}\mathbf{J}_n\\ &= \frac{1}{a}\mathbf{I} - \frac{b}{a(a+bn)}\mathbf{J}_n.\end{aligned}$$

In the other case, i.e., if $b = -a/n$, then $a\mathbf{I} + b\mathbf{J}_n = a(\mathbf{I} - \frac{1}{n}\mathbf{J})$, and because $\mathbf{I}-\frac{1}{n}\mathbf{J}_n$ is idempotent, $\frac{1}{a}(\mathbf{I}-\frac{1}{n}\mathbf{J})$ is a generalized inverse by the solutions to parts (a) and (c) of this exercise. Another generalized inverse in this case is $(1/a)\mathbf{I}_n$.
(j) $\mathbf{A}^{-1}$ is a generalized inverse of $\mathbf{A} + \mathbf{b}\mathbf{d}^T$ because

$$\begin{aligned}(\mathbf{A}+\mathbf{b}\mathbf{d}^T)\mathbf{A}^{-1}(\mathbf{A}+\mathbf{b}\mathbf{d}^T) &= \mathbf{A}\mathbf{A}^{-1}\mathbf{A} + \mathbf{b}\mathbf{d}^T\mathbf{A}^{-1}\mathbf{A} + \mathbf{A}\mathbf{A}^{-1}\mathbf{b}\mathbf{d}^T + \mathbf{b}\mathbf{d}^T\mathbf{A}^{-1}\mathbf{b}\mathbf{d}^T\\ &= \mathbf{A} + \mathbf{b}\mathbf{d}^T + \mathbf{b}(1+\mathbf{d}^T\mathbf{A}^{-1}\mathbf{b})\mathbf{d}^T\\ &= \mathbf{A} + \mathbf{b}\mathbf{d}^T.\end{aligned}$$

(k) Let $\mathbf{B}_i^-$ represent a generalized inverse of $\mathbf{B}_i$ for $i = 1, \ldots, k$. Then $\oplus_{i=1}^k\mathbf{B}_i^-$ is a generalized inverse of $\mathbf{B}$ because $\left(\oplus_{i=1}^k\mathbf{B}_i\right)\left(\oplus_{i=1}^k\mathbf{B}_i^-\right)\left(\oplus_{i=1}^k\mathbf{B}_i\right) = \oplus_{i=1}^k\mathbf{B}_i\mathbf{B}_i^-\mathbf{B}_i = \oplus_{i=1}^k\mathbf{B}_i = \mathbf{B}$.
(l) Let $\mathbf{A}^-$ and $\mathbf{B}^-$ represent generalized inverses of $\mathbf{A}$ and $\mathbf{B}$, respectively. Then $\mathbf{A}^-\otimes\mathbf{B}^-$ is a generalized inverse of $\mathbf{A}\otimes\mathbf{B}$ because, using Theorem 2.17.5,

$$(\mathbf{A}\otimes\mathbf{B})(\mathbf{A}^-\otimes\mathbf{B}^-)(\mathbf{A}\otimes\mathbf{B}) = (\mathbf{A}\mathbf{A}^-\mathbf{A})\otimes(\mathbf{B}\mathbf{B}^-\mathbf{B}) = \mathbf{A}\otimes\mathbf{B}.$$

▶ **Exercise 2** Prove Theorem 3.1.1: Let $\mathbf{A}$ represent an arbitrary matrix of rank r, and let $\mathbf{P}$ and $\mathbf{Q}$ represent nonsingular matrices such that

$$\mathbf{PAQ} = \begin{pmatrix} \mathbf{I}_r & \mathbf{0} \\ \mathbf{0} & \mathbf{0} \end{pmatrix}.$$

(Such matrices $\mathbf{P}$ and $\mathbf{Q}$ exist by Theorem 2.8.10.) Then the matrix

$$\mathbf{G} = \mathbf{Q} \begin{pmatrix} \mathbf{I}_r & \mathbf{F} \\ \mathbf{H} & \mathbf{B} \end{pmatrix} \mathbf{P},$$

where $\mathbf{F}$, $\mathbf{H}$, and $\mathbf{B}$ are arbitrary matrices of appropriate dimensions, is a generalized inverse of $\mathbf{A}$.

Solution Because $\mathbf{A} = \mathbf{P}^{-1} \begin{pmatrix} \mathbf{I}_r & \mathbf{0} \\ \mathbf{0} & \mathbf{0} \end{pmatrix} \mathbf{Q}^{-1}$, we obtain

$$\begin{aligned} \mathbf{AQ} \begin{pmatrix} \mathbf{I}_r & \mathbf{F} \\ \mathbf{H} & \mathbf{B} \end{pmatrix} \mathbf{PA} &= \mathbf{P}^{-1} \begin{pmatrix} \mathbf{I}_r & \mathbf{0} \\ \mathbf{0} & \mathbf{0} \end{pmatrix} \mathbf{Q}^{-1}\mathbf{Q} \begin{pmatrix} \mathbf{I}_r & \mathbf{F} \\ \mathbf{H} & \mathbf{B} \end{pmatrix} \mathbf{PP}^{-1} \begin{pmatrix} \mathbf{I}_r & \mathbf{0} \\ \mathbf{0} & \mathbf{0} \end{pmatrix} \mathbf{Q}^{-1} \\ &= \mathbf{P}^{-1} \begin{pmatrix} \mathbf{I}_r & \mathbf{F} \\ \mathbf{0} & \mathbf{0} \end{pmatrix} \begin{pmatrix} \mathbf{I}_r & \mathbf{0} \\ \mathbf{0} & \mathbf{0} \end{pmatrix} \mathbf{Q}^{-1} \\ &= \mathbf{P}^{-1} \begin{pmatrix} \mathbf{I}_r & \mathbf{0} \\ \mathbf{0} & \mathbf{0} \end{pmatrix} \mathbf{Q}^{-1} \\ &= \mathbf{A}, \end{aligned}$$

which establishes that $\mathbf{Q} \begin{pmatrix} \mathbf{I}_r & \mathbf{F} \\ \mathbf{H} & \mathbf{B} \end{pmatrix} \mathbf{P}$ is a generalized inverse of $\mathbf{A}$.

▶ **Exercise 3** Let $\mathbf{A}$ represent any square matrix.

(a) Prove that $\mathbf{A}$ has a nonsingular generalized inverse. (Hint: Use Theorem 3.1.1, which was stated in the previous exercise.)
(b) Let $\mathbf{G}$ represent a nonsingular generalized inverse of $\mathbf{A}$. Show, by giving a counterexample, that $\mathbf{G}^{-1}$ need not equal $\mathbf{A}$.

Solution

(a) Let n be the number of rows (and columns) of $\mathbf{A}$, and let $\mathbf{P}$ and $\mathbf{Q}$ represent nonsingular matrices such that

$$\mathbf{PAQ} = \begin{pmatrix} \mathbf{I}_r & \mathbf{0} \\ \mathbf{0} & \mathbf{0} \end{pmatrix},$$

where $r = \text{rank}(\mathbf{A})$. By Theorem 3.1.1,

$$\mathbf{Q}\begin{pmatrix} \mathbf{I}_r & \mathbf{F} \\ \mathbf{H} & \mathbf{B} \end{pmatrix}\mathbf{P},$$

where $\mathbf{F}$, $\mathbf{H}$, and $\mathbf{B}$ are arbitrary matrices of appropriate dimensions, is a generalized inverse of $\mathbf{A}$. Take $\mathbf{B} = \mathbf{I}_{n-r}$, $\mathbf{H} = \mathbf{0}_{(n-r)\times r}$, and $\mathbf{F} = \mathbf{0}_{r\times(n-r)}$. Then this generalized inverse is $\mathbf{QP}$, which is nonsingular by Theorem 2.8.9.

(b) Take $\mathbf{A} = \mathbf{J}_2$, for which $(1/2)\mathbf{I}_2$ is a nonsingular generalized inverse [because $\mathbf{J}_2(1/2)\mathbf{I}_2\mathbf{J}_2 = (1/2)\cdot 2\mathbf{J}_2 = \mathbf{J}_2$]. But the inverse of this nonsingular generalized inverse is $2\mathbf{I}_2$, which does not equal $\mathbf{J}_2$.

▶ **Exercise 4** Show, by giving a counterexample, that if the system of equations $\mathbf{Ax} = \mathbf{b}$ is not consistent, then $\mathbf{AA}^{-}\mathbf{b}$ need not equal $\mathbf{b}$.

Solution Consider the system of equations

$$\begin{pmatrix} 1 & 1 \\ 1 & 1 \end{pmatrix}\begin{pmatrix} x_1 \\ x_2 \end{pmatrix} = \begin{pmatrix} 0 \\ 1 \end{pmatrix},$$

which clearly are not consistent. Here $\mathbf{A} = \mathbf{J}_2$, for which one generalized inverse is $\begin{pmatrix} 1 & 0 \\ 0 & 0 \end{pmatrix}$. Then

$$\mathbf{AA}^{-}\mathbf{b} = \begin{pmatrix} 1 & 1 \\ 1 & 1 \end{pmatrix}\begin{pmatrix} 1 & 0 \\ 0 & 0 \end{pmatrix}\begin{pmatrix} 0 \\ 1 \end{pmatrix} = \begin{pmatrix} 0 \\ 0 \end{pmatrix} \neq \mathbf{b}.$$

▶ **Exercise 5** Let $\mathbf{X}$ represent a matrix such that $\mathbf{X}^T\mathbf{X}$ is nonsingular, and let k represent an arbitrary real number. Consider generalized inverses of the matrix $\mathbf{I} + k\mathbf{X}(\mathbf{X}^T\mathbf{X})^{-1}\mathbf{X}^T$. For each $k \in \mathbb{R}$, determine S_k, where

$$S_k = \{c \in \mathbb{R} : \mathbf{I} + c\mathbf{X}(\mathbf{X}^T\mathbf{X})^{-1}\mathbf{X}^T \text{ is a generalized inverse of } \mathbf{I} + k\mathbf{X}(\mathbf{X}^T\mathbf{X})^{-1}\mathbf{X}^T\}.$$

Solution

$$\begin{aligned}
&[\mathbf{I} + k\mathbf{X}(\mathbf{X}^T\mathbf{X})^{-1}\mathbf{X}^T][\mathbf{I} + c\mathbf{X}(\mathbf{X}^T\mathbf{X})^{-1}\mathbf{X}^T][\mathbf{I} + k\mathbf{X}(\mathbf{X}^T\mathbf{X})^{-1}\mathbf{X}^T] \\
&= [\mathbf{I} + (k + c + kc)\mathbf{X}(\mathbf{X}^T\mathbf{X})^{-1}\mathbf{X}^T][\mathbf{I} + k\mathbf{X}(\mathbf{X}^T\mathbf{X})^{-1}\mathbf{X}^T] \\
&= \mathbf{I} + (2k + k^2 + k^2c + 2kc + c)\mathbf{X}(\mathbf{X}^T\mathbf{X})^{-1}\mathbf{X}^T.
\end{aligned}$$

Thus $\mathbf{I} + c\mathbf{X}(\mathbf{X}^T\mathbf{X})^{-1}\mathbf{X}^T$ is a generalized inverse of $\mathbf{I} + k\mathbf{X}(\mathbf{X}^T\mathbf{X})^{-1}\mathbf{X}^T$ if and only if $2k+k^2+c(k+1)^2 = k$, i.e., if and only if $c(k+1)^2 = -k(k+1)$. If $k = -1$, then c is arbitrary; if $k \neq 1$, then $c = -k/(k+1)$. So $S_k = \mathbb{R}$ if $k = -1$; $S_k = \{-k/(k+1)\}$ otherwise.

▶ **Exercise 6** Let $\mathbf{A}$ represent any $m \times n$ matrix, and let $\mathbf{B}$ represent any $n \times q$ matrix. Prove that for any choices of generalized inverses $\mathbf{A}^-$ and $\mathbf{B}^-$, $\mathbf{B}^-\mathbf{A}^-$ is a generalized inverse of $\mathbf{AB}$ if and only if $\mathbf{A}^-\mathbf{ABB}^-$ is idempotent.

Solution If $\mathbf{B}^-\mathbf{A}^-$ is a generalized inverse of $\mathbf{AB}$, then $\mathbf{ABB}^-\mathbf{A}^-\mathbf{AB} = \mathbf{AB}$, implying further (upon pre-multiplying both sides of the matrix equation by $\mathbf{A}^-$ and post-multiplying by $\mathbf{B}^-$) that $\mathbf{A}^-\mathbf{ABB}^-\mathbf{A}^-\mathbf{ABB}^- = \mathbf{A}^-\mathbf{ABB}^-$, i.e., $\mathbf{A}^-\mathbf{ABB}^-$ is idempotent. Conversely, if $\mathbf{A}^-\mathbf{ABB}^-$ is idempotent, then $\mathbf{A}^-\mathbf{ABB}^-\mathbf{A}^-\mathbf{ABB}^- = \mathbf{A}^-\mathbf{ABB}^-$, implying further (upon pre-multiplying both sides of the matrix equation by $\mathbf{A}$ and post-multiplying by $\mathbf{B}$) that $\mathbf{ABB}^-\mathbf{A}^-\mathbf{AB} = \mathbf{AB}$, i.e., $\mathbf{B}^-\mathbf{A}^-$ is a generalized inverse of $\mathbf{AB}$.

▶ **Exercise 7** Prove Theorem 3.3.5: For any $n \times m$ matrix $\mathbf{A}$ of rank r, $\mathbf{AA}^-$ and $\mathbf{A}^-\mathbf{A}$ are idempotent matrices of rank r, and $\mathbf{I}_n - \mathbf{AA}^-$ and $\mathbf{I}_m - \mathbf{A}^-\mathbf{A}$ are idempotent matrices of ranks $n - r$ and $m - r$, respectively.

Solution $(\mathbf{AA}^-)(\mathbf{AA}^-) = (\mathbf{AA}^-\mathbf{A})\mathbf{A}^- = \mathbf{AA}^-$ and $(\mathbf{A}^-\mathbf{A})(\mathbf{A}^-\mathbf{A}) = \mathbf{A}^-(\mathbf{AA}^-\mathbf{A}) = \mathbf{A}^-\mathbf{A}$, which establishes that $\mathbf{AA}^-$ and $\mathbf{A}^-\mathbf{A}$ are idempotent. That $\mathbf{I}_n - \mathbf{AA}^-$ and $\mathbf{I}_m - \mathbf{A}^-\mathbf{A}$ likewise are idempotent follows immediately by Theorem 2.12.1. Now by Theorem 2.8.4,

$$\text{rank}(\mathbf{A}) = \text{rank}(\mathbf{AA}^-\mathbf{A}) \leq \text{rank}(\mathbf{AA}^-) \leq \text{rank}(\mathbf{A});$$

thus $\text{rank}(\mathbf{AA}^-) = \text{rank}(\mathbf{A}) = r$. By the idempotency of $\mathbf{I}_n - \mathbf{AA}^-$ and $\mathbf{AA}^-$ and by Theorems 2.12.2 and 2.10.1b,

$$\text{rank}(\mathbf{I}_n - \mathbf{AA}^-) = \text{tr}(\mathbf{I}_n - \mathbf{AA}^-) = \text{tr}(\mathbf{I}_n) - \text{tr}(\mathbf{AA}^-) = n - \text{rank}(\mathbf{AA}^-) = n - r.$$

Similar arguments yield $\text{rank}(\mathbf{A}^-\mathbf{A}) = r$ and $\text{rank}(\mathbf{I}_m - \mathbf{A}^-\mathbf{A}) = m - r$.

▶ **Exercise 8** Prove Theorem 3.3.7: Let $\mathbf{M} = \begin{pmatrix} \mathbf{A} & \mathbf{B} \\ \mathbf{C} & \mathbf{D} \end{pmatrix}$ represent a partitioned matrix.

(a) If $\mathcal{C}(\mathbf{B})$ is a subspace of $\mathcal{C}(\mathbf{A})$ and $\mathcal{R}(\mathbf{C})$ is a subspace of $\mathcal{R}(\mathbf{A})$ (as would be the case, e.g., if $\mathbf{A}$ was nonsingular), then the partitioned matrix

$$\begin{pmatrix} \mathbf{A}^- + \mathbf{A}^-\mathbf{BQ}^-\mathbf{CA}^- & -\mathbf{A}^-\mathbf{BQ}^- \\ -\mathbf{Q}^-\mathbf{CA}^- & \mathbf{Q}^- \end{pmatrix},$$

where $\mathbf{Q} = \mathbf{D} - \mathbf{CA}^-\mathbf{B}$, is a generalized inverse of $\mathbf{M}$.

(b) If $\mathcal{C}(\mathbf{C})$ is a subspace of $\mathcal{C}(\mathbf{D})$ and $\mathcal{R}(\mathbf{B})$ is a subspace of $\mathcal{R}(\mathbf{D})$ (as would be the case, e.g., if $\mathbf{D}$ was nonsingular), then

$$\begin{pmatrix} \mathbf{P}^- & -\mathbf{P}^-\mathbf{B}\mathbf{D}^- \\ -\mathbf{D}^-\mathbf{C}\mathbf{P}^- & \mathbf{D}^- + \mathbf{D}^-\mathbf{C}\mathbf{P}^-\mathbf{B}\mathbf{D}^- \end{pmatrix},$$

where $\mathbf{P} = \mathbf{A} - \mathbf{B}\mathbf{D}^-\mathbf{C}$, is a generalized inverse of $\mathbf{M}$.

Solution We prove part (a) only; the proof of part (b) is very similar. The given condition $\mathcal{C}(\mathbf{B}) \subseteq \mathcal{C}(\mathbf{A})$ implies that $\mathbf{B} = \mathbf{AF}$ for some $\mathbf{F}$, or equivalently that $\mathbf{AA}^-\mathbf{B} = \mathbf{B}$. Similarly, $\mathcal{R}(\mathbf{C}) \subseteq \mathcal{R}(\mathbf{A})$ implies that $\mathbf{C} = \mathbf{HA}$ for some $\mathbf{H}$, or equivalently that $\mathbf{CA}^-\mathbf{A} = \mathbf{C}$. Then

$$\begin{aligned}
&\begin{pmatrix} \mathbf{A} & \mathbf{B} \\ \mathbf{C} & \mathbf{D} \end{pmatrix} \begin{pmatrix} \mathbf{A}^- + \mathbf{A}^-\mathbf{B}\mathbf{Q}^-\mathbf{C}\mathbf{A}^- & -\mathbf{A}^-\mathbf{B}\mathbf{Q}^- \\ -\mathbf{Q}^-\mathbf{C}\mathbf{A}^- & \mathbf{Q}^- \end{pmatrix} \begin{pmatrix} \mathbf{A} & \mathbf{B} \\ \mathbf{C} & \mathbf{D} \end{pmatrix} \\
&= \begin{pmatrix} \mathbf{AA}^- + \mathbf{AA}^-\mathbf{BQ}^-\mathbf{CA}^- - \mathbf{BQ}^-\mathbf{CA}^- & -\mathbf{AA}^-\mathbf{BQ}^- + \mathbf{BQ}^- \\ \mathbf{CA}^- + \mathbf{CA}^-\mathbf{BQ}^-\mathbf{CA}^- - \mathbf{DQ}^-\mathbf{CA}^- & -\mathbf{CA}^-\mathbf{BQ}^- + \mathbf{DQ}^- \end{pmatrix} \begin{pmatrix} \mathbf{A} & \mathbf{B} \\ \mathbf{C} & \mathbf{D} \end{pmatrix} \\
&= \begin{pmatrix} \mathbf{K}_{11} & \mathbf{K}_{12} \\ \mathbf{K}_{21} & \mathbf{K}_{22} \end{pmatrix},
\end{aligned}$$

say, where

$$\begin{aligned}
\mathbf{K}_{11} &= \mathbf{AA}^-\mathbf{A} + \mathbf{AA}^-\mathbf{BQ}^-\mathbf{CA}^-\mathbf{A} - \mathbf{BQ}^-\mathbf{CA}^-\mathbf{A} - \mathbf{AA}^-\mathbf{BQ}^-\mathbf{C} + \mathbf{BQ}^-\mathbf{C} \\
&= \mathbf{A} + \mathbf{BQ}^-\mathbf{C} - \mathbf{BQ}^-\mathbf{C} - \mathbf{BQ}^-\mathbf{C} + \mathbf{BQ}^-\mathbf{C} \\
&= \mathbf{A},
\end{aligned}$$

$$\begin{aligned}
\mathbf{K}_{12} &= \mathbf{AA}^-\mathbf{B} + \mathbf{AA}^-\mathbf{BQ}^-\mathbf{CA}^-\mathbf{B} - \mathbf{BQ}^-\mathbf{CA}^-\mathbf{B} - \mathbf{AA}^-\mathbf{BQ}^-\mathbf{D} + \mathbf{BQ}^-\mathbf{D} \\
&= \mathbf{B} + \mathbf{BQ}^-\mathbf{CA}^-\mathbf{B} - \mathbf{BQ}^-\mathbf{CA}^-\mathbf{B} - \mathbf{BQ}^-\mathbf{D} + \mathbf{BQ}^-\mathbf{D} \\
&= \mathbf{B},
\end{aligned}$$

$$\begin{aligned}
\mathbf{K}_{21} &= \mathbf{CA}^-\mathbf{A} + \mathbf{CA}^-\mathbf{BQ}^-\mathbf{CA}^-\mathbf{A} - \mathbf{DQ}^-\mathbf{CA}^-\mathbf{A} - \mathbf{CA}^-\mathbf{BQ}^-\mathbf{C} + \mathbf{DQ}^-\mathbf{C} \\
&= \mathbf{C} + \mathbf{CA}^-\mathbf{BQ}^-\mathbf{C} - \mathbf{DQ}^-\mathbf{C} - \mathbf{CA}^-\mathbf{BQ}^-\mathbf{C} + \mathbf{DQ}^-\mathbf{C} \\
&= \mathbf{C},
\end{aligned}$$

and

$$\begin{aligned}
\mathbf{K}_{22} &= \mathbf{CA}^-\mathbf{B} + \mathbf{CA}^-\mathbf{BQ}^-\mathbf{CA}^-\mathbf{B} - \mathbf{DQ}^-\mathbf{CA}^-\mathbf{B} - \mathbf{CA}^-\mathbf{BQ}^-\mathbf{D} + \mathbf{DQ}^-\mathbf{D} \\
&= \mathbf{CA}^-\mathbf{B} - (\mathbf{D} - \mathbf{CA}^-\mathbf{B})\mathbf{Q}^-\mathbf{CA}^-\mathbf{B} + (\mathbf{D} - \mathbf{CA}^-\mathbf{B})\mathbf{Q}^-\mathbf{D}
\end{aligned}$$

$$= \mathbf{CA^-B} + \mathbf{QQ^-(D - CA^-B)}$$
$$= \mathbf{CA^-B + Q}$$
$$= \mathbf{D}.$$

▶ **Exercise 9** Prove Theorem 3.3.10: Let $\mathbf{M} = \mathbf{A} + \mathbf{H}$, where $\mathcal{C}(\mathbf{H})$ and $\mathcal{R}(\mathbf{H})$ are subspaces of $\mathcal{C}(\mathbf{A})$ and $\mathcal{R}(\mathbf{A})$, respectively (as would be the case, e.g., if $\mathbf{A}$ was nonsingular), and take $\mathbf{B}$, $\mathbf{C}$, and $\mathbf{D}$ to represent any three matrices such that $\mathbf{H} = \mathbf{BCD}$. Then, the matrix $\mathbf{A^- - A^-BC(C + CDA^-BC)^-CDA^-}$ is a generalized inverse of $\mathbf{M}$.

Solution Define $\mathbf{Q} = \mathbf{C} + \mathbf{CDA^-BC}$. Because $\mathcal{C}(\mathbf{H}) \subseteq \mathcal{C}(\mathbf{A})$ and $\mathcal{R}(\mathbf{H}) \subseteq \mathcal{R}(\mathbf{A})$, matrices $\mathbf{F}$ and $\mathbf{K}$ exist such that $\mathbf{H} = \mathbf{AF}$ and $\mathbf{H} = \mathbf{KA}$. It follows that $\mathbf{AA^-H} = \mathbf{AA^-AF} = \mathbf{AF} = \mathbf{H}$ and $\mathbf{HA^-A} = \mathbf{KAA^-A} = \mathbf{KA} = \mathbf{H}$. Therefore

$$\begin{aligned}
\mathbf{(A+H)(A^- - A^-BCQ^-CDA^-)(A+H)} &= \mathbf{(A+H)A^-(A+H)} \\
&\quad - \mathbf{(A+H)(A^-BCQ^-CDA^-)(A+H)} \\
&= \mathbf{A + H + H + HA^-H - R},
\end{aligned}$$

where

$$\begin{aligned}
\mathbf{R} &= \mathbf{AA^-BCQ^-CDA^-A + AA^-BCQ^-CDA^-H + HA^-BCQ^-CDA^-A} \\
&\quad \mathbf{+ HA^-BCQ^-CDA^-H} \\
&= \mathbf{AA^-BCQ^-CDA^-A + AA^-BCQ^-CDA^-HAA^-} \\
&\quad \mathbf{+ AA^-HA^-BCQ^-CDA^-A + AA^-HA^-BCQ^-CDA^-HA^-A} \\
&= \mathbf{AA^-BCQ^-CDA^-A + AA^-BCQ^-CDA^-BCDA^-A} \\
&\quad \mathbf{+ AA^-BCDA^-BCQ^-CDA^-BCDA^-A} \\
&= \mathbf{AA^-B[CQ^-C + CQ^-CDA^-BC + CDA^-BCQ^-C} \\
&\quad \mathbf{+ CDA^-BCQ^-CDA^-BC]DA^-A} \\
&= \mathbf{AA^-B[(C + CDA^-BC)Q^-(C + CDA^-BC)]DA^-A} \\
&= \mathbf{AA^-B(C + CDA^-BC)DA^-A} \\
&= \mathbf{AA^-HA^-A + AA^-HA^-HA^-A} \\
&= \mathbf{H + HA^-H}.
\end{aligned}$$

Substituting this expression for $\mathbf{R}$ into the expression for $\mathbf{(A + H)(A^- - A^-BCQ^-CDA^-)(A + H)}$, we obtain

$$\mathbf{(A + H)(A^- - A^-BCQ^-CDA^-)(A + H) = A + H},$$

which establishes that $\mathbf{A^- - A^-BCQ^-CDA^-}$ is a generalized inverse of $\mathbf{A + H}$.

▶ **Exercise 10** Prove Theorem 3.3.11: Let $\mathbf{A} = \begin{pmatrix} \mathbf{A}_{11} & \mathbf{A}_{12} \\ \mathbf{A}_{21} & \mathbf{A}_{22} \end{pmatrix}$, where $\mathbf{A}_{11}$ is of dimensions $n \times m$, represent a partitioned matrix and let $\mathbf{G} = \begin{pmatrix} \mathbf{G}_{11} & \mathbf{G}_{12} \\ \mathbf{G}_{21} & \mathbf{G}_{22} \end{pmatrix}$, where $\mathbf{G}_{11}$ is of dimensions $m \times n$, represent any generalized inverse of $\mathbf{A}$. If $\mathcal{R}(\mathbf{A}_{21}) \cap \mathcal{R}(\mathbf{A}_{11}) = \{\mathbf{0}\}$ and $\mathcal{C}(\mathbf{A}_{12}) \cap \mathcal{C}(\mathbf{A}_{11}) = \{\mathbf{0}\}$, then $\mathbf{G}_{11}$ is a generalized inverse of $\mathbf{A}_{11}$.

Solution

$$\begin{pmatrix} \mathbf{A}_{11} & \mathbf{A}_{12} \\ \mathbf{A}_{21} & \mathbf{A}_{22} \end{pmatrix} = \begin{pmatrix} \mathbf{A}_{11} & \mathbf{A}_{12} \\ \mathbf{A}_{21} & \mathbf{A}_{22} \end{pmatrix} \begin{pmatrix} \mathbf{G}_{11} & \mathbf{G}_{12} \\ \mathbf{G}_{21} & \mathbf{G}_{22} \end{pmatrix} \begin{pmatrix} \mathbf{A}_{11} & \mathbf{A}_{12} \\ \mathbf{A}_{21} & \mathbf{A}_{22} \end{pmatrix}$$
$$= \begin{pmatrix} \mathbf{A}_{11}\mathbf{G}_{11} + \mathbf{A}_{12}\mathbf{G}_{21} & \mathbf{A}_{11}\mathbf{G}_{12} + \mathbf{A}_{12}\mathbf{G}_{22} \\ \mathbf{A}_{21}\mathbf{G}_{11} + \mathbf{A}_{22}\mathbf{G}_{21} & \mathbf{A}_{21}\mathbf{G}_{12} + \mathbf{A}_{22}\mathbf{G}_{22} \end{pmatrix} \begin{pmatrix} \mathbf{A}_{11} & \mathbf{A}_{12} \\ \mathbf{A}_{21} & \mathbf{A}_{22} \end{pmatrix},$$

implying (by considering only the upper left blocks of the block matrices on the two sides of the matrix equation) that

$$\mathbf{A}_{11}\mathbf{G}_{11}\mathbf{A}_{11} + \mathbf{A}_{12}\mathbf{G}_{21}\mathbf{A}_{11} + \mathbf{A}_{11}\mathbf{G}_{12}\mathbf{A}_{21} + \mathbf{A}_{12}\mathbf{G}_{22}\mathbf{A}_{21} = \mathbf{A}_{11},$$

i.e.,

$$\mathbf{A}_{12}(\mathbf{G}_{21}\mathbf{A}_{11} + \mathbf{G}_{22}\mathbf{A}_{21}) = \mathbf{A}_{11}(\mathbf{I} - \mathbf{G}_{11}\mathbf{A}_{11} - \mathbf{G}_{12}\mathbf{A}_{21}).$$

Because the column spaces of $\mathbf{A}_{12}$ and $\mathbf{A}_{11}$ are essentially disjoint,

$$\mathbf{A}_{11}(\mathbf{I} - \mathbf{G}_{11}\mathbf{A}_{11} - \mathbf{G}_{12}\mathbf{A}_{21}) = \mathbf{0},$$

i.e.,

$$(\mathbf{I} - \mathbf{A}_{11}\mathbf{G}_{11})\mathbf{A}_{11} = \mathbf{A}_{11}\mathbf{G}_{12}\mathbf{A}_{21}.$$

Because the row spaces of $\mathbf{A}_{11}$ and $\mathbf{A}_{21}$ are essentially disjoint,

$$(\mathbf{I} - \mathbf{A}_{11}\mathbf{G}_{11})\mathbf{A}_{11} = \mathbf{0},$$

i.e.,

$$\mathbf{A}_{11}\mathbf{G}_{11}\mathbf{A}_{11} = \mathbf{A}_{11},$$

i.e., $\mathbf{G}_{11}$ is a generalized inverse of $\mathbf{A}_{11}$.

Exercise 11

(a) Prove Theorem 3.3.12: Let $\mathbf{A}$ and $\mathbf{B}$ represent two $m \times n$ matrices. If $\mathcal{C}(\mathbf{A}) \cap \mathcal{C}(\mathbf{B}) = \{\mathbf{0}\}$ and $\mathcal{R}(\mathbf{A}) \cap \mathcal{R}(\mathbf{B}) = \{\mathbf{0}\}$, then any generalized inverse of $\mathbf{A} + \mathbf{B}$ is also a generalized inverse of $\mathbf{A}$.

(b) Show that the converse of Theorem 3.3.12 is false by constructing a counterexample based on the matrices

$$\mathbf{A} = \begin{pmatrix} 1 & 0 \\ 0 & 0 \end{pmatrix} \quad \text{and} \quad \mathbf{B} = \begin{pmatrix} 0 & 0 \\ 0 & 1 \end{pmatrix}.$$

Solution

(a) $\mathbf{A} + \mathbf{B} = (\mathbf{A} + \mathbf{B})(\mathbf{A} + \mathbf{B})^-(\mathbf{A} + \mathbf{B})$, implying that

$$\mathbf{A} + \mathbf{B} = \mathbf{A}(\mathbf{A} + \mathbf{B})^-\mathbf{A} + \mathbf{B}(\mathbf{A} + \mathbf{B})^-\mathbf{A} + \mathbf{A}(\mathbf{A} + \mathbf{B})^-\mathbf{B} + \mathbf{B}(\mathbf{A} + \mathbf{B})^-\mathbf{B},$$

implying further that

$$\mathbf{A} - \mathbf{A}(\mathbf{A} + \mathbf{B})^-\mathbf{A} - \mathbf{A}(\mathbf{A} + \mathbf{B})^-\mathbf{B} = \mathbf{B}(\mathbf{A} + \mathbf{B})^-\mathbf{A} + \mathbf{B}(\mathbf{A} + \mathbf{B})^-\mathbf{B} - \mathbf{B}.$$

The columns of the matrix on the left of this last equality are elements of $\mathcal{C}(\mathbf{A})$, while those of the matrix on the right are elements of $\mathcal{C}(\mathbf{B})$. Because $\mathcal{C}(\mathbf{A}) \cap \mathcal{C}(\mathbf{B}) = \{\mathbf{0}\}$,

$$\mathbf{A} - \mathbf{A}(\mathbf{A} + \mathbf{B})^-\mathbf{A} - \mathbf{A}(\mathbf{A} + \mathbf{B})^-\mathbf{B} = \mathbf{0},$$

i.e.,

$$\mathbf{A} - \mathbf{A}(\mathbf{A} + \mathbf{B})^-\mathbf{A} = \mathbf{A}(\mathbf{A} + \mathbf{B})^-\mathbf{B}.$$

The rows of the matrix on the left of this last equality are elements of $\mathcal{R}(\mathbf{A})$, while those of the matrix on the right are elements of $\mathcal{R}(\mathbf{B})$. Because $\mathcal{R}(\mathbf{A}) \cap \mathcal{R}(\mathbf{B}) = \{\mathbf{0}\}$, $\mathbf{A} - \mathbf{A}(\mathbf{A} + \mathbf{B})^-\mathbf{A} = \mathbf{0}$. Thus $\mathbf{A} = \mathbf{A}(\mathbf{A} + \mathbf{B})^-\mathbf{A}$, showing that any generalized inverse of $\mathbf{A} + \mathbf{B}$ is also a generalized inverse of $\mathbf{A}$. This proves the theorem.

(b) Let $\mathbf{A}$ and $\mathbf{B}$ be as defined in the theorem. Then, $\mathbf{A} + \mathbf{B} = \mathbf{I}_2$, which is nonsingular so its only generalized inverse is its ordinary inverse, $\mathbf{I}_2$. It is easily verified that $\mathbf{AIA} = \mathbf{A}$, hence $\mathbf{I}_2$ is also a generalized inverse of $\mathbf{A}$. However, $\mathcal{C}(\mathbf{A}) = \left\{c\begin{pmatrix} 1 \\ 0 \end{pmatrix} : c \in \mathbb{R}\right\} \neq \mathcal{C}(\mathbf{B}) = \left\{c\begin{pmatrix} 0 \\ 1 \end{pmatrix} : c \in \mathbb{R}\right\}$. Thus $\mathcal{C}(\mathbf{A}) \cap \mathcal{C}(\mathbf{B}) = \{0\}$, and by the symmetry of $\mathbf{A}$ and $\mathbf{B}$, $\mathcal{R}(\mathbf{A}) \cap \mathcal{R}(\mathbf{B}) = \{0\}$ also.

Exercise 12 Determine Moore–Penrose inverses of each of the matrices listed in Exercise 3.1 except those in parts (c) and (j).

Solution

(a) It is easy to show that $(1/c)\mathbf{A}^+$ satisfies the four Moore–Penrose conditions.

(b) It was shown in Exercise 3.1b that $\frac{1}{\|\mathbf{a}\|^2\|\mathbf{b}\|^2}\mathbf{b}\mathbf{a}^T$ is a generalized inverse of $\mathbf{a}\mathbf{b}^T$. Observe that

$$\left(\frac{1}{\|\mathbf{a}\|^2\|\mathbf{b}\|^2}\mathbf{b}\mathbf{a}^T\right)\mathbf{a}\mathbf{b}^T\left(\frac{1}{\|\mathbf{a}\|^2\|\mathbf{b}\|^2}\mathbf{b}\mathbf{a}^T\right) = \frac{1}{\|\mathbf{a}\|^2\|\mathbf{b}\|^2}\mathbf{b}\mathbf{a}^T,$$

so this generalized inverse is reflexive. Furthermore,

$$\left(\frac{1}{\|\mathbf{a}\|^2\|\mathbf{b}\|^2}\mathbf{b}\mathbf{a}^T\right)\mathbf{a}\mathbf{b}^T = \frac{1}{\|\mathbf{b}\|^2}\mathbf{b}\mathbf{b}^T$$

and

$$\mathbf{a}\mathbf{b}^T\left(\frac{1}{\|\mathbf{a}\|^2\|\mathbf{b}\|^2}\mathbf{b}\mathbf{a}^T\right) = \frac{1}{\|\mathbf{a}\|^2}\mathbf{a}\mathbf{a}^T,$$

which are both symmetric; thus $\frac{1}{\|\mathbf{a}\|^2\|\mathbf{b}\|^2}\mathbf{b}\mathbf{a}^T$ is the Moore–Penrose inverse.

(d) It was shown in Exercise 3.1d that $\mathbf{C} = \text{diag}(c_1, \ldots, c_n)$, where

$$c_k = \begin{cases} 1/d_k, & \text{if } d_k \neq 0 \\ 0, & \text{if } d_k = 0 \end{cases}$$

(for $k = 1, \ldots, n$) is a generalized inverse of $\mathbf{D}$. It is easily verified that $\mathbf{CDC} = \mathbf{C}$ and that $\mathbf{CD}$ and $\mathbf{DC}$ are both diagonal and hence symmetric. Thus $\mathbf{C}$ is the Moore–Penrose inverse of $\mathbf{D}$.

(e) With the elements of $\mathbf{C}$ defined via

$$\mathbf{C} = \begin{pmatrix} & & c_1 \\ & \cdot^{\cdot^{\cdot}} & \\ c_n & & \end{pmatrix},$$

it was shown in Exercise 3.1e that

$$\mathbf{D} = \begin{pmatrix} & & d_n \\ & \cdot^{\cdot^{\cdot}} & \\ d_1 & & \end{pmatrix},$$

where

$$d_k = \begin{cases} 1/c_k, & \text{if } c_k \neq 0 \\ 0, & \text{if } c_k = 0 \end{cases}$$

(for $k = 1, \ldots, n$) is a generalized inverse of $\mathbf{C}$. It is easily verified that $\mathbf{DCD} = \mathbf{D}$ and that $\mathbf{DC}$ and $\mathbf{CD}$ are both diagonal and hence symmetric. Thus $\mathbf{D}$ is the Moore–Penrose inverse of $\mathbf{C}$.

(f) By Exercise 3.1f, $\mathbf{Q}^T\mathbf{A}^+\mathbf{P}^T$ is a generalized inverse of $\mathbf{PAQ}$ because $\mathbf{A}^+$ is a generalized inverse of $\mathbf{A}$. Furthermore, $(\mathbf{Q}^T\mathbf{A}^+\mathbf{P}^T)\mathbf{PAQ}(\mathbf{Q}^T\mathbf{A}^+\mathbf{P}^T) = \mathbf{Q}^T\mathbf{A}^+\mathbf{AA}^+\mathbf{P}^T = \mathbf{Q}^T\mathbf{A}^+\mathbf{P}^T$. Finally observe that $(\mathbf{Q}^T\mathbf{A}^+\mathbf{P}^T)\mathbf{PAQ} = \mathbf{Q}^T\mathbf{A}^+\mathbf{AQ}$ and $\mathbf{PAQ}(\mathbf{Q}^T\mathbf{A}^+\mathbf{P}^T) = \mathbf{PAA}^+\mathbf{P}^T$, which are both symmetric because $\mathbf{A}^+\mathbf{A}$ and $\mathbf{AA}^+$ are symmetric. Thus $\mathbf{Q}^T\mathbf{A}^+\mathbf{P}^T$ is the Moore–Penrose inverse of $\mathbf{PAQ}$.

(g) It is easy to show that $\mathbf{P}^T$ satisfies the four Moore–Penrose conditions.

(h) It is easy to show that $(1/mn)\mathbf{J}_{n\times m}$ satisfies the four Moore–Penrose conditions.

(i) If $\mathbf{b} \neq -a/n$, then $\frac{1}{a}\mathbf{I}_n - \frac{b}{a(a+bn)}\mathbf{J}_n$ is the ordinary inverse and hence the Moore–Penrose inverse. If $b = -a/n$, then the matrix under consideration is $a(\mathbf{I}_n - \frac{1}{n}\mathbf{J}_n)$. By Exercise 3.1i, $\frac{1}{a}(\mathbf{I}_n - \frac{1}{n}\mathbf{J}_n)$ is a generalized inverse of this matrix. Observe that

$$\left[\frac{1}{a}(\mathbf{I}_n - \frac{1}{n}\mathbf{J}_n)\right] a(\mathbf{I}_n - \frac{1}{n}\mathbf{J}_n) \left[\frac{1}{a}(\mathbf{I}_n - \frac{1}{n}\mathbf{J}_n)\right] = \frac{1}{a}(\mathbf{I}_n - \frac{1}{n}\mathbf{J}_n).$$

Finally, $[\frac{1}{a}(\mathbf{I}_n - \frac{1}{n}\mathbf{J}_n)]a(\mathbf{I}_n - \frac{1}{n}\mathbf{J}_n) = \mathbf{I}_n - \frac{1}{n}\mathbf{J}_n$ and $a(\mathbf{I}_n - \frac{1}{n}\mathbf{J}_n)[\frac{1}{a}(\mathbf{I}_n - \frac{1}{n}\mathbf{J}_n)] = \mathbf{I}_n - \frac{1}{n}\mathbf{J}_n$, which are both symmetric. Thus $\frac{1}{a}(\mathbf{I}_n - \frac{1}{n}\mathbf{J}_n)$ is the Moore–Penrose inverse in this case.

(k) It is easy to show that $\oplus_{i=1}^k \mathbf{B}_i^+$ satisfies the four Moore–Penrose conditions.

(l) It is easy to show, with the aid of Theorem 2.17.5, that $\mathbf{A}^+ \otimes \mathbf{B}^+$ satisfies the four Moore–Penrose conditions.

▶ **Exercise 13** Prove Theorem 3.4.2: For any matrix $\mathbf{A}$, $(\mathbf{A}^T)^+ = (\mathbf{A}^+)^T$.

Solution It suffices to show that $(\mathbf{A}^+)^T$ satisfies the four Moore–Penrose conditions for the Moore–Penrose inverse of $\mathbf{A}^T$. Using the cyclic property of the trace and the fact that $\mathbf{A}^+$ satisfies the four Moore–Penrose conditions for the Moore–Penrose inverse of $\mathbf{A}$, we obtain:

(i) $\mathbf{A}^T(\mathbf{A}^+)^T\mathbf{A}^T = (\mathbf{AA}^+\mathbf{A})^T = \mathbf{A}^T$,
(ii) $(\mathbf{A}^+)^T\mathbf{A}^T(\mathbf{A}^+)^T = (\mathbf{A}^+\mathbf{AA}^+)^T = (\mathbf{A}^+)^T$,
(iii) $[(\mathbf{A}^+)^T\mathbf{A}^T]^T = \mathbf{AA}^+ = (\mathbf{AA}^+)^T = (\mathbf{A}^+)^T\mathbf{A}^T$,
(iv) $[\mathbf{A}^T(\mathbf{A}^+)^T]^T = \mathbf{A}^+\mathbf{A} = (\mathbf{A}^+\mathbf{A})^T = \mathbf{A}^T(\mathbf{A}^+)^T$.

▶ **Exercise 14** Let $\mathbf{A}$ represent any matrix. Prove that $\text{rank}(\mathbf{A}^+) = \text{rank}(\mathbf{A})$. Does this result hold for an arbitrary generalized inverse of $\mathbf{A}$? Prove or give a counterexample.

Solution By the first two Moore–Penrose conditions for $\mathbf{A}^+$ and Theorem 2.8.4,

$$\text{rank}(\mathbf{A}^+) = \text{rank}(\mathbf{A}^+\mathbf{A}\mathbf{A}^+) \leq \text{rank}(\mathbf{A}) = \text{rank}(\mathbf{A}\mathbf{A}^+\mathbf{A}) \leq \text{rank}(\mathbf{A}^+).$$

Thus $\text{rank}(\mathbf{A}^+) = \text{rank}(\mathbf{A})$. This result does not hold for an arbitrary generalized inverse of $\mathbf{A}$. Consider, for example, $\mathbf{A} = \mathbf{J}_2$, for which $0.5\mathbf{I}_2$ is a generalized inverse as noted in (3.1). But $\text{rank}(\mathbf{J}_2) = 1$, whereas $\text{rank}(0.5\mathbf{I}_2) = 2$.

Moments of a Random Vector and of Linear and Quadratic Forms in a Random Vector

4

This chapter presents exercises on moments of a random vector and linear and quadratic forms in a random vector and provides solutions to those exercises.

▶ **Exercise 1** Prove Theorem 4.1.1: A variance–covariance matrix is nonnegative definite. Conversely, any nonnegative definite matrix is the variance–covariance matrix of some random vector.

Solution Let $\mathbf{x}$ represent a random k-vector such that $\mathrm{E}(\mathbf{x}) = \boldsymbol{\mu}$, and let $\mathbf{a} \in \mathbb{R}^k$. Then $\mathbf{a}^T \mathrm{var}(\mathbf{x})\mathbf{a} = \mathbf{a}^T \mathrm{E}[(\mathbf{x}-\boldsymbol{\mu})(\mathbf{x}-\boldsymbol{\mu})^T]\mathbf{a} = \mathrm{E}[\mathbf{a}^T(\mathbf{x}-\boldsymbol{\mu})(\mathbf{x}-\boldsymbol{\mu})^T\mathbf{a}] = \mathrm{E}\{[\mathbf{a}^T(\mathbf{x}-\boldsymbol{\mu})]^2\} \geq 0$. Conversely, let $\mathbf{z}$ represent a vector of independent random variables with common variance one, so that $\mathrm{var}(\mathbf{z}) = \mathbf{I}$, and let $\mathbf{A}$ represent any nonnegative definite matrix. By Theorem 2.15.12, a nonnegative definite matrix $\mathbf{A}^{\frac{1}{2}}$ exists such that $\mathbf{A} = \mathbf{A}^{\frac{1}{2}}\mathbf{A}^{\frac{1}{2}}$. Then $\mathrm{var}(\mathbf{A}^{\frac{1}{2}}\mathbf{z}) = \mathbf{A}^{\frac{1}{2}}\mathbf{I}\mathbf{A}^{\frac{1}{2}} = \mathbf{A}$.

▶ **Exercise 2** Prove Theorem 4.1.2: If the variances of all variables in a random vector $\mathbf{x}$ exist, and those variables are pairwise independent, then $\mathrm{var}(\mathbf{x})$ is a diagonal matrix.

Solution Without loss of generality consider $\mathrm{cov}(x_1, x_2)$, where x_1 and x_2 are the first two elements of $\mathbf{x}$. Because x_1 and x_2 are independent,

$$\mathrm{cov}(x_1, x_2) = \mathrm{E}[(x_1 - \mathrm{E}(x_1))(x_2 - \mathrm{E}(x_2))] = \mathrm{E}[x_1 - \mathrm{E}(x_1)] \cdot \mathrm{E}[x_2 - \mathrm{E}(x_2)] = 0 \cdot 0 = 0.$$

Thus all off-diagonal elements of $\mathrm{var}(\mathbf{x})$ equal 0, i.e., $\mathrm{var}(\mathbf{x})$ is diagonal.

▶ **Exercise 3** Prove that the second patterned variance–covariance matrix described in Example 4.1-1, i.e.,

D. L. Zimmerman, *Linear Model Theory*,
https://doi.org/10.1007/978-3-030-52074-8_4

$$\mathbf{\Sigma} = \sigma^2 \begin{pmatrix} 1 & 1 & \cdots & 1 & 1 \\ 1 & 2 & \cdots & 2 & 2 \\ \vdots & \vdots & \ddots & \vdots & \vdots \\ 1 & 2 & \cdots & n-1 & n-1 \\ 1 & 2 & \cdots & n-1 & n \end{pmatrix},$$

where $\sigma^2 > 0$, is positive definite.

Solution First, note that $(1/\sigma^2)\mathbf{\Sigma}$ is nonnegative definite, as shown in Example 4.1-1. Now, let $\mathbf{x}$ represent any vector for which $\mathbf{x}^T[(1/\sigma^2)\mathbf{\Sigma}]\mathbf{x} = 0$. Then

$$\begin{aligned} \mathbf{x}^T[(1/\sigma^2)\mathbf{\Sigma}]\mathbf{x} &= \mathbf{x}^T\mathbf{1}_n\mathbf{1}_n^T\mathbf{x} + \mathbf{x}^T\begin{pmatrix} 0 \\ \mathbf{1}_{n-1} \end{pmatrix}\begin{pmatrix} 0 \\ \mathbf{1}_{n-1} \end{pmatrix}^T\mathbf{x} + \mathbf{x}^T\begin{pmatrix} \mathbf{0}_2 \\ \mathbf{1}_{n-2} \end{pmatrix}\begin{pmatrix} \mathbf{0}_2 \\ \mathbf{1}_{n-2} \end{pmatrix}^T\mathbf{x} + \cdots \\ &\quad + \mathbf{x}^T\begin{pmatrix} \mathbf{0}_{n-1} \\ 1 \end{pmatrix}\begin{pmatrix} \mathbf{0}_{n-1} \\ 1 \end{pmatrix}^T\mathbf{x} \\ &= \left(\sum_{i=1}^n x_i\right)^2 + \left(\sum_{i=2}^n x_i\right)^2 + \left(\sum_{i=3}^n x_i\right)^2 + \cdots + x_n^2 \\ &= 0. \end{aligned}$$

This implies that

$$x_n = 0, \ x_{n-1} + x_n = 0, \ \ldots, \ \sum_{i=2}^n x_i = 0, \ \sum_{i=1}^n x_i = 0,$$

i.e., that $x_n = x_{n-1} = x_{n-2} = \cdots = x_1 = 0$, i.e., that $\mathbf{x} = \mathbf{0}$. Thus, $\mathbf{\Sigma}$ is positive definite.

▶ **Exercise 4** Prove that the third patterned variance–covariance matrix described in Example 4.1-1, i.e.,

$$\mathbf{\Sigma} = (\oplus_{i=1}^n \sqrt{\sigma_{ii}}) \begin{pmatrix} 1 & \rho_1 & \rho_1\rho_2 & \rho_1\rho_2\rho_3 & \cdots & \prod_{i=1}^{n-2}\rho_i & \prod_{i=1}^{n-1}\rho_i \\ & 1 & \rho_2 & \rho_2\rho_3 & \cdots & \prod_{i=2}^{n-2}\rho_i & \prod_{i=2}^{n-1}\rho_i \\ & & 1 & \rho_3 & \cdots & \prod_{i=3}^{n-2}\rho_i & \prod_{i=3}^{n-1}\rho_i \\ & & & 1 & \cdots & \prod_{i=4}^{n-2}\rho_i & \prod_{i=4}^{n-1}\rho_i \\ & & & & \ddots & \vdots & \vdots \\ & & & & & 1 & \rho_{n-1} \\ & & & & & & 1 \end{pmatrix} (\oplus_{i=1}^n \sqrt{\sigma_{ii}}),$$

where $\sigma_{ii} > 0$ and $-1 < \rho_i < 1$ for all i, is positive definite.

Solution For $k = 1, \dots, n$, define the $k \times k$ matrix

$$\mathbf{R}_k = \begin{pmatrix} 1 & \rho_1 & \rho_1\rho_2 & \rho_1\rho_2\rho_3 & \cdots & \prod_{i=1}^{k-2}\rho_i & \prod_{i=1}^{k-1}\rho_i \\ & 1 & \rho_2 & \rho_2\rho_3 & \cdots & \prod_{i=2}^{k-2}\rho_i & \prod_{i=2}^{k-1}\rho_i \\ & & 1 & \rho_3 & \cdots & \prod_{i=3}^{k-2}\rho_i & \prod_{i=3}^{k-1}\rho_i \\ & & & 1 & \cdots & \prod_{i=4}^{k-2}\rho_i & \prod_{i=4}^{k-1}\rho_i \\ & & & & \ddots & \vdots & \vdots \\ & & & & & 1 & \rho_{k-1} \\ & & & & & & 1 \end{pmatrix}.$$

Observe that the third patterned variance–covariance matrix is equal to $\mathbf{R}_n$, apart from pre- and post-multiplication by a diagonal matrix with positive main diagonal elements. Therefore, it suffices to show that $\mathbf{R}_n$ is positive definite. Also observe that for $k = 2, \dots, n$, $\mathbf{R}_k$ may be written in partitioned form as

$$\mathbf{R}_k = \begin{pmatrix} \mathbf{R}_{k-1} & \mathbf{r}_{k-1} \\ \mathbf{r}_{k-1}^T & 1 \end{pmatrix},$$

where $\mathbf{r}_{k-1} = (\prod_{i=1}^{k-1}\rho_i, \prod_{i=2}^{k-1}\rho_i, \dots, \rho_{k-1})^T$. By Theorem 2.15.11, $\mathbf{R}_n$ is positive definite if and only if $|\mathbf{R}_1| > 0$, $|\mathbf{R}_2| > 0, \dots, |\mathbf{R}_n| > 0$. We will use "proof by induction" to show that all of these inequalities are satisfied. First observe that $|\mathbf{R}_1| = |1| = 1 > 0$. Thus, it suffices to show that if $|\mathbf{R}_k| > 0$, then $|\mathbf{R}_{k+1}| > 0$. So suppose that $|\mathbf{R}_k| > 0$. By Theorem 2.11.7,

$$|\mathbf{R}_{k+1}| = |\mathbf{R}_k|(1 - \mathbf{r}_k^T\mathbf{R}_k^{-1}\mathbf{r}_k).$$

Crucially, $\mathbf{r}_k$ is equal to ρ_k times the last column of $\mathbf{R}_k$, implying that $\mathbf{R}_k^{-1}\mathbf{r}_k = \rho_k\mathbf{u}_k^{(k)}$ and $\mathbf{r}_k = \rho_k\begin{pmatrix} \mathbf{r}_{k-1} \\ 1 \end{pmatrix}$. Therefore

$$|\mathbf{R}_{k+1}| = |\mathbf{R}_k|[1 - \rho_k\begin{pmatrix} \mathbf{r}_{k-1}^T & 1 \end{pmatrix}(\rho_k\mathbf{u}_k^{(k)})] = |\mathbf{R}_k|(1 - \rho_k^2) > 0$$

because $-1 < \rho_k < 1$.

▶ **Exercise 5** Prove Theorem 4.1.3: Suppose that the skewness matrix $\mathbf{\Lambda} = (\lambda_{ijk})$ of a random n-vector $\mathbf{x}$ exists.

(a) If the elements of $\mathbf{x}$ are triple-wise independent (meaning that each trivariate marginal cdf of $\mathbf{x}$ factors into the product of its univariate marginal cdfs), then for some constants $\lambda_1, \dots, \lambda_n$,

$$\lambda_{ijk} = \begin{cases} \lambda_i & \text{if } i = j = k, \\ 0 & \text{otherwise,} \end{cases}$$

or equivalently,

$$\mathbf{\Lambda} = (\lambda_1 \mathbf{u}_1 \mathbf{u}_1^T, \lambda_2 \mathbf{u}_2 \mathbf{u}_2^T, \ldots, \lambda_n \mathbf{u}_n \mathbf{u}_n^T),$$

where $\mathbf{u}_i$ is the ith unit n-vector;

(b) If the distribution of $\mathbf{x}$ is symmetric in the sense that the distributions of $[\mathbf{x} - \mathrm{E}(\mathbf{x})]$ and $-[\mathbf{x} - \mathrm{E}(\mathbf{x})]$ are identical, then $\mathbf{\Lambda} = \mathbf{0}$.

Solution

(a) For $i = j = k$, $\lambda_{ijk} = \lambda_{iii} \equiv \lambda_i$. For any other case of i, j, and k, at least one is not equal to the other two. Without loss of generality assume that $i \neq j$ and $i \neq k$. Then

$$\lambda_{ijk} = \mathrm{E}[(x_i - \mu_i)] \cdot \mathrm{E}[(x_j - \mu_j)(x_k - \mu_k)] = 0 \cdot \sigma_{jk} = 0.$$

(b) Because two random vectors having the same distribution (in this case $\mathbf{x} - \boldsymbol{\mu}$ and $-(\mathbf{x} - \boldsymbol{\mu})$) necessarily have the same moments,

$$\begin{aligned}\lambda_{ijk} &= \mathrm{E}[(x_i - \mu_i)(x_j - \mu_j)(x_k - \mu_k)] \\ &= \mathrm{E}\{[-(x_i - \mu_i)][-(x_j - \mu_j)][-(x_k - \mu_k)]\} \\ &= -\mathrm{E}[(x_i - \mu_i)(x_j - \mu_j)(x_k - \mu_k)] = -\lambda_{ijk}\end{aligned}$$

for all i, j, k. Thus $\lambda_{ijk} = 0$ for all i, j, k, i.e., $\mathbf{\Lambda} = \mathbf{0}$.

▶ **Exercise 6** Prove Theorem 4.1.4: If the kurtosis matrix $\mathbf{\Gamma} = (\gamma_{ijkl})$ of a random n-vector $\mathbf{x}$ exists and the elements of $\mathbf{x}$ are quadruple-wise independent (meaning that each quadrivariate marginal cdf of $\mathbf{x}$ factors into the product of its univariate marginal cdfs), then for some constants $\gamma_1, \ldots, \gamma_n$,

$$\gamma_{ijkl} = \begin{cases} \gamma_i & \text{if } i = j = k = l, \\ \sigma_{ii}\sigma_{kk} & \text{if } i = j \neq k = l, \\ \sigma_{ii}\sigma_{jj} & \text{if } i = k \neq j = l \text{ or } i = l \neq j = k, \\ 0 & \text{otherwise,} \end{cases}$$

where $\mathrm{var}(\mathbf{x}) = (\sigma_{ij})$.

Solution If $i = j = k = l$, then $\gamma_{ijkl} = \gamma_{iiii} \equiv \gamma_i$. If $i = j \neq k = l$, then $\gamma_{ijkl} = \gamma_{iikk} = \mathrm{E}[(x_i - \mu_i)^2(x_k - \mu_k)^2] = \mathrm{E}[(x_i - \mu_i)^2]\mathrm{E}[(x_k - \mu_k)^2] = \sigma_{ii}\sigma_{kk}$. By exactly the same argument, $\gamma_{ijkl} = \sigma_{ii}\sigma_{jj}$ if $i = k \neq j = l$ or $i = l \neq j = k$. For any other case, at least one of i, j, k, l is not equal to any of the other three. Without loss of generality assume that $i \neq j$, $i \neq k$, and $i \neq l$. Then $\gamma_{ijkl} = \mathrm{E}[(x_i - \mu_i)]\mathrm{E}[(x_j - \mu_j)(x_k - \mu_k)(x_l - \mu_l)] = 0 \cdot \lambda_{ijk} = 0$.

▶ **Exercise 7** Prove Theorem 4.2.1: Let $\mathbf{x}$ represent a random n-vector with mean $\boldsymbol{\mu}$, let $\mathbf{A}$ represent a $t \times n$ matrix of constants, and let $\mathbf{a}$ represent a t-vector of constants. Then

$$\mathrm{E}(\mathbf{A}\mathbf{x} + \mathbf{a}) = \mathbf{A}\boldsymbol{\mu} + \mathbf{a}.$$

Solution Represent $\mathbf{A}$ by its rows as

$$\begin{pmatrix} \mathbf{a}_1^T \\ \vdots \\ \mathbf{a}_n^T \end{pmatrix}$$

and let $\mathbf{a} = (a_i)$. Then

$$\mathrm{E}(\mathbf{A}\mathbf{x} + \mathbf{a}) = \mathrm{E}\left[\begin{pmatrix} \mathbf{a}_1^T \\ \vdots \\ \mathbf{a}_n^T \end{pmatrix}\mathbf{x} + \begin{pmatrix} a_1 \\ \vdots \\ a_n \end{pmatrix}\right] = \mathrm{E}\begin{pmatrix} \mathbf{a}_1^T\mathbf{x} + a_1 \\ \vdots \\ \mathbf{a}_n^T\mathbf{x} + a_n \end{pmatrix}$$

$$= \begin{pmatrix} \mathrm{E}(\mathbf{a}_1^T\mathbf{x} + a_1) \\ \vdots \\ \mathrm{E}(\mathbf{a}_n^T\mathbf{x} + a_n) \end{pmatrix} = \begin{pmatrix} \mathbf{a}_1^T\boldsymbol{\mu} + a_1 \\ \vdots \\ \mathbf{a}_n^T\boldsymbol{\mu} + a_n \end{pmatrix} = \mathbf{A}\boldsymbol{\mu} + \mathbf{a}.$$

▶ **Exercise 8** Prove Theorem 4.2.2: Let $\mathbf{x}$ and $\mathbf{y}$ represent random n-vectors, and let $\mathbf{z}$ and $\mathbf{w}$ represent random m-vectors. If $\mathrm{var}(\mathbf{x})$, $\mathrm{var}(\mathbf{y})$, $\mathrm{var}(\mathbf{z})$, and $\mathrm{var}(\mathbf{w})$ exist, then:

(a) $\mathrm{cov}(\mathbf{x} + \mathbf{y}, \mathbf{z} + \mathbf{w}) = \mathrm{cov}(\mathbf{x}, \mathbf{z}) + \mathrm{cov}(\mathbf{x}, \mathbf{w}) + \mathrm{cov}(\mathbf{y}, \mathbf{z}) + \mathrm{cov}(\mathbf{y}, \mathbf{w})$;
(b) $\mathrm{var}(\mathbf{x} + \mathbf{y}) = \mathrm{var}(\mathbf{x}) + \mathrm{var}(\mathbf{y}) + \mathrm{cov}(\mathbf{x}, \mathbf{y}) + [\mathrm{cov}(\mathbf{x}, \mathbf{y})]^T$.

Solution

(a)

$$\begin{aligned}
\mathrm{cov}(\mathbf{x} + \mathbf{y}, \mathbf{z} + \mathbf{w}) &= \mathrm{E}[(\mathbf{x} + \mathbf{y} - \mathrm{E}(\mathbf{x} + \mathbf{y}))(\mathbf{z} + \mathbf{w} - \mathrm{E}(\mathbf{z} + \mathbf{w}))^T] \\
&= \mathrm{E}[(\mathbf{x} - \mathrm{E}(\mathbf{x}))(\mathbf{z} - \mathrm{E}(\mathbf{z}))^T + (\mathbf{x} - \mathrm{E}(\mathbf{x}))(\mathbf{w} - \mathrm{E}(\mathbf{w}))^T \\
&\quad + (\mathbf{y} - \mathrm{E}(\mathbf{y}))(\mathbf{z} - \mathrm{E}(\mathbf{z}))^T \\
&\quad + (\mathbf{y} - \mathrm{E}(\mathbf{y}))(\mathbf{w} - \mathrm{E}(\mathbf{w}))^T] \\
&= \mathrm{cov}(\mathbf{x}, \mathbf{z}) + \mathrm{cov}(\mathbf{x}, \mathbf{w}) + \mathrm{cov}(\mathbf{y}, \mathbf{z}) + \mathrm{cov}(\mathbf{y}, \mathbf{w}).
\end{aligned}$$

(b) Using part (a),

$$\begin{aligned}\text{var}(\mathbf{x}+\mathbf{y}) &= \text{cov}(\mathbf{x}+\mathbf{y}, \mathbf{x}+\mathbf{y}) \\ &= \text{cov}(\mathbf{x}, \mathbf{x}) + \text{cov}(\mathbf{x}, \mathbf{y}) + \text{cov}(\mathbf{y}, \mathbf{x}) + \text{cov}(\mathbf{y}, \mathbf{y}) \\ &= \text{var}(\mathbf{x}) + \text{var}(\mathbf{y}) + \text{cov}(\mathbf{x}, \mathbf{y}) + [\text{cov}(\mathbf{x}, \mathbf{y})]^T.\end{aligned}$$

▶ **Exercise 9** Prove Theorem 4.2.3: Let $\mathbf{x}$ and $\mathbf{z}$ represent a random n-vector and a random m-vector, respectively, with matrix of covariances $\mathbf{\Phi}$; let $\mathbf{A}$ and $\mathbf{B}$ represent $t \times n$ and $u \times m$ matrices of constants, respectively; and let $\mathbf{a}$ and $\mathbf{b}$ represent a t-vector and u-vector of constants, respectively. Then

$$\text{cov}(\mathbf{A}\mathbf{x}+\mathbf{a}, \mathbf{B}\mathbf{z}+\mathbf{b}) = \mathbf{A}\mathbf{\Phi}\mathbf{B}^T.$$

Solution

$$\begin{aligned}\text{cov}(\mathbf{A}\mathbf{x}+\mathbf{a}, \mathbf{B}\mathbf{z}+\mathbf{b}) &= \text{E}[(\mathbf{A}\mathbf{x}+\mathbf{a}-\text{E}(\mathbf{A}\mathbf{x}+\mathbf{a}))(\mathbf{B}\mathbf{z}+\mathbf{b}-\text{E}(\mathbf{B}\mathbf{z}+\mathbf{b}))^T] \\ &= \text{E}[(\mathbf{A}\mathbf{x}+\mathbf{a}-\mathbf{A}\text{E}(\mathbf{x})-\mathbf{a})(\mathbf{B}\mathbf{z}+\mathbf{b}-\mathbf{B}\text{E}(\mathbf{z})-\mathbf{b})^T] \\ &= \text{E}[\mathbf{A}(\mathbf{x}-\text{E}(\mathbf{x}))(\mathbf{z}-\text{E}(\mathbf{z}))^T\mathbf{B}^T] \\ &= \mathbf{A}\text{E}[(\mathbf{x}-\text{E}(\mathbf{x}))(\mathbf{z}-\text{E}(\mathbf{z}))^T]\mathbf{B}^T \\ &= \mathbf{A}\mathbf{\Phi}\mathbf{B}^T.\end{aligned}$$

▶ **Exercise 10** Let $x_1, \ldots, x_n$ be uncorrelated random variables with common variance σ^2. Determine $\text{cov}(x_i - \bar{x}, x_j - \bar{x})$ for $i \leq j = 1, \ldots, n$.

Solution $x_i - \bar{x} = \mathbf{u}_i^T\mathbf{x} - (1/n)\mathbf{1}^T\mathbf{x}$ and $x_j - \bar{x} = \mathbf{u}_j^T\mathbf{x} - (1/n)\mathbf{1}^T\mathbf{x}$. Thus

$$\begin{aligned}\text{cov}(x_i - \bar{x}, x_j - \bar{x}) &= \text{cov}[(\mathbf{u}_i - \frac{1}{n}\mathbf{1})^T\mathbf{x}, (\mathbf{u}_j - \frac{1}{n}\mathbf{1})^T\mathbf{x}] \\ &= (\mathbf{u}_i - \frac{1}{n}\mathbf{1})^T(\sigma^2\mathbf{I})(\mathbf{u}_j - \frac{1}{n}\mathbf{1}) \\ &= \sigma^2(\mathbf{u}_i^T\mathbf{u}_j - \frac{1}{n}\mathbf{1}^T\mathbf{u}_j - \frac{1}{n}\mathbf{u}_i^T\mathbf{1} + \frac{1}{n^2}\mathbf{1}^T\mathbf{1}) \\ &= \begin{cases}\sigma^2(1-\frac{1}{n}) & \text{if } i = j, \\ -\frac{\sigma^2}{n} & \text{if } i < j.\end{cases}\end{aligned}$$

▶ **Exercise 11** Let $\mathbf{x}$ be a random n-vector with mean $\boldsymbol{\mu}$ and positive definite variance–covariance matrix $\mathbf{\Sigma}$. Define $\mathbf{\Sigma}^{-\frac{1}{2}} = (\mathbf{\Sigma}^{\frac{1}{2}})^{-1}$.

(a) Show that $\mathbf{\Sigma}^{-\frac{1}{2}}(\mathbf{x}-\boldsymbol{\mu})$ has mean $\mathbf{0}$ and variance–covariance matrix $\mathbf{I}$.

(b) Determine $\mathrm{E}[(\mathbf{x}-\boldsymbol{\mu})^T\boldsymbol{\Sigma}^{-1}(\mathbf{x}-\boldsymbol{\mu})]$ and $\mathrm{var}[(\mathbf{x}-\boldsymbol{\mu})^T\boldsymbol{\Sigma}^{-1}(\mathbf{x}-\boldsymbol{\mu})]$, assuming for the latter that the skewness matrix $\boldsymbol{\Lambda}$ and excess kurtosis matrix $\boldsymbol{\Omega}$ of $\mathbf{x}$ exist.

Solution

(a) By Theorem 4.2.1, $\mathrm{E}[\boldsymbol{\Sigma}^{-\frac{1}{2}}(\mathbf{x}-\boldsymbol{\mu})] = \mathrm{E}(\boldsymbol{\Sigma}^{-\frac{1}{2}}\mathbf{x} - \boldsymbol{\Sigma}^{-\frac{1}{2}}\boldsymbol{\mu}) = \boldsymbol{\Sigma}^{-\frac{1}{2}}\boldsymbol{\mu} - \boldsymbol{\Sigma}^{-\frac{1}{2}}\boldsymbol{\mu} = \mathbf{0}$. By Corollary 4.2.3.1 and the symmetry of $\boldsymbol{\Sigma}^{\frac{1}{2}}$, $\mathrm{var}[\boldsymbol{\Sigma}^{-\frac{1}{2}}(\mathbf{x}-\boldsymbol{\mu})] = \boldsymbol{\Sigma}^{-\frac{1}{2}}\boldsymbol{\Sigma}\boldsymbol{\Sigma}^{-\frac{1}{2}} = \mathbf{I}$.

(b) Using Theorem 4.2.4,

$$\mathrm{E}[(\mathbf{x}-\boldsymbol{\mu})^T\boldsymbol{\Sigma}^{-1}(\mathbf{x}-\boldsymbol{\mu})] = \mathbf{0}^T\boldsymbol{\Sigma}^{-1}\mathbf{0} + \mathrm{tr}(\boldsymbol{\Sigma}^{-1}\boldsymbol{\Sigma}) = \mathrm{tr}(\mathbf{I}_n) = n.$$

Furthermore, using Theorem 4.2.6,

$$\begin{aligned}\mathrm{var}[(\mathbf{x}-\boldsymbol{\mu})^T\boldsymbol{\Sigma}^{-1}(\mathbf{x}-\boldsymbol{\mu})] &= [\mathrm{vec}(\boldsymbol{\Sigma}^{-1})]^T\boldsymbol{\Omega}[\mathrm{vec}(\boldsymbol{\Sigma}^{-1})] + 4\mathbf{0}^T\boldsymbol{\Sigma}^{-1}\boldsymbol{\Lambda}\mathrm{vec}(\boldsymbol{\Sigma}^{-1}) \\ &\quad + 2\mathrm{tr}(\boldsymbol{\Sigma}^{-1}\boldsymbol{\Sigma}\boldsymbol{\Sigma}^{-1}\boldsymbol{\Sigma}) + 4\mathbf{0}^T\boldsymbol{\Sigma}^{-1}\boldsymbol{\Sigma}\boldsymbol{\Sigma}^{-1}\mathbf{0} \\ &= [\mathrm{vec}(\boldsymbol{\Sigma}^{-1})]^T\boldsymbol{\Omega}[\mathrm{vec}(\boldsymbol{\Sigma}^{-1})] + 2\mathrm{tr}(\mathbf{I}) \\ &= [\mathrm{vec}(\boldsymbol{\Sigma}^{-1})]^T\boldsymbol{\Omega}[\mathrm{vec}(\boldsymbol{\Sigma}^{-1})] + 2n.\end{aligned}$$

▶ **Exercise 12** Let $\mathbf{x}$ be a random n-vector with variance–covariance matrix $\sigma^2\mathbf{I}$, where $\sigma^2 > 0$, and let $\mathbf{Q}$ represent an $n \times n$ orthogonal matrix. Determine the variance–covariance matrix of $\mathbf{Qx}$.

Solution Because $\mathbf{Q}$ is orthogonal, $\mathbf{Q}^T\mathbf{Q} = \mathbf{Q}\mathbf{Q}^T = \mathbf{I}$. Therefore, by Corollary 4.2.3.1, $\mathrm{var}(\mathbf{Qx}) = \mathbf{Q}(\sigma^2\mathbf{I})\mathbf{Q}^T = \sigma^2\mathbf{I}$.

▶ **Exercise 13** Let $\mathbf{x} = \begin{pmatrix}\mathbf{x}_1\\ \mathbf{x}_2\end{pmatrix}$ have mean vector $\boldsymbol{\mu} = \begin{pmatrix}\boldsymbol{\mu}_1\\ \boldsymbol{\mu}_2\end{pmatrix}$ and variance–covariance matrix $\boldsymbol{\Sigma} = \begin{pmatrix}\boldsymbol{\Sigma}_{11} & \boldsymbol{\Sigma}_{12}\\ \boldsymbol{\Sigma}_{21} & \boldsymbol{\Sigma}_{22}\end{pmatrix}$, where $\mathbf{x}_1$ and $\boldsymbol{\mu}_1$ are m-vectors and $\boldsymbol{\Sigma}_{11}$ is $m \times m$. Suppose that $\boldsymbol{\Sigma}$ is positive definite, in which case both $\boldsymbol{\Sigma}_{22}$ and $\boldsymbol{\Sigma}_{11} - \boldsymbol{\Sigma}_{12}\boldsymbol{\Sigma}_{22}^{-1}\boldsymbol{\Sigma}_{21}$ are positive definite by Theorem 2.15.7a.

(a) Determine the mean vector and variance–covariance matrix of $\mathbf{x}_{1\cdot 2} = \mathbf{x}_1 - \boldsymbol{\mu}_1 - \boldsymbol{\Sigma}_{12}\boldsymbol{\Sigma}_{22}^{-1}(\mathbf{x}_2 - \boldsymbol{\mu}_2)$.
(b) Show that $\mathrm{cov}(\mathbf{x}_1 - \boldsymbol{\Sigma}_{12}\boldsymbol{\Sigma}_{22}^{-1}\mathbf{x}_2, \mathbf{x}_2) = \mathbf{0}$.
(c) Determine $\mathrm{E}\{\mathbf{x}_{1\cdot 2}^T[\mathrm{var}(\mathbf{x}_{1\cdot 2})]^{-1}\mathbf{x}_{1\cdot 2}\}$.

Solution

(a) By Theorem 4.2.1, $\mathrm{E}[\mathbf{x}_1 - \boldsymbol{\mu}_1 - \boldsymbol{\Sigma}_{12}\boldsymbol{\Sigma}_{22}^{-1}(\mathbf{x}_2 - \boldsymbol{\mu}_2)] = \mathrm{E}(\mathbf{x}_1) - \boldsymbol{\mu}_1 - \boldsymbol{\Sigma}_{12}\boldsymbol{\Sigma}_{22}^{-1}[\mathrm{E}(\mathbf{x}_2) - \boldsymbol{\mu}_2] = \mathbf{0}$. By Theorem 4.2.2 and Corollary 4.2.3.1,

$$\begin{aligned}
\mathrm{var}[\mathbf{x}_1 - \boldsymbol{\mu}_1 - \boldsymbol{\Sigma}_{12}\boldsymbol{\Sigma}_{22}^{-1}(\mathbf{x}_2 - \boldsymbol{\mu}_2)] &= \mathrm{var}(\mathbf{x}_1) + \mathrm{var}(\boldsymbol{\Sigma}_{12}\boldsymbol{\Sigma}_{22}^{-1}\mathbf{x}_2) \\
&\quad - \mathrm{cov}(\mathbf{x}_1, \boldsymbol{\Sigma}_{12}\boldsymbol{\Sigma}_{22}^{-1}\mathbf{x}_2) \\
&\quad - [\mathrm{cov}(\mathbf{x}_1, \boldsymbol{\Sigma}_{12}\boldsymbol{\Sigma}_{22}^{-1}\mathbf{x}_2)]^T \\
&= \boldsymbol{\Sigma}_{11} + \boldsymbol{\Sigma}_{12}\boldsymbol{\Sigma}_{22}^{-1}\boldsymbol{\Sigma}_{22}(\boldsymbol{\Sigma}_{22}^{-1})^T\boldsymbol{\Sigma}_{21} \\
&\quad - \boldsymbol{\Sigma}_{12}(\boldsymbol{\Sigma}_{22}^{-1})^T\boldsymbol{\Sigma}_{21} - \boldsymbol{\Sigma}_{12}\boldsymbol{\Sigma}_{22}^{-1}\boldsymbol{\Sigma}_{21} \\
&= \boldsymbol{\Sigma}_{11} - \boldsymbol{\Sigma}_{12}\boldsymbol{\Sigma}_{22}^{-1}\boldsymbol{\Sigma}_{21}.
\end{aligned}$$

(b) By Corollary 4.2.3.1,

$$\begin{aligned}
\mathrm{cov}(\mathbf{x}_1 - \boldsymbol{\Sigma}_{12}\boldsymbol{\Sigma}_{22}^{-1}\mathbf{x}_2, \mathbf{x}_2) &= \mathrm{cov}[(\mathbf{I}, -\boldsymbol{\Sigma}_{12}\boldsymbol{\Sigma}_{22}^{-1})\mathbf{x}, (\mathbf{0}, \mathbf{I})\mathbf{x}] \\
&= (\mathbf{I}, -\boldsymbol{\Sigma}_{12}\boldsymbol{\Sigma}_{22}^{-1})\begin{pmatrix} \boldsymbol{\Sigma}_{11} & \boldsymbol{\Sigma}_{12} \\ \boldsymbol{\Sigma}_{21} & \boldsymbol{\Sigma}_{22} \end{pmatrix}\begin{pmatrix} \mathbf{0} \\ \mathbf{I} \end{pmatrix} \\
&= \left(\boldsymbol{\Sigma}_{11} - \boldsymbol{\Sigma}_{12}\boldsymbol{\Sigma}_{22}^{-1}\boldsymbol{\Sigma}_{21}, \boldsymbol{\Sigma}_{12} - \boldsymbol{\Sigma}_{12}\boldsymbol{\Sigma}_{22}^{-1}\boldsymbol{\Sigma}_{22}\right)\begin{pmatrix} \mathbf{0} \\ \mathbf{I} \end{pmatrix} \\
&= \mathbf{0}.
\end{aligned}$$

(c) By part (a) and Theorem 4.2.4, this expectation is equal to

$$\begin{aligned}
&\mathbf{0}^T(\boldsymbol{\Sigma}_{11} - \boldsymbol{\Sigma}_{12}\boldsymbol{\Sigma}_{22}^{-1}\boldsymbol{\Sigma}_{21})^{-1}\mathbf{0} + \mathrm{tr}[(\boldsymbol{\Sigma}_{11} - \boldsymbol{\Sigma}_{12}\boldsymbol{\Sigma}_{22}^{-1}\boldsymbol{\Sigma}_{21})^{-1}(\boldsymbol{\Sigma}_{11} - \boldsymbol{\Sigma}_{12}\boldsymbol{\Sigma}_{22}^{-1}\boldsymbol{\Sigma}_{21})] \\
&= 0 + \mathrm{tr}(\mathbf{I}_m) = m.
\end{aligned}$$

▶ **Exercise 14** Suppose that observations $x_1, \ldots, x_n$ have common mean μ, common variance σ^2, and common correlation ρ among pairs, so that $\boldsymbol{\mu} = \mu\mathbf{1}_n$ and $\boldsymbol{\Sigma} = \sigma^2[(1-\rho)\mathbf{I}_n + \rho\mathbf{J}_n]$ for $\rho \in [-1/(n-1), 1]$. Determine $\mathrm{E}(\bar{x})$, $\mathrm{var}(\bar{x})$, and $\mathrm{E}(s^2)$.

Solution Let $\mathbf{x} = (x_1, \ldots, x_n)^T$, so that $\mathrm{E}(\mathbf{x}) = \boldsymbol{\mu} = \mu\mathbf{1}_n$ and $\mathrm{var}(\mathbf{x}) = \sigma^2[(1-\rho)\mathbf{I}_n + \rho\mathbf{J}_n]$. Then by Theorems 4.2.1 and 4.2.4 and Corollary 4.2.3.1,

$$\begin{aligned}
\mathrm{E}(\bar{x}) &= \mathrm{E}[(1/n)\mathbf{1}^T\mathbf{x}] = (1/n)\mathbf{1}^T(\mu\mathbf{1}) = \mu; \\
\mathrm{var}(\bar{x}) &= \mathrm{var}[(1/n)\mathbf{1}^T\mathbf{x}] = (\sigma^2/n^2)\mathbf{1}^T[(1-\rho)\mathbf{I}_n + \rho\mathbf{J}_n]\mathbf{1} \\
&= (\sigma^2/n^2)[(1-\rho)n + \rho n^2] = (\sigma^2/n)[1 + \rho(n-1)];
\end{aligned}$$

$$\begin{aligned}
\mathrm{E}(s^2) &= \mathrm{E}\left(\frac{1}{n-1}\mathbf{x}^T[\mathbf{I}_n - (1/n)\mathbf{J}_n]\mathbf{x}\right) \\
&= \frac{1}{n-1}\{(\mu\mathbf{1})^T[\mathbf{I}_n - (1/n)\mathbf{J}_n](\mu\mathbf{1}\} + \frac{\sigma^2}{n-1}\mathrm{tr}\{[\mathbf{I} - (1/n)\mathbf{J}_n][(1-\rho)\mathbf{I}_n + \rho\mathbf{J}_n]\} \\
&= 0 + \frac{\sigma^2}{n-1}\mathrm{tr}\{(1-\rho)[\mathbf{I}_n - (1/n)\mathbf{J}_n]\} \\
&= \sigma^2(1-\rho).
\end{aligned}$$

► **Exercise 15** Prove that if $\mathbf{x}$ is a random vector for which $\boldsymbol{\mu} = \mathbf{0}$ and $\boldsymbol{\Lambda} = \mathbf{0}$, then any linear form $\mathbf{b}^T\mathbf{x}$ is uncorrelated with any quadratic form $\mathbf{x}^T\mathbf{A}\mathbf{x}$.

Solution By Corollary 4.2.5.1, $\mathrm{cov}(\mathbf{b}^T\mathbf{x}, \mathbf{x}^T\mathbf{A}\mathbf{x}) = 2\mathbf{b}^T\boldsymbol{\Sigma}\mathbf{A}\mathbf{0} = 0$.

► **Exercise 16** Determine how Theorems 4.2.1 and 4.2.4–4.2.6 specialize when $\boldsymbol{\mu}$ is orthogonal to every row of $\mathbf{A}$, i.e., when $\mathbf{A}\boldsymbol{\mu} = \mathbf{0}$.

Solution When $\boldsymbol{\mu}$ is orthogonal to every row of $\mathbf{A}$, i.e., when $\mathbf{A}\boldsymbol{\mu} = \mathbf{0}$, then $\mathrm{E}(\mathbf{A}\mathbf{x} + \mathbf{a}) = \mathbf{a}$, $\mathrm{E}(\mathbf{x}^T\mathbf{A}\mathbf{x}) = \mathrm{tr}(\mathbf{A}\boldsymbol{\Sigma})$, $\mathrm{cov}(\mathbf{B}\mathbf{x}, \mathbf{x}^T\mathbf{A}\mathbf{x}) = \mathbf{B}\boldsymbol{\Lambda}\mathrm{vec}(\mathbf{A})$, and $\mathrm{cov}(\mathbf{x}^T\mathbf{A}\mathbf{x}, \mathbf{x}^T\mathbf{B}\mathbf{x}) = [\mathrm{vec}(\mathbf{A})]^T\boldsymbol{\Omega}\mathrm{vec}(\mathbf{B}) + 2\boldsymbol{\mu}^T\mathbf{B}\boldsymbol{\Lambda}\mathrm{vec}(\mathbf{A}) + 2\mathrm{tr}(\mathbf{A}\boldsymbol{\Sigma}\mathbf{B}\boldsymbol{\Sigma})$.

► **Exercise 17** Let $\mathbf{x}$ be a random n-vector with mean $\boldsymbol{\mu}$ and variance–covariance matrix $\boldsymbol{\Sigma}$ of rank r. Find $\mathrm{E}(\mathbf{x}^T\boldsymbol{\Sigma}^-\mathbf{x})$ and simplify it as much as possible.

Solution By Theorems 4.2.4, 3.3.5, and 2.12.2,

$$\mathrm{E}(\mathbf{x}^T\boldsymbol{\Sigma}^-\mathbf{x}) = \boldsymbol{\mu}^T\boldsymbol{\Sigma}^-\boldsymbol{\mu} + \mathrm{tr}(\boldsymbol{\Sigma}^-\boldsymbol{\Sigma}) = \boldsymbol{\mu}^T\boldsymbol{\Sigma}^-\boldsymbol{\mu} + r.$$

► **Exercise 18** Determine the covariance between the sample mean and sample variance of observations whose joint distribution is symmetric with common mean μ, variance–covariance matrix $\boldsymbol{\Sigma}$, and finite skewness matrix $\boldsymbol{\Lambda}$.

Solution By Theorem 4.1.3b and Corollary 4.2.5.1, $\mathrm{cov}(\bar{x}, s^2) = 2(1/n)\mathbf{1}^T\boldsymbol{\Sigma}\{[\mathbf{I} - (1/n)\mathbf{J}]/(n-1)\}(\mu\mathbf{1}) = (2/n)\mathbf{1}^T\boldsymbol{\Sigma}\mathbf{0} = 0$.

Types of Linear Models

5

This chapter presents exercises on types of linear models, and provides solutions to those exercises.

▶ **Exercise 1** Give as concise a representation of the model matrix $\mathbf{X}$ as possible for a two-way main effects model when the data are balanced, and specialize further to the case $r = 1$.

Solution This model is a special case of (5.6) with $n_{ij} = r$. Thus, we have $n_{i\cdot} = mr$ and $n_{\cdot j} = qr$ for all i, j. Then by (5.7), the model matrix for a balanced two-way main effects model is given by

$$\mathbf{X} = \begin{pmatrix} \mathbf{1}_{mr} & \mathbf{1}_{mr} & \mathbf{0}_{mr} & \cdots & \mathbf{0}_{mr} & \oplus_{j=1}^{m} \mathbf{1}_r \\ \mathbf{1}_{mr} & \mathbf{0}_{mr} & \mathbf{1}_{mr} & \cdots & \mathbf{0}_{mr} & \oplus_{j=1}^{m} \mathbf{1}_r \\ \vdots & \vdots & \vdots & \ddots & \vdots & \vdots \\ \mathbf{1}_{mr} & \mathbf{0}_{mr} & \mathbf{0}_{mr} & \cdots & \mathbf{1}_{mr} & \oplus_{j=1}^{m} \mathbf{1}_r \end{pmatrix} = \left(\mathbf{1}_{qmr},\ \mathbf{I}_q \otimes \mathbf{1}_{mr},\ \mathbf{1}_q \otimes \mathbf{I}_m \otimes \mathbf{1}_r \right).$$

Furthermore, for the case that $r = 1$, we have

$$\mathbf{X} = \left(\mathbf{1}_{qm},\ \mathbf{I}_q \otimes \mathbf{1}_m,\ \mathbf{1}_q \otimes \mathbf{I}_m \right).$$

▶ **Exercise 2** Give a concise representation of the model matrix $\mathbf{X}$ for the cell-means representation of the two-way model with interaction. Is this $\mathbf{X}$ full rank? Specialize to the case of balanced data.

Solution $\mathbf{X} = \oplus_{i=1}^{q} \oplus_{j=1}^{m} \mathbf{1}_{n_{ij}}$, which has full rank qm. In the case of balanced data with $n_{ij} = r$ for all i and j, $\mathbf{X} = \oplus_{i=1}^{q} \oplus_{j=1}^{m} \mathbf{1}_r = \mathbf{I}_{qm} \otimes \mathbf{1}_r$.

D. L. Zimmerman, *Linear Model Theory*,
https://doi.org/10.1007/978-3-030-52074-8_5

▶ **Exercise 3** Give the model matrix $\mathbf{X}$ for a three-way main effects model in which each factor has two levels and each cell of the $2 \times 2 \times 2$ layout has exactly one observation, i.e., for the model

$$y_{ijk} = \mu + \alpha_i + \gamma_j + \delta_k + e_{ijk} \qquad (i = 1, 2;\ j = 1, 2;\ k = 1, 2).$$

Solution

$$\mathbf{X} = \begin{pmatrix} 1\,1\,0\,1\,0\,1\,0 \\ 1\,1\,0\,1\,0\,0\,1 \\ 1\,1\,0\,0\,1\,1\,0 \\ 1\,1\,0\,0\,1\,0\,1 \\ 1\,0\,1\,1\,0\,1\,0 \\ 1\,0\,1\,1\,0\,0\,1 \\ 1\,0\,1\,0\,1\,1\,0 \\ 1\,0\,1\,0\,1\,0\,1 \end{pmatrix} = \left(\mathbf{1}_8,\ \mathbf{I}_2 \otimes \mathbf{1}_4,\ \mathbf{1}_2 \otimes \mathbf{I}_2 \otimes \mathbf{1}_2,\ \mathbf{1}_4 \otimes \mathbf{I}_2\right).$$

▶ **Exercise 4** Give a concise representation of the model matrix $\mathbf{X}$ for the two-factor nested model, and specialize it to the case of balanced data.

Solution Assuming that the elements of $\mathbf{y}$ are ordered lexicographically,

$$\mathbf{X} = \begin{pmatrix} \mathbf{1}_{n_{1\cdot}} & \mathbf{1}_{n_{1\cdot}} & \mathbf{0}_{n_{1\cdot}} & \cdots & \mathbf{0}_{n_{1\cdot}} & \oplus_{j=1}^{m_1}\mathbf{1}_{n_{1j}} & & \\ \mathbf{1}_{n_{2\cdot}} & \mathbf{0}_{n_{2\cdot}} & \mathbf{1}_{n_{2\cdot}} & \cdots & \mathbf{0}_{n_{2\cdot}} & & \oplus_{j=1}^{m_2}\mathbf{1}_{n_{2j}} & \\ \vdots & \vdots & \vdots & \ddots & \vdots & & & \ddots & \\ \mathbf{1}_{n_{q\cdot}} & \mathbf{0}_{n_{q\cdot}} & \mathbf{0}_{n_{q\cdot}} & \cdots & \mathbf{1}_{n_{q\cdot}} & & & & \oplus_{j=1}^{m_q}\mathbf{1}_{n_{qj}} \end{pmatrix}.$$

For balanced data, $m_i = m$ and $n_{ij} = r$ for all i and j, so the model matrix may be simplified to $\mathbf{X} = \left(\mathbf{1}_{qmr},\ \mathbf{I}_q \otimes \mathbf{1}_{mr},\ \mathbf{I}_{qm} \otimes \mathbf{1}_r\right)$.

▶ **Exercise 5** Give a concise representation of the model matrix $\mathbf{X}$ for the balanced two-factor partially crossed model

$$y_{ijk} = \mu + \alpha_i - \gamma_j + e_{ijk} \quad (i \neq j = 1, \ldots, q;\ k = 1, \ldots, r).$$

Solution

$$\mathbf{X} = \left(\mathbf{1}_{rq(q-1)},\ \mathbf{I}_q \otimes \mathbf{1}_{r(q-1)},\ -(\mathbf{1}_r \mathbf{u}_j^{(q)T})_{j \neq i = 1,\ldots,q}\right).$$

▶ **Exercise 6** Give a concise representation of the model matrix $\mathbf{X}$ for a one-factor, factor-specific-slope, analysis-of-covariance model

$$y_{ij} = \mu + \alpha_i + \gamma_i x_{ij} + e_{ij} \qquad (i = 1, \ldots, q;\ j = 1, \ldots, n_i).$$

Solution

$$\mathbf{X} = \left(\mathbf{1}_{n.},\ \oplus_{i=1}^{q}\mathbf{1}_{n_i},\ \oplus_{i=1}^{q}\mathbf{x}_i\right),$$

where $n. = \sum_{i=1}^{q} n_i$ and $\mathbf{x}_i = (x_{i1}, \ldots, x_{in_i})^T$.

▶ **Exercise 7** Determine the variance–covariance matrix for each of the following two-factor mixed effects models:

(a) $y_{ij} = \mu + \alpha_i + b_j + d_{ijk}$ $(i = 1, \ldots, q;\ j = 1, \ldots, m;\ k = 1, \ldots, n_{ij})$, where $\mathrm{E}(b_j) = 0$ and $\mathrm{E}(d_{ijk}) = 0$ for all i, j, k, the b_j's and d_{ijk}'s are all uncorrelated, $\mathrm{var}(b_j) = \sigma_b^2$ for all j, and $\mathrm{var}(d_{ijk}) = \sigma^2$ for all i, j, k; with parameter space $\{\mu, \sigma_b^2, \sigma^2 : \mu \in \mathbb{R},\ \sigma_b^2 > 0,\ \sigma^2 > 0\}$.
(b) $y_{ij} = \mu + a_i + \gamma_j + d_{ijk}$ $(i = 1, \ldots, q;\ j = 1, \ldots, m;\ k = 1, \ldots, n_{ij})$, where $\mathrm{E}(a_i) = 0$ and $\mathrm{E}(d_{ijk}) = 0$ for all i, j, k, the a_i's and d_{ijk}'s are all uncorrelated, $\mathrm{var}(a_i) = \sigma_a^2$ for all i, and $\mathrm{var}(d_{ijk}) = \sigma^2$ for all i, j, k; with parameter space $\{\mu, \sigma_a^2, \sigma^2 : \mu \in \mathbb{R},\ \sigma_a^2 > 0,\ \sigma^2 > 0\}$.
(c) $y_{ij} = \mu + \alpha_i + b_j + c_{ij} + d_{ijk}$ $(i = 1, \ldots, q;\ j = 1, \ldots, m;\ k = 1, \ldots, n_{ij})$, where $\mathrm{E}(b_j) = \mathrm{E}(c_{ij}) = 0$ and $\mathrm{E}(d_{ijk}) = 0$ for all i, j, k, the b_j's, c_{ij}'s, and d_{ijk}'s are all uncorrelated, $\mathrm{var}(b_j) = \sigma_b^2$ for all j, $\mathrm{var}(c_{ij}) = \sigma_c^2$ for all i and j, and $\mathrm{var}(d_{ijk}) = \sigma^2$ for all i, j, k; with parameter space $\{\mu, \sigma_b^2, \sigma_c^2, \sigma^2 : \mu \in \mathbb{R},\ \sigma_b^2 > 0,\ \sigma_c^2 > 0,\ \sigma^2 > 0\}$.
(d) $y_{ij} = \mu + a_i + \gamma_j + c_{ij} + d_{ijk}$ $(i = 1, \ldots, q;\ j = 1, \ldots, m;\ k = 1, \ldots, n_{ij})$, where $\mathrm{E}(a_i) = \mathrm{E}(c_{ij}) = 0$ and $\mathrm{E}(d_{ijk}) = 0$ for all i, j, k, the a_i's, c_{ij}'s, and d_{ijk}'s are all uncorrelated, $\mathrm{var}(a_i) = \sigma_a^2$ for all i, $\mathrm{var}(c_{ij}) = \sigma_c^2$ for all i and j, and $\mathrm{var}(d_{ijk}) = \sigma^2$ for all i, j, k; with parameter space $\{\mu, \sigma_a^2, \sigma_c^2, \sigma^2 : \mu \in \mathbb{R},\ \sigma_a^2 > 0,\ \sigma_c^2 > 0,\ \sigma^2 > 0\}$.
(e) $y_{ij} = \mu + a_i + b_j + c_{ij} + d_{ijk}$ $(i = 1, \ldots, q;\ j = 1, \ldots, m;\ k = 1, \ldots, n_{ij})$, where $\mathrm{E}(a_i) = \mathrm{E}(b_j) = \mathrm{E}(c_{ij}) = 0$ and $\mathrm{E}(d_{ijk}) = 0$ for all i, j, k, the a_i's, b_j's, c_{ij}'s, and d_{ijk}'s are all uncorrelated, $\mathrm{var}(a_i) = \sigma_a^2$ for all i, $\mathrm{var}(b_j) = \sigma_b^2$ for all j, $\mathrm{var}(c_{ij}) = \sigma_c^2$ for all i and j, and $\mathrm{var}(d_{ijk}) = \sigma^2$ for all i, j, k; with parameter space $\{\mu, \sigma_a^2, \sigma_b^2, \sigma_c^2, \sigma^2 : \mu \in \mathbb{R},\ \sigma_a^2 > 0,\ \sigma_b^2 > 0,\ \sigma_c^2 > 0,\ \sigma^2 > 0\}$.

Solution

(a) This model may be written as

$$\mathbf{y} = \mathbf{X}\boldsymbol{\beta} + \mathbf{Z}\mathbf{b} + \mathbf{d},$$

where

$$\mathbf{Z} = \begin{pmatrix} \oplus_{j=1}^{m}\mathbf{1}_{n_{1j}} \\ \oplus_{j=1}^{m}\mathbf{1}_{n_{2j}} \\ \vdots \\ \oplus_{j=1}^{m}\mathbf{1}_{n_{qj}} \end{pmatrix}, \quad \mathbf{b} = \begin{pmatrix} b_1 \\ \vdots \\ b_m \end{pmatrix}, \quad \begin{pmatrix} d_{111} \\ d_{112} \\ \vdots \\ d_{qmn_{qm}} \end{pmatrix}.$$

Then

$$\text{var}(\mathbf{y}) = \mathbf{Z}\text{var}(\mathbf{b})\mathbf{Z}^T + \text{var}(\mathbf{d})$$

$$= \sigma_b^2 \begin{pmatrix} \oplus_{j=1}^m \mathbf{J}_{n_{1j}\times n_{1j}} & \oplus_{j=1}^m \mathbf{J}_{n_{1j}\times n_{2j}} & \cdots & \oplus_{j=1}^m \mathbf{J}_{n_{1j}\times n_{qj}} \\ \oplus_{j=1}^m \mathbf{J}_{n_{2j}\times n_{1j}} & \oplus_{j=1}^m \mathbf{J}_{n_{2j}\times n_{2j}} & \cdots & \oplus_{j=1}^m \mathbf{J}_{n_{2j}\times n_{qj}} \\ \vdots & \vdots & \ddots & \vdots \\ \oplus_{j=1}^m \mathbf{J}_{n_{qj}\times n_{1j}} & \oplus_{j=1}^m \mathbf{J}_{n_{qj}\times n_{2j}} & \cdots & \oplus_{j=1}^m \mathbf{J}_{n_{qj}\times n_{qj}} \end{pmatrix} + \sigma^2\mathbf{I}_n.$$

(b) This model may be written as

$$\mathbf{y} = \mathbf{X}\boldsymbol{\beta} + \mathbf{Z}\mathbf{b} + \mathbf{d},$$

where

$$\mathbf{Z} = \oplus_{i=1}^q \mathbf{1}_{n_{i\cdot}}, \quad \mathbf{b} = (a_1, \ldots, a_q)^T, \quad \mathbf{d} = (d_{111}, \ldots, d_{qmn_{qm}})^T.$$

Then

$$\text{var}(\mathbf{y}) = \mathbf{Z}\text{var}(\mathbf{b})\mathbf{Z}^T + \text{var}(\mathbf{d}) = \sigma_a^2 \oplus_{i=1}^q \mathbf{J}_{n_{i\cdot}} + \sigma^2\mathbf{I}_n.$$

(c) This model may be written as

$$\mathbf{y} = \mathbf{X}\boldsymbol{\beta} + \mathbf{Z}_1\mathbf{b}_1 + \mathbf{Z}_2\mathbf{b}_2 + \mathbf{d},$$

where

$$\mathbf{Z}_1 = \begin{pmatrix} \oplus_{j=1}^m \mathbf{1}_{n_{1j}} \\ \oplus_{j=1}^m \mathbf{1}_{n_{2j}} \\ \vdots \\ \oplus_{j=1}^m \mathbf{1}_{n_{qj}} \end{pmatrix}, \quad \mathbf{b} = \begin{pmatrix} b_1 \\ \vdots \\ b_m \end{pmatrix}, \quad \mathbf{Z}_2 = \oplus_{i=1}^q \oplus_{j=1}^m \mathbf{1}_{n_{ij}},$$

$$\mathbf{b}_2 = \begin{pmatrix} c_{11} \\ \vdots \\ c_{qm} \end{pmatrix}, \quad \mathbf{d} = \begin{pmatrix} d_{111} \\ \vdots \\ d_{qmn_{qm}} \end{pmatrix}.$$

Then

$$\begin{aligned} \text{var}(\mathbf{y}) &= \mathbf{Z}_1\text{var}(\mathbf{b}_1)\mathbf{Z}_1^T + \mathbf{Z}_2\text{var}(\mathbf{b}_2)\mathbf{Z}_2^T + \text{var}(\mathbf{d}) \\ &= \sigma_b^2 \begin{pmatrix} \oplus_{j=1}^m \mathbf{J}_{n_{1j}\times n_{1j}} & \oplus_{j=1}^m \mathbf{J}_{n_{1j}\times n_{2j}} & \cdots & \oplus_{j=1}^m \mathbf{J}_{n_{1j}\times n_{qj}} \\ \oplus_{j=1}^m \mathbf{J}_{n_{2j}\times n_{1j}} & \oplus_{j=1}^m \mathbf{J}_{n_{2j}\times n_{2j}} & \cdots & \oplus_{j=1}^m \mathbf{J}_{n_{2j}\times n_{qj}} \\ \vdots & \vdots & \ddots & \vdots \\ \oplus_{j=1}^m \mathbf{J}_{n_{qj}\times n_{1j}} & \oplus_{j=1}^m \mathbf{J}_{n_{qj}\times n_{2j}} & \cdots & \oplus_{j=1}^m \mathbf{J}_{n_{qj}\times n_{qj}} \end{pmatrix} \\ &\quad + \sigma_c^2 \oplus_{i=1}^q \oplus_{j=1}^m \mathbf{J}_{n_{ij}} + \sigma^2\mathbf{I}_n. \end{aligned}$$

(d) This model may be written as

$$\mathbf{y} = \mathbf{X}\boldsymbol{\beta} + \mathbf{Z}_1\mathbf{b}_1 + \mathbf{Z}_2\mathbf{b}_2 + \mathbf{d},$$

where

$$\mathbf{Z}_1 = \oplus_{i=1}^{q}\mathbf{1}_{n_{i\cdot}}, \quad \mathbf{b}_1 = \begin{pmatrix} a_1 \\ \vdots \\ a_q \end{pmatrix}, \quad \mathbf{Z}_2 = \oplus_{i=1}^{q} \oplus_{j=1}^{m} \mathbf{1}_{n_{ij}},$$

$$\mathbf{b}_2 = \begin{pmatrix} c_{11} \\ \vdots \\ c_{qm} \end{pmatrix}, \quad \mathbf{d} = \begin{pmatrix} d_{111} \\ \vdots \\ d_{qmn_{qm}} \end{pmatrix}.$$

Then

$$\begin{aligned} \text{var}(\mathbf{y}) &= \mathbf{Z}_1\text{var}(\mathbf{b}_1)\mathbf{Z}_1^T + \mathbf{Z}_2\text{var}(\mathbf{b}_2)\mathbf{Z}_2^T + \text{var}(\mathbf{d}) \\ &= \sigma_a^2 \oplus_{i=1}^{q} \mathbf{J}_{n_{i\cdot}} + \sigma_c^2 \oplus_{i=1}^{q} \oplus_{j=1}^{m}\mathbf{J}_{n_{ij}} + \sigma^2\mathbf{I}_n. \end{aligned}$$

(e) This model may be written as

$$\mathbf{y} = \mathbf{X}\boldsymbol{\beta} + \mathbf{Z}_1\mathbf{b}_1 + \mathbf{Z}_2\mathbf{b}_2 + \mathbf{Z}_3\mathbf{b}_3 + \mathbf{d},$$

where

$$\mathbf{Z}_1 = \oplus_{i=1}^{q}\mathbf{1}_{n_{i\cdot}}, \quad \mathbf{b}_1 = \begin{pmatrix} a_1 \\ \vdots \\ a_q \end{pmatrix}, \quad \mathbf{Z}_2 = \begin{pmatrix} \oplus_{j=1}^{m}\mathbf{1}_{n_{1j}} \\ \oplus_{j=1}^{m}\mathbf{1}_{n_{2j}} \\ \vdots \\ \oplus_{j=1}^{m}\mathbf{1}_{n_{qj}} \end{pmatrix}, \quad \mathbf{b}_2 = \begin{pmatrix} b_1 \\ \vdots \\ b_m \end{pmatrix},$$

$$\mathbf{Z}_3 = \oplus_{i=1}^{q} \oplus_{j=1}^{m} \mathbf{1}_{n_{ij}}, \quad \mathbf{b}_3 = (c_{11}, \ldots, c_{qm})^T, \quad \mathbf{d} = (d_{111}, \ldots, d_{qmn_{qm}})^T.$$

Then

$$\begin{aligned} \text{var}(\mathbf{y}) &= \mathbf{Z}_1\text{var}(\mathbf{b}_1)\mathbf{Z}_1^T + \mathbf{Z}_2\text{var}(\mathbf{b}_2)\mathbf{Z}_2^T + \mathbf{Z}_3\text{var}(\mathbf{b}_3)\mathbf{Z}_3^T + \text{var}(\mathbf{d}) \\ &= \sigma_a^2 \oplus_{i=1}^{q} \mathbf{J}_{n_{i\cdot}} + \sigma_b^2 \begin{pmatrix} \oplus_{j=1}^{m}\mathbf{J}_{n_{1j}\times n_{1j}} & \oplus_{j=1}^{m}\mathbf{J}_{n_{1j}\times n_{2j}} & \cdots & \oplus_{j=1}^{m}\mathbf{J}_{n_{1j}\times n_{qj}} \\ \oplus_{j=1}^{m}\mathbf{J}_{n_{2j}\times n_{1j}} & \oplus_{j=1}^{m}\mathbf{J}_{n_{2j}\times n_{2j}} & \cdots & \oplus_{j=1}^{m}\mathbf{J}_{n_{2j}\times n_{qj}} \\ \vdots & \vdots & \ddots & \vdots \\ \oplus_{j=1}^{m}\mathbf{J}_{n_{qj}\times n_{1j}} & \oplus_{j=1}^{m}\mathbf{J}_{n_{qj}\times n_{2j}} & \cdots & \oplus_{j=1}^{m}\mathbf{J}_{n_{qj}\times n_{qj}} \end{pmatrix} \\ &\quad + \sigma_c^2 \oplus_{i=1}^{q} \oplus_{j=1}^{m}\mathbf{J}_{n_{ij}} + \sigma^2\mathbf{I}_n. \end{aligned}$$

▶ **Exercise 8** Consider the stationary autoregressive model of order one, given by

$$y_i = \mu + \rho(y_{i-1} - \mu) + u_i \qquad (i = 1, \ldots, n),$$

where $u_1, \ldots, u_n$ are uncorrelated random variables with common mean 0, $\text{var}(u_1) = \sigma^2/(1-\rho^2)$, $\text{var}(u_i) = \sigma^2$ for $i = 2, \ldots, n$, and $y_0 \equiv \mu$, and the parameter space is $\{\mu, \rho, \sigma^2 : \mu \in \mathbb{R}, -1 < \rho < 1, \sigma^2 > 0\}$. Verify that this model has variance–covariance matrix given by

$$\frac{\sigma^2}{1-\rho^2}\begin{pmatrix} 1 & \rho & \rho^2 & \rho^3 & \cdots & \rho^{n-1} \\ & 1 & \rho & \rho^2 & \cdots & \rho^{n-2} \\ & & 1 & \rho & \cdots & \rho^{n-3} \\ & & & 1 & \cdots & \rho^{n-4} \\ & & & & \ddots & \vdots \\ & & & & & 1 \end{pmatrix}.$$

Solution For a fixed $i \in \{1, \ldots, n\}$ let $j \in \{0, 1, \ldots, n-1\}$. Subtracting μ from both sides of the model equation and multiplying both sides of the resulting equation by $y_{i-j} - \mu$ yields

$$(y_{i-j} - \mu)(y_i - \mu) = \rho(y_{i-j} - \mu)(y_{i-1} - \mu) + (y_{i-j} - \mu)u_i.$$

Taking expectations of both sides yields the recursive equation

$$\text{cov}(y_{i-j}, y_i) = \rho\text{cov}(y_{i-j}, y_{i-1}) + \sigma^2 I_{\{j=0\}}.$$

Then writing σ_{ij} for the (i, j)th element of the variance–covariance matrix, we obtain

$$\begin{aligned}
\sigma_{11} &= \sigma^2/(1-\rho^2), \\
\sigma_{12} &= \rho\sigma_{11} = \rho\sigma^2/(1-\rho^2), \\
\sigma_{22} &= \rho\sigma_{12} + \sigma^2 = \rho^2\sigma^2/(1-\rho^2) + \sigma^2 = \sigma^2/(1-\rho^2), \\
\sigma_{13} &= \rho\sigma_{12} = \rho^2\sigma^2/(1-\rho^2), \\
\sigma_{23} &= \rho\sigma_{22} = \rho\sigma^2/(1-\rho^2), \\
\sigma_{33} &= \rho\sigma_{23} + \sigma^2 = \rho^2\sigma^2/(1-\rho^2) + \sigma^2 = \sigma^2/(1-\rho^2).
\end{aligned}$$

The method of induction may be used to verify that the remaining elements of the variance–covariance matrix coincide with those given by (5.13).

▶ **Exercise 9** Determine the variance–covariance matrix for $\mathbf{y} = (y_1, y_2, \ldots, y_n)^T$ where

$$y_i = \mu + \sum_{j=1}^{i} u_j \qquad (i = 1, \ldots, n),$$

and the u_j's are uncorrelated random variables with mean 0 and variance $\sigma^2 > 0$.

Solution By Theorems 4.2.2 and 4.2.3, $\text{cov}(y_i, y_k) = \text{cov}\left(\mu + \sum_{j=1}^{i} u_j, \mu + \sum_{l=1}^{k} u_l\right) = \sum_{j=1}^{i} \sum_{l=1}^{k} \text{cov}(u_j, u_l) = \sum_{j=1}^{\min(i,k)} \text{var}(u_j) = \sigma^2 \min(i, k)$. Thus, the variance–covariance matrix is

$$\sigma^2 \begin{pmatrix} 1 & 1 & 1 & \cdots & 1 \\ 1 & 2 & 2 & \cdots & 2 \\ 1 & 2 & 3 & \cdots & 3 \\ \vdots & \vdots & \vdots & \ddots & \vdots \\ 1 & 2 & 3 & \cdots & n \end{pmatrix}.$$

▶ **Exercise 10** Determine the variance–covariance matrix for $\mathbf{y} = (y_1, y_2, y_3)^T$ where

$$y_i = \mu_i + \phi_{i-1}(y_{i-1} - \mu_{i-1}) + u_i \qquad (i = 1, 2, 3),$$

$y_0 \equiv \mu_0$, and u_1, u_2, u_3 are uncorrelated random variables with common mean 0 and $\text{var}(u_i) = \sigma_i^2$ for $i = 1, 2, 3$; the parameter space is $\{\mu_0, \mu_1, \mu_2, \mu_3, \phi_0, \phi_1, \phi_2, \sigma_1^2, \sigma_2^2, \sigma_3^2 : \mu_i \in \mathbb{R}$ for all i, $\phi_i \in \mathbb{R}$ for all i, and $\sigma_i^2 > 0$ for all $i\}$. This model is an extension of the first-order stationary autoregressive model (for three observations) called the *first-order antedependence model.*

Solution For a fixed $i \in \{1, 2, 3\}$ let $j \in \{0, 1, 2\}$. Subtracting μ_i from both sides of the model equation and multiplying both sides of the resulting equation by $y_{i-j} - \mu_{i-j}$ yields

$$(y_{i-j} - \mu_{i-j})(y_i - \mu_i) = \phi_{i-1}(y_{i-j} - \mu_{i-j})(y_{i-1} - \mu_{i-1}) + (y_{i-j} - \mu_{i-j})u_i.$$

Taking expectations of both sides yields the recursive equation

$$\text{cov}(y_{i-j}, y_i) = \phi_{i-1}\text{cov}(y_{i-j}, y_{i-1}) + \sigma_i^2 I_{\{j=0\}}.$$

Thus writing σ_{ij} for the (i, j)th element of the variance matrix, we obtain

$$\begin{aligned}
\sigma_{11} &= \sigma_1^2,\\
\sigma_{12} &= \phi_1\sigma_{11} = \phi_1\sigma_1^2,\\
\sigma_{22} &= \phi_1\sigma_{12} + \sigma_2^2 = \phi_1^2\sigma_1^2 + \sigma_2^2,\\
\sigma_{13} &= \phi_2\sigma_{12} = \phi_2\phi_1\sigma_1^2,\\
\sigma_{23} &= \phi_2\sigma_{22} = \phi_2\phi_1^2\sigma_1^2 + \phi_2\sigma_2^2,\\
\sigma_{33} &= \phi_2\sigma_{23} + \sigma_3^2 = \phi_2^2\phi_1^2\sigma_1^2 + \phi_2^2\sigma_2^2 + \sigma_3^2.
\end{aligned}$$

Estimability

6

This chapter presents exercises on estimability of linear functions of $\boldsymbol{\beta}$ in linear models, and provides solutions to those exercises.

▶ **Exercise 1** A linear estimator $t(\mathbf{y}) = t_0 + \mathbf{t}^T\mathbf{y}$ of $\mathbf{c}^T\boldsymbol{\beta}$ associated with the model $\{\mathbf{y}, \mathbf{X}\boldsymbol{\beta}\}$ is said to be *location equivariant* if $t(\mathbf{y} + \mathbf{X}\mathbf{d}) = t(\mathbf{y}) + \mathbf{c}^T\mathbf{d}$ for all $\mathbf{d}$.

(a) Prove that a linear estimator $t_0 + \mathbf{t}^T\mathbf{y}$ of $\mathbf{c}^T\boldsymbol{\beta}$ is location equivariant if and only if $\mathbf{t}^T\mathbf{X} = \mathbf{c}^T$.
(b) Is a location equivariant estimator of $\mathbf{c}^T\boldsymbol{\beta}$ necessarily unbiased, or vice versa? Explain.

Solution

(a) $t(\mathbf{y}+\mathbf{X}\mathbf{d}) = t_0+\mathbf{t}^T(\mathbf{y}+\mathbf{X}\mathbf{d}) = t_0+\mathbf{t}^T\mathbf{y}+\mathbf{t}^T\mathbf{X}\mathbf{d}$, and $t(\mathbf{y})+\mathbf{c}^T\mathbf{d} = t_0+\mathbf{t}^T\mathbf{y}+\mathbf{c}^T\mathbf{d}$. These two quantities are equal for all $\mathbf{d}$ if and only if $\mathbf{t}^T\mathbf{X}\mathbf{d} = \mathbf{c}^T\mathbf{d}$ for all $\mathbf{d}$, i.e., if and only if $\mathbf{t}^T\mathbf{X} = \mathbf{c}^T$ (by Theorem 2.1.1).
(b) Comparing part (a) with Theorem 6.1.1, we see that a location equivariant estimator is not necessarily unbiased, but an unbiased estimator is necessarily location equivariant.

▶ **Exercise 2** Prove Corollary 6.1.2.1: If $\mathbf{c}_1^T\boldsymbol{\beta}, \mathbf{c}_2^T\boldsymbol{\beta}, \ldots, \mathbf{c}_k^T\boldsymbol{\beta}$ are estimable under the model $\{\mathbf{y}, \mathbf{X}\boldsymbol{\beta}\}$, then so is any linear combination of them; that is, the set of estimable functions under a given unconstrained model is a linear space.

Solution Because $\mathbf{c}_1^T\boldsymbol{\beta},\ldots,\mathbf{c}_k^T\boldsymbol{\beta}$ are estimable, $\mathbf{c}_i^T \in \mathcal{R}(\mathbf{X})$ for all i. Because $\mathcal{R}(\mathbf{X})$ is a linear space, any linear combination of $\mathbf{c}_1^T\boldsymbol{\beta},\ldots,\mathbf{c}_k^T\boldsymbol{\beta}$ is an element of $\mathcal{R}(\mathbf{X})$, hence estimable.

D. L. Zimmerman, *Linear Model Theory*,
https://doi.org/10.1007/978-3-030-52074-8_6

▶ **Exercise 3** Prove Corollary 6.1.2.2: The elements of $\mathbf{X}\boldsymbol{\beta}$ are estimable under the model $\{\mathbf{y}, \mathbf{X}\boldsymbol{\beta}\}$; in fact, those elements span the space of estimable functions for that model.

Solution The ith element of $\mathbf{X}\boldsymbol{\beta}$ can be expressed as $\mathbf{x}_i^T\boldsymbol{\beta}$, where $\mathbf{x}_i^T$ is the ith row of $\mathbf{X}$. Clearly, $\mathbf{x}_i^T \in \mathcal{R}(\mathbf{X})$, so the ith element of $\mathbf{X}\boldsymbol{\beta}$ is estimable. Because $\text{span}(\mathbf{x}_1^T, \ldots, \mathbf{x}_n^T) = \mathcal{R}(\mathbf{X})$, the elements of $\mathbf{X}\boldsymbol{\beta}$ span the space of estimable functions.

▶ **Exercise 4** Prove Corollary 6.1.2.3: A set of p^* [$= \text{rank}(\mathbf{X})$] linearly independent estimable functions under the model $\{\mathbf{y}, \mathbf{X}\boldsymbol{\beta}\}$ exists such that any estimable function under that model can be written as a linear combination of functions in this set.

Solution Let $\mathbf{r}_1^T, \ldots, \mathbf{r}_{p^*}^T$ represent any basis for $\mathcal{R}(\mathbf{X})$. Then $\mathbf{r}_1^T\boldsymbol{\beta}, \ldots, \mathbf{r}_{p^*}^T\boldsymbol{\beta}$ satisfy the requirements of the definition of a basis for the set of all estimable functions.

▶ **Exercise 5** Prove Corollary 6.1.2.4: $\mathbf{c}^T\boldsymbol{\beta}$ is estimable for *every* vector $\mathbf{c}$ under the model $\{\mathbf{y}, \mathbf{X}\boldsymbol{\beta}\}$ if and only if $p^* = p$.

Solution $\mathbf{c}^T\boldsymbol{\beta}$ is estimable for all $\mathbf{c}^T \in \mathbb{R}^p$ if and only if $\mathcal{R}(\mathbf{X}) = \mathbb{R}^p$, i.e., if and only if $\mathcal{R}(\mathbf{X}) = \mathcal{R}(\mathbf{I}_p)$, i.e., if and only if $\text{rank}(\mathbf{X}) = \text{rank}(\mathbf{I}_p)$, i.e., if and only if $p^* = p$.

▶ **Exercise 6** Prove Corollary 6.1.2.5: $\mathbf{c}^T\boldsymbol{\beta}$ is estimable under the model $\{\mathbf{y}, \mathbf{X}\boldsymbol{\beta}\}$ if and only if $\mathbf{c} \perp \mathcal{N}(\mathbf{X})$.

Solution By Theorem 2.6.2, $\mathcal{R}(\mathbf{X})$ is the orthogonal complement of $\mathcal{N}(\mathbf{X})$. Thus $\mathbf{c}^T \in \mathcal{R}(\mathbf{X})$ if and only if $\mathbf{c} \perp \mathcal{N}(\mathbf{X})$.

▶ **Exercise 7** Prove Corollary 6.1.2.6: $\mathbf{c}^T\boldsymbol{\beta}$ is estimable under the model $\{\mathbf{y}, \mathbf{X}\boldsymbol{\beta}\}$ if and only if $\mathbf{c}^T \in \mathcal{R}(\mathbf{X}^T\mathbf{X})$.

Solution The result follows immediately from Theorems 6.1.2 and 6.1.3.

▶ **Exercise 8** For each of the following two questions, answer "yes" or "no." If the answer is "yes," give an example; if the answer is "no," prove it.

(a) Can the sum of an estimable function and a nonestimable function be estimable?
(b) Can the sum of two nonestimable functions be estimable?
(c) Can the sum of two linearly independent nonestimable functions be nonestimable?

Solution

(a) No. Suppose that $\mathbf{c}_1^T\boldsymbol{\beta}$ is estimable and $\mathbf{c}_2^T\boldsymbol{\beta}$ is nonestimable. Suppose further that their sum, $\mathbf{c}_1^T\boldsymbol{\beta}+\mathbf{c}_2^T\boldsymbol{\beta}$ is estimable. Now, $\mathbf{c}_2^T\boldsymbol{\beta} = (\mathbf{c}_1^T\boldsymbol{\beta}+\mathbf{c}_2^T\boldsymbol{\beta})-\mathbf{c}_1^T\boldsymbol{\beta}$, so by Corollary 6.1.2.1, $\mathbf{c}_2^T\boldsymbol{\beta}$ is estimable. But this contradicts the initial assumption that $\mathbf{c}_2^T\boldsymbol{\beta}$ is nonestimable. Thus $\mathbf{c}_1^T\boldsymbol{\beta}+\mathbf{c}_2^T\boldsymbol{\beta}$ cannot be estimable.

(b) Yes. An example occurs in the one-factor model, under which both μ and α_1 are nonestimable but their sum, $\mu+\alpha_1$, is estimable.

(c) Yes. An example occurs in the two-way main effects model, under which both α_1 and γ_1 are nonestimable and so is their sum.

► **Exercise 9** Determine $\mathcal{N}(\mathbf{X})$ for the one-factor model and for the two-way main effects model with no empty cells, and then use Corollary 6.1.2.5 to obtain the same characterizations of estimable functions given in Examples 6.1-2 and 6.1-3 for these two models.

Solution For the one-factor model, the matrix of distinct rows of $\mathbf{X}$ is $(\mathbf{1}_q, \mathbf{I}_q)$, which implies that

$$\begin{aligned}\mathcal{N}(\mathbf{X}) &= \{\mathbf{v} : (\mathbf{1}_q, \mathbf{I}_q)\mathbf{v} = \mathbf{0}\} \\ &= \{\mathbf{v} = (v_1, \mathbf{v}_2^T)^T : v_1\mathbf{1}_q + \mathbf{v}_2 = \mathbf{0}\} \\ &= \{\mathbf{v} = (v_1, \mathbf{v}_2^T)^T : \mathbf{v}_2 = -v_1\mathbf{1}_q\} \\ &= \left\{\mathbf{v} : \mathbf{v} = a\begin{pmatrix}-1 \\ \mathbf{1}_q\end{pmatrix}, a \in \mathbb{R}\right\}.\end{aligned}$$

By Corollary 6.1.2.5, $\mathbf{c}^T\boldsymbol{\beta}$ [where $\mathbf{c} = (c_i)$] is estimable if and only if $\mathbf{c} \perp \mathcal{N}(\mathbf{X})$, i.e., if and only if $\mathbf{c}^T a\begin{pmatrix}-1 \\ \mathbf{1}_q\end{pmatrix} = 0$ for all $a \in \mathbb{R}$, i.e., if and only if $c_1 = \sum_{i=2}^{q+1} c_i$.

For the two-way main effects model with no empty cells, the matrix of distinct rows of $\mathbf{X}$ is $(\mathbf{1}_{qm}, \mathbf{I}_q \otimes \mathbf{1}_m, \mathbf{1}_q \otimes \mathbf{I}_m)$, which implies that $\mathcal{N}(\mathbf{X}) = \{\mathbf{v} : (\mathbf{1}_{qm}, \mathbf{I}_q \otimes \mathbf{1}_m, \mathbf{1}_q \otimes \mathbf{I}_m)\mathbf{v} = \mathbf{0}\}$. Let $\mathbf{v} = (v_1, \mathbf{v}_2^T, \mathbf{v}_3^T)^T$ with $v_1 \in \mathbb{R}$, $\mathbf{v}_2 \in \mathbb{R}^q$, and $\mathbf{v}_3 \in \mathbb{R}^m$. Also denote the elements of $\mathbf{v}_2$ by $(v_{21}, v_{22}, \ldots, v_{2q})^T$. Then $(\mathbf{1}_{qm}, \mathbf{I}_q \otimes \mathbf{1}_m, \mathbf{1}_q \otimes \mathbf{I}_m)\mathbf{v} = \mathbf{0}$ if and only if

$$v_1\mathbf{1}_m + v_{2i}\mathbf{1}_m + \mathbf{v}_3 = \mathbf{0} \quad \text{for all } i = 1, \ldots, q,$$

i.e., if and only if $v_{2i}\mathbf{1}_m = -v_1\mathbf{1}_m + \mathbf{v}_3$ for all $i = 1, \ldots, q$, i.e., if and only if $v_{21} = v_{22} = \cdots = v_{2q} \equiv b$, say, and $\mathbf{v}_3 = -(v_1 + b)\mathbf{1}_m$. Therefore,

$$\mathcal{N}(\mathbf{X}) = \{\mathbf{v} : \mathbf{v} = (a, b\mathbf{1}_q^T, -(a+b)\mathbf{1}_m^T)^T, \ a, b \in \mathbb{R}\}.$$

By Corollary 6.1.2.5, $\mathbf{c}^T\boldsymbol{\beta}$ [where $\mathbf{c} = (c_1, c_2, \ldots, c_{q+1}, c_{q+2}, \ldots, c_{q+m+1})^T$] is estimable if and only if $\mathbf{c} \perp \mathcal{N}(\mathbf{X})$, i.e., if and only if

$$\mathbf{c}^T \begin{pmatrix} a \\ b\mathbf{1}_q \\ -(a+b)\mathbf{1}_m \end{pmatrix} = 0 \quad \text{for all } a, b \in \mathbb{R},$$

i.e., if and only if

$$ac_1 + b\sum_{i=2}^{q+1} c_i - (a+b)\sum_{i=q+2}^{q+m+1} c_i = 0 \quad \text{for all } a, b \in \mathbb{R}. \tag{6.1}$$

Putting $a = 1$ and $b = -1$ yields

$$c_1 = \sum_{i=2}^{q+1} c_i,$$

and putting $a = 1$ and $b = 0$ yields

$$c_1 = \sum_{i=q+2}^{q+m+1} c_i.$$

And, it may be easily verified that if both of these last two equations hold, then so does Eq. (6.1). Therefore, $\mathbf{c}^T\boldsymbol{\beta}$ is estimable if and only if $c_1 = \sum_{i=2}^{q+1} c_i = \sum_{i=q+2}^{q+m+1} c_i$.

▶ **Exercise 10** Prove that a two-way layout that is row-connected is also column-connected, and vice versa.

Solution Suppose that the two-way layout is row-connected, and consider $\gamma_j - \gamma_{j'}$ where $j \neq j'$. Because γ_j appears in the model for at least one observation and so does $\gamma_{j'}$, there exist i and i' such that the cell means $\mu + \alpha_i + \gamma_j$ and $\mu + \alpha_{i'} + \gamma_{j'}$ are estimable. Because the layout is row-connected, $\alpha_i - \alpha_{i'}$ is estimable. Consequently, $\gamma_j - \gamma_{j'} = (\mu + \alpha_i + \gamma_j) - (\mu + \alpha_{i'} + \gamma_{j'}) - (\alpha_i - \alpha_{i'})$, i.e., $\gamma_j - \gamma_{j'}$ is a linear combination of estimable functions, implying (by Corollary 6.1.2.1) that $\gamma_j - \gamma_{j'}$ is estimable. Because j and j' were arbitrary, the layout is column-connected. Showing that column-connectedness implies row-connectedness may be shown similarly.

▶ **Exercise 11** Prove Theorem 6.1.4: Under the two-way main effects model for a $q \times m$ layout, each of the following four conditions implies the other three:

(a) the layout is connected;
(b) $\text{rank}(\mathbf{X}) = q + m - 1$;
(c) $\mu + \alpha_i + \gamma_j$ is estimable for all i and j;
(d) all α-contrasts and γ-contrasts are estimable.

Solution

(a) ⇒ (b) Because the layout is connected, by definition $\alpha_i - \alpha_{i'}$ is estimable for all i and i' and $\gamma_j - \gamma_{j'}$ is estimable for all j and j' under the main effects model. For $j = 1$, there exists $i \in \{1, 2, \ldots, q\}$ such that $\mu + \alpha_i + \gamma_1$ is estimable (otherwise all cells in the first column of the $q \times m$ layout would be empty, rendering it a $q \times (m-1)$ design). Without loss of generality, assume that $\mu + \alpha_1 + \gamma_1$ is estimable. Then $\{\mu + \alpha_1 + \gamma_1, \alpha_2 - \alpha_1, \ldots, \alpha_q - \alpha_1, \gamma_2 - \gamma_1, \ldots, \gamma_m - \gamma_1\}$ is a set of $q + m - 1$ linearly independent estimable functions. Therefore $\text{rank}(\mathbf{X}) \geq q + m - 1$. On the other hand, $\text{rank}(\mathbf{X}) \leq q + m - 1$ because the main effects model with no empty cells (with model matrix $\mathbf{X}^*$, say) has rank $q + m - 1$ and $\mathcal{R}(\mathbf{X}) \subseteq \mathcal{R}(\mathbf{X}^*)$ because $\mathbf{X}$ can be obtained by deleting rows corresponding to empty cells (if any) in $\mathbf{X}^*$. Thus, $\text{rank}(\mathbf{X}) = q + m - 1$.

(b) ⇒ (c) If $\text{rank}(\mathbf{X}) = q + m - 1$, then for a $q \times m$ layout main effects model with no empty cells (with model matrix $\mathbf{X}^*$), it is the case that $\mathcal{R}(\mathbf{X}) \subseteq \mathcal{R}(\mathbf{X}^*)$ and $\text{rank}(\mathbf{X}) = q + m - 1 = \text{rank}(\mathbf{X}^*)$. Therefore, $\mathcal{R}(\mathbf{X}) = \mathcal{R}(\mathbf{X}^*)$ by Theorem 2.8.5. Because $\mu + \alpha_i + \beta_j$ is estimable for all i and j in the main effects model with no empty cells, by Theorem 6.1.2 it is also estimable in the main effects model with model matrix $\mathbf{X}$.

(c) ⇒ (d) Suppose that $\mu + \alpha_i + \gamma_j$ is estimable for all i and j under the main effects model. Then for any α-contrast $\sum_{i=1}^q d_i\alpha_i$ where $\sum_{i=1}^q d_i = 0$, it is the case that

$$\sum_{i=1}^q d_i\alpha_i = \sum_{i=1}^q d_i(\mu + \alpha_i + \gamma_1),$$

which is a linear combination of estimable functions, hence estimable by Corollary 6.1.2.1. Similarly, for any γ-contrast $\sum_{j=1}^m g_j\gamma_j$ where $\sum_{j=1}^m g_j = 0$, it is the case that

$$\sum_{j=1}^m g_j\gamma_j = \sum_{j=1}^m g_j(\mu + \alpha_1 + \gamma_j),$$

which is a linear combination of estimable functions, hence estimable.

(d) ⇒ (a) If all α-contrasts and γ-contrasts are estimable, then $\alpha_i - \alpha_{i'}$ is estimable for all i and i', and $\gamma_j - \gamma_{j'}$ is estimable for all j and j' because they are particular types of α- and γ-contrasts. Thus by definition, the layout is connected.

▶ **Exercise 12** For a two-way main effects model with q levels of Factor A and m levels of Factor B, what is the smallest possible value of rank($\mathbf{X}$)? Explain.

Solution $\text{rank}(\mathbf{X}) = q + m - s$ where s is the number of disconnected subarrays. Thus rank($\mathbf{X}$) is minimized by maximizing s. The maximum of s is $\min(q, m)$, so the minimum of rank($\mathbf{X}$) is $q + m - \min(q, m) = \max(q, m)$.

▶ **Exercise 13** Consider the following two-way layout, where the number in each cell indicates how many observations are in that cell:

	Levels of B									
Levels of A	1	2	3	4	5	6	7	8	9	10
1			1				2			
2									1	
3				4		1				1
4					2					2
5							1	1		
6					1					
7				1	2					
8		1							3	
9	2									

Consider the main effects model

$$y_{ijk} = \mu + \alpha_i + \gamma_j + e_{ijk} \qquad (i = 1, \ldots, 9;\ j = 1, \ldots, 10)$$

for these observations (where k indexes the observations, if any, within cell (i, j) of the table).

(a) Which α-contrasts and γ-contrasts are estimable?
(b) Give a basis for the set of all estimable functions.
(c) What is the rank of the associated model matrix $\mathbf{X}$?

Solution

(a) The "3+e" procedure, followed by an appropriate rearrangement of rows and columns, results in the following set of disconnected subarrays:

	Levels of B									
Levels of A	4	5	6	10	3	7	8	2	9	1
3	4	e	1	1						
4	e	2	e	2						
6	e	1	e	e						
7	1	2	e	e						
1					1	2	e			
5					e	1	1			
2								e	1	
8								1	3	
9										2

From this we determine that the estimable α-contrasts are of the form $\sum_{i=1}^{9} d_i\alpha_i$ where $\sum_{i\in\{3,4,6,7\}} d_i = \sum_{i\in\{1,5\}} d_i = \sum_{i\in\{2,8\}} d_i = d_9 = 0$ and that the estimable γ-contrasts are of the form $\sum_{j=1}^{10} g_j\gamma_j$ where $\sum_{j\in\{4,5,6,10\}} g_j = \sum_{j\in\{3,7,8\}} g_j = \sum_{j\in\{2,9\}} g_j = g_1 = 0$.

(b) $\{\mu+\alpha_3+\gamma_4, \alpha_4-\alpha_3, \alpha_6-\alpha_3, \alpha_7-\alpha_3, \gamma_5-\gamma_4, \gamma_6-\gamma_4, \gamma_{10}-\gamma_4, \mu+\alpha_1+\gamma_3, \alpha_5-\alpha_1, \gamma_7-\gamma_3, \gamma_8-\gamma_3, \mu+\alpha_2+\gamma_2, \alpha_8-\alpha_2, \gamma_9-\gamma_2, \mu+\alpha_9+\gamma_1\}$

(c) $\text{rank}(\mathbf{X}) = 15$ by part (b) and Corollary 6.1.2.3.

▶ **Exercise 14** Consider the following two-way layout, where the number in each cell indicates how many observations are in that cell:

	Levels of B							
Levels of A	1	2	3	4	5	6	7	8
1	1				4	1		
2	3					1		
3								1
4		1					2	
5		1						
6			2	1				1
7						3		
8		2					1	
9	1				1			

Consider the main effects model

$$y_{ijk} = \mu + \alpha_i + \gamma_j + e_{ijk} \qquad (i = 1, \ldots, 9;\ j = 1, \ldots, 8)$$

for these observations (where k indexes the observations, if any, within cell (i, j) of the table).

(a) Which α-contrasts and γ-contrasts are estimable?
(b) Give a basis for the set of all estimable functions.
(c) What is the rank of the associated model matrix $\mathbf{X}$?
(d) Suppose that it was possible to take more observations in any cells of the table, but that each additional observation in cell (i, j) of the table will cost the investigator

$$(\$20 \times i) + [\$25 \times (8 - j)].$$

How many more observations are necessary for all α-contrasts and all γ-contrasts to be estimable, and in which cells should those observations be taken to minimize the additional cost to the investigator?

Solution

(a) The following two-way layout is obtained after applying the "3+e" procedure and appropriately rearranging rows and columns:

	Levels of B							
Levels of A	1	5	6	3	4	8	2	7
1	1	4	2					
2	3	e	1					
7	e	e	3					
9	1	1	e					
3				e	e	1		
6				2	1	1		
4							1	2
5							1	e
8							2	1

Thus, the estimable α-contrasts are $\sum_{i=1}^{9} d_i\alpha_i$ where $\sum_{i\in\{1,2,7,9\}} d_i = \sum_{i\in\{3,6\}} d_i = \sum_{i\in\{4,5,8\}} d_i = 0$, and the estimable γ-contrasts are $\sum_{j=1}^{8} g_j\gamma_j$ where $\sum_{j\in\{1,5,6\}} g_j = \sum_{j\in\{3,4,8\}} g_j = \sum_{j\in\{2,7\}} g_j = 0$.

(b) $\{\mu + \alpha_1 + \gamma_1, \alpha_2 - \alpha_1, \alpha_7 - \alpha_1, \alpha_9 - \alpha_1, \gamma_5 - \gamma_1, \gamma_6 - \gamma_1, \mu + \alpha_3 + \gamma_3, \alpha_6 - \alpha_3, \gamma_4 - \gamma_3, \gamma_8 - \gamma_3, \mu + \alpha_4 + \gamma_2, \alpha_5 - \alpha_4, \alpha_8 - \alpha_4, \gamma_7 - \gamma_2\}$.
(c) $\text{rank}(\mathbf{X}) = 14$ by part (b) and Corollary 6.1.2.3.
(d) Two more observations will suffice, and the additional cost will be minimized by taking them in cells $(1, 8)$ and $(1, 7)$.

Exercise 15 Consider the two-way partially crossed model introduced in Example 5.1.4-1, i.e.,

$$y_{ijk} = \mu + \alpha_i - \alpha_j + e_{ijk} \quad (i \neq j = 1, \ldots, q;\ j = 1, \ldots, n_i)$$

for those combinations of i and j ($i \neq j$) for which $n_{ij} \geq 1$.

(a) For the case of no empty cells, what conditions must the elements of $\mathbf{c}$ satisfy for $\mathbf{c}^T\boldsymbol{\beta}$ to be estimable? In particular, are μ and all differences $\alpha_i - \alpha_j$ estimable? Determine the rank of the model matrix $\mathbf{X}$ for this case and find a basis for the set of all estimable functions.
(b) Now consider the case where all cells below the main diagonal of the two-way layout are empty, but no cells above the main diagonal are empty. How, if at all, would your answers to the same questions in part (a) change?
(c) Finally, consider the case where there are empty cells in arbitrary locations. Describe a method for determining which cell means are estimable. Under what conditions on the locations of the empty cells are all functions estimable that were estimable in part (a)?

Solution

(a) Let us agree to label the distinct rows of $\mathbf{X}$ by subscripts ij, where $j \neq i = 1, \ldots, q$. Then the matrix of distinct rows of $\mathbf{X}$ is given by $[\mathbf{1}_{q(q-1)}, \mathbf{x}_1, \mathbf{x}_2, \ldots, \mathbf{x}_q]$ where the ijth element of $\mathbf{x}_k$ is equal to 1 if $i = k$, -1 if $j = k$, and 0 otherwise. Any linear combination $\mathbf{a}^T\mathbf{X}$ may therefore be expressed as $[a_{\cdot\cdot}, a_{1\cdot} - a_{\cdot 1}, a_{2\cdot} - a_{\cdot 2}, \ldots, a_{q\cdot} - a_{\cdot q}]$, where $\mathbf{a}^T = (a_{12}, a_{13}, \ldots, a_{1q}, a_{21}, a_{23}, a_{24}, \ldots, a_{2q}, a_{31}, \ldots, a_{q,q-1})$. Thus, $\mathbf{c}^T\boldsymbol{\beta}$ is estimable if and only if the coefficients on the α_i's sum to 0; there are no restrictions on the coefficient on μ. Since $\sum_{k=1}^{q} \mathbf{x}_k = \mathbf{0}$ but $\mathbf{1} \notin \mathcal{C}(\mathbf{x}_1, \ldots, \mathbf{x}_q)$, the rank of $\mathbf{X}$ is q. A basis for the set of all estimable functions is $\{\mu, \alpha_1 - \alpha_2, \alpha_1 - \alpha_3, \ldots, \alpha_1 - \alpha_q\}$.
(b) If all cells below the main diagonal are empty, then the matrix of distinct rows of $\mathbf{X}$ is obtained from the one in part (a) by deleting those rows ij for which $i > j$. Let $\mathbf{X}^* = (\mathbf{1}_{q(q-1)/2}, \mathbf{x}_1^*, \ldots, \mathbf{x}_q^*)$ denote the resulting matrix. Then, $\mathbf{a}^T\mathbf{X}^* = (a_{\cdot\cdot}, a_{1\cdot}, a_{2\cdot} - a_{12}, a_{3\cdot} - (a_{13} + a_{23}), \ldots, a_{q-1,q} - (a_{1q} + \cdots + a_{q-1,q}))$ where $\mathbf{a}^T = (a_{12}, a_{13}, \ldots, a_{1q}, a_{23}, a_{24}, \ldots, a_{2q}, a_{34}, \ldots, a_{q-1,q})$. It is easily verified that the last q elements of $\mathbf{a}^T\mathbf{X}^*$ sum to 0 and $\mathbf{1} \notin \mathcal{C}(\mathbf{x}_1^*, \ldots, \mathbf{x}_q^*)$ as before, so the row space of $\mathbf{X}^*$ is the same as that of $\mathbf{X}$. Thus, the answers to part (a) do not change.
(c) Follow this procedure:
   (i) Since $(\mu + \alpha_{i'} - \alpha_{j'}) = (\mu + \alpha_i - \alpha_{j'}) + (\mu + \alpha_{i'} - \alpha_j) - (\mu + \alpha_i - \alpha_j)$, we can start with the $3 + e$ procedure used to determine which cell means are estimable in the two-way main effects model, with one difference. Specifically, put an "o" (for "observation") in each cell for which $n_{ij} > 0$.

Then examine the 2×2 subarray formed as the Cartesian product of any two rows and any two columns, provided that the subarray does not include a cell on the main diagonal. If any three of the four cells of this subarray are occupied by an observation but the fourth cell is empty, put an "e" (for "estimable") in the empty cell. Repeat this procedure for every 2×2 subarray, putting an "e" in every empty fourth cell if the other three cells are occupied by either an "o" or an "e."

(ii) Next, note that $[(\mu+\alpha_i-\alpha_j)+(\mu+\alpha_j-\alpha_i)]/2 = \mu$; thus, if, for any $i \neq j$, cell (i, j) and its reflection cell (j, i) are occupied by an "o" or "e," then μ is estimable, and an "e" may be put in every cell which is the reflection of a cell occupied by an "o" or "e."

(iii) Examine each 2×2 subarray formed as the Cartesian product of any two rows and any two columns such that exactly one of the four cells lies on the main diagonal. For such a subarray, subtracting the cell mean for the cell opposite to the one on the main diagonal from the sum of cell means for the other two cells yields μ. Thus, if the three off-diagonal cells for such an array are occupied by an "o" or and "e," then μ is estimable, and an "e" may be put in every cell that is the reflection of a cell occupied by an "o" or "e."

(iv) If, while carrying out steps (ii) and (iii), it was determined that μ is estimable, then re-examine every 2×2 subarray previously examined in step (iii), and if any two of the three off-diagonal cells are occupied by an "o" or and "e," put an "e" in the remaining off-diagonal cell.

(v) Cycle through the previous four steps until an "e" may not be put in any more cells.

After carrying out this procedure, the only cell means that are estimable are those corresponding to cells occupied by an "o" or an "e." If μ was not determined to be estimable in steps (ii) or (iii), then μ is not estimable and neither is any difference $\alpha_i - \alpha_{i'}$. If, however, μ was determined to be estimable, then all differences $\alpha_i - \alpha_{i'}$ corresponding to the estimable cell means are estimable.

Exercise 16 For a 3×3 layout, consider a two-way model with interaction, i.e.,

$$y_{ijk} = \mu + \alpha_i + \gamma_j + \xi_{ij} + e_{ijk} \qquad (i = 1, 2, 3;\ j = 1, 2, 3;\ k = 1, \ldots, n_{ij})$$

where some of the nine cells may be empty.

(a) Give a basis for the interaction contrasts, $\psi_{iji'j'} \equiv \xi_{ij} - \xi_{ij'} - \xi_{i'j} + \xi_{i'j'}$, that are estimable if no cells are empty.

(b) If only one of the nine cells is empty, how many interaction contrasts are there in a basis for the estimable interaction contrasts? What is rank($\mathbf{X}$)? Does the answer depend on which cell is empty?

(c) If exactly two of the nine cells are empty, how many interaction contrasts are there in a basis for the estimable interaction contrasts? What is rank($\mathbf{X}$)? Does the answer depend on which cells are empty?

(d) What is the maximum number of empty cells possible for a 3×3 layout that has at least one estimable interaction contrast? What is rank($\mathbf{X}$) in this case?

Solution

(a) Any $\psi_{iji'j'} \equiv \xi_{ij} - \xi_{ij'} - \xi_{i'j} + \xi_{i'j'} = \mu_{ij} - \mu_{ij'} - \mu_{i'j} + \mu_{i'j'}$ is estimable if the four cell means included in its definition are estimable. Thus, if there are no empty cells, all $\psi_{iji'j'}$ for $i \neq i'$ and $j \neq j'$ are estimable. A basis for this set of functions is $\{\psi_{1122}, \psi_{1123}, \psi_{1132}, \psi_{1133}, \psi_{1223}, \psi_{1233}, \psi_{2132}, \psi_{2133}, \psi_{2233}\}$. There are nine functions in the basis, so rank($\mathbf{X}$) $= 9$.
(b) The location of the empty cell determines which interaction contrasts are estimable but does not determine the number of linearly independent interaction contrasts because we can always perform row and column permutations to put the empty cell in any row and column we desire. So without loss of generality assume that the empty cell is $(3, 3)$. Then a basis for the estimable interaction contrasts is $\{\xi_{1122}, \xi_{1123}, \xi_{1132}, \xi_{1223}, \xi_{2132}\}$. The basis will change depending on which cell is empty, but its dimension is 5 regardless of which cell is empty. Thus, rank($\mathbf{X}$) $= 5$.
(c) Either the two empty cells are in the same row or column, or they are in different rows and columns. The "same row" case is equivalent to the "same column" case. For this case, again by considering row and column permutations, without loss of generality we may assume cells $(2, 3)$ and $(3, 3)$ are empty. Then a basis for the estimable interaction contrasts is $\{\xi_{1122}, \xi_{1123}, \xi_{1223}\}$. The basis will change depending on which two cells are empty, but its dimension is 3 regardless, so rank($\mathbf{X}$) $= 3$. On the other hand, if the empty cells lie in different rows and columns, which without loss of generality we may take to be cells $(1, 1)$ and $(3, 3)$, then a basis for the estimable interaction contrasts is $\{\xi_{1223}, \xi_{2132}\}$. Again, the basis will change depending on which two cells are empty, but its dimension is 2 regardless so rank($\mathbf{X}$) $= 2$.
(d) No entire row and no entire column in the 3×3 layout can be empty. The maximum allowable number of empty cells for this to be true and for there to be at least one estimable interaction contrast is 4. Without loss of generality assume that cells $(1, 3)$, $(2, 3)$, $(3, 1)$, and $(3, 2)$ are empty. Then the only estimable interaction contrast is ξ_{1122}. The lone estimable contrast will change, of course, depending on which four cells are empty, but there is always one such contrast, and the rank of $\mathbf{X}$ is 1.

▶ **Exercise 17** Consider the three-way main effects model

$$y_{ijkl} = \mu + \alpha_i + \gamma_j + \delta_k + e_{ijkl} \quad (i = 1, \ldots, q;\ j = 1, \ldots, m;\ k = 1, \ldots, s;\ l = 1, \ldots, n_{ijk}),$$

and suppose that there are no empty cells (i.e., $n_{ijk} > 0$ for all i, j, k).

(a) What conditions must the elements of $\mathbf{c}$ satisfy for $\mathbf{c}^T\boldsymbol{\beta}$ to be estimable?
(b) Find a basis for the set of all estimable functions.
(c) What is the rank of the model matrix $\mathbf{X}$?

Solution

(a) The matrix of distinct rows of $\mathbf{X}$ is

$$\left(\mathbf{1}_{qms},\ \mathbf{I}_q \otimes \mathbf{1}_{ms},\ \mathbf{1}_q \otimes \mathbf{I}_m \otimes \mathbf{1}_s,\ \mathbf{1}_{qm} \otimes \mathbf{I}_s\right).$$

An arbitrary element of $\mathcal{R}(\mathbf{X})$ may therefore be expressed as

$$\mathbf{a}^T\left(\mathbf{1}_{qms},\ \mathbf{I}_q \otimes \mathbf{1}_{ms},\ \mathbf{1}_q \otimes \mathbf{I}_m \otimes \mathbf{1}_s,\ \mathbf{1}_{qm} \otimes \mathbf{I}_s\right),$$

or equivalently as $(a_{\cdots}, a_{1\cdot\cdot}, \ldots, a_{q\cdot\cdot}, a_{\cdot 1\cdot}, \ldots, a_{\cdot m\cdot}, a_{\cdot\cdot 1}, \ldots, a_{\cdot\cdot s})$, where $\mathbf{a}^T = (a_{111}, a_{112}, \ldots, a_{11s}, a_{121}, \ldots, a_{12s}, \ldots, a_{qms})$. Thus, by Theorem 6.1.2, $\mathbf{c}^T\boldsymbol{\beta}$ is estimable if and only if the coefficients on the α_i's sum to the coefficient on μ and the coefficients on the γ_j's and δ_k's do likewise.

(b) A basis for the set of all estimable functions is $\{\mu+\alpha_1+\gamma_1+\delta_1, \alpha_2-\alpha_1, \alpha_3-\alpha_1, \ldots, \alpha_q-\alpha_1, \gamma_2-\gamma_1, \ldots, \gamma_m-\gamma_1, \delta_2-\delta_1, \ldots, \delta_s-\delta_1\}$.

(c) By part (b) and Corollary 6.1.2.3, rank$(\mathbf{X}) = 1+(q-1)+(m-1)+(s-1) = q+m+s-2$.

▶ **Exercise 18** Consider a three-factor crossed classification in which Factors A, B, and C each have two levels, and suppose that there is exactly one observation in the following four (of the eight) cells: 111, 122, 212, 221. The other cells are empty. Consider the following main effects model for these observations:

$$y_{ijk} = \mu + \alpha_i + \gamma_j + \delta_k + e_{ijk},$$

where $(i, j, k) \in \{(1, 1, 1),\ (1, 2, 2),\ (2, 1, 2),\ (2, 2, 1)\}$.

(a) Which of the functions $\alpha_2 - \alpha_1$, $\gamma_2 - \gamma_1$, and $\delta_2 - \delta_1$ are estimable?
(b) Find a basis for the set of all estimable functions.
(c) What is the rank of the model matrix $\mathbf{X}$?

Note This exercise and the next one are relevant to the issue of "confounding" in a 2^2 factorial design in blocks of size two.

Solution

(a) All of them are estimable by Corollary 6.1.2.1 because:

- $-\frac{1}{2}(\mu+\alpha_1+\gamma_1+\delta_1) - \frac{1}{2}(\mu+\alpha_1+\gamma_2+\delta_2) + \frac{1}{2}(\mu+\alpha_2+\gamma_1+\delta_2) + \frac{1}{2}(\mu+\alpha_2+\gamma_2+\delta_1) = \alpha_2-\alpha_1$;
- $-\frac{1}{2}(\mu+\alpha_1+\gamma_1+\delta_1) + \frac{1}{2}(\mu+\alpha_1+\gamma_2+\delta_2) - \frac{1}{2}(\mu+\alpha_2+\gamma_1+\delta_2) + \frac{1}{2}(\mu+\alpha_2+\gamma_2+\delta_1) = \gamma_2-\gamma_1$;

- $-\frac{1}{2}(\mu+\alpha_1+\gamma_1+\delta_1)+\frac{1}{2}(\mu+\alpha_1+\gamma_2+\delta_2)+\frac{1}{2}(\mu+\alpha_2+\gamma_1+\delta_2)-\frac{1}{2}(\mu+\alpha_2+\gamma_2+\delta_1)=\delta_2-\delta_1$.

Thus, if Factor C represents blocks, so that the layout represents a 2^2 factorial design in (incomplete) blocks of size two, Factor A and Factor B are not confounded with blocks.

(b) A basis for the set of estimable functions is $\{\mu+\alpha_1+\gamma_1+\delta_1, \alpha_2-\alpha_1, \gamma_2-\gamma_1, \delta_2-\delta_1\}$.

(c) $p^*=4$.

▶ **Exercise 19** Consider a three-factor crossed classification in which Factors A, B, and C each have two levels, and suppose that there is exactly one observation in the following four (of the eight) cells: 111, 211, 122, 222. The other cells are empty. Consider the following main effects model for these observations:

$$y_{ijk}=\mu+\alpha_i+\gamma_j+\delta_k+e_{ijk},$$

where $(i, j, k) \in \{(1,1,1),\ (2,1,1),\ (1,2,2),\ (2,2,2)\}$.

(a) Which of the functions $\alpha_2-\alpha_1$, $\gamma_2-\gamma_1$, and $\delta_2-\delta_1$ are estimable?
(b) Find a basis for the set of all estimable functions.
(c) What is the rank of the model matrix $\mathbf{X}$?

Solution

(a) By Corollary 6.1.2.1, $\alpha_2-\alpha_1$ is estimable because $-\frac{1}{2}(\mu+\alpha_1+\gamma_1+\delta_1)+\frac{1}{2}(\mu+\alpha_2+\gamma_1+\delta_1)-\frac{1}{2}(\mu+\alpha_1+\gamma_2+\delta_2)+\frac{1}{2}(\mu+\alpha_2+\gamma_2+\delta_2)=\alpha_2-\alpha_1$. So is $\gamma_2-\gamma_1+\delta_2-\delta_1$ because $-\frac{1}{2}(\mu+\alpha_1+\gamma_1+\delta_1)-\frac{1}{2}(\mu+\alpha_2+\gamma_1+\delta_1)+\frac{1}{2}(\mu+\alpha_1+\gamma_2+\delta_2)+\frac{1}{2}(\mu+\alpha_2+\gamma_2+\delta_2)=\gamma_2-\gamma_1+\delta_2-\delta_1$. However, no linear combination of the same four functions yields either $\gamma_2-\gamma_1$ or $\delta_2-\delta_1$, so neither $\gamma_2-\gamma_1$ nor $\delta_2-\delta_1$ is estimable. If Factor C represents blocks, so that the layout represents a 2^2 factorial design in (incomplete) blocks of size two, Factor B is confounded with blocks.

(b) A basis for the set of estimable functions is $\{\mu+\alpha_1+\gamma_1+\delta_1, \alpha_2-\alpha_1, \gamma_2-\gamma_1+\delta_2-\delta_1\}$.

(c) $p^*=3$.

▶ **Exercise 20** Consider a three-factor crossed classification in which Factors A and B have two levels and Factor C has three levels (i.e. a $2\times2\times3$ layout), but some cells may be empty. Consider the following main effects model for this situation:

$$y_{ijkl}=\mu+\alpha_i+\gamma_j+\delta_k+e_{ijkl},$$

where $i=1,2$; $j=1,2$; $k=1,2,3$; and $l=1,\ldots,n_{ijk}$ for the nonempty cells.

(a) What is the fewest number of observations that, if they are placed in appropriate cells, will make all functions of the form $\mu + \alpha_i + \gamma_j + \delta_k$ estimable?
(b) If all functions of the form $\mu + \alpha_i + \gamma_j + \delta_k$ are estimable, what is the rank of the model matrix?
(c) Let a represent the correct answer to part (a). If a observations are placed in *any* cells (one observation per cell), will all functions of the form $\mu + \alpha_i + \gamma_j + \delta_k$ necessarily be estimable? Explain.
(d) If $a + 1$ observations are placed in *any* cells (one observation per cell), will all functions of the form $\mu + \alpha_i + \gamma_j + \delta_k$ necessarily be estimable? Explain.
(e) Suppose there is exactly one observation in the following cells: 211, 122, 123, 221. The other cells are empty. Determine p^* and find a basis for the set of all estimable functions.

Solution

(a) Let $\mathbf{X}$ be the model matrix for the case of this model/layout where no cell is empty. Then obviously all functions of the form $\mu + \alpha_i + \gamma_j + \delta_k$ are estimable, and by Exercise 6.17c the rank of $\mathbf{X}$ is 5. Now let $\tilde{\mathbf{X}}$ be the model matrix for any $2 \times 2 \times 3$ layout for which all functions of the form $\mu + \alpha_i + \gamma_j + \delta_k$ are estimable. Then $\text{rank}(\tilde{\mathbf{X}}) \geq 5$ because spanning the set of all functions of the form $\mu + \alpha_i + \gamma_j + \delta_k$ requires at least five vectors, and $\text{rank}(\tilde{\mathbf{X}}) \leq 5$ because $\mathcal{R}(\tilde{\mathbf{X}}) \subseteq \mathcal{R}(\mathbf{X})$. Hence $\text{rank}(\tilde{\mathbf{X}}) = 5$ and it follows that $\tilde{\mathbf{X}}$ must have at least five rows, i.e., at least five observations are required. Now consider the particular case

$$\tilde{\mathbf{X}} = \begin{pmatrix} 1\,1\,0\,1\,0\,1\,0\,0 \\ 1\,1\,0\,0\,1\,1\,0\,0 \\ 1\,0\,1\,1\,0\,1\,0\,0 \\ 1\,1\,0\,1\,0\,0\,1\,0 \\ 1\,1\,0\,1\,0\,0\,0\,1 \end{pmatrix},$$

for which the corresponding design is depicted in the top panel of Fig. 6.1. Then it can be easily verified that all functions of the form $\mu + \alpha_i + \gamma_j + \delta_k$ are estimable. Therefore, it is possible to make all functions of the form $\mu + \alpha_i + \gamma_j + \delta_k$ estimable with five observations, but no fewer than 5.
(b) $\text{rank}(\mathbf{X}) = 5$, as explained in part (a).
(c) No. Suppose you put observations in cells 111, 121, 212, 222, 213, as depicted in the left bottom panel in Fig. 6.1. Then the model matrix is

$$\mathbf{X} = \begin{pmatrix} 1\,1\,0\,1\,0\,1\,0\,0 \\ 1\,1\,0\,0\,1\,1\,0\,0 \\ 1\,0\,1\,1\,0\,0\,1\,0 \\ 1\,0\,1\,0\,1\,0\,1\,0 \\ 1\,0\,1\,1\,0\,0\,0\,1 \end{pmatrix}.$$

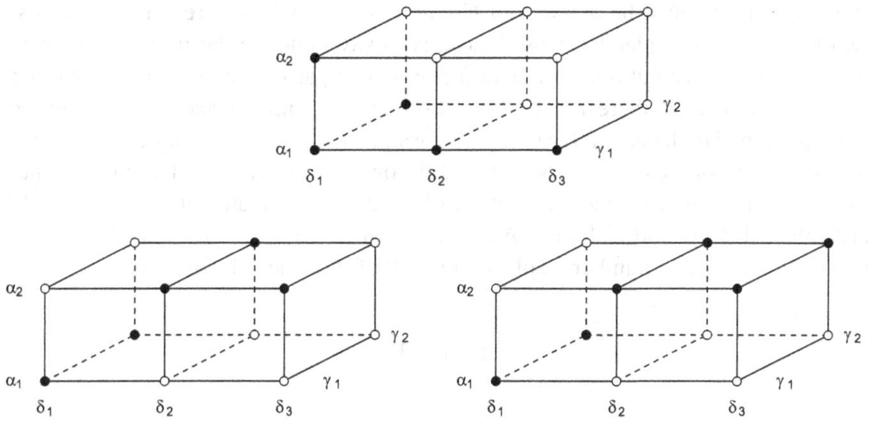

Fig. 6.1 Depictions of designs for Exercise 6.20. Open circles indicate empty cells; black circles indicate nonempty cells

Denoting the columns of $\mathbf{X}$ by $\mathbf{x}_j$, $j = 1, 2, \ldots, 8$, we observe that $\mathbf{x}_4, \mathbf{x}_5, \mathbf{x}_6, \mathbf{x}_7$ are linearly independent, and $\mathbf{x}_1 = \mathbf{x}_4 + \mathbf{x}_5$, $\mathbf{x}_2 = \mathbf{x}_6$, $\mathbf{x}_3 = \mathbf{x}_4 + \mathbf{x}_5 - \mathbf{x}_6$, and $\mathbf{x}_8 = \mathbf{x}_4 + \mathbf{x}_5 - \mathbf{x}_6 - \mathbf{x}_7$. Hence $\text{rank}(\mathbf{X}) = 4$, indicating that not all functions of the form $\mu + \alpha_i + \gamma_j + \delta_k$ are estimable.

(d) No. Suppose you put observations in the same cells as in part (c), plus cell 223; this design is shown in the bottom right panel of Fig. 6.1. Then the model matrix is

$$\mathbf{X} = \begin{pmatrix} 1\,1\,0\,1\,0\,1\,0\,0 \\ 1\,1\,0\,0\,1\,1\,0\,0 \\ 1\,0\,1\,1\,0\,0\,1\,0 \\ 1\,0\,1\,0\,1\,0\,1\,0 \\ 1\,0\,1\,1\,0\,0\,0\,1 \\ 1\,0\,1\,0\,1\,0\,0\,1 \end{pmatrix}.$$

Then columns four through seven of $\mathbf{X}$ are linearly independent here just as they were in part (b), and the other four columns of $\mathbf{X}$ here may be expressed as the same linear combinations of columns of $\mathbf{X}$ given in part (b). Hence the rank of $\mathbf{X}$ is still 4, so not all functions of the form $\mu + \alpha_i + \gamma_j + \delta_k$ are estimable.

(e) Because the cell means of those four nonempty cells are estimable and linearly independent, $p^* = 4$ and a basis for the set of all estimable functions is $\{\mu + \alpha_2 + \gamma_1 + \delta_1, \mu + \alpha_1 + \gamma_2 + \delta_2, \mu + \alpha_1 + \gamma_2 + \delta_3, \mu + \alpha_2 + \gamma_2 + \delta_1\}$.

▶ **Exercise 21** A "Latin square" design is an experimental design with three partially crossed factors, called "Rows," "Columns," and "Treatments." The number of levels of each factor is common across factors, i.e., if there are q levels of

Row, then there must be q levels of Column and q levels of Treatment. The first two factors are completely crossed, so a two-way table can be used to represent their combinations; but the Treatment factor is only partially crossed with the other two factors because there is only one observational unit for each Row×Column combination. The levels of Treatment are assigned to Row×Column combinations in such a way that each level occurs exactly once in each row and column (rather than once in each combination, as would be the case if Treatment was completely crossed with Row and Column). An example of a 3×3 "Latin square" design is depicted below: The number displayed in each of the nine cells of this layout is the

2	1	3
3	2	1
1	3	2

Treatment level (i.e., the Treatment label) which the observation in that cell received.

For observations in a Latin square design with q treatments, consider the following model:

$$y_{ijk} = \mu + \alpha_i + \gamma_j + \tau_k + e_{ijk} \qquad (i = 1, \ldots, q;\ j = 1, \ldots, q;\ k = 1, \ldots, q),$$

where y_{ijk} is the observation in row i and column j (and k is the treatment assigned to that cell); μ is an overall effect; α_i is the effect of Row i; γ_j is the effect of Column j; τ_k is the effect of Treatment k; and e_{ijk} is the error corresponding to y_{ijk}. These errors are uncorrelated with mean 0 and variance σ^2.

(a) Show that $\tau_k - \tau_{k'}$ is estimable for all k and k', and give an unbiased estimator for it.
(b) What condition(s) must the elements of $\mathbf{c}$ satisfy for $\mathbf{c}^T\boldsymbol{\beta}$ to be estimable?
(c) Find a basis for the set of all estimable functions.
(d) What is the rank of the model matrix $\mathbf{X}$?

Solution

(a)

$$\begin{aligned}\tau_k - \tau_{k'} = {} & [(\mu + \alpha_1 + \gamma_{j_1(k)} + \tau_k) - (\mu + \alpha_1 + \gamma_{j_1(k')} + \tau_{k'}) \\ & + (\mu + \alpha_2 + \gamma_{j_2(k)} + \tau_k) - (\mu + \alpha_2 + \gamma_{j_2(k')} + \tau_{k'}) \\ & + \cdots + (\mu + \alpha_q + \gamma_{j_q(k)} + \tau_k) - (\mu + \alpha_q + \gamma_{j_q(k')} + \tau_{k'})]/q,\end{aligned}$$

where $\{j_1(k), j_2(k), \ldots, j_q(k)\}$ and $\{j_1(k'), j_2(k'), \ldots, j_q(k')\}$ are permutations of $\{1, 2, \ldots, q\}$ corresponding to the columns to which treatments k

and k' were assigned in rows $1, 2, \ldots, q$. Because all of the cell means in this expression for $\tau_k - \tau_{k'}$ are estimable, every treatment difference is estimable. Obviously, the same linear combination of observations as the linear combination of cell means in $\tau_k - \tau_{k'}$, i.e.,

$$[(y_{1j_1(k)} - y_{1j_1(k')}) + (y_{2j_2(k)} - y_{2j_2(k')}) + \cdots + (y_{qj_q(k)} - y_{qj_q(k')})]/q = \bar{y}_{\cdot\cdot k} - \bar{y}_{\cdot\cdot k'}$$

is an unbiased estimator of $\tau_k - \tau_{k'}$.

(b) Given a Latin square with q treatments, let $\mathbf{X}$ be the corresponding model matrix and let $S = \{(i, j, k) :$ Treatment k is assigned to the cell in Row i and Column $j\}$. Note that there are q^2 elements in S. Then any linear combination of the rows of $\mathbf{X}$ may be written as $\mathbf{a}^T\mathbf{X}$ where $\mathbf{a} \in \mathbb{R}^{q^2}$, and let us agree to index the elements of $\mathbf{a}$ by a_{ijk}, where $(i, j, k) \in S$, arranged in lexicographic order. Then $\mathbf{a}^T\mathbf{X} = (a_{\cdots}, a_{1\cdot\cdot}, a_{2\cdot\cdot}, \ldots, a_{q\cdot\cdot}, a_{\cdot 1\cdot}, \ldots, a_{\cdot q\cdot}, a_{\cdot\cdot 1}, \ldots, a_{\cdot\cdot q})$. Therefore, $\mathbf{c}^T\boldsymbol{\beta}$ is estimable if and only if the coefficients on the α_i's sum to the coefficient on μ, and the coefficients on the γ_j's and τ_k's do likewise.

(c) Let $j_1(1)$ be defined as in part (a). Consider the set of functions $\{\mu + \alpha_1 + \gamma_{j_1(1)} + \tau_1, \alpha_2 - \alpha_1, \ldots, \alpha_q - \alpha_1, \gamma_2 - \gamma_1, \ldots, \gamma_q - \gamma_1, \tau_2 - \tau_1, \ldots, \tau_q - \tau_1\}$. There are $3q - 2$ functions in this set, all of which are estimable by part (b), and they are linearly independent. Furthermore, $\mathcal{R}(\mathbf{X}) \subseteq \mathcal{R}(\tilde{\mathbf{X}})$ where $\tilde{\mathbf{X}}$ is the model matrix of a $q \times q \times q$ three-way main effects model with no empty cells; and because $\text{rank}(\tilde{\mathbf{X}}) = 3q - 2$ by the result of Exercise 6.17c, any basis for the set of estimable functions in a Latin square with q treatments can contain no more than $3q - 2$ functions. Thus, the set in question is a basis for the set of all estimable functions.

(d) By part (c) and Corollary 6.1.2.3, $\text{rank}(\mathbf{X}) = 3q - 2$.

▶ **Exercise 22** An experiment is being planned that involves two continuous explanatory variables, x_1 and x_2, and a response variable y. The model relating y to the explanatory variables is

$$y_i = \beta_1 + \beta_2 x_{i1} + \beta_3 x_{i2} + e_i \qquad (i = 1, \ldots, n).$$

Suppose that only three observations can be taken (i.e. $n = 3$), and that these three observations must be taken at distinct values of (x_{i1}, x_{i2}) such that $x_{i1} = 1, 2,$ or 3, and $x_{i2} = 1, 2,$ or 3 (see Fig. 6.2 for a depiction of the nine allowable (x_{i1}, x_{i2})-pairs, which are labelled as A, B, ..., I for easier reference). A three-observation design can thus be represented by three letters: for example, ACI is the design consisting of the three points A, C, and I. For some of the three-observation designs, all parameters in the model's mean structure (β_1, β_2, and β_3) are estimable. However, there are some three-observation designs for which not all of these parameters are estimable. List eight designs of the latter type, and justify your list.

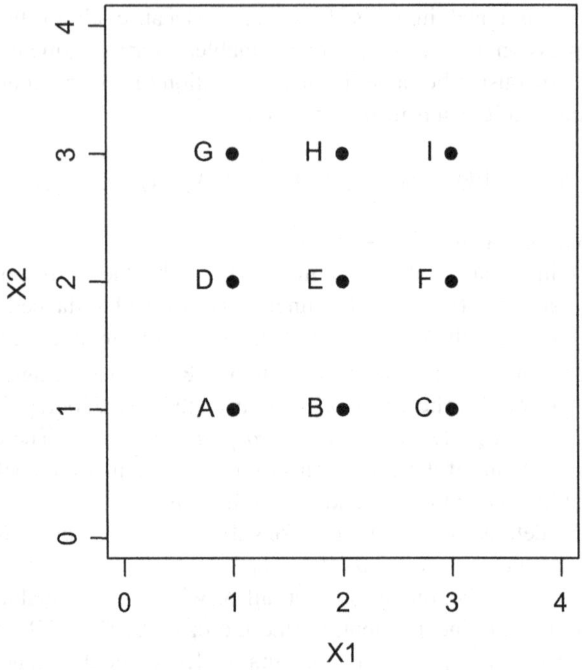

Fig. 6.2 Experimental layout for Exercise 6.22

Solution For all elements of $\boldsymbol{\beta}$ (β_1, β_2, and β_3) to be estimable, rank($\mathbf{X}$) must equal 3. But for designs ABC, DEF, GHI, GDA, HEB, and IFC, $\mathbf{X} = (\mathbf{1}, \mathbf{x}_2, \mathbf{x}_3)$ where either $\mathbf{x}_2$ or $\mathbf{x}_3$ is equal to $c\mathbf{1}$ where $c \in \{1, 2, 3\}$, so rank($\mathbf{X}$) $= 2$ for these six designs. Furthermore, for design AEI, $\mathbf{x}_2 = \mathbf{x}_3 = (1, 2, 3)^T$, so rank($\mathbf{X}$) $= 2$ for this design. Finally, for design GEC, $\mathbf{x}_2 = (1, 2, 3)^T = 4\mathbf{1} - \mathbf{x}_3$ so rank($\mathbf{X}$) $= 2$ for this design also.

▶ **Exercise 23** Suppose that $\mathbf{y}$ follows a Gauss–Markov model in which

$$\mathbf{X} = \begin{pmatrix} 1 & 1 & t \\ 1 & 2 & 2t \\ \vdots & \vdots & \vdots \\ 1 & t & t^2 \\ 1 & 1 & s \\ 1 & 2 & 2s \\ \vdots & \vdots & \vdots \\ 1 & s & s^2 \end{pmatrix}, \quad \boldsymbol{\beta} = \begin{pmatrix} \beta_1 \\ \beta_2 \\ \beta_3 \end{pmatrix}.$$

Here, t and s are integers such that $t \geq 2$ and $s \geq 2$. [Note: The corresponding model is called the "conditional linear model" and has been used for data that are informatively right-censored; see, e.g., Wu and Bailey (1989).]

(a) Determine which of the following linear functions of the elements of $\boldsymbol{\beta}$ are always estimable, with no further assumptions on t and s.
   (i) β_1,
   (ii) $\beta_1 + \beta_2 t + \beta_3 s$,
   (iii) $\beta_2 + \beta_3(s+t)/2$.

(b) Determine necessary and sufficient conditions on t and s for all linear functions of the elements of $\boldsymbol{\beta}$ to be estimable.

Solution

(a) (i) $\beta_1 = (1, 0, 0)\boldsymbol{\beta} = [2(1, 1, t) - (1, 2, 2t)]\boldsymbol{\beta}$, so β_1 is always estimable.
   (ii) Let $\mathbf{d}^T = (\mathbf{a}^T, \mathbf{b}^T) = (a_1, \ldots, a_t, b_1, \ldots, b_s)$. Then $\mathbf{d}^T\mathbf{X} = (\sum_{i=1}^t a_i + \sum_{j=1}^s b_j,\ \sum_{i=1}^t a_i i + \sum_{j=1}^s b_j j,\ t\sum_{i=1}^t a_i i + s\sum_{j=1}^s b_j j) \equiv (A_1,\ A_2 + A_3,\ tA_2 + sA_3)$, say. Now, $\beta_1 + \beta_2 t + \beta_3 s = \mathbf{c}^T\boldsymbol{\beta}$ where $\mathbf{c}^T = (1, t, s)$. For this function to be estimable, there must exist A_1, A_2, A_3 such that $A_1 = 1$, $A_2 + A_3 = t$, and $tA_2 + sA_3 = s$. The last two equalities cannot simultaneously hold if $t = s \neq 1$. So $\beta_1 + \beta_2 t + \beta_3 s$ is not always estimable.
   (iii) $\beta_2 + \beta_3\left(\frac{s+t}{2}\right) = (0,\ 1,\ (s+t)/2)\boldsymbol{\beta} = \{(1/2)[(1, 2, 2t) - (1, 1, t)] + (1/2)[(1, 2, 2s) - (1, 1, s)]\}\boldsymbol{\beta}$, so it is always estimable.

(b) A necessary and sufficient condition is $s \neq t$. To show sufficiency, let $s \neq t$ and let us show that the unit vectors in three-dimensions form a basis for $\mathcal{R}(\mathbf{X})$. It was shown in part (a) that $(1, 0, 0) \in \mathcal{R}(\mathbf{X})$. Hence it is also the case that $(1, 1, t) - (1, 0, 0) = (0, 1, t) \in \mathcal{R}(\mathbf{X})$ and $(1, 1, s) - (1, 0, 0) = (0, 1, s) \in \mathcal{R}(\mathbf{X})$. Thus $[(0, 1, t) - (0, 1, s)]/(t - s) = (0, 0, 1) \in \mathcal{R}(\mathbf{X})$ and $[(1/t)(0, 1, t) - (1/s)(0, 1, s)]/[(1/t) - (1/s)] = (0, 1, 0) \in \mathcal{R}(\mathbf{X})$. It follows that $\mathcal{R}(\mathbf{X}) = \mathbb{R}^3$ and that all linear functions of $\boldsymbol{\beta}$ are estimable. To show necessity, suppose that $s = t$. Then $\{(1, 0, 0), (0, 1, t)\}$ is a basis for $\mathcal{R}(\mathbf{X})$, implying that the dimensionality of $\mathcal{R}(\mathbf{X})$ equals 2 and hence that not all linear functions of $\boldsymbol{\beta}$ are estimable.

▶ **Exercise 24** Consider three linear models labelled I, II, and III, which all have model matrix

$$\mathbf{X} = \begin{pmatrix} 1 & 1 & 1 & 1 \\ 0 & 1 & 0 & 1 \\ 1 & 0 & 1 & 0 \\ 1 & 1 & 1 & 1 \\ 1 & 0 & 1 & 0 \end{pmatrix}.$$

The three models differ only with respect to their parameter spaces for $\boldsymbol{\beta} = [\beta_1, \beta_2, \beta_3, \beta_4]^T$, as indicated below:

$$\begin{aligned}
&\text{Parameter space for Model I} : \{\boldsymbol{\beta} : \boldsymbol{\beta} \in \mathbb{R}^4\}, \\
&\text{Parameter space for Model II} : \{\boldsymbol{\beta} : \beta_1 + \beta_4 = 5\}, \\
&\text{Parameter space for Model III} : \{\boldsymbol{\beta} : \beta_1 + \beta_4 = 0\}.
\end{aligned}$$

Under which, if any, of these models does an unbiased estimator of $\beta_2 + \beta_3$ exist which is of the form $\mathbf{t}^T\mathbf{y}$? Justify your answer.

Solution

(a) By Theorem 6.2.1, $\mathbf{t}^T\mathbf{y}$ is unbiased for $\beta_2 + \beta_3 = (0, 1, 1, 0)\boldsymbol{\beta}$ if and only if a scalar g exists such that $gh = 0$ and $\mathbf{t}^T\mathbf{X} + g\mathbf{A} = (0, 1, 1, 0)$. Under Model I, $\mathbf{A} = \mathbf{0}$ so $\mathbf{t}^T\mathbf{y}$ is unbiased for $\beta_2 + \beta_3$ if and only if $\mathbf{t}^T\mathbf{X} = (0, 1, 1, 0)$. But $(0, 1, 1, 0) \notin \mathcal{R}(\mathbf{X}) = \text{span}\{(1, 0, 1, 0), (0, 1, 0, 1)\}$, so $\mathbf{t}^T\mathbf{y}$ cannot be unbiased under Model I. Under Model II, $\mathbf{A} = (1, 0, 0, 1)$ and $\mathbf{t}^T\mathbf{y}$ is unbiased for $\beta_2 + \beta_3$ if and only if a scalar g exists such that $g \cdot 5 = 0$ and $\mathbf{t}^T\mathbf{X} + g(1, 0, 0, 1) = (0, 1, 1, 0)$, or equivalently, if and only if $\mathbf{t}^T\mathbf{X} = (0, 1, 1, 0)$. Thus, as previously, $\mathbf{t}^T\mathbf{y}$ cannot be unbiased under Model II. Under Model III, $\mathbf{t}^T\mathbf{y}$ is unbiased for $\beta_2 + \beta_3 = (0, 1, 1, 0)\boldsymbol{\beta}$ if and only if a scalar g exists such that $g \cdot 0 = 0$ and $\mathbf{t}^T\mathbf{X} + g\mathbf{A} = (0, 1, 1, 0)$. Choosing $g = -1$ and $\mathbf{t}^T = (1, 0, 0, 0, 0)$ satisfies the requirements, so $\mathbf{t}^T\mathbf{y}$ is unbiased under Model III.

▶ **Exercise 25** Consider two linear models of the form

$$\mathbf{y} = \mathbf{X}\boldsymbol{\beta} + \mathbf{e},$$

where $\mathbf{y}$ is a 4-vector, $\boldsymbol{\beta}$ is the 3-vector $(\beta_1, \beta_2, \beta_3)^T$, and $\mathbf{X}$ is the 4×3 matrix

$$\begin{pmatrix} 1 & 3 & 5 \\ 0 & 4 & 4 \\ -2 & 6 & 2 \\ 4 & -3 & 5 \end{pmatrix}.$$

(Observe that the third column of $\mathbf{X}$ is equal to the second column plus twice the first column, but the first and the second columns are linearly independent, so the rank of $\mathbf{X}$ is 2). In the first model, $\boldsymbol{\beta}$ is unconstrained in $\mathbb{R}^3$. In the second model, $\boldsymbol{\beta}$ is constrained to the subset of $\mathbb{R}^3$ for which $\beta_1 = \beta_2$.

(a) Under which of the two models is the function $\beta_2 + \beta_3$ estimable? (Your answer may be neither model, the unconstrained model only, the constrained model only, or both models.)

(b) Under which of the two models is the function $\beta_1 + \beta_3$ estimable? (Your answer may be neither model, the unconstrained model only, the constrained model only, or both models.)

Solution

(a) $\beta_2 + \beta_3 = (0, 1, 1)\boldsymbol{\beta}$, which is estimable under both models because

$$(0, 1, 1) = (1/4)(0, 4, 4) \in \mathcal{R}(\mathbf{X}) \subseteq \mathcal{R}\begin{pmatrix} \mathbf{X} \\ \mathbf{A} \end{pmatrix}.$$

(b) $\beta_1 + \beta_3 = (1, 0, 1)\boldsymbol{\beta}$, which is estimable under the constrained model but not under the unconstrained model. To see this, observe that the second and third rows of $\mathbf{X}$ are linearly independent and span $\mathcal{R}(\mathbf{X})$, and the only linear combination of these two rows yielding a vector whose first element is 1 and second element is 0 is $(3/4)(0, 4, 4) + (-1/2)(-2, 6, 2) = (1, 0, 2)$. So $(1, 0, 1) \notin \mathcal{R}(\mathbf{X})$. However,

$$(1, 0, 1) = (1/4)(0, 4, 4) + (1, -1, 0),$$

where the first vector on the right-hand side is an element of $\mathcal{R}(\mathbf{X})$ and the second vector on the right-hand side is an element of $\mathcal{R}(\mathbf{A}) = \mathcal{R}(1, -1, 0)$. Thus $(1, 0, 1) \in \mathcal{R}\begin{pmatrix} \mathbf{X} \\ \mathbf{A} \end{pmatrix}$.

▶ **Exercise 26** Prove Corollary 6.2.2.1: If $\mathbf{c}_1^T\boldsymbol{\beta}, \mathbf{c}_2^T\boldsymbol{\beta}, \ldots, \mathbf{c}_k^T\boldsymbol{\beta}$ are estimable under the model $\{\mathbf{y}, \mathbf{X}\boldsymbol{\beta} : \mathbf{A}\boldsymbol{\beta} = \mathbf{h}\}$, then so is any linear combination of them; that is, the set of estimable functions under a given constrained model is a linear space.

Solution Because $\mathbf{c}_1^T\boldsymbol{\beta},\ldots,\mathbf{c}_k^T\boldsymbol{\beta}$ are estimable, $\mathbf{c}_i^T \in \mathcal{R}\begin{pmatrix} \mathbf{X} \\ \mathbf{A} \end{pmatrix}$ for all i. Because $\mathcal{R}\begin{pmatrix} \mathbf{X} \\ \mathbf{A} \end{pmatrix}$ is a linear space, any linear combination of $\mathbf{c}_1^T\boldsymbol{\beta},\ldots,\mathbf{c}_k^T\boldsymbol{\beta}$ is an element of $\mathcal{R}\begin{pmatrix} \mathbf{X} \\ \mathbf{A} \end{pmatrix}$, hence estimable.

▶ **Exercise 27** Prove Corollary 6.2.2.2: The elements of $\mathbf{X}\boldsymbol{\beta}$ are estimable under the model $\{\mathbf{y}, \mathbf{X}\boldsymbol{\beta} : \mathbf{A}\boldsymbol{\beta} = \mathbf{h}\}$.

Solution The ith element of $\mathbf{X}\boldsymbol{\beta}$ can be expressed as $\mathbf{x}_i^T\boldsymbol{\beta}$, where $\mathbf{x}_i^T$ is the ith row of $\mathbf{X}$. Clearly, $\mathbf{x}_i^T \in \mathcal{R}\begin{pmatrix} \mathbf{X} \\ \mathbf{A} \end{pmatrix}$, so the ith element of $\mathbf{X}\boldsymbol{\beta}$ is estimable.

▶ **Exercise 28** Prove Corollary 6.2.2.3: A set of $\text{rank}\begin{pmatrix}\mathbf{X}\\ \mathbf{A}\end{pmatrix}$ linearly independent estimable functions under the model $\{\mathbf{y}, \mathbf{X}\boldsymbol{\beta} : \mathbf{A}\boldsymbol{\beta} = \mathbf{h}\}$ exist such that any estimable function under that model can be written as a linear combination of functions in this set.

Solution Next, let $\mathbf{r}_1^T, \ldots, \mathbf{r}_s^T$ represent any basis for $\mathcal{R}\begin{pmatrix}\mathbf{X}\\ \mathbf{A}\end{pmatrix}$ where necessarily $s = \text{rank}\begin{pmatrix}\mathbf{X}\\ \mathbf{A}\end{pmatrix}$. Then $\mathbf{r}_1^T\boldsymbol{\beta}, \ldots, \mathbf{r}_s^T\boldsymbol{\beta}$ satisfy the requirements of the definition of a basis for the set of all estimable functions.

▶ **Exercise 29** Prove Corollary 6.2.2.4: $\mathbf{c}^T\boldsymbol{\beta}$ is estimable for *every* vector $\mathbf{c}$ under the model $\{\mathbf{y}, \mathbf{X}\boldsymbol{\beta} : \mathbf{A}\boldsymbol{\beta} = \mathbf{h}\}$ if and only if $\text{rank}\begin{pmatrix}\mathbf{X}\\ \mathbf{A}\end{pmatrix} = p$.

Solution $\mathbf{c}^T\boldsymbol{\beta}$ is estimable for all $\mathbf{c}^T \in \mathbb{R}^p$ if and only if $\mathcal{R}\begin{pmatrix}\mathbf{X}\\ \mathbf{A}\end{pmatrix} = \mathbb{R}^p$, i.e., if and only if $\mathcal{R}\begin{pmatrix}\mathbf{X}\\ \mathbf{A}\end{pmatrix} = \mathcal{R}(\mathbf{I}_p)$, i.e., if and only if $\text{rank}\begin{pmatrix}\mathbf{X}\\ \mathbf{A}\end{pmatrix} = \text{rank}(\mathbf{I}_p)$, i.e., if and only if $\text{rank}\begin{pmatrix}\mathbf{X}\\ \mathbf{A}\end{pmatrix} = p$.

▶ **Exercise 30** Prove Corollary 6.2.2.5: $\mathbf{c}^T\boldsymbol{\beta}$ is estimable under the model $\{\mathbf{y}, \mathbf{X}\boldsymbol{\beta} : \mathbf{A}\boldsymbol{\beta} = \mathbf{h}\}$ if and only if $\mathbf{c}^T \in \mathcal{R}\begin{pmatrix}\mathbf{X}^T\mathbf{X}\\ \mathbf{A}\end{pmatrix}$.

Solution Corollary 6.2.2.5 follows immediately from Theorems 6.2.2 and 6.2.3.

▶ **Exercise 31** Consider the constrained model $\{\mathbf{y}, \mathbf{X}\boldsymbol{\beta} : \mathbf{A}\boldsymbol{\beta} = \mathbf{h}\}$.

(a) Suppose that a function $\mathbf{c}^T\boldsymbol{\beta}$ exists that is estimable under this model but is nonestimable under the unconstrained model $\{\mathbf{y}, \mathbf{X}\boldsymbol{\beta}\}$. Under these circumstances, how do $\mathcal{R}(\mathbf{X}) \cap \mathcal{R}(\mathbf{A})$, $\mathcal{R}\begin{pmatrix}\mathbf{X}\\ \mathbf{A}\end{pmatrix}$, and $\mathcal{R}(\mathbf{X})$ compare to each other? Indicate which subset inclusions, if any, are strict.
(b) Again suppose that a function $\mathbf{c}^T\boldsymbol{\beta}$ exists that is estimable under this model but is nonestimable under the model $\{\mathbf{y}, \mathbf{X}\boldsymbol{\beta}\}$. Under these circumstances, if a vector $\mathbf{t}$ exists such that $\mathbf{t}^T\mathbf{y}$ is an unbiased estimator of $\mathbf{c}^T\boldsymbol{\beta}$ under the constrained model, what can be said about $\mathbf{A}$ and $\mathbf{h}$? Be as specific as possible.

Solution

(a) Because $\mathbf{c}^T\boldsymbol{\beta}$ is estimable under the constrained model, $\mathbf{c}^T \in \mathcal{R}\begin{pmatrix}\mathbf{X}\\ \mathbf{A}\end{pmatrix}$. And, because $\mathbf{c}^T\boldsymbol{\beta}$ is nonestimable under the unconstrained model, $\mathbf{c}^T \notin \mathcal{R}(\mathbf{X})$. Thus, $\{\mathcal{R}(\mathbf{X}) \cap \mathcal{R}(\mathbf{A})\} \subseteq \mathcal{R}(\mathbf{X}) \subseteq \mathcal{R}\begin{pmatrix}\mathbf{X}\\ \mathbf{A}\end{pmatrix}$, where the latter inclusion is strict.

(b) The situation described implies, by Theorem 6.2.1, that a vector $\mathbf{g}$ exists such that $\mathbf{g}^T\mathbf{h} = 0$, $\mathbf{t}^T\mathbf{X} + \mathbf{g}^T\mathbf{A} = \mathbf{c}^T$, and $\mathbf{g}^T\mathbf{A} \notin \mathcal{R}(\mathbf{X})$.

Reference

Wu, M. C. & Bailey, K. R. (1989). Estimation and comparison of changes in the presence of informative right censoring: conditional linear model. *Biometrics, 45*, 939–955.

Least Squares Estimation for the Gauss–Markov Model

7

This chapter presents exercises on least squares estimation for the Gauss–Markov model, and provides solutions to those exercises.

▶ **Exercise 1** The proof of the consistency of the normal equations (Theorem 7.1.1) relies upon an identity (Theorem 3.3.3c) involving a generalized inverse of $\mathbf{X}^T\mathbf{X}$. Provide an alternate proof of the consistency of the normal equations that does not rely on this identity, by showing that the normal equations are compatible.

Solution Let $\mathbf{v}$ represent any vector such that $\mathbf{v}^T\mathbf{X}^T\mathbf{X} = \mathbf{0}^T$. Post-multiplication of both sides of this equation by $\mathbf{v}$ results in another equation $\mathbf{v}^T\mathbf{X}^T\mathbf{X}\mathbf{v} = 0$, implying (by Corollary 2.10.4.1) that $\mathbf{X}\mathbf{v} = \mathbf{0}$, hence that $\mathbf{v}^T\mathbf{X}^T\mathbf{y} = (\mathbf{X}\mathbf{v})^T\mathbf{y} = 0$. This establishes that the normal equations are compatible, hence consistent by Theorem 3.2.1.

▶ **Exercise 2** For observations following a certain linear model, the normal equations are

$$\begin{pmatrix} 7 & -2 & -5 \\ -2 & 3 & -1 \\ -5 & -1 & 6 \end{pmatrix} \begin{pmatrix} \beta_1 \\ \beta_2 \\ \beta_3 \end{pmatrix} = \begin{pmatrix} 17 \\ 34 \\ -51 \end{pmatrix}.$$

Observe that the three rows of the coefficient matrix sum to $\mathbf{0}_3^T$.

(a) Obtain two distinct solutions, say $\hat{\boldsymbol{\beta}}_1$ and $\hat{\boldsymbol{\beta}}_2$, to the normal equations.
(b) Characterize, as simply as possible, the collection of linear functions $\mathbf{c}^T\boldsymbol{\beta} = c_1\beta_1 + c_2\beta_2 + c_3\beta_3$ that are estimable.
(c) List one nonzero estimable function, $\mathbf{c}_E^T\boldsymbol{\beta}$, and one nonestimable function, $\mathbf{c}_{NE}^T\boldsymbol{\beta}$. (Give numerical entries for $\mathbf{c}_E^T$ and $\mathbf{c}_{NE}^T$.)

D. L. Zimmerman, *Linear Model Theory*,
https://doi.org/10.1007/978-3-030-52074-8_7

(d) Determine numerically whether the least squares estimator of your estimable function from part (c) is the same for both of your solutions from part (a), i.e., determine whether $\mathbf{c}_E^T\hat{\boldsymbol{\beta}}_1 = \mathbf{c}_E^T\hat{\boldsymbol{\beta}}_2$. Similarly, determine whether $\mathbf{c}_{NE}^T\hat{\boldsymbol{\beta}}_1 = \mathbf{c}_{NE}^T\hat{\boldsymbol{\beta}}_2$. Which theorem does this exemplify?

Solution

(a) The coefficient matrix has dimensions 3×3 but rank 2, so by Theorems 3.2.3 and 3.3.6 two solutions to the normal equations are

$$\hat{\boldsymbol{\beta}}_1 = \begin{pmatrix} \frac{3}{17} & \frac{2}{17} & 0 \\ \frac{2}{17} & \frac{7}{17} & 0 \\ 0 & 0 & 0 \end{pmatrix} \begin{pmatrix} 17 \\ 34 \\ -51 \end{pmatrix} = \begin{pmatrix} 7 \\ 16 \\ 0 \end{pmatrix} \text{ and } \hat{\boldsymbol{\beta}}_2 = \begin{pmatrix} 0 & 0 & 0 \\ 0 & \frac{6}{17} & \frac{1}{17} \\ 0 & \frac{1}{17} & \frac{3}{17} \end{pmatrix} \begin{pmatrix} 17 \\ 34 \\ -51 \end{pmatrix} = \begin{pmatrix} 0 \\ 9 \\ -7 \end{pmatrix}.$$

(b) $\mathcal{R}(\mathbf{X}) = \{\mathbf{c}^T \in \mathbb{R}^3 : \mathbf{c}^T\mathbf{1} = 0\}$.

(c) $\mathbf{c}_E^T\boldsymbol{\beta} = \beta_1 - 2\beta_2 + \beta_3$, $\mathbf{c}_{NE}^T\boldsymbol{\beta} = \beta_1 + \beta_2 + \beta_3$.

(d) $\mathbf{c}_E^T\hat{\boldsymbol{\beta}}_1 = (1, -2, 1)\begin{pmatrix} 7 \\ 16 \\ 0 \end{pmatrix} = -25 = (1, -2, 1)\begin{pmatrix} 0 \\ 9 \\ -7 \end{pmatrix} = \mathbf{c}_E^T\hat{\boldsymbol{\beta}}_2$, but $\mathbf{c}_{NE}^T\hat{\boldsymbol{\beta}}_1 =$

$(1, 1, 1)\begin{pmatrix} 7 \\ 16 \\ 0 \end{pmatrix} = 23 \neq 2 = (1, 1, 1)\begin{pmatrix} 0 \\ 9 \\ -7 \end{pmatrix} = \mathbf{c}_{NE}^T\hat{\boldsymbol{\beta}}_2$. This exemplifies Theorem 7.1.3.

▶ **Exercise 3** Consider a linear model for which

$$\mathbf{y} = \begin{pmatrix} y_1 \\ y_2 \\ y_3 \\ y_4 \\ y_5 \\ y_6 \\ y_7 \\ y_8 \end{pmatrix}, \quad \mathbf{X} = \begin{pmatrix} 1 & 1 & 1 & -1 \\ 1 & 1 & 1 & -1 \\ 1 & 1 & -1 & 1 \\ 1 & 1 & -1 & 1 \\ 1 & -1 & 1 & 1 \\ 1 & -1 & 1 & 1 \\ -1 & 1 & 1 & 1 \\ -1 & 1 & 1 & 1 \end{pmatrix}, \quad \boldsymbol{\beta} = \begin{pmatrix} \beta_1 \\ \beta_2 \\ \beta_3 \\ \beta_4 \end{pmatrix}.$$

(a) Obtain the normal equations for this model and solve them.
(b) Are all functions $\mathbf{c}^T\boldsymbol{\beta}$ estimable? Justify your answer.
(c) Obtain the least squares estimator of $\beta_1 + \beta_2 + \beta_3 + \beta_4$.

Solution

(a) The normal equations are

$$\begin{pmatrix} 8 & 0 & 0 & 0 \\ 0 & 8 & 0 & 0 \\ 0 & 0 & 8 & 0 \\ 0 & 0 & 0 & 8 \end{pmatrix} \boldsymbol{\beta} = \begin{pmatrix} y_1 + y_2 + y_3 + y_4 + y_5 + y_6 - y_7 - y_8 \\ y_1 + y_2 + y_3 + y_4 - y_5 - y_6 + y_7 + y_8 \\ y_1 + y_2 - y_3 - y_4 + y_5 + y_6 + y_7 + y_8 \\ -y_1 - y_2 + y_3 + y_4 + y_5 + y_6 + y_7 + y_8 \end{pmatrix}$$

and their unique solution is

$$\hat{\boldsymbol{\beta}} = \begin{pmatrix} (y_1 + y_2 + y_3 + y_4 + y_5 + y_6 - y_7 - y_8)/8 \\ (y_1 + y_2 + y_3 + y_4 - y_5 - y_6 + y_7 + y_8)/8 \\ (y_1 + y_2 - y_3 - y_4 + y_5 + y_6 + y_7 + y_8)/8 \\ (-y_1 - y_2 + y_3 + y_4 + y_5 + y_6 + y_7 + y_8)/8 \end{pmatrix}.$$

(b) Because $\mathbf{X}^T\mathbf{X} = 8\mathbf{I}$ has full rank, all functions $\mathbf{c}^T\boldsymbol{\beta}$ are estimable by Corollary 6.1.2.4.

(c) $\mathbf{c}^T\hat{\boldsymbol{\beta}} = \mathbf{1}_4^T\hat{\boldsymbol{\beta}} = (2y_1 + 2y_2 + 2y_3 + 2y_4 + 2y_5 + 2y_6 + 2y_7 + 2y_8)/8 = 2\bar{y}$.

▶ **Exercise 4** Prove Corollary 7.1.4.1: If $\mathbf{C}^T\boldsymbol{\beta}$ is estimable under the model $\{\mathbf{y}, \mathbf{X}\boldsymbol{\beta}\}$ and $\mathbf{L}^T$ is a matrix of constants having the same number of columns as $\mathbf{C}^T$ has rows, then $\mathbf{L}^T\mathbf{C}^T\boldsymbol{\beta}$ is estimable and its least squares estimator (associated with the specified model) is $\mathbf{L}^T(\mathbf{C}^T\hat{\boldsymbol{\beta}})$.

Solution Because $\mathbf{C}^T\boldsymbol{\beta}$ is estimable under the specified model, $\mathbf{C}^T = \mathbf{A}^T\mathbf{X}$ for some $\mathbf{A}$. Then $\mathbf{L}^T\mathbf{C}^T = \mathbf{L}^T\mathbf{A}^T\mathbf{X} = (\mathbf{A}\mathbf{L})^T\mathbf{X}$, implying that every row of $\mathbf{L}^T\mathbf{C}^T$ is an element of $\mathcal{R}(\mathbf{X})$. Hence $\mathbf{L}^T\mathbf{C}^T\boldsymbol{\beta}$ is estimable, and by Theorem 7.1.4 its least squares estimator is $\mathbf{L}^T\mathbf{C}^T(\mathbf{X}^T\mathbf{X})^-\mathbf{X}^T\mathbf{y} = \mathbf{L}^T[\mathbf{C}^T(\mathbf{X}^T\mathbf{X})^-\mathbf{X}^T\mathbf{y}] = \mathbf{L}^T\mathbf{C}^T\hat{\boldsymbol{\beta}}$.

▶ **Exercise 5** Prove that under the Gauss–Markov model $\{\mathbf{y}, \mathbf{X}\boldsymbol{\beta}, \sigma^2\mathbf{I}\}$, the least squares estimator of an estimable function $\mathbf{c}^T\boldsymbol{\beta}$ associated with that model has uniformly (in $\boldsymbol{\beta}$ and σ^2) smaller mean squared error than any other linear location equivariant estimator of $\mathbf{c}^T\boldsymbol{\beta}$. (Refer back to Exercise 6.1 for the definition of a location equivariant estimator.)

Solution Let $t_0 + \mathbf{t}^T\mathbf{y}$ represent any linear location equivariant estimator of $\mathbf{c}^T\boldsymbol{\beta}$ associated with the specified model. Then by Exercise 6.1, $\mathbf{t}^T\mathbf{X} = \mathbf{c}^T$, implying that $\mathbf{t}^T\mathbf{y}$ is unbiased for $\mathbf{c}^T\boldsymbol{\beta}$. Now,

$$\begin{aligned} \mathrm{MSE}(t_0 + \mathbf{t}^T\mathbf{y}) &= \mathrm{var}(t_0 + \mathbf{t}^T\mathbf{y}) + [\mathrm{E}(t_0 + \mathbf{t}^T\mathbf{y}) - \mathbf{c}^T\boldsymbol{\beta}]^2 \\ &= \mathrm{var}(\mathbf{t}^T\mathbf{y}) + [t_0 + \mathbf{t}^T\mathbf{X}\boldsymbol{\beta} - \mathbf{c}^T\boldsymbol{\beta}]^2 \\ &= \mathrm{var}(\mathbf{t}^T\mathbf{y}) + t_0^2 \\ &\geq \mathrm{var}(\mathbf{c}^T\hat{\boldsymbol{\beta}}), \end{aligned}$$

where the inequality holds by Theorem 7.2.3, with equality holding if and only if $t_0 = 0$ and $\mathbf{t}^T\mathbf{y} = \mathbf{c}^T\hat{\boldsymbol{\beta}}$ with probability one.

▶ **Exercise 6** Consider the model $\{\mathbf{y}, \mathbf{X}\boldsymbol{\beta}\}$ with observations partitioned into two groups, so that the model may be written alternatively as $\left\{\begin{pmatrix}\mathbf{y}_1\\ \mathbf{y}_2\end{pmatrix}, \begin{pmatrix}\mathbf{X}_1\\ \mathbf{X}_2\end{pmatrix}\boldsymbol{\beta}\right\}$, where $\mathbf{y}_1$ is an n_1-vector. Suppose that $\mathbf{c}^T \in \mathcal{R}(\mathbf{X}_1)$ and $\mathcal{R}(\mathbf{X}_1) \cap \mathcal{R}(\mathbf{X}_2) = \{\mathbf{0}\}$. Prove that the least squares estimator of $\mathbf{c}^T\boldsymbol{\beta}$ associated with the model $\{\mathbf{y}_1, \mathbf{X}_1\boldsymbol{\beta}\}$ for the first n_1 observations is identical to the least squares estimator of $\mathbf{c}^T\boldsymbol{\beta}$ associated with the model for all the observations; i.e., the additional observations do not affect the least squares estimator.

Solution First observe that $\mathbf{c}^T\boldsymbol{\beta}$ is estimable under both models. Let $\hat{\boldsymbol{\beta}}$ denote a solution to the normal equations for the model for all the observations. Then $\mathbf{X}^T\mathbf{X}\hat{\boldsymbol{\beta}} = \mathbf{X}^T\mathbf{y}$, i.e.,

$$\begin{pmatrix}\mathbf{X}_1\\ \mathbf{X}_2\end{pmatrix}^T \begin{pmatrix}\mathbf{X}_1\\ \mathbf{X}_2\end{pmatrix}\hat{\boldsymbol{\beta}} = \begin{pmatrix}\mathbf{X}_1\\ \mathbf{X}_2\end{pmatrix}^T \mathbf{y},$$

i.e.,

$$\mathbf{X}_1^T\mathbf{X}_1\hat{\boldsymbol{\beta}} + \mathbf{X}_2^T\mathbf{X}_2\hat{\boldsymbol{\beta}} = \mathbf{X}_1^T\mathbf{y} + \mathbf{X}_2^T\mathbf{y}.$$

Algebraic transposition yields

$$\mathbf{X}_1^T\mathbf{X}_1\hat{\boldsymbol{\beta}} - \mathbf{X}_1^T\mathbf{y} = -(\mathbf{X}_2^T\mathbf{X}_2\hat{\boldsymbol{\beta}} - \mathbf{X}_2^T\mathbf{y}),$$

and then matrix transposition yields

$$(\hat{\boldsymbol{\beta}}^T\mathbf{X}_1^T - \mathbf{y}^T)\mathbf{X}_1 = -(\hat{\boldsymbol{\beta}}^T\mathbf{X}_2^T - \mathbf{y}^T)\mathbf{X}_2.$$

The vector on the left-hand side of this last system of equations belongs to $\mathcal{R}(\mathbf{X}_1)$, while the vector on the right-hand side belongs to $\mathcal{R}(\mathbf{X}_2)$. Because these two row spaces are essentially disjoint, both vectors must be null, implying in particular that

$$(\hat{\boldsymbol{\beta}}^T\mathbf{X}_1^T - \mathbf{y}^T)\mathbf{X}_1 = \mathbf{0}^T,$$

or equivalently that $\mathbf{X}_1^T\mathbf{X}_1\hat{\boldsymbol{\beta}} = \mathbf{X}_1^T\mathbf{y}$. Thus $\hat{\boldsymbol{\beta}}$ also satisfies the normal equations for the model for only the first n_1 observations, implying that $\mathbf{c}^T\hat{\boldsymbol{\beta}}$ is the least squares estimator of $\mathbf{c}^T\boldsymbol{\beta}$ under both models.

Exercise 7 For an arbitrary model $\{\mathbf{y}, \mathbf{X}\boldsymbol{\beta}\}$:

(a) Prove that $\bar{x}_1\beta_1 + \bar{x}_2\beta_2 + \cdots + \bar{x}_p\beta_p$ is estimable.
(b) Prove that if one of the columns of $\mathbf{X}$ is a column of ones, then the least squares estimator of the function in part (a) is $\bar{y}$.

Solution

(a) $\bar{x}_1\beta_1 + \bar{x}_2\beta_2 + \cdots + \bar{x}_p\beta_p = \mathbf{c}^T\boldsymbol{\beta}$ with $\mathbf{c}^T = (\bar{x}_1, \bar{x}_2, \ldots, \bar{x}_p)$. But this $\mathbf{c}^T = (1/n)\mathbf{1}_n^T\mathbf{X}$ which is an element of $\mathcal{R}(\mathbf{X})$, so $\mathbf{c}^T\boldsymbol{\beta}$ is estimable.
(b) Because $\mathbf{1}_n \in \mathcal{C}(\mathbf{X})$, $\mathbf{1}_n = \mathbf{Xa}$ for some p-vector $\mathbf{a}$. Then, using part (a) and Theorem 7.1.4, the least squares estimator of $\mathbf{c}^T\boldsymbol{\beta}$ is $(1/n)\mathbf{1}^T\mathbf{X}(\mathbf{X}^T\mathbf{X})^-\mathbf{X}^T\mathbf{y} = (1/n)\mathbf{a}^T\mathbf{X}^T\mathbf{X}(\mathbf{X}^T\mathbf{X})^-\mathbf{X}^T\mathbf{y} = (1/n)\mathbf{a}^T\mathbf{X}^T\mathbf{y} = (1/n)\mathbf{1}^T\mathbf{y} = \bar{y}$.

Exercise 8 For the centered simple linear regression model

$$y_i = \beta_1 + \beta_2(x_i - \bar{x}) + e_i \quad (i = 1, \ldots, n),$$

obtain the least squares estimators of β_1 and β_2 and their variance–covariance matrix (under Gauss–Markov assumptions).

Solution

$$\begin{pmatrix} \hat{\beta}_1 \\ \hat{\beta}_2 \end{pmatrix} = [(\mathbf{1}, \mathbf{x} - \bar{x}\mathbf{1})^T(\mathbf{1}, \mathbf{x} - \bar{x}\mathbf{1})]^{-1}(\mathbf{1}, \mathbf{x} - \bar{x}\mathbf{1})^T\mathbf{y}$$

$$= \begin{pmatrix} n & 0 \\ 0 & SXX \end{pmatrix}^{-1} \begin{pmatrix} \sum_{i=1}^n y_i \\ SXY \end{pmatrix} = \begin{pmatrix} \bar{y} \\ SXX/SXY \end{pmatrix},$$

and $\mathrm{var}\begin{pmatrix} \hat{\beta}_1 \\ \hat{\beta}_2 \end{pmatrix} = \sigma^2[(\mathbf{1}, \mathbf{x} - \bar{x}\mathbf{1})^T(\mathbf{1}, \mathbf{x} - \bar{x}\mathbf{1})]^{-1} = \sigma^2\begin{pmatrix} 1/n & 0 \\ 0 & 1/SXX \end{pmatrix}.$

Exercise 9 For the no-intercept simple linear regression model, obtain the least squares estimator of the slope and its variance (under Gauss–Markov assumptions).

Solution $\hat{\beta} = (\mathbf{x}^T\mathbf{x})^{-1}\mathbf{x}^T\mathbf{y} = \sum_{i=1}^n x_iy_i / \sum_{i=1}^n x_i^2$, and $\mathrm{var}(\hat{\beta}) = \sigma^2(\mathbf{x}^T\mathbf{x})^{-1} = \sigma^2/\sum_{i=1}^n x_i^2$.

Exercise 10 For the two-way main effects model with cell frequencies specified by the last 2×2 layout of Example 7.1-3, obtain the variance–covariance matrix (under Gauss–Markov assumptions) of the least squares estimators of:

(a) the cell means;
(b) the Factor A and B differences.

Solution

(a) $$\text{var}\begin{pmatrix} \widehat{\mu+\alpha_1+\gamma_1} \\ \widehat{\mu+\alpha_1+\gamma_2} \\ \widehat{\mu+\alpha_2+\gamma_1} \\ \widehat{\mu+\alpha_2+\gamma_2} \end{pmatrix} = \sigma^2 \begin{pmatrix} 0.5 & 0 & 0 & -0.5 \\ 0 & 1 & 0 & 1 \\ 0 & 0 & 1 & 1 \\ -0.5 & 1 & 1 & 2.5 \end{pmatrix}.$$

(b) $$\text{var}\begin{pmatrix} \widehat{\alpha_1-\alpha_2} \\ \widehat{\gamma_1-\gamma_2} \end{pmatrix} = \sigma^2 \begin{pmatrix} 1.5 & 0.5 \\ 0.5 & 1.5 \end{pmatrix}.$$

▶ **Exercise 11** For the two-way main effects model with equal cell frequencies r, obtain the least squares estimators, and their variance–covariance matrix (under Gauss–Markov assumptions), of:

(a) the cell means;
(b) the Factor A and Factor B differences.

Solution

(a) The model matrix is $\mathbf{X} = (\mathbf{1}_{qmr}, \mathbf{I}_q \otimes \mathbf{1}_{mr}, \mathbf{1}_q \otimes \mathbf{I}_m \otimes \mathbf{1}_r)$ and

$$\mathbf{X}^T\mathbf{X} = \begin{pmatrix} \mathbf{A} & \mathbf{B} \\ \mathbf{C} & \mathbf{D} \end{pmatrix},$$

where

$$\mathbf{A} = \begin{pmatrix} qmr & mr\mathbf{1}_q^T \\ mr\mathbf{1}_q & mr\mathbf{I}_q \end{pmatrix}, \quad \mathbf{B} = \begin{pmatrix} qr\mathbf{1}_m^T \\ r\mathbf{J}_{q\times m} \end{pmatrix}$$
$$\mathbf{C} = \left(qr\mathbf{1}_m, r\mathbf{J}_{m\times q}\right), \quad \mathbf{D} = qr\mathbf{I}_m.$$

By Theorem 3.3.6, one generalized inverse of $\mathbf{A}$ is

$$\mathbf{A}^- = \begin{pmatrix} 0 & \mathbf{0}_q^T \\ \mathbf{0}_q & (1/mr)\mathbf{I}_q \end{pmatrix}.$$

Because each column of $\mathbf{B}$ is equal to $(1/m)$ times the first column of $\mathbf{A}$, $\mathcal{C}(\mathbf{B}) \subseteq \mathcal{C}(\mathbf{A})$, and similarly $\mathcal{R}(\mathbf{C}) \subseteq \mathcal{R}(\mathbf{A})$. Let $\mathbf{Q} = \mathbf{D} - \mathbf{C}\mathbf{A}^-\mathbf{B} = qr\mathbf{I}_m - (qr/m)\mathbf{J}_m$. One generalized inverse of $\mathbf{Q}$ is $\mathbf{Q}^- = (1/qr)\mathbf{I}_m - (1/qmr)\mathbf{J}_m$, and $\mathbf{B}\mathbf{Q}^- = \begin{pmatrix} qr\mathbf{1}_m^T[(1/qr)\mathbf{I}_m - (1/qmr)\mathbf{J}_m] \\ r\mathbf{J}_{q\times m}[(1/qr)\mathbf{I}_m - (1/qmr)\mathbf{J}_m] \end{pmatrix} = \mathbf{0}$. Therefore, by Theorem 3.3.7, one generalized inverse of $\mathbf{X}^T\mathbf{X}$ is

$$\begin{pmatrix} \mathbf{A}^- & \mathbf{0} \\ \mathbf{0} & \mathbf{Q}^- \end{pmatrix}.$$

Finally, let $\mathbf{C}^T = (\mathbf{1}_{qm},\ \mathbf{I}_q \otimes \mathbf{1}_m,\ \mathbf{1}_q \otimes \mathbf{I}_m)$ denote the matrix consisting of all distinct rows of $\mathbf{X}$. Then the least squares estimator of the vector of cell means is

$$\begin{pmatrix} \widehat{\mu + \alpha_1 + \gamma_1} \\ \widehat{\mu + \alpha_1 + \gamma_2} \\ \vdots \\ \widehat{\mu + \alpha_q + \gamma_m} \end{pmatrix} = \mathbf{C}^T(\mathbf{X}^T\mathbf{X})^-\mathbf{X}^T\mathbf{y}$$

$$= (\mathbf{1}_{qm},\ \mathbf{I}_q \otimes \mathbf{1}_m,\ \mathbf{1}_q \otimes \mathbf{I}_m)$$

$$\times \begin{pmatrix} 0 & \mathbf{0}_q^T & \mathbf{0} \\ \mathbf{0}_q & (1/mr)\mathbf{I}_q & \mathbf{0} \\ \mathbf{0} & \mathbf{0} & (1/qr)\mathbf{I}_m - (1/qmr)\mathbf{J}_m \end{pmatrix} \begin{pmatrix} y_{\cdots} \\ y_{1\cdot\cdot} \\ \vdots \\ y_{q\cdot\cdot} \\ y_{\cdot 1 \cdot} \\ \vdots \\ y_{\cdot m \cdot} \end{pmatrix}$$

$$= \Big(\mathbf{0}_{qm},\ (1/mr)\mathbf{I}_q \otimes \mathbf{1}_m,\ (1/qr)\mathbf{1}_q \otimes [\mathbf{I}_m - (1/m)\mathbf{J}_m]\Big) \begin{pmatrix} y_{\cdots} \\ y_{1\cdot\cdot} \\ \vdots \\ y_{q\cdot\cdot} \\ y_{\cdot 1 \cdot} \\ \vdots \\ y_{\cdot m \cdot} \end{pmatrix}$$

$$= \begin{pmatrix} \bar{y}_{1\cdot\cdot} + \bar{y}_{\cdot 1 \cdot} - \bar{y}_{\cdots} \\ \bar{y}_{1\cdot\cdot} + \bar{y}_{\cdot 2 \cdot} - \bar{y}_{\cdots} \\ \vdots \\ \bar{y}_{q\cdot\cdot} + \bar{y}_{\cdot m \cdot} - \bar{y}_{\cdots} \end{pmatrix}.$$

By Theorem 7.2.2,

$$\text{var}\begin{pmatrix} \bar{y}_{1\cdot\cdot} + \bar{y}_{\cdot 1 \cdot} - \bar{y}_{\cdots} \\ \bar{y}_{1\cdot\cdot} + \bar{y}_{\cdot 2 \cdot} - \bar{y}_{\cdots} \\ \vdots \\ \bar{y}_{q\cdot\cdot} + \bar{y}_{\cdot m \cdot} - \bar{y}_{\cdots} \end{pmatrix} = \sigma^2\mathbf{C}^T(\mathbf{X}^T\mathbf{X})^-\mathbf{C}$$

$$= \sigma^2 \left(\mathbf{0}_{qm}, (1/mr)\mathbf{I}_q \otimes \mathbf{1}_m, (1/qr)\mathbf{1}_q \otimes [\mathbf{I}_m - (1/m)\mathbf{J}_m] \right) \begin{pmatrix} \mathbf{1}_{qm}^T \\ \mathbf{I}_q \otimes \mathbf{1}_m^T \\ \mathbf{1}_q^T \otimes \mathbf{I}_m \end{pmatrix}$$

$$= \sigma^2 \{(1/mr)(\mathbf{I}_q \otimes \mathbf{J}_m) + (1/qr)\mathbf{J}_q \otimes [\mathbf{I}_m - (1/m)\mathbf{J}_m]\}.$$

(b) By part (a) and Corollary 7.1.4.1, the least squares estimators of the Factor A differences and Factor B differences are

$$\widehat{\alpha_i - \alpha_{i'}} = \bar{y}_{i\cdot\cdot} - \bar{y}_{i'\cdot\cdot} \qquad \text{and} \qquad \widehat{\gamma_j - \gamma_{j'}} = \bar{y}_{\cdot j\cdot} - \bar{y}_{\cdot j'\cdot},$$

for $i \neq i' = 1, \dots, q$, $j \neq j' = 1, \dots, m$. To compute the variance–covariance matrix of these estimators, we use an approach similar to that used in Example 7.2-2. First we obtain

$$\mathrm{cov}(\bar{y}_{i\cdot\cdot}, \bar{y}_{i'\cdot\cdot}) = \begin{cases} \frac{\sigma^2}{mr} & \text{if } i = i', \\ 0 & \text{if } i \neq i'; \end{cases}$$

$$\mathrm{cov}(\bar{y}_{\cdot j\cdot}, \bar{y}_{\cdot j'\cdot}) = \begin{cases} \frac{\sigma^2}{qr} & \text{if } j = j', \\ 0 & \text{if } j \neq j'; \end{cases}$$

$$\mathrm{cov}(\bar{y}_{i\cdot\cdot}, \bar{y}_{\cdot j\cdot}) = \frac{\sigma^2}{qmr} \qquad \text{for all } i, j.$$

Therefore, entries of the variance–covariance matrix obtained using Theorem 4.2.2a are as follows: for $i \leq s$, $i' > i = 1, \dots, q$, and $s' > s = 1, \dots, q$,

$$\begin{aligned} \mathrm{cov}(\bar{y}_{i\cdot\cdot} - \bar{y}_{i'\cdot\cdot}, \bar{y}_{s\cdot\cdot} - \bar{y}_{s'\cdot\cdot}) &= \mathrm{cov}(\bar{y}_{i\cdot\cdot}, \bar{y}_{s\cdot\cdot}) - \mathrm{cov}(\bar{y}_{i'\cdot\cdot}, \bar{y}_{s\cdot\cdot}) \\ &\quad - \mathrm{cov}(\bar{y}_{i\cdot\cdot}, \bar{y}_{s'\cdot\cdot}) + \mathrm{cov}(\bar{y}_{i'\cdot\cdot}, \bar{y}_{s'\cdot\cdot}) \\ &= \begin{cases} \frac{2\sigma^2}{mr} & \text{if } i = s \text{ and } i' = s', \\ \frac{\sigma^2}{mr} & \text{if } i = s \text{ and } i' \neq s', \text{ or } i \neq s \text{ and } i' = s', \\ -\frac{\sigma^2}{mr} & \text{if } i' = s, \\ 0, & \text{otherwise}; \end{cases} \end{aligned}$$

for $j \leq t$, $j' > j = 1, \dots, m$, and $t' > t = 1, \dots, m$,

$$\begin{aligned} \mathrm{cov}(\bar{y}_{\cdot j\cdot} - \bar{y}_{\cdot j'\cdot}, \bar{y}_{\cdot t\cdot} - \bar{y}_{\cdot t'\cdot}) &= \mathrm{cov}(\bar{y}_{\cdot j\cdot}, \bar{y}_{\cdot t\cdot}) - \mathrm{cov}(\bar{y}_{\cdot j'\cdot}, \bar{y}_{\cdot t'\cdot}) \\ &\quad - \mathrm{cov}(\bar{y}_{\cdot j\cdot}, \bar{y}_{\cdot t'\cdot}) + \mathrm{cov}(\bar{y}_{\cdot j'\cdot}, \bar{y}_{\cdot t'\cdot}) \\ &= \begin{cases} \frac{2\sigma^2}{qr} & \text{if } j = t \text{ and } j' = t', \\ \frac{\sigma^2}{qr} & \text{if } j = t \text{ and } j' \neq t', \text{ or } j \neq t \text{ and } j' = t', \\ -\frac{\sigma^2}{qr} & \text{if } j' = t, \\ 0, & \text{otherwise}; \end{cases} \end{aligned}$$

and for $i' > i = 1, \ldots, q$ and $j' > j = 1, \ldots, m$,

$$\begin{aligned}\operatorname{cov}(\bar{y}_{i\cdot\cdot} - \bar{y}_{i'\cdot\cdot}, \bar{y}_{\cdot j\cdot} - \bar{y}_{\cdot j'\cdot}) &= \operatorname{cov}(\bar{y}_{i\cdot\cdot}, \bar{y}_{\cdot j\cdot}) - \operatorname{cov}(\bar{y}_{i'\cdot\cdot}, \bar{y}_{\cdot j\cdot}) \\ &\quad - \operatorname{cov}(\bar{y}_{i\cdot\cdot}, \bar{y}_{\cdot j'\cdot}) + \operatorname{cov}(\bar{y}_{i'\cdot\cdot}, \bar{y}_{\cdot j'\cdot}) \\ &= \frac{\sigma^2}{qmr} - \frac{\sigma^2}{qmr} - \frac{\sigma^2}{qmr} + \frac{\sigma^2}{qmr} \\ &= 0.\end{aligned}$$

▶ **Exercise 12** Consider the connected 6×4 layout displayed in Example 6.1-4. Suppose that the model for the "data" in the occupied cells is the Gauss–Markov two-way main effects model

$$y_{ij} = \mu + \alpha_i + \gamma_j + e_{ij},$$

where $(i, j) \in \{(1, 1), (1, 2), (2, 1), (2, 3), (3, 1), (3, 4), (4, 2), (4, 3), (5, 2), (5, 4), (6, 3), (6, 4)\}$ (the occupied cells). Obtain specialized expressions for the least squares estimators of $\gamma_j - \gamma_{j'}$ $(j' > j = 1, 2, 3, 4)$, and show that their variances are equal. The homoscedasticity of the variances of these differences is a nice property of balanced incomplete block designs such as this one.

Solution The model matrix is (only the nonzero elements are shown)

$$\mathbf{X} = \begin{pmatrix} 1 & 1 & & & & & & 1 & & & \\ 1 & 1 & & & & & & & 1 & & \\ 1 & & 1 & & & & & 1 & & & \\ 1 & & 1 & & & & & & & 1 & \\ 1 & & & 1 & & & & 1 & & & \\ 1 & & & 1 & & & & & & & 1 \\ 1 & & & & 1 & & & & 1 & & \\ 1 & & & & 1 & & & & & 1 & \\ 1 & & & & & 1 & & & 1 & & \\ 1 & & & & & 1 & & & & & 1 \\ 1 & & & & & & 1 & & & 1 & \\ 1 & & & & & & 1 & & & & 1 \end{pmatrix}$$

and the coefficient matrix of the normal equations is the symmetric matrix

$$
\mathbf{X}^T\mathbf{X} = \begin{pmatrix}
12 & 2 & 2 & 2 & 2 & 2 & 2 & 3 & 3 & 3 & 3 \\
 & 2 & 0 & 0 & 0 & 0 & 0 & 1 & 1 & 0 & 0 \\
 & & 2 & 0 & 0 & 0 & 0 & 1 & 0 & 1 & 0 \\
 & & & 2 & 0 & 0 & 0 & 1 & 0 & 0 & 1 \\
 & & & & 2 & 0 & 0 & 0 & 1 & 1 & 0 \\
 & & & & & 2 & 0 & 0 & 1 & 0 & 1 \\
 & & & & & & 2 & 0 & 0 & 1 & 1 \\
 & & & & & & & 3 & 0 & 0 & 0 \\
 & & & & & & & & 3 & 0 & 0 \\
 & & & & & & & & & 3 & 0 \\
 & & & & & & & & & & 3
\end{pmatrix}.
$$

Clearly $\mathbf{X}^T\mathbf{X}$ satisfies the conditions of Theorem 3.3.7a with $\mathbf{A}$ in that theorem taken to be the upper left 7×7 block of $\mathbf{X}^T\mathbf{X}$. Furthermore, by Theorem 3.3.6 one generalized inverse of the matrix $\mathbf{A}$ so defined is $\begin{pmatrix} 0 & \mathbf{0}_6^T \\ \mathbf{0}_6 & \frac{1}{2}\mathbf{I}_6 \end{pmatrix}$. Thus, one generalized inverse of $\mathbf{X}^T\mathbf{X}$ is given by a matrix of the form specified in Theorem 3.3.7a, with

$$
\begin{aligned}
\mathbf{Q} &= 3\mathbf{I}_4 - \begin{pmatrix}
3 & 1 & 1 & 1 & 0 & 0 & 0 \\
3 & 1 & 0 & 0 & 1 & 1 & 0 \\
3 & 0 & 1 & 0 & 1 & 0 & 1 \\
3 & 0 & 0 & 1 & 0 & 1 & 1
\end{pmatrix}
\begin{pmatrix} 0 & \mathbf{0}_6^T \\ \mathbf{0}_6 & \frac{1}{2}\mathbf{I}_6 \end{pmatrix}
\begin{pmatrix}
3 & 3 & 3 & 3 \\
1 & 1 & 0 & 0 \\
1 & 0 & 1 & 0 \\
1 & 0 & 0 & 1 \\
0 & 1 & 1 & 0 \\
0 & 1 & 0 & 1 \\
0 & 0 & 1 & 1
\end{pmatrix} \\
&= 3\mathbf{I}_4 - \frac{1}{2}\begin{pmatrix}
3 & 1 & 1 & 1 \\
1 & 3 & 1 & 1 \\
1 & 1 & 3 & 1 \\
1 & 1 & 1 & 3
\end{pmatrix} \\
&= 2(\mathbf{I}_4 - \frac{1}{4}\mathbf{J}_4).
\end{aligned}
$$

Either by first principles or by the results of Exercise 3.1a, c, we find that one generalized inverse of $\mathbf{Q}$ is

$$
\mathbf{Q}^- = \frac{1}{2}(\mathbf{I}_4 - \frac{1}{4}\mathbf{J}_4) = \begin{pmatrix}
\frac{3}{8} & -\frac{1}{8} & -\frac{1}{8} & -\frac{1}{8} \\
-\frac{1}{8} & \frac{3}{8} & -\frac{1}{8} & -\frac{1}{8} \\
-\frac{1}{8} & -\frac{1}{8} & \frac{3}{8} & -\frac{1}{8} \\
-\frac{1}{8} & -\frac{1}{8} & -\frac{1}{8} & \frac{3}{8}
\end{pmatrix}.
$$

Using further notation from Theorem 3.3.7a,

$$\mathbf{Q}^{-}\mathbf{C}\mathbf{A}^{-} = \frac{1}{2}(\mathbf{I}_4 - \frac{1}{4}\mathbf{J}_4)\begin{pmatrix} 3 & 1 & 1 & 1 & 0 & 0 & 0 \\ 3 & 1 & 0 & 0 & 1 & 1 & 0 \\ 3 & 0 & 1 & 0 & 1 & 0 & 1 \\ 3 & 0 & 0 & 1 & 0 & 1 & 1 \end{pmatrix}\begin{pmatrix} 0 & \mathbf{0}_6^T \\ \mathbf{0}_6 & \frac{1}{2}\mathbf{I}_6 \end{pmatrix} = \begin{pmatrix} -\frac{1}{4} & -\frac{1}{4} & -\frac{1}{4} & -\frac{1}{2} & -\frac{1}{2} & -\frac{1}{2} \\ -\frac{1}{4} & -\frac{1}{2} & -\frac{1}{2} & -\frac{1}{4} & -\frac{1}{4} & -\frac{1}{2} \\ -\frac{1}{2} & -\frac{1}{4} & -\frac{1}{2} & -\frac{1}{4} & -\frac{1}{2} & -\frac{1}{4} \\ -\frac{1}{2} & -\frac{1}{2} & -\frac{1}{4} & -\frac{1}{2} & -\frac{1}{4} & -\frac{1}{4} \end{pmatrix}.$$

Thus, the last four elements of $(\mathbf{X}^T\mathbf{X})^{-}\mathbf{X}^T\mathbf{y}$ (which are the only elements needed to obtain least squares estimators of the γ-differences) are

$$\begin{pmatrix} \frac{1}{4} & \frac{1}{4} & \frac{1}{4} & \frac{1}{2} & \frac{1}{2} & \frac{1}{2} & \frac{3}{8} & -\frac{1}{8} & -\frac{1}{8} & -\frac{1}{8} \\ \frac{1}{4} & \frac{1}{2} & \frac{1}{2} & \frac{1}{4} & \frac{1}{4} & \frac{1}{2} & -\frac{1}{8} & \frac{3}{8} & -\frac{1}{8} & -\frac{1}{8} \\ \frac{1}{2} & \frac{1}{4} & \frac{1}{2} & \frac{1}{4} & \frac{1}{2} & \frac{1}{4} & -\frac{1}{8} & -\frac{1}{8} & \frac{3}{8} & -\frac{1}{8} \\ \frac{1}{2} & \frac{1}{2} & \frac{1}{4} & \frac{1}{2} & \frac{1}{4} & \frac{1}{4} & -\frac{1}{8} & -\frac{1}{8} & -\frac{1}{8} & \frac{3}{8} \end{pmatrix}\begin{pmatrix} y_{1\cdot} \\ y_{2\cdot} \\ y_{3\cdot} \\ y_{4\cdot} \\ y_{5\cdot} \\ y_{6\cdot} \\ y_{\cdot 1} \\ y_{\cdot 2} \\ y_{\cdot 3} \\ y_{\cdot 4} \end{pmatrix}.$$

Thus, the least squares estimators of the γ-differences are

$$\hat{\gamma}_1 - \hat{\gamma}_2 = \frac{1}{4}(y_{4\cdot} + y_{5\cdot}) - \frac{1}{4}(y_{2\cdot} + y_{3\cdot}) + \frac{1}{2}(y_{\cdot 1} - y_{\cdot 2}),$$

$$\hat{\gamma}_1 - \hat{\gamma}_3 = \frac{1}{4}(y_{4\cdot} + y_{6\cdot}) - \frac{1}{4}(y_{1\cdot} + y_{3\cdot}) + \frac{1}{2}(y_{\cdot 1} - y_{\cdot 3}),$$

$$\hat{\gamma}_1 - \hat{\gamma}_4 = \frac{1}{4}(y_{5\cdot} + y_{6\cdot}) - \frac{1}{4}(y_{1\cdot} + y_{2\cdot}) + \frac{1}{2}(y_{\cdot 1} - y_{\cdot 4}),$$

$$\hat{\gamma}_2 - \hat{\gamma}_3 = \frac{1}{4}(y_{1\cdot} + y_{2\cdot}) - \frac{1}{4}(y_{5\cdot} + y_{6\cdot}) + \frac{1}{2}(y_{\cdot 2} - y_{\cdot 3}),$$

$$\hat{\gamma}_2 - \hat{\gamma}_4 = \frac{1}{4}(y_{3\cdot} + y_{6\cdot}) - \frac{1}{4}(y_{1\cdot} + y_{4\cdot}) + \frac{1}{2}(y_{\cdot 2} - y_{\cdot 4}),$$

$$\hat{\gamma}_3 - \hat{\gamma}_4 = \frac{1}{4}(y_{3\cdot} + y_{5\cdot}) - \frac{1}{4}(y_{2\cdot} + y_{4\cdot}) + \frac{1}{2}(y_{\cdot 3} - y_{\cdot 4}).$$

Written in terms of the individual observations, the first of these estimators is

$$\hat{\gamma}_1 - \hat{\gamma}_2 = \frac{1}{2}(y_{11} - y_{12}) + \frac{1}{4}(y_{21} - y_{23} + y_{31} - y_{34} - y_{42} + y_{43} - y_{52} + y_{54}),$$

which has variance $(\frac{1}{2})^2(2\sigma^2) + (\frac{1}{4})^2(8\sigma^2) = \sigma^2$. The variances of the other five estimators may likewise be shown to equal σ^2.

▶ **Exercise 13** For the two-factor nested model, obtain the least squares estimators, and their variance–covariance matrix (under Gauss–Markov assumptions), of:

(a) the cell means;
(b) the Factor B differences (within levels of Factor A).

Solution

(a) For the full-rank reparameterized model

$$\mathbf{y} = \breve{\mathbf{X}} \begin{pmatrix} \mu + \alpha_1 + \gamma_{11} \\ \mu + \alpha_1 + \gamma_{12} \\ \vdots \\ \mu + \alpha_q + \gamma_{qm_q} \end{pmatrix} + \mathbf{e}$$

where $\breve{\mathbf{X}} = \oplus_{i=1}^{q} \oplus_{j=1}^{m_i} \mathbf{1}_{n_{ij}}$, least squares estimators of the cell means (the elements of $\boldsymbol{\beta}$ in this parameterization) may be obtained as follows:

$$\begin{pmatrix} \widehat{\mu + \alpha_1 + \gamma_{11}} \\ \widehat{\mu + \alpha_1 + \gamma_{12}} \\ \vdots \\ \widehat{\mu + \alpha_q + \gamma_{qm_q}} \end{pmatrix} = \left[\left(\oplus_{i=1}^{q} \oplus_{j=1}^{m_i} \mathbf{1}_{n_{ij}}^T\right)\left(\oplus_{i=1}^{q} \oplus_{j=1}^{m_i} \mathbf{1}_{n_{ij}}\right)\right]^{-1} \left(\oplus_{i=1}^{q} \oplus_{j=1}^{m_i} \mathbf{1}_{n_{ij}}^T\right)\mathbf{y}$$

$$= \left(\oplus_{i=1}^{q} \oplus_{j=1}^{m_i} (1/n_{ij})\right) \begin{pmatrix} y_{11\cdot} \\ y_{12\cdot} \\ \vdots \\ y_{qm_q\cdot} \end{pmatrix} = \begin{pmatrix} \bar{y}_{11\cdot} \\ \bar{y}_{12\cdot} \\ \vdots \\ \bar{y}_{qm_q\cdot} \end{pmatrix}.$$

From the elements of

$$\text{var}\begin{pmatrix} \widehat{\mu + \alpha_1 + \gamma_{11}} \\ \widehat{\mu + \alpha_1 + \gamma_{12}} \\ \vdots \\ \widehat{\mu + \alpha_q + \gamma_{qm_q}} \end{pmatrix} = \sigma^2(\breve{\mathbf{X}}^T\breve{\mathbf{X}})^{-1} = \sigma^2\left(\oplus_{i=1}^{q} \oplus_{j=1}^{m_i} (1/n_{ij})\right),$$

we obtain

$$\text{cov}(\bar{y}_{ij\cdot}, \bar{y}_{i'j'\cdot}) = \begin{cases} \frac{\sigma^2}{n_{ij}} & \text{if } i = i' \text{ and } j = j', \\ 0 & \text{otherwise.} \end{cases}$$

(b) By part (a) and Corollary 7.1.4.1, the least squares estimators of $\{\gamma_{ij} - \gamma_{ij'} : i = 1, \ldots, q;\ j' > j = 1, \ldots, m_i\}$ are

$$\widehat{\gamma_{ij} - \gamma_{ij'}} = \bar{y}_{ij\cdot} - \bar{y}_{ij'\cdot}.$$

Furthermore, using Theorem 4.2.2a and the variances and covariances among the estimated cell means determined in part (a), entries of the variance–covariance matrix of the least squares estimators are as follows, for $i \leq s$, $j' > j = 1, \ldots, m_i$ and $t' > t = 1, \ldots, m_s$:

$$\begin{aligned} \text{cov}(\bar{y}_{ij\cdot} - \bar{y}_{ij'\cdot}, \bar{y}_{st\cdot} - \bar{y}_{st'\cdot}) &= \text{cov}(\bar{y}_{ij\cdot}, \bar{y}_{st\cdot}) - \text{cov}(\bar{y}_{ij'\cdot}, \bar{y}_{st\cdot}) \\ &\quad - \text{cov}(\bar{y}_{ij\cdot}, \bar{y}_{st'\cdot}) + \text{cov}(\bar{y}_{ij'\cdot}, \bar{y}_{st'\cdot}) \\ &= \begin{cases} \sigma^2(\frac{1}{n_{ij}} + \frac{1}{n_{ij'}}) & \text{if } i = s,\ j = t,\ j' = t', \\ \frac{\sigma^2}{n_{ij}} & \text{if } i = s,\ j = t,\ j' \neq t', \\ \frac{\sigma^2}{n_{ij'}} & \text{if } i = s,\ j \neq t,\ j' = t', \\ -\frac{\sigma^2}{n_{ij}} & \text{if } i = s \text{ and } j = t', \\ -\frac{\sigma^2}{n_{ij'}} & \text{if } i = s \text{ and } j' = t, \\ 0 & \text{otherwise.} \end{cases} \end{aligned}$$

▶ **Exercise 14** Consider the two-way partially crossed model introduced in Example 5.1.4-1, with one observation per cell, i.e.,

$$y_{ij} = \mu + \alpha_i - \alpha_j + e_{ij} \qquad (i \neq j = 1, \ldots, q),$$

where the e_{ij}'s satisfy Gauss–Markov assumptions.

(a) Obtain expressions for the least squares estimators of those functions in the set $\{\mu, \alpha_i - \alpha_j : i \neq j\}$ that are estimable.
(b) Obtain expressions for the variances and covariances (under Gauss–Markov assumptions) of the least squares estimators obtained in part (a).

Solution

(a) Recall from Example 5.1.4-1 that $\mathbf{X} = (\mathbf{1}_{q(q-1)}, (\mathbf{v}_{ij}^T)_{i \neq j = 1, \ldots, q})$, where $\mathbf{v}_{ij} = \mathbf{u}_i^{(q)} - \mathbf{u}_j^{(q)}$. Now observe that

$$\mathbf{1}_{q(q-1)}^T \mathbf{1}_{q(q-1)} = q(q-1),$$

$$\begin{aligned} &\mathbf{1}_{q(q-1)}^T (\mathbf{u}_i^{(q)} - \mathbf{u}_j^{(q)})_{i \neq j = 1, \ldots, q} \\ &= \big((1)(q-1) + (-1)(q-1), \ldots, (1)(q-1) + (-1)(q-1)\big)_{1 \times q} = \mathbf{0}_q^T, \end{aligned}$$

and

$$\begin{aligned}
[(\mathbf{v}_{ij}^T)_{i\neq j=1,\ldots,q}]^T[(\mathbf{v}_{ij}^T)_{i\neq j=1,\ldots,q}] &= \sum_{i=1}^{q}\sum_{j\neq i}^{q}(\mathbf{u}_i^{(q)}-\mathbf{u}_j^{(q)})(\mathbf{u}_i^{(q)}-\mathbf{u}_j^{(q)})^T \\
&= (q-1)\sum_{i=1}^{q}\mathbf{u}_i^{(q)}\mathbf{u}_i^{(q)T}+(q-1)\sum_{j=1}^{q}\mathbf{u}_j^{(q)}\mathbf{u}_j^{(q)T} \\
&\quad -\sum_{i=1}^{q}\sum_{j\neq i}(\mathbf{u}_i^{(q)}\mathbf{u}_j^{(q)T}+\mathbf{u}_j^{(q)}\mathbf{u}_i^{(q)T}) \\
&= 2q\mathbf{I}_q-2\mathbf{J}_q \\
&= 2q[\mathbf{I}_q-(1/q)\mathbf{J}_q].
\end{aligned}$$

One generalized inverse of this last matrix is $(1/2q)\mathbf{I}_q$, so by Theorem 3.3.7 one generalized inverse of $\mathbf{X}^T\mathbf{X}$ is

$$(\mathbf{X}^T\mathbf{X})^- = \begin{pmatrix} 1/[q(q-1)] & \mathbf{0}_q^T \\ \mathbf{0}_q & (1/2q)\mathbf{I}_q \end{pmatrix}.$$

Now

$$\mathbf{X}^T\mathbf{y} = \begin{pmatrix} \sum_{i\neq j} y_{ij} \\ y_{1\cdot}-y_{\cdot 1} \\ y_{2\cdot}-y_{\cdot 2} \\ \vdots \\ y_{q\cdot}-y_{\cdot q} \end{pmatrix},$$

where $y_{i\cdot}=\sum_{j\neq i} y_{ij}$ and $y_{\cdot j}=\sum_{i\neq j} y_{ij}$. Thus, one solution to the normal equations is

$$\hat{\boldsymbol{\beta}} = \begin{pmatrix} 1/[q(q-1)] & \mathbf{0}_q^T \\ \mathbf{0}_q & (1/2q)\mathbf{I}_q \end{pmatrix}\begin{pmatrix} \sum_{i\neq j} y_{ij} \\ y_{1\cdot}-y_{\cdot 1} \\ y_{2\cdot}-y_{\cdot 2} \\ \vdots \\ y_{q\cdot}-y_{\cdot q} \end{pmatrix} = \begin{pmatrix} \bar{y}_{\cdot\cdot} \\ (y_{1\cdot}-y_{\cdot 1})/2q \\ (y_{2\cdot}-y_{\cdot 2})/2q \\ \vdots \\ (y_{q\cdot}-y_{\cdot q})/2q \end{pmatrix},$$

where $\bar{y}_{\cdot\cdot}=[1/q(q-1)]\sum_{i=1}^{q}\sum_{j\neq i} y_{ij}$. By the result of Exercise 6.15a, μ and $\alpha_i-\alpha_j$ are estimable ($j>i=1,\ldots,q$), and by Theorem 7.1.4 their least squares estimators are

$$\mathbf{u}_1^{(q+1)T}\hat{\boldsymbol{\beta}}=\bar{y}_{\cdot\cdot}$$

and

$$(\mathbf{u}_{i+1}^{(q+1)} - \mathbf{u}_{j+1}^{(q+1)})^T \hat{\boldsymbol{\beta}} = (y_{i\cdot} - y_{\cdot i} - y_{j\cdot} + y_{\cdot j})/2q.$$

(b) $\text{var}(\bar{y}_{\cdot\cdot}) = \sigma^2/[q(q-1)]$,

$$\text{var}[(y_{i\cdot} - y_{\cdot i} - y_{j\cdot} + y_{\cdot j})/2q] = \sigma^2(\mathbf{u}_{i+1}^{(q+1)} - \mathbf{u}_{j+1}^{(q+1)})^T \begin{pmatrix} 1/[q(q-1)] & \mathbf{0}_q^T \\ \mathbf{0}_q & (1/2q)\mathbf{I}_q \end{pmatrix} \times (\mathbf{u}_{i+1}^{(q+1)} - \mathbf{u}_{j+1}^{(q+1)}) = \sigma^2/q,$$

$$\text{cov}[\bar{y}_{\cdot\cdot}, (y_{i\cdot} - y_{\cdot i} - y_{j\cdot} + y_{\cdot j})/2q] = \sigma^2\mathbf{u}_1^T \begin{pmatrix} 1/[q(q-1)] & \mathbf{0}_q^T \\ \mathbf{0}_q & (1/2q)\mathbf{I}_q \end{pmatrix} \times (\mathbf{u}_{i+1}^{(q+1)} - \mathbf{u}_{j+1}^{(q+1)}) = 0,$$

and for $i \le s$, $j > i = 1, \ldots, q$, and $t > s = 1, \ldots, q$,

$$\begin{aligned} &\text{cov}[(y_{i\cdot} - y_{\cdot i} - y_{j\cdot} + y_{\cdot j})/2q, (y_{s\cdot} - y_{\cdot s} - y_{t\cdot} + y_{\cdot t})/2q] \\ &= \sigma^2(\mathbf{u}_{i+1}^{(q+1)} - \mathbf{u}_{j+1}^{(q+1)})^T \begin{pmatrix} 1/[q(q-1)] & \mathbf{0}_q^T \\ \mathbf{0}_q & (1/2q)\mathbf{I}_q \end{pmatrix} (\mathbf{u}_{s+1}^{(q+1)} - \mathbf{u}_{t+1}^{(q+1)}) \\ &= \begin{cases} \sigma^2/q & \text{if } i = s \text{ and } j = t, \\ \sigma^2/2q & \text{if } i = s \text{ and } j \neq t, \text{ or } i \neq s \text{ and } j = t, \\ -\sigma^2/2q & \text{if } j = s, \\ 0 & \text{otherwise.} \end{cases} \end{aligned}$$

▶ **Exercise 15** For the Latin square design with q treatments described in Exercise 6.21:

(a) Obtain the least squares estimators of the treatment differences $\tau_k - \tau_{k'}$, where $k \neq k' = 1, \ldots, q$.
(b) Obtain expressions for the variances and covariances (under Gauss–Markov assumptions) of those least squares estimators.

Solution

(a) As in the solution to Exercise 6.21a, for each $i = 1, \ldots, q$ let $\{j_1(k), j_2(k), \ldots, j_q(k)\}$ represent a permutation of $\{1, \ldots, q\}$ such that the kth treatment in row i is assigned to column $j_i(k)$. Then the model matrix may be written as

$$\mathbf{X} = \left(\mathbf{1}_{q^2}, \mathbf{I}_q \otimes \mathbf{1}_q, \mathbf{1}_q \otimes \mathbf{I}_q, \tilde{\mathbf{I}}\right),$$

where $\tilde{\mathbf{I}} = \begin{pmatrix} \tilde{\mathbf{I}}_1 \\ \tilde{\mathbf{I}}_2 \\ \vdots \\ \tilde{\mathbf{I}}_q \end{pmatrix}$ consists of q blocks where the columns of each block are a permutation of the columns of $\mathbf{I}_q$, so that the kth column of $\tilde{\mathbf{I}}_i$ is the unit q-vector $\mathbf{u}_{j_i(k)}$. It is easy to verify that whatever the specific assignment of treatments to rows and columns,

$$\mathbf{X}^T\mathbf{X} = \begin{pmatrix} q^2 & q\mathbf{1}_q^T & q\mathbf{1}_q^T & q\mathbf{1}_q^T \\ q\mathbf{1}_q & q\mathbf{I}_q & \mathbf{J}_q & \mathbf{J}_q \\ q\mathbf{1}_q & \mathbf{J}_q & q\mathbf{I}_q & \mathbf{J}_q \\ q\mathbf{1}_q & \mathbf{J}_q & \mathbf{J}_q & q\mathbf{I}_q \end{pmatrix} \equiv \begin{pmatrix} \mathbf{A} & \mathbf{B} \\ \mathbf{C} & \mathbf{D} \end{pmatrix},$$

say, where $\mathbf{A} = q^2$, $\mathbf{B} = q\mathbf{1}_{3q}^T = \mathbf{C}^T$, and $\mathbf{D} = q\{\mathbf{I}_3 \otimes [\mathbf{I}_q - (1/q)\mathbf{J}_q] + \mathbf{J}_3 \otimes (1/q)\mathbf{J}_q\}$. Observe that $\mathbf{D}\mathbf{1}_{3q} = (3q)\mathbf{1}_{3q}$, so $\mathcal{C}(\mathbf{C}) \subseteq \mathcal{C}(\mathbf{D})$ and $\mathcal{R}(\mathbf{B}) \subseteq \mathcal{R}(\mathbf{D})$. Also, it is easily verified that one generalized inverse of $\mathbf{D}$ is

$$\mathbf{D}^- = (1/q)\{\mathbf{I}_3 \otimes [\mathbf{I}_q - (1/q)\mathbf{J}_q] + (1/9)\mathbf{J}_3 \otimes (1/q)\mathbf{J}_q\}.$$

Then by Theorem 3.3.7, one generalized inverse of $\mathbf{X}^T\mathbf{X}$ is

$$(\mathbf{X}^T\mathbf{X})^- = \begin{pmatrix} 0 & \mathbf{0}^T \\ \mathbf{0} & \mathbf{D}^- \end{pmatrix},$$

where we used the fact that $\mathbf{P} \equiv \mathbf{A} - \mathbf{B}\mathbf{D}^-\mathbf{C} = \mathbf{0}$ so $\mathbf{0}$ is a generalized inverse of $\mathbf{P}$. Now let $\mathcal{S} = \{(i, j, k) : \text{treatment } k \text{ is assigned to cell } (i, j)\}$ and let $\mathbf{y} = (y_{ijk} : (i, j, k) \in \mathcal{S})$ be ordered lexicographically. Then a solution to the normal equations is

$$\hat{\boldsymbol{\beta}} = (\mathbf{X}^T\mathbf{X})^-\mathbf{X}^T\mathbf{y} = \begin{pmatrix} 0 & \mathbf{0}^T \\ \mathbf{0} & (1/q)\{\mathbf{I}_3 \otimes [\mathbf{I}_q - (1/q)\mathbf{J}_q] + (1/9)\mathbf{J}_3 \otimes (1/q)\mathbf{J}_q\} \end{pmatrix} \begin{pmatrix} y_{\cdots} \\ y_{1\cdot\cdot} \\ \vdots \\ y_{q\cdot\cdot} \\ y_{\cdot 1\cdot} \\ \vdots \\ y_{\cdot q\cdot} \\ y_{\cdot\cdot 1} \\ \vdots \\ y_{\cdot\cdot q} \end{pmatrix}$$

$$= \begin{pmatrix} 0 \\ \bar{y}_{1\cdot\cdot} - (2/3)\bar{y}_{\cdot\cdot\cdot} \\ \bar{y}_{2\cdot\cdot} - (2/3)\bar{y}_{\cdot\cdot\cdot} \\ \vdots \\ \bar{y}_{q\cdot\cdot} - (2/3)\bar{y}_{\cdot\cdot\cdot} \\ \bar{y}_{\cdot 1\cdot} - (2/3)\bar{y}_{\cdot\cdot\cdot} \\ \vdots \\ \bar{y}_{\cdot q\cdot} - (2/3)\bar{y}_{\cdot\cdot\cdot} \\ \bar{y}_{\cdot\cdot 1} - (2/3)\bar{y}_{\cdot\cdot\cdot} \\ \vdots \\ \bar{y}_{\cdot\cdot q} - (2/3)\bar{y}_{\cdot\cdot\cdot} \end{pmatrix}.$$

It follows that the least squares estimator of $\tau_k - \tau_{k'}$ is $[\bar{y}_{\cdot\cdot k} - (2/3)\bar{y}_{\cdot\cdot\cdot}] - [\bar{y}_{\cdot\cdot k'} - (2/3)\bar{y}_{\cdot\cdot\cdot}] = \bar{y}_{\cdot\cdot k} - \bar{y}_{\cdot\cdot k'}$ $(k \neq k' = 1\ldots, q)$.

(b) For $k \leq l$, $k < k'$, and $l < l'$,

$$\text{cov}(\bar{y}_{\cdot\cdot k} - \bar{y}_{\cdot\cdot k'}, \bar{y}_{\cdot\cdot l} - \bar{y}_{\cdot\cdot l'}) = \begin{cases} 2\sigma^2/q & \text{if } k = l \text{ and } k' = l', \\ \sigma^2/q & \text{if } k = l \text{ and } k' \neq l', \text{ or } k \neq l \text{ and } k' = l', \\ -\sigma^2/q & \text{if } k' = l, \\ 0 & \text{otherwise.} \end{cases}$$

▶ **Exercise 16** Consider a simple linear regression model (with Gauss–Markov assumptions) for (x, y)-observations $\{(i, y_i) : i = 1, 2, 3, 4, 5\}$.

(a) Give a nonmatrix expression for the least squares estimator, $\hat{\beta}_2$, of the slope, in terms of the y_i's only.
(b) Give the variance of $\hat{\beta}_2$ as a multiple of σ^2.
(c) Show that each of the linear estimators

$$\hat{\gamma} = y_2 - y_1 \quad \text{and} \quad \hat{\eta} = (y_5 - y_2)/3$$

is unbiased for the slope, and determine their variances. Are these variances larger, smaller, or equal to the variance of $\hat{\beta}_2$?
(d) Determine the bias and variance of

$$c\hat{\gamma} + (1 - c)\hat{\eta}$$

(where $0 \leq c \leq 1$) as an estimator of the slope.
(e) Find the choice of c that minimizes the variance in part (d), and determine this minimized variance. Is this variance larger, smaller, or equal to the variance of $\hat{\beta}_2$?

Solution

(a) $\hat{\beta}_2 = \frac{\sum_{i=1}^{5}(x_i-\bar{x})y_i}{\sum_{i=1}^{5}(x_i-\bar{x})^2} = \frac{y_4+2y_5-y_2-2y_1}{10}$

(b) $\text{var}(\hat{\beta}_2) = \sigma^2/10$

(c) $\text{E}(y_2 - y_1) = \beta_1 + 2\beta_2 - (\beta_1 + \beta_2) = \beta_2$, $\text{E}[(y_5 - y_2)/3] = [\beta_1 + 5\beta_2 - (\beta_1 + 2\beta_2)]/3 = \beta_2$, $\text{var}(y_2 - y_1) = \text{var}(y_2) + \text{var}(y_1) = 2\sigma^2$, $\text{var}[(y_5 - y_2)/3] = [\text{var}(y_5) + \text{var}(y_2)]/9 = 2\sigma^2/9$. These variances are larger than $\text{var}(\hat{\beta}_2)$.

(d) $\text{E}[c\hat{\gamma} + (1-c)\hat{\eta}] = c\text{E}(\hat{\gamma}) + (1-c)\text{E}(\hat{\eta}) = c\beta_2 + (1-c)\beta_2 = \beta_2$, so the bias is 0. Because $c\hat{\gamma} + (1-c)\hat{\eta} = c(y_2 - y_1) + [(1-c)/3](y_5 - y_2) = [(1-c)/3]y_5 + [(4c-1)/3]y_2 - cy_1$,

$$\begin{aligned}\text{var}(c\hat{\gamma} + (1-c)\hat{\eta}) &= \sigma^2\{[(1-c)^2/9] + [(4c-1)^2/9] + [9c^2/9]\}\\ &= (\sigma^2/9)(1 - 2c + c^2 + 16c^2 - 8c + 1 + 9c^2)\\ &= (\sigma^2/9)(26c^2 - 10c + 2).\end{aligned}$$

(e) Define $f(c) = 26c^2 - 10c + 2$. Taking the first derivative of f and setting it equal to 0 yields an equation whose solution is 5/26. The second derivative of f is 52, which is positive, so the stationary point of f at $c = 5/26$ is a point of minimum. The minimized variance is $\text{var}[(5/26)\hat{\gamma} + (21/26)\hat{\eta}] = (\sigma^2/9)[26(5/26)^2 - 10(5/26) + 2] = 3\sigma^2/26$. This variance is larger than $\text{var}(\hat{\beta}_2)$.

▶ **Exercise 17** Consider the cell-means parameterization of the two-way model with interaction, i.e.,

$$y_{ijk} = \mu_{ij} + e_{ijk} \qquad (i = 1, \ldots, q;\ j = 1, \ldots, m;\ k = 1, \ldots, n_{ij}).$$

Suppose that none of the cells is empty. Under Gauss–Markov assumptions, obtain expressions for the variance of the least squares estimator of $\mu_{ij} - \mu_{ij'} - \mu_{i'j} + \mu_{i'j'}$ and the covariance between the least squares estimators of every pair $\mu_{ij} - \mu_{ij'} - \mu_{i'j} + \mu_{i'j'}$ and $\mu_{st} - \mu_{st'} - \mu_{s't} + \mu_{s't'}$ of such functions.

Solution We seek an expression for $\text{cov}(\bar{y}_{ij} - \bar{y}_{ij'} - \bar{y}_{i'j} + \bar{y}_{i'j'}, \bar{y}_{st} - \bar{y}_{st'} - \bar{y}_{s't} + \bar{y}_{s't'})$, where i and i' are distinct, j and j' are distinct, s and s' are distinct, and t and t' are distinct. Without loss of generality we may take $i \leq s$. The least squares estimator of each interaction contrast of the type described in this exercise corresponds to a 2×2 array of cells. It is considerably less complicated/tedious if we systematize by the extent and nature of the overlap of these 2×2 arrays, rather than by the indices of the cells. With respect to the extent of overlap, there are four cases: no overlap, a one-cell overlap, a two-cell overlap, and a complete (four-cell overlap); it is not possible for only three cells to overlap. An instance of each of the possibilities is depicted in Fig. 7.1. Suppose initially that $i < i'$, $j < j'$, $k < k'$, and $m < m'$.

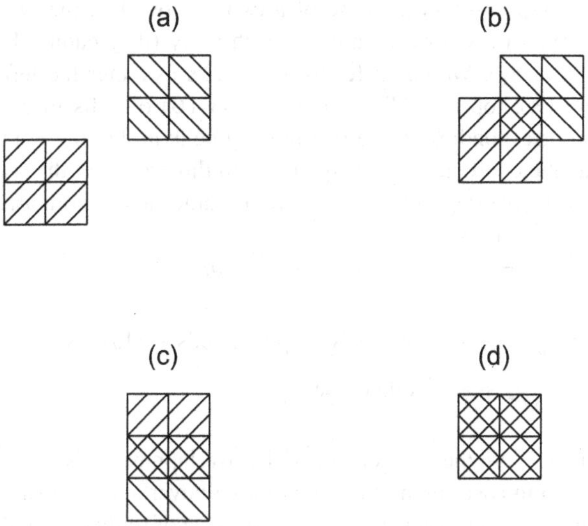

Fig. 7.1 Four cases of overlap among 2 × 2 arrays in a two-way layout

- **No overlap case:** In this case, all eight means in the two contrasts are uncorrelated, so

$$\text{cov}(\bar{y}_{ij} - \bar{y}_{ij'} - \bar{y}_{i'j} + \bar{y}_{i'j'}, \bar{y}_{st} - \bar{y}_{st'} - \bar{y}_{s't} + \bar{y}_{s't'}) = 0.$$

- **One-cell overlap case:** In order to deal with the 1-cell overlap cases, let us agree to refer to the cells in each array as the top left (TL), top right (TR), bottom left (BL), and bottom right (BR) cells. By assumption, indices ij and st refer to the TL cells of the first and second arrays, respectively. Thus, the coefficients on the four cells in any interaction contrast are $1, -1, -1, 1$ for the TL, TR, BL, and BR cells, respectively. Let TL/TL denote the situation in which the overlapping cell is the TL cell of both arrays, TL/TR the situation in which the overlapping cell is the TL cell of the first array but the TR cell of the second array, etc. Observe that the coefficients corresponding to the overlapping cell in the two interaction contrasts have the same sign if the overlapping cell is at the same corner or opposite corners in the two arrays; otherwise, those coefficients have opposite sign. Thus,

$$\text{cov}(\bar{y}_{ij} - \bar{y}_{ij'} - \bar{y}_{i'j} + \bar{y}_{i'j'}, \bar{y}_{st} - \bar{y}_{st'} - \bar{y}_{s't} + \bar{y}_{s't'})$$

$$= \begin{cases} \sigma^2\left(\frac{1}{n^*}\right) & \text{if the overlapping cell is TL/TL, TL/BR, TR/TR, TR/BL,} \\ & \text{BL/BL, BL/TR, BR/BR, BR/TL} \\ -\sigma^2\left(\frac{1}{n^*}\right) & \text{otherwise,} \end{cases}$$

where n^* is the frequency of the overlapping cell.

- **Two-cell overlap case:** In the case of a two-cell overlap, the two overlapping cells form either a row or a column in each array (they cannot lie in opposite corners of an array). We can refer to these cells as either the left column (L), right column (R), top row (T), or bottom row (B) of cells in each array. The coefficients corresponding to the overlapping cells in the first contrast have the same sign as the coefficients corresponding to those cells in the second contrast if and only if the overlapping cells are on the same side of both arrays, and have opposite signs otherwise. Thus,

$$\mathrm{cov}(\bar{y}_{ij} - \bar{y}_{ij'} - \bar{y}_{i'j} + \bar{y}_{i'j'}, \bar{y}_{st} - \bar{y}_{st'} - \bar{y}_{s't} + \bar{y}_{s't'})$$

$$= \begin{cases} \sigma^2\left(\frac{1}{n^*} + \frac{1}{n^{**}}\right) & \text{if the overlapping cells are L/L, R/R, T/T, or B/B,} \\ -\sigma^2\left(\frac{1}{n^*} + \frac{1}{n^{**}}\right) & \text{otherwise,} \end{cases}$$

  where n^* and n^{**} are the frequencies of the overlapping cells.
- **Four-cell overlap case:** In the case of a four-cell overlap, the covariance between the two estimated contrasts is merely the variance of (either) one of them:

$$\mathrm{cov}(\bar{y}_{ij} - \bar{y}_{ij'} - \bar{y}_{i'j} + \bar{y}_{i'j'}, \bar{y}_{st} - \bar{y}_{st'} - \bar{y}_{s't} + \bar{y}_{s't'}) = \sigma^2\left(\frac{1}{n_{ij}} + \frac{1}{n_{ij'}} + \frac{1}{n_{i'j}} + \frac{1}{n_{i'j'}}\right).$$

▶ **Exercise 18** This exercise generalizes Theorem 7.2.3 (the Gauss–Markov theorem) to a situation in which one desires to estimate several estimable functions and say something about their joint optimality. Let $\mathbf{C}^T\boldsymbol{\beta}$ be a k-vector of estimable functions under the Gauss–Markov model $\{\mathbf{y}, \mathbf{X}\boldsymbol{\beta}, \sigma^2\mathbf{I}\}$, and let $\mathbf{C}^T\hat{\boldsymbol{\beta}}$ be the k-vector of least squares estimators of those functions associated with that model. Furthermore, let $\mathbf{k}+\mathbf{B}^T\mathbf{y}$ be any vector of linear unbiased estimators of $\mathbf{C}^T\boldsymbol{\beta}$. Prove the following results (the last two results follow from the first). For the last result, use the following lemma, which is essentially the same as Theorem 18.1.6 of Harville (1997): If $\mathbf{M}$ is a positive definite matrix, $\mathbf{Q}$ is a nonnegative definite matrix of the same dimensions as $\mathbf{M}$, and $\mathbf{M}-\mathbf{Q}$ is nonnegative definite, then $|\mathbf{M}-\mathbf{Q}| \leq |\mathbf{M}|$.

(a) $\mathrm{var}(\mathbf{k}+\mathbf{B}^T\mathbf{y}) - \mathrm{var}(\mathbf{C}^T\hat{\boldsymbol{\beta}})$ is nonnegative definite;
(b) $\mathrm{tr}[\mathrm{var}(\mathbf{k}+\mathbf{B}^T\mathbf{y})] \geq \mathrm{tr}[\mathrm{var}(\mathbf{C}^T\hat{\boldsymbol{\beta}})]$;
(c) $|\mathrm{var}(\mathbf{k}+\mathbf{B}^T\mathbf{y})| \geq |\mathrm{var}(\mathbf{C}^T\hat{\boldsymbol{\beta}})|$.

Solution

(a) For any k-vector $\boldsymbol{\ell}$, $\boldsymbol{\ell}^T\mathbf{C}^T\boldsymbol{\beta}$ is estimable (by Corollary 6.1.2.1), and $\boldsymbol{\ell}^T\mathbf{k} + (\mathbf{B}\boldsymbol{\ell})^T\mathbf{y}$ and $(\mathbf{C}\boldsymbol{\ell})^T\hat{\boldsymbol{\beta}}$ are linear unbiased estimators of it (if $\boldsymbol{\ell}^T\mathbf{k} = 0$). Therefore, by Theorem 7.2.3,

$$\begin{aligned}0 &\leq \text{var}(\boldsymbol{\ell}^T\mathbf{k}+\boldsymbol{\ell}^T\mathbf{B}^T\mathbf{y}) - \text{var}(\boldsymbol{\ell}^T\mathbf{C}^T\hat{\boldsymbol{\beta}}) \\ &= \boldsymbol{\ell}^T[\text{var}(\mathbf{B}^T\mathbf{y}) - \text{var}(\mathbf{C}^T\hat{\boldsymbol{\beta}})]\boldsymbol{\ell} \\ &= \boldsymbol{\ell}^T[\text{var}(\mathbf{k}+\mathbf{B}^T\mathbf{y}) - \text{var}(\mathbf{C}^T\hat{\boldsymbol{\beta}})]\boldsymbol{\ell}.\end{aligned}$$

The result follows from the definition of nonnegative definiteness.

(b) This follows immediately from the result of part (a) upon taking $\boldsymbol{\ell}$ to equal each of the unit vectors, and then summing both sides of the corresponding inequalities over all such vectors.

(c) $\text{var}(\mathbf{k}+\mathbf{B}^T\mathbf{y}) = \sigma^2\mathbf{B}^T\mathbf{B}$ and, because $\mathbf{B}^T\mathbf{X} = \mathbf{C}^T$ (due to the unbiasedness of $\mathbf{k}+\mathbf{B}^T\mathbf{y}$),

$$\begin{aligned}\text{var}(\mathbf{C}^T\hat{\boldsymbol{\beta}}) &= \sigma^2\mathbf{C}^T(\mathbf{X}^T\mathbf{X})^-\mathbf{C} \\ &= \sigma^2\mathbf{B}^T\mathbf{X}(\mathbf{X}^T\mathbf{X})^-\mathbf{X}^T\mathbf{B}.\end{aligned}$$

Thus, the rank of $\text{var}(\mathbf{C}^T\hat{\boldsymbol{\beta}})$ is less than or equal to the rank of $\mathbf{B}$. Now consider the following two cases. If $|\text{var}(\mathbf{k}+\mathbf{B}^T\mathbf{y})| = 0$, then $\text{rank}(\mathbf{B}) < k$ by Theorem 2.11.2, implying that $\text{rank}[\text{var}(\mathbf{C}^T\hat{\boldsymbol{\beta}})] < k$ and hence that $|\text{var}(\mathbf{C}^T\hat{\boldsymbol{\beta}})| = 0$. Thus the result holds for this case. On the other hand, if $|\text{var}(\mathbf{k}+\mathbf{B}^T\mathbf{y})| > 0$, then by substituting $\text{var}(\mathbf{k}+\mathbf{B}^T\mathbf{y})$ and $[\text{var}(\mathbf{k}+\mathbf{B}^T\mathbf{y}) - \text{var}(\mathbf{C}^T\hat{\boldsymbol{\beta}})]$ for $\mathbf{M}$ and $\mathbf{Q}$, respectively, in the result from part (a), we obtain the desired result for this case also.

▶ **Exercise 19** For a balanced two-factor crossed classification with two levels of each factor and r observations per cell:

(a) Show that the model

$$y_{ijk} = \theta + (-1)^i\tau_1 + (-1)^j\tau_2 + e_{ijk} \quad (i = 1, 2;\ j = 1, 2;\ k = 1, \ldots, r)$$

is a reparameterization of the two-way main effects model given by (5.6).

(b) Determine how θ, τ_1, and τ_2 are related to the parameters μ, α_1, α_2, γ_1, and γ_2 for parameterization (5.6) of the two-way main effects model.

(c) What "nice" property does the model matrix for this reparameterization have? Does it still have this property if the data are unbalanced?

(d) Indicate how you might reparameterize a three-way main effects model with two levels for each factor and balanced data in a similar way, and give the corresponding model matrix.

Solution

(a) The model matrices associated with parameterization (5.6) and the parameterization introduced in this exercise may be written, respectively, as

$$\mathbf{X} = \begin{pmatrix} \mathbf{1}_r \otimes (1\ 1\ 0\ 1\ 0) \\ \mathbf{1}_r \otimes (1\ 1\ 0\ 0\ 1) \\ \mathbf{1}_r \otimes (1\ 0\ 1\ 1\ 0) \\ \mathbf{1}_r \otimes (1\ 0\ 1\ 0\ 1) \end{pmatrix} \quad \text{and} \quad \breve{\mathbf{X}} = \begin{pmatrix} \mathbf{1}_r \otimes (1\ -1\ -1) \\ \mathbf{1}_r \otimes (1\ -1\ 1) \\ \mathbf{1}_r \otimes (1\ 1\ -1) \\ \mathbf{1}_r \otimes (1\ 1\ 1) \end{pmatrix}.$$

Let $\mathbf{X}^*$ and $\breve{\mathbf{X}}^*$ be the submatrices of unique rows of $\mathbf{X}$ and $\breve{\mathbf{X}}$, respectively. Then

$$\breve{\mathbf{X}}^* = \mathbf{X}^* \begin{pmatrix} 1 & 0 & 0 \\ 0 & -1 & 0 \\ 0 & 1 & 0 \\ 0 & 0 & -1 \\ 0 & 0 & 1 \end{pmatrix} \quad \text{and} \quad \mathbf{X}^* = \breve{\mathbf{X}}^* \begin{pmatrix} 1 & 0.5 & 0.5 & 0.5 & 0.5 \\ 0 & -0.5 & 0.5 & 0 & 0 \\ 0 & 0 & 0 & -0.5 & 0.5 \end{pmatrix},$$

which establishes that $\mathcal{C}(\mathbf{X}) = \mathcal{C}(\breve{\mathbf{X}})$ and hence that the two models are reparameterizations of each other.

(b) $\begin{pmatrix} \theta \\ \tau_1 \\ \tau_2 \end{pmatrix} = \begin{pmatrix} \mu + (\alpha_1 + \alpha_2)/2 + (\gamma_1 + \gamma_2)/2 \\ (\alpha_2 - \alpha_1)/2 \\ (\gamma_2 - \gamma_1)/2 \end{pmatrix}.$

(c) The columns of $\breve{\mathbf{X}}$ are orthogonal, so $\breve{\mathbf{X}}^T\breve{\mathbf{X}} = 4r\mathbf{I}$. $\breve{\mathbf{X}}$ does not retain this property if the data are unbalanced.

(d) The analogous reparameterization is

$$y_{ijkl} = \theta + (-1)^i\tau_1 + (-1)^j\tau_2 + (-1)^k\tau_3 + e_{ijkl}$$
$$(i = 1, 2;\ j = 1, 2;\ j = 1, 2;\ l = 1, \ldots, r)$$

and the corresponding model matrix is

$$\breve{\mathbf{X}} = \begin{pmatrix} \mathbf{1}_r \otimes (1\ -1\ -1\ -1) \\ \mathbf{1}_r \otimes (1\ -1\ -1\ 1) \\ \mathbf{1}_r \otimes (1\ -1\ 1\ -1) \\ \mathbf{1}_r \otimes (1\ -1\ 1\ 1) \\ \mathbf{1}_r \otimes (1\ 1\ -1\ -1) \\ \mathbf{1}_r \otimes (1\ 1\ -1\ 1) \\ \mathbf{1}_r \otimes (1\ 1\ 1\ -1) \\ \mathbf{1}_r \otimes (1\ 1\ 1\ 1) \end{pmatrix}.$$

Exercise 20 Consider the second-order polynomial regression model

$$y_i = \beta_1 + \beta_2 x_i + \beta_3 x_i^2 + e_i, \quad (i = 1, \dots, n)$$

and the reparameterization

$$y_i = \tau_1 + \tau_2(x_i - \bar{x}) + \tau_3(x_i - \bar{x})^2 + e_i, \quad (i = 1, \dots, n).$$

The second model is obtained from the first by centering the x_i's. Suppose that at least three of the x_i's are distinct and that the errors satisfy Gauss–Markov assumptions.

(a) Let $\mathbf{X}$ and $\breve{\mathbf{X}}$ represent the model matrices corresponding to these two models. Determine the columns of $\mathbf{X}$ and $\breve{\mathbf{X}}$ and verify that the second model is a reparameterization of the first model.
(b) Let $\hat{\tau}_j$ denote the least squares estimators of τ_j $(j = 1, 2, 3)$ in the model $\mathbf{y} = \breve{\mathbf{X}}\boldsymbol{\tau} + \mathbf{e}$. Suppose that $x_i = i$ $(i = 1 \dots, n)$. Determine $\text{cov}(\hat{\tau}_1, \hat{\tau}_2)$ and $\text{cov}(\hat{\tau}_2, \hat{\tau}_3)$.
(c) Determine the variance inflation factors for the regressors in the second model.

Solution

(a) $\mathbf{X} = (\mathbf{1}, \mathbf{x}_1, \mathbf{x}_2)$ where $\mathbf{x}_1 = (x_1, \dots, x_n)^T$ and $\mathbf{x}_2 = (x_1^2, \dots, x_n^2)^T$. Furthermore, $\breve{\mathbf{X}} = (\mathbf{1}, \mathbf{x}_1 - \bar{x}\mathbf{1}, \breve{\mathbf{x}})$ where $\breve{\mathbf{x}} = ((x_1 - \bar{x})^2, \dots, (x_n - \bar{x})^2)^T = (x_1^2 - 2\bar{x}x_1 + \bar{x}^2, \dots, x_n^2 - 2\bar{x}x_n + \bar{x}^2)^T$. Thus

$$\breve{\mathbf{X}} = (\mathbf{1}, \mathbf{x}_1, \mathbf{x}_2)\begin{pmatrix} 1 & -\bar{x} & \bar{x}^2 \\ 0 & 1 & -2\bar{x} \\ 0 & 0 & 1 \end{pmatrix} = \mathbf{XT},$$

say, implying that $\mathcal{C}(\breve{\mathbf{X}}) \subseteq \mathcal{C}(\mathbf{X})$. Furthermore, $\mathbf{T}$ is nonsingular, so $\mathcal{C}(\breve{\mathbf{X}}) = \mathcal{C}(\mathbf{X})$ by Theorems 2.8.5 and 2.8.9. Thus the second model is a reparameterization of the first.

(b) Of course, $\sum_{i=1}^n (x_i - \bar{x}) = 0$; moreover, by the symmetry of the values of x_i around their mean, $\sum_{i=1}^n (x_i - \bar{x})^3 = 0$. Thus,

$$\text{var}(\hat{\boldsymbol{\tau}}) = \sigma^2(\breve{\mathbf{X}}^T\breve{\mathbf{X}})^{-1} = \sigma^2 \begin{pmatrix} n & \sum_{i=1}^n (x_i - \bar{x}) & \sum_{i=1}^n (x_i - \bar{x})^2 \\ & \sum_{i=1}^n (x_i - \bar{x})^2 & \sum_{i=1}^n (x_i - \bar{x})^3 \\ & & \sum_{i=1}^n (x_i - \bar{x})^4 \end{pmatrix}^{-1}$$

$$= \sigma^2 \begin{pmatrix} n & 0 & \sum_{i=1}^n (x_i - \bar{x})^2 \\ & \sum_{i=1}^n (x_i - \bar{x})^2 & 0 \\ & & \sum_{i=1}^n (x_i - \bar{x})^4 \end{pmatrix}^{-1}.$$

Rearranging the elements of $\hat{\boldsymbol{\tau}}$, we obtain

$$\text{var}\begin{pmatrix}\hat{\tau}_2\\ \hat{\tau}_1\\ \hat{\tau}_3\end{pmatrix} = \sigma^2\begin{pmatrix}\sum_{i=1}^n (x_i-\bar{x})^2 & 0 & 0\\ 0 & n & \sum_{i=1}^n (x_i-\bar{x})^2\\ 0 & \sum_{i=1}^n (x_i-\bar{x})^2 & \sum_{i=1}^n (x_i-\bar{x})^4\end{pmatrix}^{-1}$$

$$= \sigma^2\begin{pmatrix}1/\sum_{i=1}^n (x_i-\bar{x})^2 & 0 & 0\\ 0 & * & *\\ 0 & * & *\end{pmatrix},$$

where we have written an asterisk for elements whose exact form is unimportant for the purposes of this exercise. Thus, $\text{cov}(\hat{\tau}_1, \hat{\tau}_2) = \text{cov}(\hat{\tau}_2, \hat{\tau}_3) = 0$.

(c) Using the notation introduced in Methodological Interlude #1,

$$VIF_2 = \left\{1 - \frac{\mathbf{a}^T\mathbf{A}^{-1}\mathbf{a}}{\sum_{i=1}^n [(x_i-\bar{x}) - (1/n)\sum_{j=1}^n (x_j-\bar{x})]^2}\right\}^{-1},$$

where in this case

$$\mathbf{a} = \sum_{i=1}^n (x_i-\bar{x})[(x_i-\bar{x})^2 - (1/n)\sum_{j=1}^n (x_j-\bar{x})^2] = \sum_{i=1}^n (x_i-\bar{x})^3 = 0.$$

Thus $VIF_2 = 1$. VIF_3 is given by an analogous expression, for which some quantities are different but $\mathbf{a}$ is the same as it is for VIF_2. Thus $VIF_3 = 1$ also.

▶ **Exercise 21** Consider the Gauss–Markov model $\{\mathbf{y}, \mathbf{X}\boldsymbol{\beta}, \sigma^2\mathbf{I}\}$, and suppose that $\mathbf{X} = (x_{ij})$ $(i = 1, \ldots, n;\ j = 1, \ldots, p)$ has full column rank. Let $\hat{\beta}_j$ be the least squares estimator of β_j $(j = 1, \ldots, p)$ associated with this model.

(a) Prove that

$$\text{var}(\hat{\beta}_j) \geq \frac{\sigma^2}{\sum_{i=1}^n x_{ij}^2} \quad \text{for all } j = 1, \ldots, p,$$

with equality if and only if $\sum_{i=1}^n x_{ij}x_{ik} = 0$ for all $j \neq k = 1, \ldots, p$. (Hint: Without loss of generality take $j = 1$, and use Theorem 2.9.5.)

(b) Suppose that $n = 8$ and $p = 4$. Let $\mathcal{X}_c$ represent the set of all 8×4 model matrices $\mathbf{X} = (x_{ij})$ for which $\sum_{i=1}^8 x_{ij}^2 \leq c$ for all $j = 1, \ldots, 4$. From part (a), it follows that if an $\mathbf{X} \in \mathcal{X}_c$ exists for which $\sum_{i=1}^8 x_{ij}x_{ik} = 0$ for all $j \neq k = 1, \ldots, 4$ and $\sum_{i=1}^8 x_{ij}^2 = c$ for $j = 1, \ldots, 4$, then $\mathbf{X}$ minimizes $\text{var}(\hat{\beta}_j)$, for all

$j = 1, \ldots, 4$, over $\mathcal{X}_c$. Use this fact to show that

$$\mathbf{X} = \begin{pmatrix} 1 & -1 & -1 & -1 \\ 1 & -1 & -1 & 1 \\ 1 & -1 & 1 & -1 \\ 1 & -1 & 1 & 1 \\ 1 & 1 & -1 & -1 \\ 1 & 1 & -1 & 1 \\ 1 & 1 & 1 & -1 \\ 1 & 1 & 1 & 1 \end{pmatrix}$$

minimizes $\text{var}(\hat{\beta}_j)$ for all $j = 1, \ldots, 4$ among all 8×4 model matrices for which $-1 \leq x_{ij} \leq 1$ ($i = 1, \ldots, 8$ and $j = 1, \ldots, 4$).

(c) The model matrix $\mathbf{X}$ displayed in part (b) corresponds to what is called a 2^3 *factorial design*, i.e., an experimental design for which there are three completely crossed experimental factors, each having two levels (coded as -1 and 1), with one observation per combination of factor levels. According to part (b), a 2^3 factorial design minimizes $\text{var}(\hat{\beta}_j)$ (for all $j = 1, \ldots, 4$) among all 8×4 model matrices for which $x_{i1} \equiv 1$ and $-1 \leq x_{ij} \leq 1$ ($i = 1, \ldots, 8$ and $j = 2, 3, 4$), under the "first-order" model

$$y_i = \beta_1 + \beta_2 x_{i2} + \beta_3 x_{i3} + \beta_4 x_{i4} + e_i \quad (i = 1, \ldots, 8).$$

Now consider adding seven additional combinations of the three quantitative explanatory variables to the eight combinations in a 2^3 factorial design. These seven additional combinations are as listed in rows 9 through 15 of the following new, larger model matrix:

$$\mathbf{X} = \begin{pmatrix} 1 & -1 & -1 & -1 \\ 1 & -1 & -1 & 1 \\ 1 & -1 & 1 & -1 \\ 1 & -1 & 1 & 1 \\ 1 & 1 & -1 & -1 \\ 1 & 1 & -1 & 1 \\ 1 & 1 & 1 & -1 \\ 1 & 1 & 1 & 1 \\ 1 & 0 & 0 & 0 \\ 1 & -\alpha & 0 & 0 \\ 1 & \alpha & 0 & 0 \\ 1 & 0 & -\alpha & 0 \\ 1 & 0 & \alpha & 0 \\ 1 & 0 & 0 & -\alpha \\ 1 & 0 & 0 & \alpha \end{pmatrix}.$$

Here α is any positive number. The experimental design associated with this model matrix is known as a 2^3 *factorial central composite design*. Use the results of parts (a) and (b) to show that this design minimizes $\text{var}(\hat{\beta}_j)$ ($j = 2, 3, 4$) among all 15×4 model matrices for which $x_{i1} \equiv 1$ and $\sum_{i=1}^{15} x_{ij}^2 \leq 8 + 2\alpha^2$ ($j = 2, 3, 4$), under the "first-order" model

$$y_i = \beta_1 + \beta_2 x_{i2} + \beta_3 x_{i3} + \beta_4 x_{i4} + e_i \quad (i = 1, \ldots, 15).$$

Solution

(a) Partition $\mathbf{X} = (\mathbf{x}_1, \mathbf{X}_{-1})$, so that $(\mathbf{X}^T\mathbf{X})^{-1} = \begin{pmatrix} a & \mathbf{b}^T \\ \mathbf{b} & \mathbf{C} \end{pmatrix}$ where, by Theorem 2.9.5,

$$a = [\mathbf{x}_1^T\mathbf{x}_1 - \mathbf{x}_1^T\mathbf{X}_{-1}(\mathbf{X}_{-1}^T\mathbf{X}_{-1})^{-1}\mathbf{X}_{-1}^T\mathbf{x}_1]^{-1} \geq 1/(\mathbf{x}_1^T\mathbf{x}_1),$$

with equality if and only if $\mathbf{x}_1^T\mathbf{X}_{-1} = \mathbf{0}^T$. That is, $a \geq 1/\sum_{i=1}^n x_{i1}^2$, with equality if and only if $\sum_{i=1}^n x_{i1}x_{ik} = 0$ for all $k = 2, \ldots, p$. Because $\text{var}(\hat{\beta}_1) = \sigma^2 a$, the result holds for $j = 1$. Results for $j = 2, \ldots, p$ may be shown similarly.

(b) According to the result stated in this part, $\mathbf{X}$ minimizes $\text{var}(\hat{\beta}_j)$, $j = 1, 2, 3, 4$, overall 8×4 matrices for which $\sum_{i=1}^n x_{ij}^2 \leq 8$ for $j = 1, 2, 3, 4$. Because $\{x_{ij} : -1 \leq x_{ij} \leq 1, i = 1, \ldots, 8 \text{ and } j = 1, 2, 3, 4\}$ is a subset of that set, $\mathbf{X}$ also minimizes $\text{var}(\hat{\beta}_j)$, $j = 1, 2, 3, 4$, over this subset.

(c) By parts (a) and (b), this design minimizes $\text{var}(\hat{\beta}_j)$ ($j = 2, 3, 4$) among all 15×3 model matrices for the model

$$y_i = \beta_2 x_{i2} + \beta_3 x_{i3} + \beta_4 x_{i4} + e_i$$

for which $\sum_{i=1}^{15} x_{ij}^2 \leq 8 + 2\alpha^2$ ($j = 2, 3, 4$). Because the columns of this model are orthogonal to $\mathbf{1}_{15}$, this design also minimizes $\text{var}(\hat{\boldsymbol{\beta}}_j)$ ($j = 2, 3, 4$) under the model

$$y_i = \beta_1 + \beta_2 x_{i2} + \beta_3 x_{i3} + \beta_4 x_{i4} + e_i.$$

▶ **Exercise 22** Consider the model $\{\mathbf{y}, \mathbf{X}\boldsymbol{\beta}\}$ and a reparameterization $\{\mathbf{y}, \breve{\mathbf{X}}\boldsymbol{\tau}\}$ of it. Let matrices $\mathbf{T}$ and $\mathbf{S}^T$ be defined such that $\breve{\mathbf{X}} = \mathbf{X}\mathbf{T}$ and $\mathbf{X} = \breve{\mathbf{X}}\mathbf{S}^T$. Show that the converse of Theorem 7.3.1 is false by letting

$$\mathbf{X} = \begin{pmatrix} 1 & 2 & 1 & 2 \\ 1 & 2 & 0 & 0 \\ 1 & 2 & 0 & 0 \end{pmatrix}, \quad \breve{\mathbf{X}} = \begin{pmatrix} 1 & 1 & 2 \\ 1 & 0 & 0 \\ 1 & 0 & 0 \end{pmatrix}, \quad \mathbf{S}^T = \begin{pmatrix} 1 & 2 & 0 & 0 \\ 0 & 0 & 1 & 2 \\ 0 & 0 & 0 & 0 \end{pmatrix},$$

and finding a vector $\mathbf{b}$ such that $\mathbf{b}^T\mathbf{S}^T\boldsymbol{\beta}$ is estimable under the original model but $\mathbf{b}^T\boldsymbol{\tau}$ is not estimable under the reparameterized model.

Solution Let $\mathbf{b} = \mathbf{1}_3$. Then $\mathbf{b}^T\mathbf{S}^T = (1, 2, 1, 2) \in \mathcal{R}(\mathbf{X})$, so $\mathbf{b}^T\mathbf{S}^T\boldsymbol{\beta}$ is estimable under the original model. However, $\mathbf{b}^T \notin \mathcal{R}(\breve{\mathbf{X}})$ so $\mathbf{b}^T\boldsymbol{\tau}$ is not estimable under the reparameterized model.

▶ **Exercise 23** Prove Theorem 7.3.2: For a full-rank reparameterization of the model $\{\mathbf{y}, \mathbf{X}\boldsymbol{\beta}\}$, $\mathcal{R}(\mathbf{S}^T) = \mathcal{R}(\mathbf{X})$ and hence:

(a) $\text{rank}(\mathbf{S}^T) = p^*$ and
(b) a function $\mathbf{c}^T\boldsymbol{\beta}$ is estimable if and only if $\mathbf{c}^T \in \mathcal{R}(\mathbf{S}^T)$.

Solution Because $\mathbf{X} = \breve{\mathbf{X}}\mathbf{S}^T$, $\mathcal{R}(\mathbf{X}) \subseteq \mathcal{R}(\mathbf{S}^T)$. Furthermore,

$$\text{rank}(\mathbf{X}) = \text{rank}(\breve{\mathbf{X}}\mathbf{S}^T) \leq \text{rank}(\mathbf{S}^T) \leq \text{rank}(\mathbf{X}),$$

where the first inequality holds by Theorem 2.8.4 and the last inequality holds because $\mathbf{S}^T$ has p^* rows. Thus $\text{rank}(\mathbf{S}^T) = \text{rank}(\mathbf{X})$, proving part (a). By Theorem 2.8.5, $\mathcal{R}(\mathbf{S}^T) = \mathcal{R}(\mathbf{X})$, so part (b) follows from Theorem 6.1.2.

Reference

Harville, D. A. (1997). Matrix algebra from a statistician's perspective. New York: Springer-Verlag.

8 Least Squares Geometry and the Overall ANOVA

This chapter presents exercises on least squares geometry and the overall analysis of variance, and provides solutions to those exercises.

▶ **Exercise 1** Prove Corollary 8.1.1.1: If $\breve{\mathbf{X}}$ is any matrix for which $\mathcal{C}(\breve{\mathbf{X}}) = \mathcal{C}(\mathbf{X})$, then $\mathbf{P}_{\breve{\mathbf{X}}} = \mathbf{P}_{\mathbf{X}}$ and $\breve{\mathbf{X}}\hat{\boldsymbol{\tau}} = \mathbf{X}\hat{\boldsymbol{\beta}}$, where $\hat{\boldsymbol{\beta}}$ is any solution to $\mathbf{X}^T\mathbf{X}\boldsymbol{\beta} = \mathbf{X}^T\mathbf{y}$ and $\hat{\boldsymbol{\tau}}$ is any solution to $\breve{\mathbf{X}}^T\breve{\mathbf{X}}\boldsymbol{\tau} = \breve{\mathbf{X}}^T\mathbf{y}$.

Solution By Theorem 8.1.1, $\hat{\mathbf{y}} = \mathbf{X}\hat{\boldsymbol{\beta}} = \mathbf{X}(\mathbf{X}^T\mathbf{X})^-\mathbf{X}^T\mathbf{y} = \mathbf{P}_{\mathbf{X}}\mathbf{y}$ and $\hat{\mathbf{y}} = \breve{\mathbf{X}}\hat{\boldsymbol{\tau}} = \breve{\mathbf{X}}(\breve{\mathbf{X}}^T\breve{\mathbf{X}})^-\breve{\mathbf{X}}^T\mathbf{y} = \mathbf{P}_{\breve{\mathbf{X}}}\mathbf{y}$. Thus $\mathbf{P}_{\mathbf{X}}\mathbf{y} = \mathbf{P}_{\breve{\mathbf{X}}}\mathbf{y}$ for all $\mathbf{y}$, implying (by Theorem 2.1.1) that $\mathbf{P}_{\mathbf{X}} = \mathbf{P}_{\breve{\mathbf{X}}}$.

▶ **Exercise 2** Prove Theorem 8.1.6c–e: Let p_{ij} denote the ijth element of $\mathbf{P}_{\mathbf{X}}$. Then:

(c) $p_{ii} \leq 1/r_i$, where r_i is the number of rows of $\mathbf{X}$ that are identical to the ith row. If, in addition, $\mathbf{1}_n$ is one of the columns of $\mathbf{X}$, then:
(d) $p_{ii} \geq 1/n$;
(e) $\sum_{i=1}^n p_{ij} = \sum_{j=1}^n p_{ij} = 1$.

[Hint: To prove part (c), use Theorem 3.3.10 to show (without loss of generality) that the result holds for p_{11} by representing $\mathbf{X}$ as $\begin{pmatrix} \mathbf{1}_{r_1-1}\mathbf{x}_1^T \\ \mathbf{X}_{\boxed{2}} \end{pmatrix}$ where $\mathbf{x}_1^T$ is the first distinct row of $\mathbf{X}$ and r_1 is the number of replicates of that row, and $\mathbf{X}_{\boxed{2}}$ consists of the remaining rows of $\mathbf{X}$ including, as its first row, the last replicate of $\mathbf{x}_1^T$. To prove part (d), consider properties of $\mathbf{P}_{\mathbf{X}} - (1/n)\mathbf{J}_n$.]

Solution According to Theorem 8.1.3b, c, $\mathbf{I} - \mathbf{P}_{\mathbf{X}}$ is symmetric and idempotent. Thus, by Corollary 2.15.9.1 and Theorem 2.15.1, its diagonal elements $1 - p_{ii}$ $(i = 1, \ldots, n)$ are nonnegative, or equivalently $p_{ii} \leq 1$ $(i = 1, \ldots, n)$. Now represent $\mathbf{X}$

D. L. Zimmerman, *Linear Model Theory*,
https://doi.org/10.1007/978-3-030-52074-8_8

as

$$\mathbf{X} = \begin{pmatrix} \mathbf{1}_{r1}\mathbf{x}_1^T \\ \mathbf{1}_{r2}\mathbf{x}_2^T \\ \vdots \\ \mathbf{1}_{rm}\mathbf{x}_m^T \end{pmatrix},$$

where $\mathbf{x}_1, \ldots, \mathbf{x}_m^T$ are the distinct rows of $\mathbf{X}$ and $r_1, \ldots, r_m$ are the corresponding number of replicates of those rows. Without loss of generality, assume that $r_1 > 1$, and we will show that $p_{11} \leq 1/r_1$. The general result can then be obtained by permuting the rows of $\mathbf{X}$.

As suggested in the hint, define

$$\mathbf{X}_{\boxed{2}} = \begin{pmatrix} \mathbf{x}_1^T \\ \mathbf{1}_{r2}\mathbf{x}_2^T \\ \vdots \\ \mathbf{1}_{rm}\mathbf{x}_m^T \end{pmatrix}.$$

Then

$$\mathbf{X} = \begin{pmatrix} \mathbf{1}_{r_1-1}\mathbf{x}_1^T \\ \mathbf{X}_{\boxed{2}} \end{pmatrix}$$

and $\mathbf{X}^T\mathbf{X} = (r_1-1)\mathbf{x}_1\mathbf{x}_1^T + \mathbf{X}_{\boxed{2}}^T\mathbf{X}_{\boxed{2}}$. Now by Theorem 3.3.10 (with $\mathbf{X}_{\boxed{2}}^T\mathbf{X}_{\boxed{2}}$, $\mathbf{x}_1, r_1-1$, and $\mathbf{x}_1^T$ playing the roles of **A**, **B**, **C**, and **D** in the theorem) we obtain

$$\begin{aligned}(\mathbf{X}^T\mathbf{X})^- &= (\mathbf{X}_{\boxed{2}}^T\mathbf{X}_{\boxed{2}})^- - (\mathbf{X}_{\boxed{2}}^T\mathbf{X}_{\boxed{2}})^-\mathbf{x}_1(r_1-1)[(r_1-1) + (r_1-1)^2\mathbf{x}_1^T(\mathbf{X}_{\boxed{2}}^T\mathbf{X}_{\boxed{2}})^-\mathbf{x}_1]^- \\ &\quad \cdot (r_1-1)\mathbf{x}_1^T[(\mathbf{X}_{\boxed{2}}^T\mathbf{X}_{\boxed{2}})^-]^T.\end{aligned}$$

Let $p_{11(-1)} = \mathbf{x}_1^T(\mathbf{X}_{\boxed{2}}^T\mathbf{X}_{\boxed{2}})^-\mathbf{x}_1$ be the (1,1) element of $\mathbf{P}_{\mathbf{X}_{\boxed{2}}}$. Then by pre- and post-multiplying the expression for $(\mathbf{X}^T\mathbf{X})^-$ above by $\mathbf{x}_1^T$ and $\mathbf{x}_1$, respectively, we obtain

$$\begin{aligned}p_{11} &= p_{11(-1)} - p_{11(-1)}^2(r_1-1)^2/[(r_1-1) + (r_1-1)^2 p_{11(-1)}] \\ &= p_{11(-1)}/[1 + (r_1-1)p_{11(-1)}] \\ &= 1/[(1/p_{11(-1)}) + r_1 - 1] \\ &\leq 1/r_1,\end{aligned}$$

where the inequality holds because $p_{11(-1)} \leq 1$ which implies that $(1/p_{11(-1)}) - 1 \geq 0$. This proves part (c). Now suppose that $\mathbf{1}_n$ is one of the columns of $\mathbf{X}$. Then by Theorem 8.1.2a(i), $\mathbf{P_X 1}_n = \mathbf{1}_n$, implying that

$$(1/n)\mathbf{J}_n\mathbf{P_X} = (1/n)\mathbf{1}_n\mathbf{1}_n^T\mathbf{P_X} = (1/n)\mathbf{1}_n\mathbf{1}_n^T = (1/n)\mathbf{J}_n = (1/n)\mathbf{P_X}\mathbf{J}_n.$$

Therefore,

$$\begin{aligned}[\mathbf{P_X} - (1/n)\mathbf{J}_n][\mathbf{P_X} - (1/n)\mathbf{J}_n] &= \mathbf{P_X}\mathbf{P_X} - (1/n)\mathbf{J}_n\mathbf{P_X} - (1/n)\mathbf{P_X}\mathbf{J}_n + (1/n^2)\mathbf{J}_n\mathbf{J}_n \\ &= \mathbf{P_X} - (1/n)\mathbf{J}_n - (1/n)\mathbf{J}_n + (1/n)\mathbf{J}_n \\ &= \mathbf{P_X} - (1/n)\mathbf{J}_n.\end{aligned}$$

Thus $\mathbf{P_X} - (1/n)\mathbf{J}_n$ is idempotent, and it is clearly symmetric as well; hence by Corollary 2.15.9.1, $\mathbf{P_X} - (1/n)\mathbf{J}_n$ is nonnegative definite. Thus, by Theorem 2.15.1 its diagonal elements $p_{ii} - (1/n)$ $(i = 1, \dots, n)$ are nonnegative, or equivalently $p_{ii} \geq 1/n$ $(i = 1, \dots, n)$. This proves part (d). For part (e), consideration of each element of the equalities $\mathbf{P_X 1}_n = \mathbf{1}_n$ and $\mathbf{1}_n^T\mathbf{P_X} = \mathbf{1}_n^T$ established above yields $\sum_{j=1}^n p_{ij} = 1$ for all $i = 1, \dots, n$ and $\sum_{i=1}^n p_{ij} = 1$ for all $j = 1, \dots, n$.

▶ **Exercise 3** Find var$(\hat{\mathbf{e}} - \mathbf{e})$ and cov$(\hat{\mathbf{e}} - \mathbf{e}, \mathbf{e})$ under the Gauss–Markov model $\{\mathbf{y}, \mathbf{X}\boldsymbol{\beta}, \sigma^2\mathbf{I}\}$.

Solution

$$\text{var}(\hat{\mathbf{e}}-\mathbf{e}) = \text{var}[(\mathbf{I}-\mathbf{P_X})\mathbf{y}-\mathbf{e}] = \text{var}[(\mathbf{I}-\mathbf{P_X})\mathbf{e}-\mathbf{e}] = \text{var}(-\mathbf{P_X}\mathbf{e}) = \mathbf{P_X}(\sigma^2\mathbf{I})\mathbf{P_X}^T = \sigma^2\mathbf{P_X},$$

and

$$\text{cov}(\hat{\mathbf{e}}-\mathbf{e}, \mathbf{e}) = \text{cov}(\hat{\mathbf{e}}, \mathbf{e})-\text{var}(\mathbf{e}) = \text{cov}[(\mathbf{I}-\mathbf{P_X})\mathbf{e}, \mathbf{e}]-\sigma^2\mathbf{I} = (\mathbf{I}-\mathbf{P_X})(\sigma^2\mathbf{I})-\sigma^2\mathbf{I} = -\sigma^2\mathbf{P_X}.$$

▶ **Exercise 4** Determine $\mathbf{P_X}$ for each of the following models:

(a) the no-intercept simple linear regression model;
(b) the two-way main effects model with balanced data;
(c) the two-way model with interaction and balanced data;
(d) the two-way partially crossed model introduced in Example 5.1.4-1, with one observation per cell;
(e) the two-factor nested model.

Solution

(a) $\mathbf{P_X} = \mathbf{x}(\mathbf{x}^T\mathbf{x})^{-1}\mathbf{x}^T = \left(1/\sum_{i=1}^n x_i^2\right)\mathbf{x}\mathbf{x}^T$.
(b) Recall that for this situation, $\mathbf{X} = (\mathbf{1}_{qmr},\ \mathbf{I}_q \otimes \mathbf{1}_{mr},\ \mathbf{1}_q \otimes \mathbf{I}_m \otimes \mathbf{1}_r)$. By the same development given in the solution to Exercise 7.11, we obtain the following

generalized inverse of $\mathbf{X}^T\mathbf{X}$:

$$(\mathbf{X}^T\mathbf{X})^- = \begin{pmatrix} 0 & \mathbf{0}_q^T & \mathbf{0} \\ \mathbf{0}_q & (1/mr)\mathbf{I}_q & \mathbf{0} \\ \mathbf{0} & \mathbf{0} & (1/qr)\mathbf{I}_m - (1/qmr)\mathbf{J}_m \end{pmatrix}.$$

Thus,

$$\mathbf{P_X} = (\mathbf{1}_{qmr},\, \mathbf{I}_q \otimes \mathbf{1}_{mr},\, \mathbf{1}_q \otimes \mathbf{I}_m \otimes \mathbf{1}_r) \begin{pmatrix} 0 & \mathbf{0}_q^T & \mathbf{0} \\ \mathbf{0}_q & (1/mr)\mathbf{I}_q & \mathbf{0} \\ \mathbf{0} & \mathbf{0} & (1/qr)\mathbf{I}_m - (1/qmr)\mathbf{J}_m \end{pmatrix}$$

$$\times \begin{pmatrix} \mathbf{1}_{qmr}^T \\ \mathbf{I}_q \otimes \mathbf{1}_{mr}^T \\ \mathbf{1}_q^T \otimes \mathbf{I}_m \otimes \mathbf{1}_r^T \end{pmatrix}$$

$$= \left((1/mr)\mathbf{I}_q \otimes \mathbf{1}_{mr},\, (1/qr)\mathbf{1}_q \otimes [\mathbf{I}_m - (1/m)\mathbf{J}_m] \otimes \mathbf{1}_r \right) \begin{pmatrix} \mathbf{I}_q \otimes \mathbf{1}_{mr}^T \\ \mathbf{1}_q^T \otimes \mathbf{I}_m \otimes \mathbf{1}_r^T \end{pmatrix}$$

$$= (1/mr)\mathbf{I}_q \otimes \mathbf{J}_{mr} + (1/qr)\mathbf{J}_q \otimes [\mathbf{I}_m - (1/m)\mathbf{J}_m] \otimes \mathbf{J}_r.$$

(c) The cell-means model

$$y_{ijk} = \mu_{ij} + e_{ijk} \quad (i = 1, \ldots, q;\ j = 1, \ldots, m;\ k = 1, \ldots, r)$$

is a reparameterization of the two-way model with interaction and balanced data. Because the model matrix for this model is $\mathbf{X} = \mathbf{I}_{qm} \otimes \mathbf{1}_r$, we obtain

$$\begin{aligned} \mathbf{P_X} &= \left(\mathbf{I}_{qm} \otimes \mathbf{1}_r\right)\left[\left(\mathbf{I}_{qm} \otimes \mathbf{1}_r\right)^T \left(\mathbf{I}_{qm} \otimes \mathbf{1}_r\right)\right]^- \left(\mathbf{I}_{qm} \otimes \mathbf{1}_r\right)^T \\ &= \left(\mathbf{I}_{qm} \otimes \mathbf{1}_r\right)\left(r\mathbf{I}_{qm}\right)^- \left(\mathbf{I}_{qm} \otimes \mathbf{1}_r^T\right) \\ &= \mathbf{I}_{qm} \otimes (1/r)\mathbf{J}_r. \end{aligned}$$

(d) Recall from Example 5.1.4-1 that $\mathbf{X} = (\mathbf{1}_{q(q-1)}, (\mathbf{v}_{ij}^T)_{i \neq j=1,\ldots,q})$, where $\mathbf{v}_{ij} = \mathbf{u}_i^{(q)} - \mathbf{u}_j^{(q)}$. Now observe that

$$\mathbf{1}_{q(q-1)}^T \mathbf{1}_{q(q-1)} = q(q-1),$$

$$\begin{aligned} &\mathbf{1}_{q(q-1)}^T (\mathbf{u}_i^{(q)} - \mathbf{u}_j^{(q)})_{i \neq j=1,\ldots,q} \\ &= \left((1)(q-1) + (-1)(q-1),\, \ldots,\, (1)(q-1) + (-1)(q-1)\right)_{1 \times q} = \mathbf{0}_q^T, \end{aligned}$$

and

$$
\begin{aligned}
[(\mathbf{v}_{ij}^T)_{i\neq j=1,\ldots,q}]^T[(\mathbf{v}_{ij}^T)_{i\neq j=1,\ldots,q}] &= \sum_{i=1}^{q}\sum_{j\neq i}^{q}(\mathbf{u}_i^{(q)}-\mathbf{u}_j^{(q)})(\mathbf{u}_i^{(q)}-\mathbf{u}_j^{(q)})^T \\
&= (q-1)\sum_{i=1}^{q}\mathbf{u}_i^{(q)}\mathbf{u}_i^{(q)T} + (q-1)\sum_{j=1}^{q}\mathbf{u}_j^{(q)}\mathbf{u}_j^{(q)T} \\
&\quad -\sum_{i=1}^{q}\sum_{j\neq i}(\mathbf{u}_i^{(q)}\mathbf{u}_j^{(q)T}+\mathbf{u}_j^{(q)}\mathbf{u}_i^{(q)T}) \\
&= 2q\mathbf{I}_q - 2\mathbf{J}_q \\
&= 2q[(\mathbf{I}_q - (1/q)\mathbf{J}_q].
\end{aligned}
$$

One generalized inverse of this last matrix is $(1/2q)\mathbf{I}_q$, so by Theorem 3.3.7 one generalized inverse of $\mathbf{X}^T\mathbf{X}$ is

$$
(\mathbf{X}^T\mathbf{X})^- = \begin{pmatrix} 1/[q(q-1)] & \mathbf{0}_q^T \\ \mathbf{0}_q & (1/2q)\mathbf{I}_q \end{pmatrix}.
$$

Then

$$
\begin{aligned}
\mathbf{P_X} = \mathbf{X}(\mathbf{X}^T\mathbf{X})^-\mathbf{X}^T &= (\mathbf{1}_{q(q-1)},\ (\mathbf{v}_{ij}^T)_{i\neq j=1,\ldots,q})\begin{pmatrix} 1/[q(q-1)] & \mathbf{0}_q^T \\ \mathbf{0}_q & (1/2q)\mathbf{I}_q \end{pmatrix} \\
&\quad \times \begin{pmatrix} \mathbf{1}_{q(q-1)}^T \\ (\mathbf{v}_{ij})_{i\neq j=1,\ldots,q} \end{pmatrix} \\
&= \frac{1}{q(q-1)}\mathbf{J}_{q(q-1)} + \mathbf{P}_2,
\end{aligned}
$$

where the rows and columns of $\mathbf{P}_2 = (p_{2,iji'j'})$ are indexed by double subscripts ij and $i'j'$, respectively, and

$$
\begin{aligned}
p_{2,iji'j'} &= (1/2q)\cdot(\mathbf{u}_i^{(q)}-\mathbf{u}_j^{(q)})^T(\mathbf{u}_{i'}^{(q)}-\mathbf{u}_{j'}^{(q)}) \\
&= (1/2q)\cdot(\mathbf{u}_i^{(q)T}\mathbf{u}_{i'}^{(q)} - \mathbf{u}_i^{(q)T}\mathbf{u}_{j'}^{(q)} - \mathbf{u}_j^{(q)T}\mathbf{u}_{i'}^{(q)} + \mathbf{u}_j^{(q)T}\mathbf{u}_{j'}^{(q)} \\
&= \begin{cases} \frac{1}{q} & \text{if } i=i' \text{ and } j=j', \\ -\frac{1}{q} & \text{if } i=j' \text{ and } j=i', \\ \frac{1}{2q} & \text{if } i=i' \text{ and } j\neq j', \text{ or } i\neq i' \text{ and } j=j', \\ -\frac{1}{2q} & \text{if } i=j' \text{ and } j\neq i', \text{ or } j=i' \text{ and } i\neq j', \\ 0 & \text{if } i\neq i',\ i\neq j',\ j\neq i', \text{ and } j\neq j'. \end{cases}
\end{aligned}
$$

Thus $\mathbf{P_X} = (p_{iji'j'})$ where

$$p_{iji'j'} = \begin{cases} \frac{1}{q(q-1)} + \frac{1}{q} & \text{if } i = i' \text{ and } j = j', \\ \frac{1}{q(q-1)} - \frac{1}{q} & \text{if } i = j' \text{ and } j = i', \\ \frac{1}{q(q-1)} + \frac{1}{2q} & \text{if } i = i' \text{ and } j \neq j', \text{ or } i \neq i' \text{ and } j = j', \\ \frac{1}{q(q-1)} - \frac{1}{2q} & \text{if } i = j' \text{ and } j \neq i', \text{ or } j = i' \text{ and } i \neq j', \\ \frac{1}{q(q-1)} & \text{if } i \neq i', i \neq j', j \neq i', \text{ and } j \neq j'. \end{cases}$$

(e) By Exercise 5.4, the model matrix for the two-factor nested model may be written as $\mathbf{X} = (\mathbf{X}_1, \mathbf{X}_2, \mathbf{X}_3)$ where $\mathbf{X}_1 = \mathbf{1}$, $\mathbf{X}_2 = \oplus_{i=1}^q \mathbf{1}_{n_i}$, and $\mathbf{X}_3 = \oplus_{i=1}^q \oplus_{j=1}^{m_i} \mathbf{1}_{n_{ij}}$. Observe that $\mathbf{X}_1$ and every column of $\mathbf{X}_2$ can be written as a linear combination of the columns of $\mathbf{X}_3$, so $\mathcal{C}(\mathbf{X}_3) = \mathcal{C}(\mathbf{X})$. By Corollary 8.1.1.1,

$$\begin{aligned} \mathbf{P_X} = \mathbf{P}_{\mathbf{X}_3} &= \left(\oplus_{i=1}^q \oplus_{j=1}^{m_i} \mathbf{1}_{n_{ij}}\right) \left[\left(\oplus_{i=1}^q \oplus_{j=1}^{m_i} \mathbf{1}_{n_{ij}}^T\right) \left(\oplus_{i=1}^q \oplus_{j=1}^{m_i} \mathbf{1}_{n_{ij}}\right)\right]^- \\ &\quad \times \left(\oplus_{i=1}^q \oplus_{j=1}^{m_i} \mathbf{1}_{n_{ij}}^T\right) \\ &= \left(\oplus_{i=1}^q \oplus_{j=1}^{m_i} \mathbf{1}_{n_{ij}}\right) \left(\oplus_{i=1}^q \oplus_{j=1}^{m_i} (1/n_{ij})\right) \left(\oplus_{i=1}^q \oplus_{j=1}^{m_i} \mathbf{1}_{n_{ij}}^T\right) \\ &= \oplus_{i=1}^q \oplus_{j=1}^{m_i} (1/n_{ij})\mathbf{J}_{n_{ij}}. \end{aligned}$$

▶ **Exercise 5** Determine the overall ANOVA, including a nonmatrix expression for the model sum of squares, for each of the following models:

(a) the no-intercept simple linear regression model;
(b) the two-way main effects model with balanced data;
(c) the two-way model with interaction and balanced data;
(d) the two-way partially crossed model introduced in Example 5.1.4-1, with one observation per cell;
(e) the two-factor nested model.

Solution

(a)

Source	Rank	Sum of squares
Model	1	$\mathbf{y}^T\mathbf{x}(\mathbf{x}^T\mathbf{x})^{-1}\mathbf{x}^T\mathbf{y} = \left(\sum_{i=1}^n x_i y_i\right)^2 / \left(\sum_{i=1}^n x_i^2\right)$
Residual	$n-1$	By subtraction
Total	n	$\mathbf{y}^T\mathbf{y}$

(b) Using the solution to Exercise 8.4b,

$$\mathbf{P_X y} = \{(1/mr)\mathbf{I}_q \otimes \mathbf{J}_{mr} + (1/qr)\mathbf{J}_q \otimes [\mathbf{I}_m - (1/m)\mathbf{J}_m] \otimes \mathbf{J}_r\}\mathbf{y}$$

$$= \begin{pmatrix} \bar{y}_{1\cdot\cdot} \\ \bar{y}_{2\cdot\cdot} \\ \vdots \\ \bar{y}_{q\cdot\cdot} \end{pmatrix} \otimes \mathbf{1}_{mr} + \mathbf{1}_q \otimes \begin{pmatrix} \bar{y}_{\cdot 1\cdot} \\ \bar{y}_{\cdot 2\cdot} \\ \vdots \\ \bar{y}_{\cdot m\cdot} \end{pmatrix} \otimes \mathbf{1}_r - \bar{y}_{\cdots}\mathbf{1}_{qmr}.$$

Hence $\mathbf{y}^T\mathbf{P_X y} = (\mathbf{P_X y})^T\mathbf{P_X y} = r\sum_{i=1}^q \sum_{j=1}^m (\bar{y}_{i\cdot\cdot} + \bar{y}_{\cdot j\cdot} - \bar{y}_{\cdots})^2$. Therefore, the overall ANOVA is as follows:

Source	Rank	Sum of squares
Model	$q+m-1$	$r\sum_{i=1}^q \sum_{j=1}^m (\bar{y}_{i\cdot\cdot} + \bar{y}_{\cdot j\cdot} - \bar{y}_{\cdots})^2$
Residual	$qmr-q-m+1$	By subtraction
Total	qmr	$\mathbf{y}^T\mathbf{y}$

(c) Using the solution to Exercise 8.4c, $\mathbf{P_X y} = \left(\oplus_{i=1}^q \oplus_{j=1}^m (1/r)\mathbf{J}_r\right)\mathbf{y} = \begin{pmatrix} \bar{y}_{11\cdot} \\ \bar{y}_{12\cdot} \\ \vdots \\ \bar{y}_{qm\cdot} \end{pmatrix} \otimes \mathbf{1}_r$. Hence $\mathbf{y}^T\mathbf{P_X y} = (\mathbf{P_X y})^T\mathbf{P_X y} = r\sum_{i=1}^q \sum_{j=1}^m \bar{y}_{ij\cdot}^2$. Therefore, the overall ANOVA is as follows:

Source	Rank	Sum of squares
Model	qm	$r\sum_{i=1}^q \sum_{j=1}^m \bar{y}_{ij\cdot}^2$
Residual	$qm(r-1)$	By subtraction
Total	qmr	$\mathbf{y}^T\mathbf{y}$

(d) Using the solution to Exercise 8.4d,

$$\mathbf{P_X y} = \left(\mathbf{1}_{q(q-1)}, (\mathbf{v}_{ij}^T)_{i\neq j=1,\ldots,q}\right)\begin{pmatrix} 1/[q(q-1)] & \mathbf{0}_q^T \\ \mathbf{0}_q & (1/2q)\mathbf{I}_q \end{pmatrix}$$

$$\times \left(\mathbf{1}_{q(q-1)}, (\mathbf{v}_{ij}^T)_{i\neq j=1,\ldots,q}\right)^T \mathbf{y}$$

$$= \left(\{1/[q(q-1)]\}\mathbf{1}_{q(q-1)}, (1/2q)(\mathbf{v}_{ij}^T)_{i\neq j=1,\ldots,q}\right)\begin{pmatrix} y_{\cdot\cdot} \\ y_{1\cdot} - y_{\cdot 1} \\ y_{2\cdot} - y_{\cdot 2} \\ \vdots \\ y_{q\cdot} - y_{\cdot q} \end{pmatrix}$$

$$= \begin{pmatrix} \bar{y}_{\cdot\cdot} + [(\bar{y}_{1\cdot} - \bar{y}_{\cdot 1})/2] - [\bar{y}_{2\cdot} - \bar{y}_{\cdot 2})/2] \\ \bar{y}_{\cdot\cdot} + [(\bar{y}_{1\cdot} - \bar{y}_{\cdot 1})/2] - [\bar{y}_{3\cdot} - \bar{y}_{\cdot 3})/2] \\ \vdots \\ \bar{y}_{\cdot\cdot} + [(\bar{y}_{q-1\cdot} - \bar{y}_{\cdot q-1})/2] - [\bar{y}_{q\cdot} - \bar{y}_{\cdot q})/2] \end{pmatrix}.$$

Hence $\mathbf{y}^T\mathbf{P_X}\mathbf{y} = (\mathbf{P_X}\mathbf{y})^T\mathbf{P_X}\mathbf{y} = \sum_{i=1}^{q}\sum_{j\neq i}[\bar{y}_{\cdot\cdot} + (1/2)(\bar{y}_{i\cdot} - \bar{y}_{\cdot i}) - (1/2)(\bar{y}_{j\cdot} - \bar{y}_{\cdot j})]^2$. Therefore, the overall ANOVA is as follows:

Source	Rank	Sum of squares
Model	q	$\sum_{i=1}^{q}\sum_{j\neq i}[\bar{y}_{\cdot\cdot} + (1/2)(\bar{y}_{i\cdot} - \bar{y}_{\cdot i}) - (1/2)(\bar{y}_{j\cdot} - \bar{y}_{\cdot j})]^2$
Residual	$(q-1)^2$	By subtraction
Total	$q(q-1)$	$\mathbf{y}^T\mathbf{y}$

(e) Using the solution to Exercise 8.4e,

$$\mathbf{P_X}\mathbf{y} = \oplus_{i=1}^{q}\oplus_{j=1}^{m_i}(1/n_{ij})\mathbf{J}_{n_{ij}}\mathbf{y} = \oplus_{i=1}^{q}\oplus_{j=1}^{m_i}\mathbf{1}_{n_{ij}}\bar{y}_{ij\cdot}.$$

Hence

$$\mathbf{y}^T\mathbf{P_X}\mathbf{y} = (\mathbf{P_X}\mathbf{y})^T(\mathbf{P_X}\mathbf{y}) = \left(\oplus_{i=1}^{q}\oplus_{j=1}^{m_i}\mathbf{1}_{n_{ij}}\bar{y}_{ij\cdot}\right)^T\left(\oplus_{i=1}^{q}\oplus_{j=1}^{m_i}\mathbf{1}_{n_{ij}}\bar{y}_{ij\cdot}\right)$$
$$= \sum_{i=1}^{q}\sum_{j=1}^{m_i} n_{ij}\bar{y}_{ij\cdot}^2.$$

Therefore, the overall ANOVA is as follows:

Source	Rank	Sum of squares
Model	$\sum_{i=1}^{q} m_i$	$\sum_{i=1}^{q}\sum_{j=1}^{m_i} n_{ij}\bar{y}_{ij\cdot}^2$
Residual	$\sum_{i=1}^{q}\sum_{j=1}^{m_i} n_{ij} - \sum_{i=1}^{q} m_i$	By subtraction
Total	$\sum_{i=1}^{q}\sum_{j=1}^{m_i} n_{ij}$	$\mathbf{y}^T\mathbf{y}$

▶ **Exercise 6** For the special case of a (Gauss–Markov) simple linear regression model with even sample size n and $n/2$ observations at each of $x = -n$ and $x = n$, obtain the variances of the fitted residuals and the correlations among them.

Solution In this situation, $\bar{x} = 0$ and $SXX = \sum_{i=1}^{n} n^2 = n^3$. Specializing the general expressions for the elements of $\mathbf{P_X}$ provided in Example 8.1-2, we obtain

$$p_{ij} = \begin{cases} \frac{1}{n} + \frac{n^2}{n^3} = \frac{2}{n} \text{ if } x_i = x_j, \\ \frac{1}{n} - \frac{n^2}{n^3} = 0 \text{ if } x_i = -x_j. \end{cases}$$

So by Theorem 8.1.7d, $\mathrm{var}(\hat{e}_i) = \sigma^2[1 - (2/n)]$ and for $i \neq j$,

$$\mathrm{corr}(\hat{e}_i, \hat{e}_j) = \begin{cases} \frac{-2/n}{1-(2/n)} = \frac{-2}{n-2} & \text{if } x_i = x_j, \\ 0 & \text{if } x_i = -x_j. \end{cases}$$

▶ **Exercise 7** For the (Gauss–Markov) no-intercept simple linear regression model, obtain $\mathrm{var}(\hat{e}_i)$ and $\mathrm{corr}(\hat{e}_i, \hat{e}_j)$.

Solution $\mathbf{P_x} = \mathbf{x}(\mathbf{x}^T\mathbf{x})^{-1}\mathbf{x}^T = (1/\sum_{i=1}^{n} x_i^2)\mathbf{x}\mathbf{x}^T$, so by Theorem 8.1.7d, $\mathrm{var}(\hat{e}_i) = \sigma^2(1 - x_i^2/\sum_{j=1}^{n} x_j^2)$ and $\mathrm{corr}(\hat{e}_i, \hat{e}_j) = \frac{-x_i x_j/\sum_{k=1}^{n} x_k^2}{\sqrt{(1-x_i^2/\sum_{k=1}^{n} x_k^2)(1-x_j^2/\sum_{k=1}^{n} x_k^2)}}$.

▶ **Exercise 8** Consider the Gauss–Markov model $\{\mathbf{y}, \mathbf{X}\boldsymbol{\beta}, \sigma^2\mathbf{I}\}$. Prove that if $\mathbf{X}^T\mathbf{X} = k\mathbf{I}$ for some $k > 0$, and rows i and j of $\mathbf{X}$ are orthogonal, then $\mathrm{corr}(\hat{e}_i, \hat{e}_j) = 0$.

Solution Let $\mathbf{x}_i^T$ represent row i of $\mathbf{X}$. By Theorem 8.1.7d,

$$\begin{aligned} \mathrm{cov}(\hat{e}_i, \hat{e}_j) &= -\sigma^2\mathbf{x}_i^T(\mathbf{X}^T\mathbf{X})^-\mathbf{x}_j \\ &= -\sigma^2\mathbf{x}_i^T(k\mathbf{I})^{-1}\mathbf{x}_j = -(\sigma^2/k)\mathbf{x}_i^T\mathbf{x}_j = (\sigma^2/k)\cdot 0 = 0. \end{aligned}$$

▶ **Exercise 9** Under the Gauss–Markov model $\{\mathbf{y}, \mathbf{X}\boldsymbol{\beta}, \sigma^2\mathbf{I}\}$ with $n > p^*$ and excess kurtosis matrix $\mathbf{0}$, find the estimator of σ^2 that minimizes the mean squared error within the class of estimators of the form $\mathbf{y}^T(\mathbf{I} - \mathbf{P_X})\mathbf{y}/k$, where $k > 0$. How does the minimized mean squared error compare to the mean squared error of $\hat{\sigma}^2$ specified in Theorem 8.2.3, which is $2\sigma^4/(n - p^*)$?

Solution By Theorem 4.2.4 and Corollary 4.2.6.2,

$$E[\mathbf{y}^T(\mathbf{I}-\mathbf{P_X})\mathbf{y}] = (\mathbf{X}\boldsymbol{\beta})^T(\mathbf{I}-\mathbf{P_X})\mathbf{X}\boldsymbol{\beta} + \mathrm{tr}[(\mathbf{I}-\mathbf{P_X})(\sigma^2\mathbf{I})] = 0 + \sigma^2\mathrm{tr}(\mathbf{I}-\mathbf{P_X}) = \sigma^2(n-p^*)$$

and

$$\begin{aligned} \mathrm{var}[\mathbf{y}^T(\mathbf{I} - \mathbf{P_X})\mathbf{y}] &= 4(\mathbf{X}\boldsymbol{\beta})^T(\mathbf{I} - \mathbf{P_X})\boldsymbol{\Lambda}\mathrm{vec}(\mathbf{I} - \mathbf{P_X}) + 2\mathrm{tr}[(\mathbf{I} - \mathbf{P_X})(\sigma^2\mathbf{I})(\mathbf{I} - \mathbf{P_X})(\sigma^2\mathbf{I})] \\ &\quad + 4(\mathbf{X}\boldsymbol{\beta})^T(\mathbf{I} - \mathbf{P_X})(\sigma^2\mathbf{I})(\mathbf{I} - \mathbf{P_X})\mathbf{X}\boldsymbol{\beta} \\ &= 2\sigma^4(n - p^*). \end{aligned}$$

Therefore, writing $MSE(\hat{\sigma}^2(k))$ for the mean squared error of $\hat{\sigma}^2(k) \equiv [\mathbf{y}^T(\mathbf{I} - \mathbf{P_X})\mathbf{y}]/k$, we obtain

$$\begin{aligned}
MSE(\hat{\sigma}^2(k)) &= \mathrm{var}(\hat{\sigma}^2(k)) + [\mathrm{E}(\hat{\sigma}^2(k)) - \sigma^2]^2 \\
&= \frac{1}{k^2}[2\sigma^4(n-p^*)] + \left(\frac{\sigma^2(n-p^*)}{k} - \sigma^2\right)^2 \\
&= \frac{\sigma^4}{k^2}[2(n-p^*) + (n-p^*)^2] - \frac{2\sigma^4}{k}(n-p^*) + \sigma^4 \\
&= \sigma^4[2(n-p^*)h^2 + (n-p^*)^2h^2 - 2(n-p^*)h + 1],
\end{aligned}$$

where $h \equiv (1/k)$. Differentiating this last expression with respect to h and setting the result equal to 0, we obtain a stationary point at $h = 1/(n - p^* + 2)$. Because the second derivative is positive, this point minimizes the mean squared error. Equivalently, $MSE(\hat{\sigma}^2(k))$ is minimized at $k = n - p^* + 2$. Then,

$$\begin{aligned}
MSE(\hat{\sigma}^2(k)) &= \frac{1}{(n-p^*+2)^2}[2\sigma^4(n-p^*)] + \left(\frac{\sigma^2(n-p^*)}{n-p^*+2} - \sigma^2\right)^2 \\
&= \frac{\sigma^4}{(n-p^*+2)^2}[2(n-p^*) + 4] \\
&= \frac{2\sigma^4}{n-p^*+2}.
\end{aligned}$$

This mean squared error is smaller than that of $\hat{\sigma}^2$ by a multiplicative factor of $(n - p^*)/(n - p^* + 2)$.

▶ **Exercise 10** Define a *quadratic estimator* of σ^2 associated with the Gauss–Markov model $\{\mathbf{y}, \mathbf{X}\boldsymbol{\beta}, \sigma^2\mathbf{I}\}$ to be any quadratic form $\mathbf{y}^T\mathbf{A}\mathbf{y}$, where $\mathbf{A}$ is a positive definite matrix.

(a) Prove that a quadratic estimator $\mathbf{y}^T\mathbf{A}\mathbf{y}$ of σ^2 is unbiased under the specified Gauss–Markov model if and only if $\mathbf{AX} = \mathbf{0}$ and $\mathrm{tr}(\mathbf{A}) = 1$. (An estimator $t(\mathbf{y})$ is said to be unbiased for σ^2 under a Gauss–Markov model if $\mathrm{E}[t(\mathbf{y})] = \sigma^2$ for all $\boldsymbol{\beta} \in \mathbb{R}^p$ and all $\sigma^2 > 0$ under that model.)
(b) Prove the following extension of Theorem 8.2.2: Let $\mathbf{c}^T\hat{\boldsymbol{\beta}}$ be the least squares estimator of an estimable function associated with the model $\{\mathbf{y}, \mathbf{X}\boldsymbol{\beta}\}$, suppose that $n > p^*$, and let $\tilde{\sigma}^2$ be a quadratic unbiased estimator of σ^2 under the Gauss–Markov model $\{\mathbf{y}, \mathbf{X}\boldsymbol{\beta}, \sigma^2\mathbf{I}\}$. If $\mathbf{e}$ has skewness matrix $\mathbf{0}$, then $\mathrm{cov}(\mathbf{c}^T\hat{\boldsymbol{\beta}}, \tilde{\sigma}^2) = 0$.
(c) Determine as simple an expression as possible for the variance of a quadratic unbiased estimator of σ^2 under a Gauss–Markov model for which the skewness matrix of $\mathbf{e}$ equals $\mathbf{0}$ and the excess kurtosis matrix of $\mathbf{e}$ equals $\mathbf{0}$.

Solution

(a) By Theorem 4.2.4,

$$\mathrm{E}(\mathbf{y}^T\mathbf{A}\mathbf{y}) = (\mathbf{X}\boldsymbol{\beta})^T\mathbf{A}\mathbf{X}\boldsymbol{\beta} + \mathrm{tr}[\mathbf{A}(\sigma^2\mathbf{I})] = \boldsymbol{\beta}^T\mathbf{X}^T\mathbf{A}\mathbf{X}\boldsymbol{\beta} + \sigma^2\mathrm{tr}(\mathbf{A}).$$

Thus, $\mathrm{E}(\mathbf{y}^T\mathbf{A}\mathbf{y}) = \sigma^2$ for all $\boldsymbol{\beta} \in \mathbb{R}^p$ and all $\sigma^2 > 0$ if and only if $\mathbf{X}^T\mathbf{A}\mathbf{X} = \mathbf{0}$ and $\mathrm{tr}(\mathbf{A}) = 1$. The condition $\mathbf{X}^T\mathbf{A}\mathbf{X} = \mathbf{0}$ may be re-expressed as $\mathbf{X}^T\mathbf{C}^T\mathbf{C}\mathbf{X} = \mathbf{0}$ (where $\mathbf{A} = \mathbf{C}^T\mathbf{C}$ by the nonnegative definiteness of $\mathbf{A}$), or equivalently as either $\mathbf{C}\mathbf{X} = \mathbf{0}$ or $\mathbf{A}\mathbf{X} = \mathbf{0}$.

(b) By Corollary 4.2.5.1 and part (a),

$$\mathrm{cov}(\mathbf{c}^T\hat{\boldsymbol{\beta}}, \tilde{\sigma}^2) = 2\mathbf{c}^T(\mathbf{X}^T\mathbf{X})^-\mathbf{X}^T(\sigma^2\mathbf{I})\mathbf{A}\mathbf{X}\boldsymbol{\beta} = 0.$$

(c) By Corollary 4.2.6.3,

$$\mathrm{var}(\mathbf{y}^T\mathbf{A}\mathbf{y}) = 2\mathrm{tr}[\mathbf{A}(\sigma^2\mathbf{I})\mathbf{A}(\sigma^2\mathbf{I})] + 4(\mathbf{X}\boldsymbol{\beta})^T\mathbf{A}(\sigma^2\mathbf{I})\mathbf{A}\mathbf{X}\boldsymbol{\beta} = 2\sigma^4\mathrm{tr}(\mathbf{A}^2).$$

▶ **Exercise 11** Verify the alternative expressions for the five influence diagnostics described in Methodological Interlude #3.

Solution

$$\begin{aligned}
\hat{e}_{i,-i} &= y_i - \mathbf{x}_i^T\hat{\boldsymbol{\beta}}_{-i} = y_i - \mathbf{x}_i^T\left[\hat{\boldsymbol{\beta}} - \left(\frac{\hat{e}_i}{1-p_{ii}}\right)(\mathbf{X}^T\mathbf{X})^{-1}\mathbf{x}_i\right] \\
&= y_i - \mathbf{x}_i^T\hat{\boldsymbol{\beta}} + \left(\frac{\hat{e}_i}{1-p_{ii}}\right)\mathbf{x}_i^T(\mathbf{X}^T\mathbf{X})^{-1}\mathbf{x}_i = \hat{e}_i + \frac{p_{ii}}{1-p_{ii}}\hat{e}_i \\
&= \frac{\hat{e}_i}{1-p_{ii}},
\end{aligned}$$

$$\mathrm{DFBETAS}_{j,i} = \frac{\hat{\beta}_j - \hat{\beta}_{j,i}}{\hat{\sigma}_{-i}\sqrt{[(\mathbf{X}^T\mathbf{X})^{-1}]_{jj}}} = \left(\frac{\hat{e}_i}{1-p_{ii}}\right)\frac{[(\mathbf{X}^T\mathbf{X})^{-1}\mathbf{x}_i]_j}{\hat{\sigma}_{-i}[(\mathbf{X}^T\mathbf{X})^{-1}]_{jj}}$$

$$\begin{aligned}
\mathrm{DFFITS}_i &= \frac{\hat{y}_i - \hat{y}_{i,-i}}{\hat{\sigma}_{-i}\sqrt{p_{ii}}} = \frac{\mathbf{X}(\hat{\boldsymbol{\beta}} - \hat{\boldsymbol{\beta}}_{-i})_i}{\hat{\sigma}_{-i}\sqrt{p_{ii}}} = \left(\frac{\hat{e}_i}{1-p_{ii}}\right)\frac{[\mathbf{X}(\mathbf{X}^T\mathbf{X})^{-1}\mathbf{x}_i]_i}{\hat{\sigma}_{-i}\sqrt{p_{ii}}} \\
&= \frac{\hat{e}_i\sqrt{p_{ii}}}{\hat{\sigma}_{-i}(1-p_{ii})},
\end{aligned}$$

$$\begin{aligned}
\text{Cook's } D_i &= \frac{(\hat{\boldsymbol{\beta}} - \hat{\boldsymbol{\beta}}_{-i})^T\mathbf{X}^T\mathbf{X}(\hat{\boldsymbol{\beta}} - \hat{\boldsymbol{\beta}}_{-i})}{p\hat{\sigma}^2} = \left(\frac{\hat{e}_i}{1-p_{ii}}\right)^2\frac{\mathbf{x}_i^T(\mathbf{X}^T\mathbf{X})^{-1}\mathbf{X}^T\mathbf{X}(\mathbf{X}^T\mathbf{X})^{-1}\mathbf{x}_i}{p\hat{\sigma}^2} \\
&= \frac{\hat{e}_i^2 p_{ii}}{(1-p_{ii})^2 p\hat{\sigma}^2},
\end{aligned}$$

$$\text{COVRATIO}_i = \frac{|(\mathbf{X}_{-i}^T\mathbf{X}_{-i})^{-1}\hat{\sigma}_{-i}^2|}{|(\mathbf{X}^T\mathbf{X})^{-1}\hat{\sigma}^2|} = \left(\frac{\hat{\sigma}_{-i}^2}{\hat{\sigma}^2}\right)^p \frac{|\mathbf{X}^T\mathbf{X}|}{|\mathbf{X}_{-i}^T\mathbf{X}_{-i}|} = \left(\frac{\hat{\sigma}_{-i}^2}{\hat{\sigma}^2}\right)^p \frac{|\mathbf{X}^T\mathbf{X}|}{|\mathbf{X}^T\mathbf{X}|(1-p_{ii})}$$

$$= \frac{1}{1-p_{ii}}\left(\frac{\hat{\sigma}_{-i}^2}{\hat{\sigma}^2}\right)^p,$$

where we used Theorem 2.11.8 to obtain the penultimate expression for COVRATIO_i.

▶ **Exercise 12** Find $\text{E}(\hat{e}_{i,-i})$, $\text{var}(\hat{e}_{i,-i})$, and $\text{corr}(\hat{e}_{i,-i}, \hat{e}_{j,-j})$ (for $i \neq j$) under the Gauss–Markov model $\{\mathbf{y}, \mathbf{X}\boldsymbol{\beta}, \sigma^2\mathbf{I}\}$.

Solution Using Theorem 8.1.7b, d,

$$\text{E}(\hat{e}_{i,-i}) = \text{E}\left(\frac{\hat{e}_i}{1-p_{ii}}\right) = \frac{\text{E}(\hat{e}_i)}{1-p_{ii}} = 0,$$

$$\text{var}(\hat{e}_{i,-i}) = \text{var}\left(\frac{\hat{e}_i}{1-p_{ii}}\right) = \frac{\text{var}(\hat{e}_i)}{(1-p_{ii})^2} = \frac{\sigma^2}{1-p_{ii}},$$

and for $i \neq j$,

$$\text{corr}(\hat{e}_{i,-i}, \hat{e}_{j,-j}) = \text{corr}\left(\frac{\hat{e}_i}{1-p_{ii}}, \frac{\hat{e}_j}{1-p_{jj}}\right) = \frac{\text{cov}\left(\frac{\hat{e}_i}{1-p_{ii}}, \frac{\hat{e}_j}{1-p_{jj}}\right)}{\sqrt{\text{var}\left(\frac{\hat{e}_i}{1-p_{ii}}\right)\text{var}\left(\frac{\hat{e}_j}{1-p_{jj}}\right)}}$$

$$= \frac{\left(\frac{1}{1-p_{ii}}\right)\left(\frac{1}{1-p_{jj}}\right)(-\sigma^2 p_{ij})}{\sqrt{\left(\frac{\sigma^2}{1-p_{ii}}\right)\left(\frac{\sigma^2}{1-p_{jj}}\right)}}$$

$$= \frac{-p_{ij}}{\sqrt{(1-p_{ii})(1-p_{jj})}}.$$

Least Squares Estimation and ANOVA for Partitioned Models

9

This chapter presents exercises on least squares estimation and ANOVA for partitioned linear models and provides solutions to those exercises.

▶ **Exercise 1** Prove Theorem 9.1.1: For the orthogonal projection matrices $\mathbf{P}_{12}$ and $\mathbf{P}_1$ corresponding to the ordered two-part model $\{\mathbf{y}, \mathbf{X}_1\boldsymbol{\beta}_1 + \mathbf{X}_2\boldsymbol{\beta}_2\}$ and its submodel $\{\mathbf{y}, \mathbf{X}_1\boldsymbol{\beta}_1\}$, respectively, the following results hold:

(a) $\mathbf{P}_{12}\mathbf{P}_1 = \mathbf{P}_1$ and $\mathbf{P}_1\mathbf{P}_{12} = \mathbf{P}_1$;
(b) $\mathbf{P}_1(\mathbf{P}_{12} - \mathbf{P}_1) = \mathbf{0}$;
(c) $\mathbf{P}_1(\mathbf{I} - \mathbf{P}_{12}) = \mathbf{0}$;
(d) $(\mathbf{P}_{12} - \mathbf{P}_1)(\mathbf{I} - \mathbf{P}_{12}) = \mathbf{0}$;
(e) $(\mathbf{P}_{12} - \mathbf{P}_1)(\mathbf{P}_{12} - \mathbf{P}_1) = \mathbf{P}_{12} - \mathbf{P}_1$;
(f) $\mathbf{P}_{12} - \mathbf{P}_1$ is symmetric and $\text{rank}(\mathbf{P}_{12} - \mathbf{P}_1) = \text{rank}(\mathbf{X}_1, \mathbf{X}_2) - \text{rank}(\mathbf{X}_1)$.

Solution $\mathbf{P}_{12}\mathbf{P}_1 = \mathbf{P}_{12}\mathbf{X}_1(\mathbf{X}_1^T\mathbf{X}_1)^-\mathbf{X}_1^T = \mathbf{X}_1(\mathbf{X}_1^T\mathbf{X}_1)^-\mathbf{X}_1^T = \mathbf{P}_1$, upon which it follows that $\mathbf{P}_1 = \mathbf{P}_1^T = (\mathbf{P}_{12}\mathbf{P}_1)^T = \mathbf{P}_1^T\mathbf{P}_{12}^T = \mathbf{P}_1\mathbf{P}_{12}$. This proves part (a). Then by part (a), $\mathbf{P}_1(\mathbf{P}_{12} - \mathbf{P}_1) = \mathbf{P}_1\mathbf{P}_{12} - \mathbf{P}_1\mathbf{P}_1 = \mathbf{P}_1 - \mathbf{P}_1 = \mathbf{0}$, which proves part (b). Similarly, $\mathbf{P}_1(\mathbf{I} - \mathbf{P}_{12}) = \mathbf{P}_1 - \mathbf{P}_1\mathbf{P}_{12} = \mathbf{P}_1 - \mathbf{P}_1 = \mathbf{0}$, $(\mathbf{P}_{12} - \mathbf{P}_1)(\mathbf{I} - \mathbf{P}_{12}) = \mathbf{P}_{12} - \mathbf{P}_1 - \mathbf{P}_{12}\mathbf{P}_{12} + \mathbf{P}_1\mathbf{P}_{12} = \mathbf{P}_{12} - \mathbf{P}_1 - \mathbf{P}_{12} + \mathbf{P}_1 = \mathbf{0}$, and $(\mathbf{P}_{12} - \mathbf{P}_1)(\mathbf{P}_{12} - \mathbf{P}_1) = \mathbf{P}_{12}\mathbf{P}_{12} - \mathbf{P}_1\mathbf{P}_{12} - \mathbf{P}_{12}\mathbf{P}_1 + \mathbf{P}_1\mathbf{P}_1 = \mathbf{P}_{12} - \mathbf{P}_1 - \mathbf{P}_1 + \mathbf{P}_1 = \mathbf{P}_{12} - \mathbf{P}_1$, proving parts (c), (d), and (e). Theorem 2.7.1b implies that $\mathbf{P}_{12} - \mathbf{P}_1$ is symmetric, whereas the idempotency of $\mathbf{P}_{12}$, $\mathbf{P}_1$, and their difference together with Theorem 2.12.2 imply that $\text{rank}(\mathbf{P}_{12} - \mathbf{P}_1) = \text{tr}(\mathbf{P}_{12} - \mathbf{P}_1) = \text{tr}(\mathbf{P}_{12}) - \text{tr}(\mathbf{P}_1) = \text{rank}(\mathbf{X}_1, \mathbf{X}_2) - \text{rank}(\mathbf{X}_1)$. This establishes part (f).

D. L. Zimmerman, *Linear Model Theory*,
https://doi.org/10.1007/978-3-030-52074-8_9

▶ **Exercise 2** Prove Theorem 9.1.3: For the matrices $\mathbf{P}_{12\cdots j}$ ($j = 1, \ldots, k$) given by (9.4) and their successive differences, the following properties hold:

(a) $\mathbf{P}_{12\cdots j}$ is symmetric and idempotent, and $\text{rank}(\mathbf{P}_{12\cdots j}) = \text{rank}(\mathbf{X}_1, \ldots, \mathbf{X}_j)$ for $j = 1, \ldots, k$;
(b) $\mathbf{P}_{12\cdots j}\mathbf{X}_{j'} = \mathbf{X}_{j'}$ for $1 \leq j' \leq j$, $j = 1, \ldots, k$;
(c) $\mathbf{P}_{12\cdots j}\mathbf{P}_{12\cdots j'} = \mathbf{P}_{12\cdots j'}$ and $\mathbf{P}_{12\cdots j'}\mathbf{P}_{12\cdots j} = \mathbf{P}_{12\cdots j'}$ for $j' < j = 2, \ldots, k$;
(d) $\mathbf{P}_{1\ldots j}(\mathbf{I} - \mathbf{P}_{12\cdots k}) = \mathbf{0}$ for $j = 1, \ldots, k$ and $(\mathbf{P}_{12\cdots j} - \mathbf{P}_{12\cdots j-1})(\mathbf{I} - \mathbf{P}_{12\cdots k}) = \mathbf{0}$ for $j = 2, \ldots, k$;
(e) $\mathbf{P}_{1\ldots j-1}(\mathbf{P}_{12\cdots j} - \mathbf{P}_{12\cdots j-1}) = \mathbf{0}$ for $j = 2, \ldots, k$ and $(\mathbf{P}_{12\cdots j} - \mathbf{P}_{12\cdots j-1})(\mathbf{P}_{12\cdots j'} - \mathbf{P}_{12\cdots j'-1}) = \mathbf{0}$ for all $j \neq j'$;
(f) $(\mathbf{P}_{12\cdots j} - \mathbf{P}_{12\cdots j-1})(\mathbf{P}_{12\cdots j} - \mathbf{P}_{12\cdots j-1}) = \mathbf{P}_{12\cdots j} - \mathbf{P}_{12\cdots j-1}$ for $j = 2, \ldots, k$;
(g) $\mathbf{P}_{12\cdots j} - \mathbf{P}_{12\cdots j-1}$ is symmetric and $\text{rank}(\mathbf{P}_{12\cdots j} - \mathbf{P}_{12\cdots j-1}) = \text{rank}(\mathbf{X}_1, \ldots, \mathbf{X}_j) - \text{rank}(\mathbf{X}_1, \ldots, \mathbf{X}_{j-1})$ for $j = 2, \ldots, k$;
(h) $\mathbf{P}_{12\cdots j} - \mathbf{P}_{12\cdots j-1}$ is the orthogonal projection matrix onto $\mathcal{C}[(\mathbf{I} - \mathbf{P}_{12\cdots j-1})\mathbf{X}_{12\cdots j}]$, which is the orthogonal complement of $\mathcal{C}(\mathbf{X}_{12\cdots j-1})$ relative to $\mathcal{C}(\mathbf{X}_{12\cdots j})$ for $j = 2, \ldots, k$.

Solution Parts (a) and (b) are merely restatements of various parts of Theorem 8.1.2a as they apply to $(\mathbf{X}_1, \ldots, \mathbf{X}_j)$. Part (c) may be established using part (b) by observing that for $j' < j$,

$$\begin{aligned}
\mathbf{P}_{12\cdots j}\mathbf{P}_{12\cdots j'} &= \mathbf{P}_{12\cdots j}(\mathbf{X}_1, \ldots, \mathbf{X}_{j'})[(\mathbf{X}_1, \ldots, \mathbf{X}_{j'})^T(\mathbf{X}_1, \ldots, \mathbf{X}_{j'})]^-(\mathbf{X}_1, \ldots, \mathbf{X}_{j'})^T \\
&= (\mathbf{X}_1, \ldots, \mathbf{X}_{j'})[(\mathbf{X}_1, \ldots, \mathbf{X}_{j'})^T(\mathbf{X}_1, \ldots, \mathbf{X}_{j'})]^-(\mathbf{X}_1, \ldots, \mathbf{X}_{j'})^T \\
&= \mathbf{P}_{12\cdots j'},
\end{aligned}$$

upon which it follows that $\mathbf{P}_{12\cdots j'} = \mathbf{P}_{12\cdots j'}^T = (\mathbf{P}_{12\cdots j}\mathbf{P}_{12\cdots j'})^T = \mathbf{P}_{12\cdots j'}^T\mathbf{P}_{12\cdots j}^T = \mathbf{P}_{12\cdots j'}\mathbf{P}_{12\cdots j}$. Parts (d), (e), and (f) follow easily from part (c). We may obtain part (g) using part (a) and an argument very similar to that used to prove part (j) of Theorem 9.1.1. Finally, the proof of part (h) is exactly like that of Theorem 9.1.2, with $\mathbf{P}_{12\cdots j}$ and $\mathbf{P}_{12\cdots j-1}$, respectively, substituted for $\mathbf{P}_{12}$ and $\mathbf{P}_1$.

▶ **Exercise 3** For the ordered k-part Gauss–Markov model $\{\mathbf{y}, \sum_{l=1}^k \mathbf{X}_l\boldsymbol{\beta}_l, \sigma^2\mathbf{I}\}$, show that $\text{E}[\mathbf{y}^T(\mathbf{P}_{12\cdots j} - \mathbf{P}_{12\cdots j-1})\mathbf{y}] = (\sum_{l=j}^k \mathbf{X}_l\boldsymbol{\beta}_l)^T(\mathbf{P}_{12\cdots j} - \mathbf{P}_{12\cdots j-1})$ $\mathbf{X}(\sum_{l=j}^k \mathbf{X}_l\boldsymbol{\beta}_l) + \sigma^2[\text{rank}(\mathbf{X}_1, \ldots, \mathbf{X}_j) - \text{rank}(\mathbf{X}_1, \ldots, \mathbf{X}_{j-1})]$.

Solution By Theorem 4.2.4,

$$
\begin{aligned}
E[\mathbf{y}^T(\mathbf{P}_{12\cdots j} - \mathbf{P}_{12\cdots j-1})\mathbf{y}] &= \left(\sum_{l=1}^{k} \mathbf{X}_l \boldsymbol{\beta}_l\right)^T (\mathbf{P}_{12\cdots j} - \mathbf{P}_{12\cdots j-1}) \left(\sum_{l=1}^{k} \mathbf{X}_l \boldsymbol{\beta}_l\right) \\
&\quad + \mathrm{tr}[(\mathbf{P}_{12\cdots j} - \mathbf{P}_{12\cdots j-1})(\sigma^2 \mathbf{I})] \\
&= \left(\sum_{l=j}^{k} \mathbf{X}_l \boldsymbol{\beta}_l\right)^T (\mathbf{P}_{12\cdots j} - \mathbf{P}_{12\cdots j-1}) \left(\sum_{l=j}^{k} \mathbf{X}_l \boldsymbol{\beta}_l\right) \\
&\quad + \sigma^2[\mathrm{rank}(\mathbf{X}_1, \ldots, \mathbf{X}_j) - \mathrm{rank}(\mathbf{X}_1, \ldots, \mathbf{X}_{j-1})]
\end{aligned}
$$

where for the last equality we used Theorem 9.1.3b, e.

▶ **Exercise 4** Verify the expressions for the expected mean squares in the corrected sequential ANOVA for the two-way main effects model with one observation per cell given in Example 9.3.4-3.

Solution

$$
\bar{y}_{i\cdot} = (1/m) \sum_{j=1}^{m} y_{ij} \Rightarrow E(\bar{y}_{i\cdot}) = (1/m) \sum_{j=1}^{m} (\mu + \alpha_i + \gamma_j) = \mu + \alpha_i + \bar{\gamma}.
$$

Similarly,

$$
\bar{y}_{\cdot j} = (1/q) \sum_{i=1}^{q} y_{ij} \Rightarrow E(\bar{y}_{\cdot j}) = (1/q) \sum_{i=1}^{q} (\mu + \alpha_i + \gamma_j) = \mu + \bar{\alpha} + \gamma_j
$$

and

$$
\bar{y}_{\cdot\cdot} = (1/qm) \sum_{i=1}^{q} \sum_{j=1}^{m} y_{ij} \Rightarrow E(\bar{y}_{\cdot\cdot}) = (1/qm) \sum_{i=1}^{q} \sum_{j=1}^{m} (\mu + \alpha_i + \gamma_j) = \mu + \bar{\alpha} + \bar{\gamma}.
$$

Thus,

$$
m \sum_{i=1}^{q} [E(\bar{y}_{i\cdot}) - E(\bar{y}_{\cdot\cdot})]^2 = m \sum_{i=1}^{q} (\alpha_i - \bar{\alpha})^2
$$

and

$$q\sum_{j=1}^{m}[\mathrm{E}(\bar{y}_{\cdot j}) - \mathrm{E}(\bar{y}_{\cdot\cdot})]^2 = q\sum_{j=1}^{m}(\gamma_j - \bar{\gamma})^2,$$

so the expected mean squares for Factor A and Factor B are, respectively,

$$\sigma^2 + \frac{m}{q-1}\sum_{i=1}^{q}(\alpha_i - \bar{\alpha})^2$$

and

$$\sigma^2 + \frac{q}{m-1}\sum_{j=1}^{m}(\gamma_j - \bar{\gamma})^2.$$

▶ **Exercise 5** Obtain the corrected sequential ANOVA for a balanced (r observations per cell) two-way main effects model, with Factor A fitted first (but after the overall mean, of course). Give nonmatrix expressions for the sums of squares in this ANOVA table and for the corresponding expected mean squares (under a Gauss–Markov version of the model).

Solution By the result obtained in Example 9.3.3-2, the sum of squares for the Factor A effects in the corrected sequential ANOVA in which those effects are fitted first is $mr\sum_{i=1}^{q}(\bar{y}_{i\cdot\cdot} - \bar{y}_{\cdot\cdot\cdot})^2$. By the same result, the sum of squares for the Factor B effects in the corrected sequential ANOVA in which those effects are fitted first is $qr\sum_{j=1}^{m}(\bar{y}_{\cdot j\cdot} - \bar{y}_{\cdot\cdot\cdot})^2$. Because equal cell frequencies are a special case of proportional frequencies, by the result obtained in Example 9.3.4-2 the sums of squares corresponding to the Factor A and Factor B effects are invariant to the order in which the two factors are fitted. Thus, the corrected sequential ANOVA is

Source	Rank	Sum of squares	Expected mean square
Factor A	$q-1$	$mr\sum_{i=1}^{q}(\bar{y}_{i\cdot\cdot} - \bar{y}_{\cdot\cdot\cdot})^2$	$\sigma^2 + \frac{mr}{q-1}\sum_{i=1}^{q}(\alpha_i - \bar{\alpha})^2$
Factor B	$m-1$	$qr\sum_{j=1}^{m}(\bar{y}_{\cdot j\cdot} - \bar{y}_{\cdot\cdot\cdot})^2$	$\sigma^2 + \frac{qr}{m-1}\sum_{j=1}^{m}(\gamma_j - \bar{\gamma})^2$
Residual	$qmr-q-m+1$	By subtraction	σ^2
Corrected total	$qmr-1$	$\sum_{i=1}^{q}\sum_{j=1}^{m}\sum_{k=1}^{r}(y_{ijk} - \bar{y}_{\cdot\cdot\cdot})^2$	

► **Exercise 6** For the balanced ($r \geq 2$ observations per cell) two-way model with interaction:

(a) Obtain the corrected sequential ANOVA with Factor A fitted first (but after the overall mean, of course), then Factor B, and then the interaction effects. Give nonmatrix expressions for the sums of squares in this ANOVA table and for the corresponding expected mean squares (under a Gauss–Markov version of the model).

(b) Would the corrected sequential ANOVA corresponding to the ordered model in which the overall mean is fitted first, then Factor B, then Factor A, and then the interaction be the same (apart from order) as the sequential ANOVA of part (a)? Justify your answer.

Solution

(a) Clearly, the lines for Factor A, Factor B, and Corrected Total in this corrected sequential ANOVA coincide with those for the balanced two-way main effects model given in the solution to Exercise 9.5. The sum of squares for interaction effects may be expressed as $\mathbf{y}^T(\mathbf{P_X} - \mathbf{P}_{123})\mathbf{y}$, where

$$\mathbf{X} = (\mathbf{X}_1, \mathbf{X}_2, \mathbf{X}_3, \mathbf{X}_4) = (\mathbf{1}_{qmr}, \mathbf{I}_q \otimes \mathbf{1}_{mr}, \mathbf{1}_q \otimes \mathbf{I}_m \otimes \mathbf{1}_r, \mathbf{I}_{qm} \otimes \mathbf{1}_r)$$

and $\mathbf{P}_{123}$ is the orthogonal projection matrix onto the column space of $(\mathbf{X}_1, \mathbf{X}_2, \mathbf{X}_3)$. Using results from the solutions to Exercises 8.5b and 8.5c, we obtain

$$\begin{aligned}(\mathbf{P_X} - \mathbf{P}_{123})\mathbf{y} &= \mathbf{P_X}\mathbf{y} - \mathbf{P}_{123}\mathbf{y} \\ &= \begin{pmatrix} \bar{y}_{11\cdot} \\ \bar{y}_{12\cdot} \\ \vdots \\ \bar{y}_{qm\cdot} \end{pmatrix} \otimes \mathbf{1}_r - \begin{pmatrix} \bar{y}_{1\cdot\cdot} \\ \bar{y}_{2\cdot\cdot} \\ \vdots \\ \bar{y}_{q\cdot\cdot} \end{pmatrix} \otimes \mathbf{1}_{mr} - \mathbf{1}_q \otimes \begin{pmatrix} \bar{y}_{\cdot 1\cdot} \\ \bar{y}_{\cdot 2\cdot} \\ \vdots \\ \bar{y}_{\cdot m\cdot} \end{pmatrix} \otimes \mathbf{1}_r + \bar{y}_{\cdot\cdot\cdot}\mathbf{1}_{qmr}.\end{aligned}$$

Therefore,

$$\mathbf{y}^T(\mathbf{P_X}-\mathbf{P}_{123})\mathbf{y} = [(\mathbf{P_X}-\mathbf{P}_{123})\mathbf{y}]^T(\mathbf{P_X}-\mathbf{P}_{123})\mathbf{y} = r\sum_{i=1}^{q}\sum_{j=1}^{m}(\bar{y}_{ij\cdot}-\bar{y}_{i\cdot\cdot}-\bar{y}_{\cdot j\cdot}+\bar{y}_{\cdot\cdot\cdot})^2.$$

Also, the rank of $\mathbf{P_X} - \mathbf{P}_{123}$ is $qm - (q + m - 1) = qm - q - m + 1 = (q - 1)(m - 1)$. Therefore, the corrected sequential ANOVA table is as follows:

Source	Rank	Sum of squares
Factor A	$q - 1$	$mr \sum_{i=1}^{q} (\bar{y}_{i\cdot\cdot} - \bar{y}_{\cdots})^2$
Factor B	$m - 1$	$qr \sum_{j=1}^{m} (\bar{y}_{\cdot j\cdot} - \bar{y}_{\cdots})^2$
Interaction	$(q - 1)(m - 1)$	$r \sum_{i=1}^{q} \sum_{j=1}^{m} (\bar{y}_{ij\cdot} - \bar{y}_{i\cdot\cdot} - \bar{y}_{\cdot j\cdot} + \bar{y}_{\cdots})^2$
Residual	$qm(r - 1)$	By subtraction
Corrected total	$qmr - 1$	$\sum_{i=1}^{q} \sum_{j=1}^{m} \sum_{k=1}^{r} (y_{ijk} - \bar{y}_{\cdots})^2$

To obtain the expected mean squares, we use the same type of approach used in the solution to Exercise 9.4, obtaining

$$
\begin{aligned}
\mathrm{E}(\bar{y}_{ij\cdot}) &= \mu + \alpha_i + \gamma_j + \bar{\xi}_{ij}, \\
\mathrm{E}(\bar{y}_{i\cdot\cdot}) &= \mu + \alpha_i + \bar{\gamma} + \bar{\xi}_{i\cdot}, \\
\mathrm{E}(\bar{y}_{\cdot j\cdot}) &= \mu + \bar{\alpha} + \gamma_j + \bar{\xi}_{\cdot j}, \\
\mathrm{E}(\bar{y}_{\cdots}) &= \mu + \bar{\alpha} + \bar{\gamma} + \bar{\xi}_{\cdot\cdot}.
\end{aligned}
$$

Substituting these expressions for the corresponding sample means in the sums of squares column, simplifying, dividing the result by the corresponding degrees of freedom, and finally adding σ^2, we obtain the following table of expected mean squares:

Source	Expected mean square
Factor A	$\sigma^2 + \frac{mr}{q-1} \sum_{i=1}^{q} (\alpha_i - \bar{\alpha} + \bar{\xi}_{i\cdot} - \bar{\xi}_{\cdot\cdot})^2$
Factor B	$\sigma^2 + \frac{qr}{m-1} \sum_{j=1}^{m} (\gamma_j - \bar{\gamma} + \bar{\xi}_{\cdot j} - \bar{\xi}_{\cdot\cdot})^2$
Interaction	$\sigma^2 + \frac{r}{(q-1)(m-1)} \sum_{i=1}^{q} \sum_{j=1}^{m} (\xi_{ij} - \bar{\xi}_{i\cdot} - \bar{\xi}_{\cdot j} + \bar{\xi}_{\cdot\cdot})^2$
Residual	σ^2

(b) The only difference between this corrected sequential ANOVA and that given in part (a) is that the order of fitting Factor A and Factor B is interchanged. Thus, the answer to the question is yes because, as shown in Example 9.3.4-2, the two corrected sequential ANOVAs for a balanced two-way main effects model are identical apart from order.

▶ **Exercise 7** Obtain the corrected sequential ANOVA for a two-factor nested model in which the Factor A effects are fitted first (but after the overall mean, of course). Give nonmatrix expressions for the sums of squares in this ANOVA table and for the corresponding expected mean squares (under a Gauss–Markov version of the model).

Solution Write the model as the ordered three-part model

$$\mathbf{y} = \mathbf{X}_1\boldsymbol{\beta}_1 + \mathbf{X}_2\boldsymbol{\beta}_2 + \mathbf{X}_3\boldsymbol{\beta}_3 + \mathbf{e}$$

where $\mathbf{X}_1 = \mathbf{1}_n$, $\mathbf{X}_2 = \oplus_{i=1}^q \mathbf{1}_{n_{i\cdot}}$ where $n_{i\cdot} = \sum_{j=1}^{m_i} n_{ij}$, $\mathbf{X}_3 = \oplus_{i=1}^q \oplus_{j=1}^{m_i} \mathbf{1}_{n_{ij}}$, $\boldsymbol{\beta}_1 = \mu$, $\boldsymbol{\beta}_2 = (\alpha_1, \ldots, \alpha_q)^T$, and $\boldsymbol{\beta}_3 = (\gamma_{11}, \gamma_{12}, \ldots, \gamma_{qm_q})^T$. Clearly $\mathbf{P}_1 = (1/n)\mathbf{J}_n$ so the corrected total sum of squares is

$$\mathbf{y}^T[\mathbf{I} - (1/n)\mathbf{J}_n]\mathbf{y} = \sum_{i=1}^{q}\sum_{j=1}^{m_i}\sum_{k=1}^{n_{ij}}(y_{ijk} - \bar{y}_{\cdots})^2,$$

with rank $n - 1$. Furthermore, because $\mathcal{C}(\mathbf{X}_1) \subseteq \mathcal{C}(\mathbf{X}_2)$,

$$\begin{aligned}\mathbf{P}_{12} = \mathbf{X}_2(\mathbf{X}_2^T\mathbf{X}_2)^-\mathbf{X}_2^T &= (\oplus_{i=1}^q \mathbf{1}_{n_{i\cdot}})\left((\oplus_{i=1}^q \mathbf{1}_{n_{i\cdot}})^T(\oplus_{i=1}^q \mathbf{1}_{n_{i\cdot}})\right)^- (\oplus_{i=1}^q \mathbf{1}_{n_{i\cdot}})^T \\ &= \oplus_{i=1}^q (1/n_{i\cdot})\mathbf{J}_{n_{i\cdot}}.\end{aligned}$$

Thus,

$$\mathrm{SS}(\mathbf{X}_2|\mathbf{1}) = \mathbf{y}^T(\mathbf{P}_{12} - \mathbf{P}_1)\mathbf{y} = \mathbf{y}^T\left[\oplus_{i=1}^q (1/n_{i\cdot})\mathbf{J}_{n_{i\cdot}} - (1/n)\mathbf{J}_n\right]\mathbf{y} = \sum_{i=1}^{q} n_{i\cdot}(\bar{y}_{i\cdot\cdot} - \bar{y}_{\cdots})^2,$$

with rank $q - 1$. Also, from Exercise 8.4e, $\mathbf{P}_{123} = \mathbf{P}_\mathbf{X} = \oplus_{i=1}^q \oplus_{j=1}^{m_i} (1/n_{ij})\mathbf{J}_{n_{ij}}$. Thus,

$$\begin{aligned}\mathrm{SS}(\mathbf{X}_3|\mathbf{1}, \mathbf{X}_2) &= \mathbf{y}^T(\mathbf{P}_{123} - \mathbf{P}_{12})\mathbf{y} = \mathbf{y}^T\left[\oplus_{i=1}^q \oplus_{j=1}^{m_i} (1/n_{ij})\mathbf{J}_{n_{ij}} - \oplus_{i=1}^q (1/n_{i\cdot})\mathbf{J}_{n_{i\cdot}}\right]\mathbf{y} \\ &= \sum_{i=1}^{q}\sum_{j=1}^{m_i} n_{ij}(\bar{y}_{ij\cdot} - \bar{y}_{i\cdot\cdot})^2,\end{aligned}$$

with rank $\sum_{i=1}^q m_i - q$. Finally, the residual sum of squares is

$$\mathbf{y}^T(\mathbf{I} - \mathbf{P}_{123})\mathbf{y} = \mathbf{y}^T\left[\mathbf{I} - \oplus_{i=1}^q \oplus_{j=1}^{m_i} (1/n_{ij})\mathbf{J}_{n_{ij}}\right]\mathbf{y} = \sum_{i=1}^{q}\sum_{j=1}^{m_i}\sum_{k=1}^{n_{ij}}(y_{ijk} - \bar{y}_{ij\cdot})^2,$$

with rank $n - \sum_{i=1}^{q} m_i$. Thus, the corrected sequential ANOVA table is

Source	Rank	Sum of squares
Factor A	$q - 1$	$\sum_{i=1}^{q} n_{i\cdot}(\bar{y}_{i\cdot\cdot} - \bar{y}_{\cdot\cdot\cdot})^2$
Factor B	$\sum_{i=1}^{q} m_i - q$	$\sum_{i=1}^{q}\sum_{j=1}^{m_i} n_{ij}(\bar{y}_{ij\cdot} - \bar{y}_{i\cdot\cdot})^2$
Residual	$n - \sum_{i=1}^{q} m_i$	$\sum_{i=1}^{q}\sum_{j=1}^{m_i}\sum_{k=1}^{n_{ij}}(y_{ijk} - \bar{y}_{ij\cdot})^2$
Corrected total	$n - 1$	$\sum_{i=1}^{q}\sum_{j=1}^{m_i}\sum_{k=1}^{n_{ij}}(y_{ijk} - \bar{y}_{\cdot\cdot\cdot})^2$

▶ **Exercise 8** Suppose, in a k-part model, that $\mathcal{C}(\mathbf{X}_j) \neq \mathcal{C}(\mathbf{X}_{j'})$ for all $j \neq j'$. Prove that the sequential ANOVAs corresponding to all ordered k-part models are identical (apart from order of listing) if and only if $\mathbf{X}_j^T\mathbf{X}_{j'} = \mathbf{0}$ for all $j \neq j'$.

Solution Matrices $\mathbf{B}_1, \ldots, \mathbf{B}_k$ exist such that $\mathbf{X}_j^T\mathbf{X}_j\mathbf{B}_j = \mathbf{X}_j^T$ $(j = 1, \ldots, k)$, and $\mathbf{P}_j = \mathbf{X}_j\mathbf{B}_j$ is the orthogonal projection matrix onto $\mathcal{C}(\mathbf{X}_j)$. Now suppose that $\mathbf{X}_j^T\mathbf{X}_{j'} = \mathbf{0}$ for all $j \neq j'$. Then,

$$\begin{pmatrix} \mathbf{X}_1^T\mathbf{X}_1 & \mathbf{0} & \cdots & \mathbf{0} \\ \mathbf{0} & \mathbf{X}_2^T\mathbf{X}_2 & \cdots & \mathbf{0} \\ \vdots & \vdots & \ddots & \vdots \\ \mathbf{0} & \mathbf{0} & \cdots & \mathbf{X}_k^T\mathbf{X}_k \end{pmatrix} \begin{pmatrix} \mathbf{B}_1 \\ \mathbf{B}_2 \\ \vdots \\ \mathbf{B}_k \end{pmatrix} = \begin{pmatrix} \mathbf{X}_1^T \\ \mathbf{X}_2^T \\ \vdots \\ \mathbf{X}_k^T \end{pmatrix};$$

that is, $\mathbf{B} \equiv (\mathbf{B}_1^T, \mathbf{B}_2^T, \ldots, \mathbf{B}_k^T)^T$ satisfies the matrix equation $\mathbf{X}^T\mathbf{X}\mathbf{B} = \mathbf{X}^T$, where $\mathbf{X} = (\mathbf{X}_1, \ldots, \mathbf{X}_k)$. Hence $\mathbf{P}_{12\cdots k} = \mathbf{X}\mathbf{B} = \mathbf{X}_1\mathbf{B}_1 + \mathbf{X}_2\mathbf{B}_2 + \cdots + \mathbf{X}_k\mathbf{B}_k = \mathbf{P}_1 + \mathbf{P}_2 + \cdots + \mathbf{P}_k$. By applying the same argument to every ordered submodel of order less than k, we find that the orthogonal projection matrix onto the column space of any subset of $\{\mathbf{X}_1, \ldots, \mathbf{X}_k\}$ is equal to the sum of the orthogonal projection matrices onto the column spaces of each element of the subset. Thus, letting $\{\pi_1, \pi_2, \ldots, \pi_k\}$ represent any permutation of the first k positive integers, we have

$$\mathbf{y}^T(\mathbf{P}_{\pi_1\pi_2} - \mathbf{P}_{\pi_1})\mathbf{y} = \mathbf{y}^T\mathbf{P}_{\pi_2}\mathbf{y},$$
$$\mathbf{y}^T(\mathbf{P}_{\pi_1\pi_2\pi_3} - \mathbf{P}_{\pi_1\pi_2})\mathbf{y} = \mathbf{y}^T\mathbf{P}_{\pi_3}\mathbf{y},$$

and so on, for all $\mathbf{y}$. Then the sequential ANOVA corresponding to the ordered k-part model

$$\mathbf{y} = \mathbf{X}_{\pi_1}\boldsymbol{\beta}_{\pi_1} + \mathbf{X}_{\pi_2}\boldsymbol{\beta}_{\pi_2} + \cdots + \mathbf{X}_{\pi_k}\boldsymbol{\beta}_{\pi_k} + \mathbf{e}$$

is

Source	Rank	Sum of squares
$\mathbf{X}_{\pi_1}$	$\text{rank}(\mathbf{X}_{\pi_1})$	$\mathbf{y}^T\mathbf{P}_{\pi_1}\mathbf{y}$
$\mathbf{X}_{\pi_2}\vert\mathbf{X}_{\pi_1}$	$\text{rank}(\mathbf{X}_{\pi_1}, \mathbf{X}_{\pi_2}) - \text{rank}(\mathbf{X}_{\pi_1})$	$\mathbf{y}^T\mathbf{P}_{\pi_2}\mathbf{y}$
$\mathbf{X}_{\pi_3}\vert\mathbf{X}_{\pi_1}, \mathbf{X}_{\pi_2}$	$\text{rank}(\mathbf{X}_{\pi_1}, \mathbf{X}_{\pi_2}, \mathbf{X}_{\pi_3}) - \text{rank}(\mathbf{X}_{\pi_1}, \mathbf{X}_{\pi_2})$	$\mathbf{y}^T\mathbf{P}_{\pi_3}\mathbf{y}$
$\vdots$	$\vdots$	$\vdots$
$\mathbf{X}_{\pi_k}\vert\mathbf{X}_{\pi_1}, \ldots, \mathbf{X}_{\pi_{k-1}}$	$\text{rank}(\mathbf{X}) - \text{rank}(\mathbf{X}_{\pi_1}, \ldots, \mathbf{X}_{\pi_{k-1}})$	$\mathbf{y}^T\mathbf{P}_{\pi_k}\mathbf{y}$
Residual	$n - \text{rank}(\mathbf{X})$	$\mathbf{y}^T(\mathbf{I} - \mathbf{P_X})\mathbf{y}$
Total	n	$\mathbf{y}^T\mathbf{y}$

Because the permutation was arbitrary, this establishes that the sequential ANOVAs corresponding to all $k!$ ordered k-part models are identical, apart from order of listing, under the given condition.

Conversely, suppose that the sequential ANOVAs corresponding to all $k!$ ordered k-part models are identical, apart from order of listing. Let j and j' be distinct integers between 1 and k, inclusive. By considering only those permutations for which $\pi_1 = j$ and $\pi_2 = j'$, we see that either $\mathbf{P}_j = \mathbf{P}_{j'}$ or $\mathbf{P}_j = \mathbf{P}_{jj'} - \mathbf{P}_{j'}$. If $\mathbf{P}_j = \mathbf{P}_{j'}$, then $\mathcal{C}(\mathbf{X}_j) = \mathcal{C}(\mathbf{X}_{j'})$, which contradicts the assumption. If, instead, $\mathbf{P}_j = \mathbf{P}_{jj'} - \mathbf{P}_{j'}$, then $\mathbf{X}_j^T\mathbf{P}_j\mathbf{X}_{j'} = \mathbf{X}_j^T(\mathbf{P}_{jj'} - \mathbf{P}_{j'})\mathbf{X}_{j'} = \mathbf{X}_j^T(\mathbf{P}_{jj'}\mathbf{X}_{j'} - \mathbf{P}_{j'}\mathbf{X}_{j'}) = \mathbf{X}_j^T(\mathbf{X}_{j'} - \mathbf{X}_{j'}) = \mathbf{0}$, i.e., $\mathbf{X}_j^T\mathbf{X}_{j'} = \mathbf{0}$. Because j and j' were arbitrary, the result is established.

▶ **Exercise 9** Prove Theorem 9.3.2: Consider the two ordered three-part models

$$\mathbf{y} = \mathbf{X}_1\boldsymbol{\beta}_1 + \mathbf{X}_2\boldsymbol{\beta}_2 + \mathbf{X}_3\boldsymbol{\beta}_3 + \mathbf{e}$$

and

$$\mathbf{y} = \mathbf{X}_1\boldsymbol{\beta}_1 + \mathbf{X}_3\boldsymbol{\beta}_3 + \mathbf{X}_2\boldsymbol{\beta}_2 + \mathbf{e},$$

and suppose that $\mathcal{C}[(\mathbf{I} - \mathbf{P}_1)\mathbf{X}_2] \neq \mathcal{C}[(\mathbf{I} - \mathbf{P}_1)\mathbf{X}_3]$. A necessary and sufficient condition for the sequential ANOVAs of these models to be identical (apart from order of listing) is

$$\mathbf{X}_2^T(\mathbf{I} - \mathbf{P}_1)\mathbf{X}_3 = \mathbf{0}.$$

Solution

$$\begin{aligned}\text{SS}(\mathbf{X}_3|\mathbf{X}_1,\mathbf{X}_2) = \text{SS}(\mathbf{X}_3|\mathbf{X}_1) &\Rightarrow \mathbf{P}_{123} - \mathbf{P}_{12} = \mathbf{P}_{13} - \mathbf{P}_1\\
&\Rightarrow \mathbf{X}_2^T(\mathbf{P}_{123} - \mathbf{P}_{12})\mathbf{X}_3 = \mathbf{X}_2^T(\mathbf{P}_{13} - \mathbf{P}_1)\mathbf{X}_3\\
&\Rightarrow \mathbf{X}_2^T\mathbf{X}_3 - \mathbf{X}_2^T\mathbf{X}_3 = \mathbf{X}_2^T\mathbf{X}_3 - \mathbf{X}_2^T\mathbf{P}_1\mathbf{X}_3\\
&\Rightarrow \mathbf{X}_2^T(\mathbf{I} - \mathbf{P}_1)\mathbf{X}_3 = \mathbf{0}.\end{aligned}$$

For the converse, suppose that $\mathbf{X}_2^T(\mathbf{I}-\mathbf{P}_1)\mathbf{X}_3 = \mathbf{0}$. We aim to show that $\mathbf{P}_{123}-\mathbf{P}_{12} = \mathbf{P}_{13} - \mathbf{P}_1$, or equivalently that $\mathbf{P}_{123} - \mathbf{P}_1 = (\mathbf{P}_{12} - \mathbf{P}_1) + (\mathbf{P}_{13} - \mathbf{P}_1)$. Recall from Theorem 9.1.2 that $\mathbf{P}_{12}-\mathbf{P}_1$ is the orthogonal projection matrix onto $\mathcal{C}[(\mathbf{I}-\mathbf{P}_1)\mathbf{X}_2]$. Thus,

$$\mathbf{P}_{12} - \mathbf{P}_1 = (\mathbf{I} - \mathbf{P}_1)\mathbf{X}_2\dot{\mathbf{B}}_2,$$

where $\dot{\mathbf{B}}_2$ is any matrix satisfying $\mathbf{X}_2^T(\mathbf{I} - \mathbf{P}_1)\mathbf{X}_2\dot{\mathbf{B}}_2 = \mathbf{X}_2^T(\mathbf{I} - \mathbf{P}_1)$. Similarly,

$$\mathbf{P}_{13} - \mathbf{P}_1 = (\mathbf{I} - \mathbf{P}_1)\mathbf{X}_3\dot{\mathbf{B}}_3,$$

where $\dot{\mathbf{B}}_3$ is any matrix satisfying $\mathbf{X}_3^T(\mathbf{I} - \mathbf{P}_1)\mathbf{X}_3\dot{\mathbf{B}}_3 = \mathbf{X}_3^T(\mathbf{I} - \mathbf{P}_1)$. By similar reasoning,

$$\mathbf{P}_{123} - \mathbf{P}_1 = (\mathbf{I} - \mathbf{P}_1)(\mathbf{X}_2, \mathbf{X}_3)\begin{pmatrix}\ddot{\mathbf{B}}_2\\ \ddot{\mathbf{B}}_3\end{pmatrix}$$

where $\begin{pmatrix}\ddot{\mathbf{B}}_2\\ \ddot{\mathbf{B}}_3\end{pmatrix}$ is any matrix satisfying

$$(\mathbf{X}_2, \mathbf{X}_3)^T(\mathbf{I} - \mathbf{P}_1)(\mathbf{X}_2, \mathbf{X}_3)\begin{pmatrix}\ddot{\mathbf{B}}_2\\ \ddot{\mathbf{B}}_3\end{pmatrix} = (\mathbf{X}_2, \mathbf{X}_3)^T(\mathbf{I} - \mathbf{P}_1).$$

Expanding the last system of equations above yields

$$\begin{pmatrix}\mathbf{X}_2^T(\mathbf{I} - \mathbf{P}_1)\mathbf{X}_2 & \mathbf{X}_2^T(\mathbf{I} - \mathbf{P}_1)\mathbf{X}_3\\ \mathbf{X}_3^T(\mathbf{I} - \mathbf{P}_1)\mathbf{X}_2 & \mathbf{X}_3^T(\mathbf{I} - \mathbf{P}_1)\mathbf{X}_3\end{pmatrix}\begin{pmatrix}\ddot{\mathbf{B}}_2\\ \ddot{\mathbf{B}}_3\end{pmatrix} = \begin{pmatrix}\mathbf{X}_2^T(\mathbf{I} - \mathbf{P}_1)\\ \mathbf{X}_3^T(\mathbf{I} - \mathbf{P}_1)\end{pmatrix},$$

but by hypothesis the coefficient matrix is block diagonal, so these equations separate as follows:

$$\mathbf{X}_2^T(\mathbf{I}-\mathbf{P}_1)\mathbf{X}_2\ddot{\mathbf{B}}_2 = \mathbf{X}_2^T(\mathbf{I}-\mathbf{P}_1),$$
$$\mathbf{X}_3^T(\mathbf{I}-\mathbf{P}_1)\mathbf{X}_3\ddot{\mathbf{B}}_3 = \mathbf{X}_3^T(\mathbf{I}-\mathbf{P}_1).$$

But this is the same system of equations as the system defining $\dot{\mathbf{B}}_2$ and $\dot{\mathbf{B}}_3$; hence,

$$\begin{aligned}\mathbf{P}_{123}-\mathbf{P}_1 &= (\mathbf{I}-\mathbf{P}_1)(\mathbf{X}_2,\mathbf{X}_3)\begin{pmatrix}\dot{\mathbf{B}}_2\\ \dot{\mathbf{B}}_3\end{pmatrix}\\ &= (\mathbf{I}-\mathbf{P}_1)\mathbf{X}_2\dot{\mathbf{B}}_2 + (\mathbf{I}-\mathbf{P}_1)\mathbf{X}_3\dot{\mathbf{B}}_3\\ &= (\mathbf{P}_{12}-\mathbf{P}_1)+(\mathbf{P}_{13}-\mathbf{P}_1).\end{aligned}$$

▶ **Exercise 10** Consider the linear model specified in Exercise 7.3.

(a) Obtain the overall ANOVA (Source, Rank, and Sum of squares) for this model. Give nonmatrix expressions for the model and total sums of squares (the residual sum of squares may be obtained by subtraction).
(b) Obtain the sequential ANOVA (Source, Rank, and Sum of squares) for the ordered two-part model $\{\mathbf{y}, \mathbf{X}_1\boldsymbol{\beta}_1 + \mathbf{X}_2\boldsymbol{\beta}_2\}$, where $\mathbf{X}_1$ is the submatrix of $\mathbf{X}$ consisting of its first two columns and $\mathbf{X}_2$ is the submatrix of $\mathbf{X}$ consisting of its last two columns. (Again, give nonmatrix expressions for all sums of squares except the residual sum of squares.)
(c) Would the sequential ANOVA for the ordered two-part model $\{\mathbf{y}, \mathbf{X}_2\boldsymbol{\beta}_2+\mathbf{X}_1\boldsymbol{\beta}_1\}$ be identical to the ANOVA you obtained in part (b), apart from order of listing? Justify your answer.

Solution

(a) Here, using expressions for $\hat{\boldsymbol{\beta}}$, $\mathbf{X}$, and $\mathbf{y}$ from Exercise 7.3, Model SS $=$ $\hat{\boldsymbol{\beta}}^T\mathbf{X}^T\mathbf{y} = (y_1+y_2+y_3+y_4+y_5+y_6-y_7-y_8)^2/8 + (y_1+y_2+y_3+y_4-y_5-y_6+y_7+y_8)^2/8 + (y_1+y_2-y_3-y_4+y_5+y_6+y_7+y_8)^2/8 + (-y_1-y_2+y_3+y_4+y_5+y_6+y_7+y_8)^2/8$. So the ANOVA table is

Source	Rank	Sum of squares
Model	4	Model SS
Residual	4	By subtraction
Total	8	$\sum_{i=1}^{8} y_i^2$

(b)

Source	Rank	Sum of squares
$\mathbf{X}_1$	2	$(y_1 + y_2 + y_3 + y_4 + y_5 + y_6 - y_7 - y_8)^2/8$
		$+(y_1 + y_2 + y_3 + y_4 - y_5 - y_6 + y_7 + y_8)^2/8$
$\mathbf{X}_2\|\mathbf{X}_1$	2	$(y_1 + y_2 - y_3 - y_4 + y_5 + y_6 + y_7 + y_8)^2/8$
		$+(-y_1 - y_2 + y_3 + y_4 + y_5 + y_6 + y_7 + y_8)^2/8$
Residual	4	By subtraction
Total	8	$\sum_{i=1}^{8} y_i^2$

(c) Yes, by Theorem 9.3.1, because

$$\mathbf{X}_1^T\mathbf{X}_2 = \begin{pmatrix} 1 & 1 & 1 & 1 & 1 & 1 & -1 & -1 \\ 1 & 1 & 1 & 1 & -1 & -1 & 1 & 1 \end{pmatrix} \begin{pmatrix} 1 & -1 \\ 1 & -1 \\ -1 & 1 \\ -1 & 1 \\ 1 & 1 \\ 1 & 1 \\ 1 & 1 \\ 1 & 1 \end{pmatrix} = \begin{pmatrix} 0 & 0 \\ 0 & 0 \end{pmatrix}.$$

▶ **Exercise 11** Consider the two-way partially crossed model introduced in Example 5.1.4-1, with one observation per cell, i.e.,

$$y_{ij} = \mu + \alpha_i - \alpha_j + e_{ij} \quad (i \neq j = 1, \ldots, q).$$

(a) Would the sequential ANOVAs corresponding to the ordered two-part models in which the overall mean is fitted first and last be identical (apart from order of listing)? Justify your answer.
(b) Obtain the sequential ANOVA (Source, Rank, and Sum of squares) corresponding to the ordered two-part version of this model in which the overall mean is fitted first. Give nonmatrix expressions for the sums of squares and the corresponding expected mean squares (under a Gauss–Markov version of the model).

Solution

(a) Recall from Example 5.1.4-1 that $\mathbf{X} = (\mathbf{X}_1, \mathbf{X}_2)$, where $\mathbf{X}_1 = \mathbf{1}_{q(q-1)}$ and $\mathbf{X}_2 = (\mathbf{v}_{ij}^T)_{i \neq j=1,\ldots,q}$, where $\mathbf{v}_{ij} = \mathbf{u}_i^{(q)} - \mathbf{u}_j^{(q)}$. Now,

$$\begin{aligned}\mathbf{X}_1^T\mathbf{X}_2 &= \mathbf{1}_{q(q-1)}^T(\mathbf{u}_i^{(q)} - \mathbf{u}_j^{(q)})_{i\neq j=1,\ldots,q}^T \\ &= ((1)(q-1)+(-1)(q-1), \quad (1)(q-1)+(-1)(q-1), \ldots, \\ &\quad (1)(q-1)+(-1)(q-1))_{1\times q} \\ &= \mathbf{0}_q^T.\end{aligned}$$

Thus, by Theorem 9.3.1, the two ANOVAs would be identical (apart from order of listing).

(b) By part (a) and the argument used to prove the sufficiency of Theorem 9.3.1, $\mathbf{P_X} = \mathbf{P}_1 + \mathbf{P}_2$, where $\mathbf{P}_1 = [1/q(q-1)]\mathbf{J}_{q(q-1)}$ and $\mathbf{P}_2$ is the orthogonal projection matrix onto $\mathcal{C}[(\mathbf{v}_{ij}^T)_{i\neq j=1,\ldots,q}]$. Recall from Exercise 8.4d that $\mathbf{P_X} = (p_{iji'j'})$, where the elements of the $q(q-1) \times q(q-1)$ matrix $\mathbf{P_X}$ are ordered by row index ij and column index $i'j'$ and

$$p_{iji'j'} = \begin{cases} \frac{1}{q(q-1)} + \frac{1}{q} & \text{if } i = i' \text{ and } j = j', \\ \frac{1}{q(q-1)} - \frac{1}{q} & \text{if } i = j' \text{ and } j = i', \\ \frac{1}{q(q-1)} + \frac{1}{2q} & \text{if } i = i' \text{ and } j \neq j', \text{ or } i \neq i' \text{ and } j = j', \\ \frac{1}{q(q-1)} - \frac{1}{2q} & \text{if } i = j' \text{ and } j \neq i', \text{ or } j = i' \text{ and } i \neq j', \\ \frac{1}{q(q-1)} & \text{if } i \neq i', i \neq j', j \neq i', \text{ and } j \neq j'. \end{cases}$$

Thus, $\mathbf{P}_2 = (p_{2,iji'j'})$, where

$$p_{2,iji'j'} = \begin{cases} 1/q & \text{if } i = i' \text{ and } j = j', \\ -1/q & \text{if } i = j' \text{ and } j = i', \\ 1/2q & \text{if } i = i' \text{ and } j \neq j', \text{ or } i \neq i' \text{ and } j = j', \\ -1/2q & \text{if } i = j' \text{ and } j \neq i', \text{ or } j = i' \text{ and } i \neq j', \\ 0 & \text{if } i \neq i', i \neq j', j \neq i', \text{ and } j \neq j'. \end{cases}$$

Thus,

$$\begin{aligned}\mathbf{y}^T\mathbf{P}_2\mathbf{y} = (1/2q)\Bigg[2\Big(\sum_{i\neq j} y_{ij}^2 - \sum_{i\neq j} y_{ij}y_{ji}\Big) + \sum_{i,j,j':i\neq j,\, i\neq j'} y_{ij}y_{ij'} \\ + \sum_{i,j,i':i\neq j,\, i'\neq j} y_{ij}y_{i'j} \\ - \Bigg(\sum_{i,j,i':i\neq j,\, i'\neq i} y_{ij}y_{i'i} + \sum_{i,j,j':i\neq j,\, j\neq j'} y_{ij}y_{jj'}\Bigg)\Bigg].\end{aligned}$$

Furthermore, $\text{rank}(\mathbf{P}_2) = \text{tr}(\mathbf{P}_2) = q(q-1)(1/q) = q-1$. Thus, the sequential ANOVA table is as follows:

Source	Rank	Sum of squares
$\mathbf{1}$	1	$q(q-1)\bar{y}_{\cdot\cdot}^2$
$(\mathbf{v}_{ij})_{i \neq j}$	$q-1$	$\mathbf{y}^T\mathbf{P}_2\mathbf{y}$
Residual	$q(q-2)$	By subtraction
Total	$q(q-1)$	$\sum_{i=1}^n y_i^2$

▶ **Exercise 12** Consider the two corrected sequential ANOVAs for the analysis-of-covariance model having a single factor of classification and a single quantitative variable, which was described in Sect. 9.4.

(a) Obtain nonmatrix expressions for the expected mean squares (under a Gauss–Markov version of the model) in both ANOVAs.
(b) Obtain a necessary and sufficient condition for the two ANOVAs to be identical (apart from order of listing).

Solution

(a)

$$\bar{y}_{i\cdot} = (1/n_i)\sum_{j=1}^{n_i} y_{ij} \Rightarrow \text{E}(\bar{y}_{i\cdot}) = (1/n_i)\sum_{j=1}^{n_i}(\mu + \alpha_i + \gamma z_{ij}) = \mu + \alpha_i + \gamma \bar{z}_{i\cdot}$$

and

$$\begin{aligned}\bar{y}_{\cdot\cdot} = (1/n)\sum_{i=1}^{q}\sum_{j=1}^{n_i} y_{ij} &\Rightarrow \text{E}(\bar{y}_{\cdot\cdot}) = (1/n)\sum_{i=1}^{q}\sum_{j=1}^{n_i}(\mu + \alpha_i + \gamma z_{ij}) \\ &= \mu + \sum_{i=1}^{q}(n_i/n)\alpha_i + \gamma \bar{z}_{\cdot\cdot},\end{aligned}$$

from which we obtain

$$\begin{aligned}\text{E}(\bar{y}_{i\cdot}) - \text{E}(\bar{y}_{\cdot\cdot}) &= \alpha_i - \sum_{k=1}^{q}(n_k/n)\alpha_k + \gamma(\bar{z}_{i\cdot} - \bar{z}_{\cdot\cdot}), \\ \text{E}(y_{ij}) - \text{E}(\bar{y}_{i\cdot}) &= \gamma(z_{ij} - \bar{z}_{i\cdot}),\end{aligned}$$

$$\mathrm{E}(y_{ij}) - \mathrm{E}(\bar{y}_{\cdot\cdot}) = \mu + \alpha_i + \gamma z_{ij} - [\mu + \sum_{k=1}^{q}(n_k/n)\alpha_k + \gamma \bar{z}_{\cdot\cdot}]$$

$$= \alpha_i - \sum_{k=1}^{q}(n_k/n)\alpha_k + \gamma(z_{ij} - \bar{z}_{\cdot\cdot}).$$

Thus in the first corrected ANOVA,

$$\mathrm{EMS(Classes)} = \sigma^2 + \frac{1}{q-1}\sum_{i=1}^{q} n_i[\alpha_i - \sum_{k=1}^{q}(n_k/n)\alpha_k + \gamma(\bar{z}_{i\cdot} - \bar{z}_{\cdot\cdot})]^2$$

and

$$\mathrm{EMS}(\mathbf{z}|\mathrm{Classes}) = \sigma^2 + \gamma^2 \sum_{i=1}^{q}\sum_{j=1}^{n_i}(z_{ij} - \bar{z}_{i\cdot})^2.$$

In the second corrected ANOVA,

$$\mathrm{EMS}(\mathbf{z}) = \sigma^2 + \frac{\{\sum_{i=1}^{q}\sum_{j=1}^{n_i}[\alpha_i - \sum_{k=1}^{q}(n_k/n)\alpha_k](z_{ij} - \bar{z}_{\cdot\cdot}) + \gamma\sum_{i=1}^{q}\sum_{j=1}^{n_i}(z_{ij} - \bar{z}_{\cdot\cdot})^2\}^2}{\sum_{i=1}^{q}\sum_{j=1}^{n_i}(z_{ij} - \bar{z}_{\cdot\cdot})^2}$$

and

$$\mathrm{EMS(Classes}|\mathbf{z}) = \sigma^2 + \frac{1}{q-1}\left\{\sum_{i=1}^{q} n_i[\alpha_i - \sum_{k=1}^{q}(n_k/n)\alpha_k + \gamma(\bar{z}_{i\cdot} - \bar{z}_{\cdot\cdot})]^2 + \gamma^2\left[\sum_{i=1}^{q}\sum_{j=1}^{n_i}(z_{ij} - \bar{z}_{i\cdot})^2\right] - \frac{\{\sum_{i=1}^{q}\sum_{j=1}^{n_i}[\alpha_i - \sum_{k=1}^{q}(n_k/n)\alpha_k](z_{ij} - \bar{z}_{\cdot\cdot}) + \gamma\sum_{i=1}^{q}\sum_{j=1}^{n_i}(z_{ij} - \bar{z}_{\cdot\cdot})^2\}^2}{\sum_{i=1}^{q}\sum_{j=1}^{n_i}(z_{ij} - \bar{z}_{\cdot\cdot})^2}\right\}.$$

(b) Observe that

$$\mathbf{X}_2^T[\mathbf{I} - (1/n)\mathbf{J}]\mathbf{X}_3 = \left(\oplus_{i=1}^{q}\mathbf{1}_{n_i}\right)^T[\mathbf{I} - (1/n)\mathbf{J}]\mathbf{z} = \begin{pmatrix} \sum_{j=1}^{n_1}(z_{1j} - \bar{z}_{\cdot\cdot}) \\ \vdots \\ \sum_{j=1}^{n_q}(z_{qj} - \bar{z}_{\cdot\cdot}) \end{pmatrix}.$$

This vector, which is the left-hand side of (9.13) as it specializes to this model, will equal $\mathbf{0}$ if and only if $\bar{z}_{i\cdot} = \bar{z}_{\cdot\cdot}$ $(i = 1, \ldots, q)$. So (9.13), the necessary and sufficient condition for the two ANOVAs to be identical (by Theorem 9.3.2), is satisfied if and only if $\bar{z}_{i\cdot} = \bar{z}_{\cdot\cdot}$ $(i = 1, \ldots, q)$.

▶ **Exercise 13** Consider the analysis-of-covariance model having a single factor of classification with q levels and a single quantitative variable, i.e.,

$$y_{ij} = \mu + \alpha_i + \gamma z_{ij} + e_{ij} \qquad (i = 1, \ldots, q;\ j = 1, \ldots, n_i)$$

where the e_{ij}'s satisfy Gauss–Markov assumptions. Assume that $n_i \geq 2$ and $z_{i1} \neq z_{i2}$ for all $i = 1, \ldots, q$. In Sect. 9.4 it was shown that one solution to the normal equations for this model is $\hat{\mu} = 0$, $\hat{\alpha}_i = \bar{y}_{i\cdot} - \hat{\gamma}\bar{z}_{i\cdot}$ $(i = 1, \ldots, q)$, and

$$\hat{\gamma} = \frac{\sum_{i=1}^{q}\sum_{j=1}^{n_i}(z_{ij} - \bar{z}_{i\cdot})y_{ij}}{\sum_{i=1}^{q}\sum_{j=1}^{n_i}(z_{ij} - \bar{z}_{i\cdot})^2}.$$

Thus, the least squares estimator of $\mu + \alpha_i$ is $\bar{y}_{i\cdot} - \hat{\gamma}\bar{z}_{i\cdot}$ $(i = 1, \ldots, q)$. Obtain nonmatrix expressions for:

(a) $\text{var}(\hat{\gamma})$
(b) $\text{var}(\bar{y}_{i\cdot} - \hat{\gamma}\bar{z}_{i\cdot})$ $\quad (i = 1, \ldots, q)$
(c) $\text{cov}(\bar{y}_{i\cdot} - \hat{\gamma}\bar{z}_{i\cdot}, \bar{y}_{i'\cdot} - \hat{\gamma}\bar{z}_{i'\cdot})$ $(i' > i = 1, \ldots, q)$

Solution

(a) $\text{var}(\hat{\gamma}) = \frac{\sum_{i=1}^{q}\sum_{j=1}^{n_i}(z_{ij}-\bar{z}_{i\cdot})^2\sigma^2}{[\sum_{i=1}^{q}\sum_{j=1}^{n_i}(z_{ij}-\bar{z}_{i\cdot})^2]^2} = \frac{\sigma^2}{\sum_{i=1}^{q}\sum_{j=1}^{n_i}(z_{ij}-\bar{z}_{i\cdot})^2} = \frac{\sigma^2}{SZZ}$, say.

(b) Without loss of generality, consider the case $i = 1$. We obtain

$$\begin{aligned}\text{cov}(\bar{y}_{1\cdot}, \hat{\gamma}\bar{z}_{1\cdot}) &= \text{cov}((1/n_1)\mathbf{1}_{n_1}^T, \mathbf{0}_{n-n_1}^T)\mathbf{y}, (1/SZZ)\\ &\quad \times(\mathbf{z}_1^T - \bar{z}_{1\cdot}\mathbf{1}_{n_1}^T, \ldots, \mathbf{z}_q^T - \bar{z}_{q\cdot}\mathbf{1}_{n_q}^T)\mathbf{y}]\bar{z}_{1\cdot}\\ &= \sigma^2[(1/n_1)(1/SZZ)\mathbf{1}_{n_1}^T(\mathbf{z}_1 - \bar{z}_{1\cdot}\mathbf{1}_{n_1})]\bar{z}_{1\cdot} = 0.\end{aligned}$$

Hence $\text{var}(\bar{y}_{1\cdot} - \hat{\gamma}\bar{z}_{1\cdot}) = \text{var}(\bar{y}_{1\cdot}) + \bar{z}_{1\cdot}^2\text{var}(\hat{\gamma}) = (\sigma^2/n_1) + \bar{z}_{1\cdot}^2\sigma^2/SZZ$.

(c) By Theorem 4.2.2a and part (b) of this exercise, $\text{cov}(\bar{y}_{i\cdot} - \hat{\gamma}\bar{z}_{i\cdot}, \bar{y}_{i'\cdot} - \hat{\gamma}\bar{z}_{i'\cdot}) = \text{cov}(\bar{y}_{i\cdot}, \bar{y}_{i'\cdot}) - \bar{z}_{i\cdot}\text{cov}(\hat{\gamma}, \bar{y}_{i'\cdot}) - \bar{z}_{i'\cdot}\text{cov}(\bar{y}_{i\cdot}, \hat{\gamma}) + \bar{z}_{i\cdot}\bar{z}_{i'\cdot}\text{var}(\hat{\gamma}) = 0 - 0 - 0 + \bar{z}_{i\cdot}\bar{z}_{i'\cdot}\sigma^2/SZZ = (\bar{z}_{i\cdot}\bar{z}_{i'\cdot}/SZZ)\sigma^2$.

▶ **Exercise 14** Consider the one-factor, factor-specific-slope analysis-of-covariance model

$$y_{ij} = \mu + \alpha_i + \gamma_i(z_{ij} - \bar{z}_{i\cdot}) + e_{ij} \quad (i = 1, \ldots, q;\ j = 1, \ldots, n_i)$$

and assume that $z_{i1} \neq z_{i2}$ for all i.

(a) Obtain the least squares estimators of $\mu + \alpha_i$ and γ_i $(i = 1, \ldots, q)$.
(b) Obtain the corrected sequential ANOVA corresponding to the ordered version of this model in which $\mathbf{Z}\boldsymbol{\gamma}$ appears first (but after the overall mean, of course), where $\mathbf{Z} = \oplus_{i=1}^q(\mathbf{z}_i - \bar{z}_{i\cdot}\mathbf{1}_{n_i})$ and $\mathbf{z}_i = (z_{i1}, z_{i2}, \ldots, z_{in_i})^T$. Assume that $\mathbf{z}_i$ has at least two distinct elements for each i. Give nonmatrix expressions for the sums of squares in the ANOVA table.

Solution

(a) This model can be written in two-part form as follows:

$$\mathbf{y} = \mathbf{X}\boldsymbol{\beta} + \mathbf{Z}\boldsymbol{\gamma} + \mathbf{e},$$

where $\mathbf{X} = (\mathbf{1}_n, \oplus_{i=1}^q \mathbf{1}_{n_i})$, $\mathbf{Z} = \oplus_{i=1}^q(\mathbf{z}_i - \bar{z}_{i\cdot}\mathbf{1}_{n_i})$, $\boldsymbol{\beta} = (\mu, \alpha_1, \ldots, \alpha_q)^T$, and $\boldsymbol{\gamma} = (\gamma_1, \ldots, \gamma_q)^T$. Because

$$\begin{aligned}
\mathbf{X}^T\mathbf{Z} &= \begin{pmatrix} \mathbf{1}_n^T \\ \oplus_{i=1}^q \mathbf{1}_{n_i}^T \end{pmatrix} \oplus_{i=1}^q (\mathbf{z}_i - \bar{z}_{i\cdot}\mathbf{1}_{n_i}) \\
&= \begin{pmatrix} z_{1\cdot} - n_1\bar{z}_{1\cdot} & z_{2\cdot} - n_2\bar{z}_{2\cdot} & \cdots & z_{q\cdot} - n_q\bar{z}_{q\cdot} \\ z_{1\cdot} - n_1\bar{z}_{1\cdot} & 0 & \cdots & 0 \\ 0 & z_{2\cdot} - n_2\bar{z}_{2\cdot} & \cdots & 0 \\ \vdots & \vdots & \ddots & \vdots \\ 0 & 0 & \cdots & z_{q\cdot} - n_q\bar{z}_{q\cdot} \end{pmatrix} \\
&= \mathbf{0},
\end{aligned}$$

the corresponding two-part normal equations are

$$\begin{pmatrix} n & \mathbf{n}^T \\ \mathbf{n} & \oplus_{i=1}^q n_i \end{pmatrix} \boldsymbol{\beta} + \mathbf{0}\boldsymbol{\gamma} = \begin{pmatrix} y_{\cdot\cdot} \\ y_{1\cdot} \\ \vdots \\ y_{q\cdot} \end{pmatrix},$$

$$\mathbf{0}\boldsymbol{\beta} + \left(\oplus_{i=1}^q \left(\textstyle\sum_{j=1}^{n_i}(z_{ij} - \bar{z}_{i\cdot})^2\right)\right)\boldsymbol{\gamma} = \begin{pmatrix} \sum_{j=1}^{n_1}(z_{1j} - \bar{z}_{1\cdot})y_{1j} \\ \vdots \\ \sum_{j=1}^{n_q}(z_{qj} - \bar{z}_{q\cdot})y_{qj} \end{pmatrix},$$

where $\mathbf{n} = (n_1, \ldots, n_q)^T$. A solution for $\boldsymbol{\gamma}$ can be easily obtained from the "bottom" subset of equations:

$$\hat{\boldsymbol{\gamma}} = \begin{pmatrix} \hat{\gamma}_1 \\ \vdots \\ \hat{\gamma}_q \end{pmatrix} = \begin{pmatrix} \frac{\sum_{j=1}^{n_1}(z_{1j}-\bar{z}_{1\cdot})y_{1j}}{\sum_{j=1}^{n_1}(z_{1j}-\bar{z}_{1\cdot})^2} \\ \vdots \\ \frac{\sum_{j=1}^{n_q}(z_{qj}-\bar{z}_{q\cdot})y_{qj}}{\sum_{j=1}^{n_1}(z_{qj}-\bar{z}_{q\cdot})^2} \end{pmatrix}$$

Because one generalized inverse of the coefficient matrix of $\boldsymbol{\beta}$ in the "top" subset of equations is

$$\begin{pmatrix} 0 & \mathbf{0}^T \\ \mathbf{0} & \oplus_{i=1}^q (1/n_i) \end{pmatrix},$$

one solution for $\boldsymbol{\beta}$ is

$$\hat{\boldsymbol{\beta}} = \begin{pmatrix} \hat{\mu} \\ \hat{\alpha}_1 \\ \vdots \\ \hat{\alpha}_q \end{pmatrix} = \begin{pmatrix} 0 & \mathbf{0}^T \\ \mathbf{0} & \oplus_{i=1}^q (1/n_i) \end{pmatrix} \begin{pmatrix} y_{\cdot\cdot} \\ y_{1\cdot} \\ \vdots \\ y_{q\cdot} \end{pmatrix} = \begin{pmatrix} 0 \\ \bar{y}_{1\cdot} \\ \vdots \\ \bar{y}_{q\cdot} \end{pmatrix}.$$

Hence, the least squares estimator of $(\mu + \alpha_1, \ldots, \mu + \alpha_q)^T$ is

$$\begin{pmatrix} \widehat{\mu + \alpha_1} \\ \vdots \\ \widehat{\mu + \alpha_q} \end{pmatrix} = \begin{pmatrix} \bar{y}_{1\cdot} \\ \vdots \\ \bar{y}_{q\cdot} \end{pmatrix}.$$

(b) The corrected total sum of squares is

$$\mathbf{y}^T(\mathbf{I} - \frac{1}{n}\mathbf{J}_n)\mathbf{y} = \sum_{i=1}^{q}\sum_{j=1}^{n_i}(y_{ij} - \bar{y}_{\cdot\cdot})^2$$

with rank $n - 1$, and

$$\text{SS}(\mathbf{1}) = \mathbf{y}^T(\frac{1}{n}\mathbf{J}_n)\mathbf{y} = n\bar{y}_{\cdot\cdot}^2.$$

Now consider the orthogonal projection of $\mathbf{y}$ onto $\mathcal{C}(\mathbf{1}, \mathbf{Z})$. The two-part normal equations corresponding to this submodel are

$$\begin{pmatrix} n & \mathbf{0}^T \\ \mathbf{0} & \oplus_{i=1}^{q}\left(\sum_{j=1}^{n_i}(z_{ij}-\bar{z}_{i\cdot})^2\right) \end{pmatrix} \begin{pmatrix} \mu \\ \boldsymbol{\gamma} \end{pmatrix} = \begin{pmatrix} \mathbf{1}_n^T\mathbf{y} \\ \mathbf{Z}^T\mathbf{y} \end{pmatrix}.$$

Because the coefficient matrix is diagonal and nonsingular, the unique solution is

$$\hat{\mu} = \bar{y}_{\cdot\cdot}, \quad \hat{\boldsymbol{\gamma}} = \begin{pmatrix} \hat{\gamma}_1 \\ \vdots \\ \hat{\gamma}_q \end{pmatrix} = \begin{pmatrix} \frac{\sum_{j=1}^{n_1}(z_{1j}-\bar{z}_{1\cdot})y_{1j}}{\sum_{j=1}^{n_1}(z_{1j}-\bar{z}_{1\cdot})^2} \\ \vdots \\ \frac{\sum_{j=1}^{n_q}(z_{qj}-\bar{z}_{q\cdot})y_{qj}}{\sum_{j=1}^{n_1}(z_{qj}-\bar{z}_{q\cdot})^2} \end{pmatrix}.$$

Therefore,

$$\begin{aligned} \mathrm{SS}(\mathbf{1}, \mathbf{Z}) &= \left(\hat{\mu},\, \hat{\boldsymbol{\gamma}}^T\right)\begin{pmatrix} \mathbf{1}_n^T\mathbf{y} \\ \mathbf{Z}^T\mathbf{y} \end{pmatrix} \\ &= \hat{\mu}\mathbf{1}_n^T\mathbf{y} + \hat{\boldsymbol{\gamma}}^T\mathbf{Z}^T\mathbf{y} \\ &= n\bar{y}_{\cdot\cdot}^2 + \sum_{i=1}^{q}\frac{\left(\sum_{j=1}^{n_i} z_{ij}-\bar{z}_{i\cdot})y_{ij}\right)^2}{\sum_{j=1}^{n_i}(z_{ij}-\bar{z}_{i\cdot})^2} \end{aligned}$$

and thus,

$$\begin{aligned} \mathrm{SS}(\mathbf{Z} \mid \mathbf{1}) &= \mathrm{SS}(\mathbf{1}, \mathbf{Z}) - \mathrm{SS}(\mathbf{1}) \\ &= \sum_{i=1}^{q}\frac{\left(\sum_{j=1}^{n_i}(z_{ij}-\bar{z}_{i\cdot})y_{ij}\right)^2}{\sum_{j=1}^{n_i}(z_{ij}-\bar{z}_{i\cdot})^2} \end{aligned}$$

with rank $(q+1)-1=q$.

Denote $\mathbf{X}^* = \oplus_{i=1}^{q}\mathbf{1}_{n_i}$.

Now,

$$\mathrm{SS}(\mathbf{X}^*, \mathbf{Z}, \mathbf{1}) = \left(\hat{\boldsymbol\beta}^T, \hat{\boldsymbol\gamma}^T\right)\begin{pmatrix}\mathbf{X}^T \\ \mathbf{Z}^T\end{pmatrix}\mathbf{y} = \hat{\boldsymbol\beta}^T\mathbf{X}^T\mathbf{y} + \hat{\boldsymbol\gamma}^T\mathbf{Z}^T\mathbf{y}$$

$$= \begin{pmatrix} 0 \\ \bar{y}_{1\cdot} \\ \vdots \\ \bar{y}_{q\cdot}\end{pmatrix}^T \begin{pmatrix} y_{\cdot\cdot} \\ y_{1\cdot} \\ \vdots \\ y_{q\cdot}\end{pmatrix} + \begin{pmatrix} \frac{\sum_{j=1}^{n_1}(z_{1j}-\bar{z}_{1\cdot})y_{1j}}{\sum_{j=1}^{n_1}(z_{1j}-\bar{z}_{1\cdot})^2} \\ \vdots \\ \frac{\sum_{j=1}^{n_q}(z_{qj}-\bar{z}_{q\cdot})y_{qj}}{\sum_{j=1}^{n_1}(z_{qj}-\bar{z}_{q\cdot})^2}\end{pmatrix}^T \begin{pmatrix} \sum_{j=1}^{n_1}(z_{1j}-\bar{z}_{1\cdot})y_{1j} \\ \vdots \\ \sum_{j=1}^{n_q}(z_{qj}-\bar{z}_{q\cdot})y_{qj}\end{pmatrix}$$

$$= \sum_{i=1}^q n_i\bar{y}_{i\cdot}^2 + \sum_{i=1}^q \frac{\left(\sum_{j=1}^{n_i}(z_{ij}-\bar{z}_{i\cdot})y_{ij}\right)^2}{\sum_{j=1}^{n_i}(z_{ij}-\bar{z}_{i\cdot})^2}.$$

Therefore,

$$\begin{aligned}
\mathrm{SS}(\mathbf{X}^* \mid \mathbf{Z}, \mathbf{1}) &= \mathrm{SS}(\mathbf{X}^*, \mathbf{Z}, \mathbf{1}) - \mathrm{SS}(\mathbf{1}, \mathbf{Z}) \\
&= \left(\sum_{i=1}^q n_i\bar{y}_{i\cdot}^2 + \sum_{i=1}^q \frac{\left(\sum_{j=1}^{n_i}(z_{ij}-\bar{z}_{i\cdot})y_{ij}\right)^2}{\sum_{j=1}^{n_i}(z_{ij}-\bar{z}_{i\cdot})^2}\right) \\
&\quad - \left(n\bar{y}_{\cdot\cdot}^2 + \sum_{i=1}^q \frac{\left(\sum_{j=1}^{n_i}(z_{ij}-\bar{z}_{i\cdot})y_{ij}\right)^2}{\sum_{j=1}^{n_i}(z_{ij}-\bar{z}_{i\cdot})^2}\right) \\
&= \sum_{i=1}^q n_i\bar{y}_{i\cdot}^2 - n\bar{y}_{\cdot\cdot}^2
\end{aligned}$$

with rank $2q - q - 1 = q - 1$.

Finally, the corrected sequential ANOVA is as follows:

Source	Rank	Sum of squares
$\mathbf{Z}$	q	$\mathrm{SS}(\mathbf{Z} \mid \mathbf{1})$
Classes $\mid \mathbf{Z}$	$q-1$	$\mathrm{SS}(\mathbf{X}^* \mid \mathbf{Z}, \mathbf{1})$
Residual	$n-2q$	By subtraction
Corrected Total	$n-1$	$\sum_{i=1}^q\sum_{j=1}^{n_i}(y_{ij}-\bar{y}_{\cdot\cdot})^2$

▶ **Exercise 15** Consider the balanced two-way main effects analysis-of-covariance model with one observation per cell, i.e.,

$$y_{ij} = \mu + \alpha_i + \gamma_j + \xi z_{ij} + e_{ij} \quad (i = 1, \ldots, q;\ j = 1, \ldots, m).$$

(a) Obtain expressions for the least squares estimators of $\alpha_i - \alpha_{i'}$, $\gamma_j - \gamma_{j'}$, and ξ.
(b) Obtain the corrected sequential ANOVAs corresponding to the two ordered four-part models in which $\mathbf{1}\mu$ appears first and $\mathbf{z}\xi$ appears second.

Solution

(a) To obtain a solution to the normal equations, let us first formulate the model as a special case of the two-part model

$$\mathbf{y} = \mathbf{X}_1\boldsymbol{\beta}_1 + \mathbf{X}_2\boldsymbol{\beta}_2 + \mathbf{e}$$

where $\mathbf{X}_1 = (\mathbf{1}_{qm}, \mathbf{I}_q \otimes \mathbf{1}_m, \mathbf{1}_q \otimes \mathbf{I}_m)$, $\boldsymbol{\beta}_1 = (\mu, \alpha_1, \ldots, \alpha_q, \gamma_1, \ldots, \gamma_m)^T$, $\mathbf{X}_2 = \mathbf{z} = (z_{11}, z_{12}, \ldots, z_{qm})^T$, and $\boldsymbol{\beta}_2 = \xi$. The two-part normal equations corresponding to this model are easily shown to be

$$\begin{pmatrix} mq & m\mathbf{1}_q^T & q\mathbf{1}_m^T \\ m\mathbf{1}_q & m\mathbf{I}_q & \mathbf{J}_{q\times m} \\ q\mathbf{1}_m & \mathbf{J}_{m\times q} & q\mathbf{I}_m \end{pmatrix} \boldsymbol{\beta}_1 + \begin{pmatrix} z_{\cdot\cdot} \\ z_{1\cdot} \\ \vdots \\ z_{q\cdot} \\ z_{\cdot 1} \\ \vdots \\ z_{\cdot m} \end{pmatrix} \xi = \begin{pmatrix} y_{\cdot\cdot} \\ y_{1\cdot} \\ \vdots \\ y_{q\cdot} \\ y_{\cdot 1} \\ \vdots \\ y_{\cdot m} \end{pmatrix},$$

$$(z_{\cdot\cdot}, z_{1\cdot}, \ldots, z_{q\cdot}, z_{\cdot 1}, \ldots, z_{\cdot m})\boldsymbol{\beta}_1 + \sum_{i=1}^{q}\sum_{j=1}^{m} z_{ij}^2 \xi = \sum_{i=1}^{q}\sum_{j=1}^{m} z_{ij} y_{ij}.$$

Next observe that

$$\begin{aligned}
\mathbf{P}_1 &= \mathbf{X}_1(\mathbf{X}_1^T\mathbf{X}_1)^-\mathbf{X}_1^T \\
&= (\mathbf{1}_{qm}, \mathbf{I}_q \otimes \mathbf{1}_m, \mathbf{1}_q \otimes \mathbf{I}_m) \begin{pmatrix} mq & m\mathbf{1}_q^T & q\mathbf{1}_m^T \\ m\mathbf{1}_q & m\mathbf{I}_q & \mathbf{J}_{q\times m} \\ q\mathbf{1}_m & \mathbf{J}_{m\times q} & q\mathbf{I}_m \end{pmatrix}^- \\
&\quad \times (\mathbf{1}_{qm}, \mathbf{I}_q \otimes \mathbf{1}_m, \mathbf{1}_q \otimes \mathbf{I}_m)^T \\
&= (\mathbf{1}_{qm}, \mathbf{I}_q \otimes \mathbf{1}_m, \mathbf{1}_q \otimes \mathbf{I}_m) \begin{pmatrix} 0 & \mathbf{0}_q^T & \mathbf{0}_m^T \\ \mathbf{0}_q & (1/m)\mathbf{I}_q & \mathbf{0}_{q\times m} \\ \mathbf{0}_m & \mathbf{0}_{m\times q} & (1/q)\mathbf{I}_m - (1/qm)\mathbf{J}_m \end{pmatrix} \\
&\quad \times \begin{pmatrix} \mathbf{1}_{qm}^T \\ \mathbf{I}_q \otimes \mathbf{1}_m^T \\ \mathbf{1}_q^T \otimes \mathbf{I}_m \end{pmatrix}
\end{aligned}$$

$$= (\mathbf{0}_{qm}, \mathbf{I}_q \otimes (1/m)\mathbf{1}_m, \mathbf{1}_q \otimes [(1/q)\mathbf{I}_m - (1/qm)\mathbf{J}_m] \begin{pmatrix} \mathbf{1}_{qm}^T \\ \mathbf{I}_q \otimes \mathbf{1}_m^T \\ \mathbf{1}_q^T \otimes \mathbf{I}_m \end{pmatrix}$$

$$= \mathbf{I}_q \otimes (1/m)\mathbf{J}_m + \mathbf{J}_q \otimes [(1/q)\mathbf{I}_m - (1/qm)\mathbf{J}_m].$$

Thus, the reduced normal equation for ξ is

$$\begin{aligned} &\mathbf{z}^T\{\mathbf{I}_{qm} - [\mathbf{I}_q \otimes (1/m)\mathbf{J}_m] - \mathbf{J}_q \otimes [(1/q)\mathbf{I}_m - (1/qm)\mathbf{J}_m)]\}\mathbf{z}\xi \\ &\quad = \mathbf{z}^T\{\mathbf{I}_{qm} - [\mathbf{I}_q \otimes (1/m)\mathbf{J}_m] - \mathbf{J}_q \otimes [(1/q)\mathbf{I}_m - (1/qm)\mathbf{J}_m)]\}\mathbf{y}. \end{aligned}$$

The unique solution to this equation (which is also the least squares estimator of ξ) is

$$\begin{aligned} \hat{\xi} &= [\mathbf{z}^T\{\mathbf{I}_{qm} - [\mathbf{I}_q \otimes (1/m)\mathbf{J}_m] - \mathbf{J}_q \otimes [(1/q)\mathbf{I}_m - (1/qm)\mathbf{J}_m)]\}\mathbf{z}]^{-1} \\ &\quad \times \mathbf{z}^T\{\mathbf{I}_{qm} - [\mathbf{I}_q \otimes (1/m)\mathbf{J}_m] - \mathbf{J}_q \otimes [(1/q)\mathbf{I}_m - (1/qm)\mathbf{J}_m)]\}\mathbf{y} \\ &= \frac{\sum_{i=1}^q \sum_{j=1}^m (z_{ij} - \bar{z}_{i\cdot} - \bar{z}_{\cdot j} + \bar{z}_{\cdot\cdot})(y_{ij} - \bar{y}_{i\cdot} - \bar{y}_{\cdot j} + \bar{y}_{\cdot\cdot})}{\sum_{i=1}^q \sum_{j=1}^m (z_{ij} - \bar{z}_{i\cdot} - \bar{z}_{\cdot j} + \bar{z}_{\cdot\cdot})^2}. \end{aligned}$$

Back-solving for $\boldsymbol{\beta}_1$, we obtain

$$\hat{\boldsymbol{\beta}}_1 = \begin{pmatrix} \hat{\mu} \\ \hat{\alpha}_1 \\ \vdots \\ \hat{\alpha}_q \\ \hat{\gamma}_1 \\ \vdots \\ \hat{\gamma}_m \end{pmatrix} = \begin{pmatrix} mq & m\mathbf{1}_q^T & q\mathbf{1}_m^T \\ m\mathbf{1}_q & m\mathbf{I}_q & \mathbf{J}_{q\times m} \\ q\mathbf{1}_m & \mathbf{J}_{m\times q} & q\mathbf{I}_m \end{pmatrix}^- \left[\begin{pmatrix} y_{\cdot\cdot} \\ y_{1\cdot} \\ \vdots \\ y_{q\cdot} \\ y_{\cdot 1} \\ \vdots \\ y_{\cdot m} \end{pmatrix} - \begin{pmatrix} z_{\cdot\cdot} \\ z_{1\cdot} \\ \vdots \\ z_{q\cdot} \\ z_{\cdot 1} \\ \vdots \\ z_{\cdot m} \end{pmatrix} \hat{\xi} \right]$$

$$= \begin{pmatrix} 0 & \mathbf{0}_q^T & \mathbf{0}_m^T \\ \mathbf{0}_q & (1/m)\mathbf{I}_q & \mathbf{0}_{q\times m} \\ \mathbf{0}_m & \mathbf{0}_{m\times q} & (1/q)\mathbf{I}_m - (1/qm)\mathbf{J}_m \end{pmatrix} \begin{pmatrix} y_{\cdot\cdot} - \hat{\xi} z_{\cdot\cdot} \\ y_{1\cdot} - \hat{\xi} z_{1\cdot} \\ \vdots \\ y_{q\cdot} - \hat{\xi} z_{q\cdot} \\ y_{\cdot 1} - \hat{\xi} z_{\cdot 1} \\ \vdots \\ y_{\cdot m} - \hat{\xi} z_{\cdot m} \end{pmatrix}$$

$$= \begin{pmatrix} 0 \\ \bar{y}_{1\cdot} - \hat{\xi}\bar{z}_{1\cdot} \\ \vdots \\ \bar{y}_{q\cdot} - \hat{\xi}\bar{z}_{q\cdot} \\ \bar{y}_{\cdot 1} - \hat{\xi}\bar{z}_{\cdot 1} - (\bar{y}_{\cdot\cdot} - \hat{\xi}\bar{z}_{\cdot\cdot}) \\ \vdots \\ \bar{y}_{\cdot m} - \hat{\xi}\bar{z}_{\cdot m} - (\bar{y}_{\cdot\cdot} - \hat{\xi}\bar{z}_{\cdot\cdot}) \end{pmatrix}.$$

So the least squares estimator of $\alpha_i - \alpha_{i'}$ is $\bar{y}_{i\cdot} - \bar{y}_{i'\cdot} - \hat{\xi}(\bar{z}_{i\cdot} - \bar{z}_{i'\cdot})$ $(i \neq i' = 1, \ldots, q)$, and the least squares estimator of $\gamma_j - \gamma_{j'}$ is $\bar{y}_{\cdot j} - \bar{y}_{\cdot j'} - \hat{\xi}(\bar{z}_{\cdot j} - \bar{z}_{\cdot j'})$ $(j \neq j' = 1, \ldots, m)$.

(b) First consider the ordered four-part model in which $\mathbf{1}\mu$ appears first, $\mathbf{z}\xi$ appears second, $\mathbf{X}_3\boldsymbol{\alpha}$ appears third, and $\mathbf{X}_4\boldsymbol{\gamma}$ appears last, where $\mathbf{X}_3 = \mathbf{I}_q \otimes \mathbf{1}_m$, $\boldsymbol{\alpha} = (\alpha_1, \ldots, \alpha_q)^T$, $\mathbf{X}_4 = \mathbf{1}_q \otimes \mathbf{I}_m$, and $\boldsymbol{\gamma} = (\gamma_1, \ldots, \gamma_m)^T$. (Here the Factor A effects are fit prior to the Factor B effects.) For this model, we may exploit results for the second corrected sequential ANOVA of the one-way analysis of covariance model presented in Sect. 9.4 to obtain

$$\text{SS}(\mathbf{z}|\mathbf{1}) = \frac{\left[\sum_{i=1}^q \sum_{j=1}^m (y_{ij} - \bar{y}_{\cdot\cdot})(z_{ij} - \bar{z}_{\cdot\cdot})\right]^2}{\sum_{i=1}^q \sum_{j=1}^m (z_{ij} - \bar{z}_{\cdot\cdot})^2}$$

and

$$\text{SS(A Classes}|\mathbf{z}, \mathbf{1}) = \sum_{i=1}^q m(\bar{y}_{i\cdot} - \bar{y}_{\cdot\cdot})^2 + \hat{\xi}\sum_{i=1}^q \sum_{j=1}^m (z_{ij} - \bar{z}_{i\cdot})(y_{ij} - \bar{y}_{i\cdot})$$
$$- \frac{\left[\sum_{i=1}^q \sum_{j=1}^m (y_{ij} - \bar{y}_{\cdot\cdot})(z_{ij} - \bar{z}_{\cdot\cdot})\right]^2}{\sum_{i=1}^q \sum_{j=1}^m (z_{ij} - \bar{z}_{\cdot\cdot})^2}.$$

To obtain SS(B Classes|A Classes, $\mathbf{z}, \mathbf{1}$), we may obtain the Model sum of squares for the overall ANOVA and subtract SS(A Classes|$\mathbf{z}, \mathbf{1}$) from it. The former is, using results from part (a),

$$\hat{\boldsymbol{\beta}}^T\mathbf{X}^T\mathbf{y} = \hat{\boldsymbol{\beta}}_1\mathbf{X}_1^T\mathbf{y} + \hat{\xi}\mathbf{z}^T\mathbf{y}$$

$$= \begin{pmatrix} 0 \\ \bar{y}_{1\cdot} - \hat{\xi}\bar{z}_{1\cdot} \\ \vdots \\ \bar{y}_{q\cdot} - \hat{\xi}\bar{z}_{q\cdot} \\ \bar{y}_{\cdot 1} - \hat{\xi}\bar{z}_{\cdot 1} - (\bar{y}_{\cdot\cdot} - \hat{\xi}\bar{z}_{\cdot\cdot}) \\ \vdots \\ \bar{y}_{\cdot m} - \hat{\xi}\bar{z}_{\cdot m} - (\bar{y}_{\cdot\cdot} - \hat{\xi}\bar{z}_{\cdot\cdot}) \end{pmatrix}^T \begin{pmatrix} y_{\cdot\cdot} \\ y_{1\cdot} \\ \vdots \\ y_{q\cdot} \\ y_{\cdot 1} \\ \vdots \\ y_{\cdot m} \end{pmatrix}$$

$$= \sum_{i=1}^{q}(\bar{y}_{i\cdot} - \hat{\xi}\bar{z}_{i\cdot})y_{i\cdot} + \sum_{j=1}^{m}[\bar{y}_{\cdot j} - \hat{\xi}\bar{z}_{\cdot j} - (\bar{y}_{\cdot\cdot} - \hat{\xi}\bar{z}_{\cdot\cdot})]y_{\cdot j}$$

$$+\hat{\xi}\sum_{i=1}^{q}\sum_{j=1}^{m}(z_{ij} - \bar{z}_{i\cdot} - \bar{z}_{\cdot j} + \bar{z}_{\cdot\cdot})(y_{ij} - \bar{y}_{i\cdot} - \bar{y}_{\cdot j} + \bar{y}_{\cdot\cdot}).$$

Thus, the corrected sequential ANOVA for this model is

Source	Rank	Sum of squares
$\mathbf{z}$	1	$\frac{\left[\sum_{i=1}^{q}\sum_{j=1}^{m}(y_{ij}-\bar{y}_{\cdot\cdot})(z_{ij}-\bar{z}_{\cdot\cdot})\right]^2}{\sum_{i=1}^{q}\sum_{j=1}^{m}(z_{ij}-\bar{z}_{\cdot\cdot})^2}$
A classes$\vert\mathbf{z}$	$q-1$	$\sum_{i=1}^{q} m(\bar{y}_{i\cdot} - \bar{y}_{\cdot\cdot})^2 + \hat{\xi}\sum_{i=1}^{q}\sum_{j=1}^{m}(z_{ij} - \bar{z}_{i\cdot})(y_{ij} - \bar{y}_{i\cdot})$ $-\frac{\left[\sum_{i=1}^{q}\sum_{j=1}^{m}(y_{ij}-\bar{y}_{\cdot\cdot})(z_{ij}-\bar{z}_{\cdot\cdot})\right]^2}{\sum_{i=1}^{q}\sum_{j=1}^{m}(z_{ij}-\bar{z}_{\cdot\cdot})^2}$
B classes$\vert$A classes,$\mathbf{z}$	$m-1$	$\sum_{i=1}^{q}(\bar{y}_{i\cdot} - \hat{\xi}\bar{z}_{i\cdot})y_{i\cdot} + \sum_{j=1}^{m}[\bar{y}_{\cdot j} - \hat{\xi}\bar{z}_{\cdot j} - (\bar{y}_{\cdot\cdot} - \hat{\xi}\bar{z}_{\cdot\cdot})]y_{\cdot j}$ $+\hat{\xi}\sum_{i=1}^{q}\sum_{j=1}^{m}(z_{ij} - \bar{z}_{i\cdot} - \bar{z}_{\cdot j} + \bar{z}_{\cdot\cdot})(y_{ij} - \bar{y}_{i\cdot} - \bar{y}_{\cdot j} + \bar{y}_{\cdot\cdot})$ $-\sum_{i=1}^{q} m(\bar{y}_{i\cdot} - \bar{y}_{\cdot\cdot})^2 - \hat{\xi}\sum_{i=1}^{q}\sum_{j=1}^{m}(z_{ij} - \bar{z}_{i\cdot})(y_{ij} - \bar{y}_{i\cdot})$ $+\frac{\left[\sum_{i=1}^{q}\sum_{j=1}^{m}(y_{ij}-\bar{y}_{\cdot\cdot})(z_{ij}-\bar{z}_{\cdot\cdot})\right]^2}{\sum_{i=1}^{q}\sum_{j=1}^{m}(z_{ij}-\bar{z}_{\cdot\cdot})^2}$
Residual	$qm-q-m-1$	By subtraction
Corrected total	$qm-1$	$\sum_{i=1}^{q}\sum_{j=1}^{m}(y_{ij} - \bar{y}_{\cdot\cdot})^2$

By the "symmetry" of the situation with Factor A and Factor B effects, the corrected sequential ANOVA for the model in which $\mathbf{1}\mu$ appears first, $\mathbf{z}\xi$ appears second, $\mathbf{X}_4\boldsymbol{\gamma}$ appears third, and $\mathbf{X}_3\boldsymbol{\alpha}$ appears last is as follows:

Source	Rank	Sum of squares
$\mathbf{z}$	1	$\frac{\left[\sum_{i=1}^q \sum_{j=1}^m (y_{ij}-\bar{y}_{\cdot\cdot})(z_{ij}-\bar{z}_{\cdot\cdot})\right]^2}{\sum_{i=1}^q \sum_{j=1}^m (z_{ij}-\bar{z}_{\cdot\cdot})^2}$
B classes|$\mathbf{z}$	$m-1$	$\sum_{y=1}^m q(\bar{y}_{\cdot j}-\bar{y}_{\cdot\cdot})^2 + \hat{\xi}\sum_{i=1}^q \sum_{j=1}^m (z_{ij}-\bar{z}_{\cdot j})(y_{ij}-\bar{y}_{\cdot j})$
		$-\frac{\left[\sum_{i=1}^q \sum_{j=1}^m (y_{ij}-\bar{y}_{\cdot\cdot})(z_{ij}-\bar{z}_{\cdot\cdot})\right]^2}{\sum_{i=1}^q \sum_{j=1}^m (z_{ij}-\bar{z}_{\cdot\cdot})^2}$
A classes|B classes,$\mathbf{z}$	$q-1$	$\sum_{i=1}^q (\bar{y}_{i\cdot}-\hat{\xi}\bar{z}_{i\cdot})y_{i\cdot} + \sum_{j=1}^m [\bar{y}_{\cdot j}-\hat{\xi}\bar{z}_{\cdot j}-(\bar{y}_{\cdot\cdot}-\hat{\xi}\bar{z}_{\cdot\cdot})]y_{\cdot j}$
		$+\hat{\xi}\sum_{i=1}^q \sum_{j=1}^m (z_{ij}-\bar{z}_{i\cdot}-\bar{z}_{\cdot j}+\bar{z}_{\cdot\cdot})(y_{ij}-\bar{y}_{i\cdot}-\bar{y}_{\cdot j}+\bar{y}_{\cdot\cdot})$
		$-\sum_{j=1}^m q(\bar{y}_{\cdot j}-\bar{y}_{\cdot\cdot})^2 - \hat{\xi}\sum_{i=1}^q \sum_{j=1}^m (z_{ij}-\bar{z}_{\cdot j})(y_{ij}-\bar{y}_{\cdot j})$
		$+\frac{\left[\sum_{i=1}^q \sum_{j=1}^m (y_{ij}-\bar{y}_{\cdot\cdot})(z_{ij}-\bar{z}_{\cdot\cdot})\right]^2}{\sum_{i=1}^q \sum_{j=1}^m (z_{ij}-\bar{z}_{\cdot\cdot})^2}$
Residual	$qm-q-m-1$	By subtraction
Corrected total	$qm-1$	$\sum_{i=1}^q \sum_{j=1}^m (y_{ij}-\bar{y}_{\cdot\cdot})^2$

▶ **Exercise 16** Consider an analysis-of-covariance extension of the model in Exercise 9.6, written in cell-means form, as

$$y_{ijk} = \mu_{ij} + \gamma z_{ijk} + e_{ijk} \quad (i = 1, \ldots, q;\ j = 1, \ldots, m;\ k = 1, \ldots, r).$$

(a) Give the reduced normal equation for γ (in either matrix or nonmatrix form), and assuming that the inverse of its coefficient "matrix" exists, obtain the unique solution to the reduced normal equation in nonmatrix form.
(b) Back-solve to obtain a solution for the μ_{ij}'s.
(c) Using the overall ANOVA from Exercise 9.15 as a starting point, give the sequential ANOVA (sums of squares and ranks of the corresponding matrices), uncorrected for the mean, for the ordered two-part model $\{\mathbf{y}, \mathbf{X}_1\boldsymbol{\beta}_1 + \mathbf{z}\gamma\}$, where $\mathbf{X}_1$ is the model matrix without the covariate and $\mathbf{z}$ is the vector of covariates.

Solution

(a)

$$\mathbf{z}^T[\mathbf{I}_{qmr} - (\mathbf{I}_{qm} \otimes \frac{1}{r}\mathbf{J}_r)]\mathbf{z}\gamma = \mathbf{z}^T[\mathbf{I}_{qmr} - (\mathbf{I}_{qm} \otimes \frac{1}{r}\mathbf{J}_r)]\mathbf{y},$$

so

$$\hat{\gamma} = \frac{\sum_{i=1}^{q}\sum_{j=1}^{m}\sum_{k=1}^{r} z_{ijk}(y_{ijk} - \bar{y}_{ij\cdot})}{\sum_{i=1}^{q}\sum_{j=1}^{m}\sum_{k=1}^{r}(z_{ijk} - \bar{z}_{ij\cdot})^2}.$$

(b) Observe that $\mathbf{X}^T\mathbf{X} = r\mathbf{I}_{qm}$, so the "top" subset of equations in the two-part normal equations is

$$r\begin{pmatrix}\mu_{11}\\ \mu_{12}\\ \vdots\\ \mu_{qm}\end{pmatrix} + \begin{pmatrix}z_{11\cdot}\\ z_{12\cdot}\\ \vdots\\ z_{qm\cdot}\end{pmatrix}\gamma = \begin{pmatrix}y_{11\cdot}\\ y_{12\cdot}\\ \vdots\\ y_{qm\cdot}\end{pmatrix}.$$

Back-solving yields $\hat{\mu}_{ij} = \bar{y}_{ij\cdot} - \hat{\gamma}\bar{z}_{ij\cdot}$ $(i = 1, \ldots, q;\ j = 1, \ldots, m)$.

(c)

Source	Rank	Sum of squares
$\mathbf{X}_1$	qm	$r\sum_{i=1}^{q}\sum_{j=1}^{m}\bar{y}_{ij\cdot}^2$
$\mathbf{z}\vert\mathbf{X}_1$	1	$\hat{\gamma}\sum_{i=1}^{q}\sum_{j=1}^{m}\sum_{k=1}^{r} z_{ijk}(y_{ijk} - \bar{y}_{ij\cdot})$
Residual	$qm(r-1)-1$	By subtraction
Total	qmr	$\sum_{i=1}^{q}\sum_{j=1}^{m}\sum_{k=1}^{r} y_{ijk}^2$

▶ **Exercise 17** Prove Theorem 9.5.1: For the $n \times n$ matrix $\bar{\mathbf{J}}$ defined in Sect 9.5, the following results hold:

(a) $\bar{\mathbf{J}}$ is symmetric and idempotent, and $\text{rank}(\bar{\mathbf{J}}) = m$;
(b) $\mathbf{I} - \bar{\mathbf{J}}$ is symmetric and idempotent, and $\text{rank}(\mathbf{I} - \bar{\mathbf{J}}) = n - m$;
(c) $\bar{\mathbf{J}}\mathbf{X} = \mathbf{X}$;
(d) $\bar{\mathbf{J}}\mathbf{P}_{\mathbf{X}} = \mathbf{P}_{\mathbf{X}}$;
(e) $(\mathbf{I} - \bar{\mathbf{J}})(\bar{\mathbf{J}} - \mathbf{P}_{\mathbf{X}}) = \mathbf{0}$;
(f) $\bar{\mathbf{J}} - \mathbf{P}_{\mathbf{X}}$ is symmetric and idempotent, and $\text{rank}(\bar{\mathbf{J}} - \mathbf{P}_{\mathbf{X}}) = m - p^*$.

Solution The symmetry of $\bar{\mathbf{J}} = \oplus_{i=1}^{m}(1/n_i)\mathbf{J}_{n_i}$ follows immediately from that of each $\mathbf{J}_{n_i}$, and the idempotency holds because

$$\begin{aligned}\bar{\mathbf{J}}\bar{\mathbf{J}} &= \left(\oplus_{i=1}^{m}(1/n_i)\mathbf{J}_{n_i}\right)\left(\oplus_{i=1}^{m}(1/n_i)\mathbf{J}_{n_i}\right) = \oplus_{i=1}^{m}(1/n_i)\mathbf{J}_{n_i}(1/n_i)\mathbf{J}_{n_i}\\ &= \oplus_{i=1}^{m}(1/n_i)\mathbf{J}_{n_i} = \bar{\mathbf{J}}.\end{aligned}$$

Thus, by Theorem 2.12.2, $\text{rank}(\bar{\mathbf{J}}) = \text{tr}(\bar{\mathbf{J}}) = \sum_{i=1}^{m}(1/n_i)n_i = m$. This proves part (a). The symmetry and idempotency of $\mathbf{I} - \bar{\mathbf{J}}$ follow from that of $\bar{\mathbf{J}}$ (the latter by

Theorem 2.12.1), so again by Theorem 2.12.2, rank$(\mathbf{I} - \bar{\mathbf{J}}) = \text{tr}(\mathbf{I} - \bar{\mathbf{J}}) = \text{tr}(\mathbf{I}) - \text{tr}(\bar{\mathbf{J}}) = n - m$, establishing part (b). As for part (c),

$$\bar{\mathbf{J}}\mathbf{X} = \left[\oplus_{i=1}^{m}(1/n_i)\mathbf{J}_{n_i}\right]\begin{pmatrix} \mathbf{1}_{n_1}\mathbf{x}_1^T \\ \vdots \\ \mathbf{1}_{n_m}\mathbf{x}_m^T \end{pmatrix} = \begin{pmatrix} (1/n_1)\mathbf{J}_{n_1}\mathbf{1}_{n_1}\mathbf{x}_1^T \\ \vdots \\ (1/n_m)\mathbf{J}_{n_m}\mathbf{1}_{n_m}\mathbf{x}_m^T \end{pmatrix} = \begin{pmatrix} \mathbf{1}_{n_1}\mathbf{x}_1^T \\ \vdots \\ \mathbf{1}_{n_m}\mathbf{x}_m^T \end{pmatrix} = \mathbf{X}.$$

By part (c), $\bar{\mathbf{J}}\mathbf{P}_\mathbf{X} = \bar{\mathbf{J}}\mathbf{X}(\mathbf{X}^T\mathbf{X})^-\mathbf{X}^T = \mathbf{X}(\mathbf{X}^T\mathbf{X})^-\mathbf{X}^T = \mathbf{P}_\mathbf{X}$, proving part (d). By parts (a) and (d), $(\mathbf{I}-\bar{\mathbf{J}})(\bar{\mathbf{J}}-\mathbf{P}_\mathbf{X}) = \mathbf{I}\bar{\mathbf{J}} - \mathbf{I}\mathbf{P}_\mathbf{X} - \bar{\mathbf{J}}\bar{\mathbf{J}} + \bar{\mathbf{J}}\mathbf{P}_\mathbf{X} = \bar{\mathbf{J}} - \mathbf{P}_\mathbf{X} - \bar{\mathbf{J}} + \mathbf{P}_\mathbf{X} = \mathbf{0}$, proving part (e). As for part (f), the symmetry of $\bar{\mathbf{J}} - \mathbf{P}_\mathbf{X}$ follows easily from that of $\bar{\mathbf{J}}$ (part (a)) and $\mathbf{P}_\mathbf{X}$. Also, $(\bar{\mathbf{J}} - \mathbf{P}_\mathbf{X})(\bar{\mathbf{J}} - \mathbf{P}_\mathbf{X}) = \bar{\mathbf{J}}\bar{\mathbf{J}} - \bar{\mathbf{J}}\mathbf{P}_\mathbf{X} - \mathbf{P}_\mathbf{X}\bar{\mathbf{J}} + \mathbf{P}_\mathbf{X}\mathbf{P}_\mathbf{X} = \bar{\mathbf{J}} - \mathbf{P}_\mathbf{X} - \mathbf{P}_\mathbf{X} + \mathbf{P}_\mathbf{X} = \bar{\mathbf{J}} - \mathbf{P}_\mathbf{X}$, where we have used parts (a) and (d) and the fact that $\mathbf{P}_\mathbf{X}\bar{\mathbf{J}} = (\bar{\mathbf{J}}\mathbf{P}_\mathbf{X})^T = \mathbf{P}_\mathbf{X}^T = \mathbf{P}_\mathbf{X}$. Thus $\bar{\mathbf{J}} - \mathbf{P}_\mathbf{X}$ is symmetric and idempotent, so by Theorem 2.12.2 once more, rank$(\bar{\mathbf{J}} - \mathbf{P}_\mathbf{X}) = \text{tr}(\bar{\mathbf{J}} - \mathbf{P}_\mathbf{X}) = \text{tr}(\bar{\mathbf{J}}) - \text{tr}(\mathbf{P}_\mathbf{X}) = \text{rank}(\bar{\mathbf{J}}) - \text{rank}(\mathbf{P}_\mathbf{X}) = m - p^*$.

10 Constrained Least Squares Estimation and ANOVA

This chapter presents exercises on least squares estimation and ANOVA for constrained linear models and provides solutions to those exercises.

▶ **Exercise 1** Consider the constrained one-factor models featured in Example 10.1-1.

(a) Under the model with constraint $\alpha_q = h$, verify that (10.2) is the constrained least squares estimator of an arbitrary estimable function.
(b) Under the model with constraint $\sum_{i=1}^{q} \alpha_i = h$, obtain a general expression for the constrained least squares estimator of an arbitrary estimable function and verify that the estimators of level means and differences coincide with their counterparts under the unconstrained model.
(c) Under the model with constraint $\mu+\alpha_1 = h$, verify that (10.3) is the constrained least squares estimator of an arbitrary estimable function. Also verify the expressions for the constrained least squares estimators of estimable functions and for $\text{var}(\mathbf{C}^T\breve{\boldsymbol{\beta}})$ that immediately follow (10.3).

Solution

(a) Using Theorem 2.9.5b and the fact that $n-\mathbf{n}_{-q}^T(\oplus_{i=1}^{q-1} n_i^{-1})\mathbf{n}_{-q} = n_q$, the inverse of the upper left $q \times q$ block of the constrained normal equations coefficient matrix displayed in the second paragraph of Example 10.1-1 may be expressed as

$$\begin{pmatrix} n & \mathbf{n}_{-q}^T \\ \mathbf{n}_{-q} & \oplus_{i=1}^{q-1} n_i \end{pmatrix}^{-1} = \begin{pmatrix} n_q^{-1} & -n_q^{-1}\mathbf{1}_{q-1}^T \\ -n_q^{-1}\mathbf{1}_{q-1} & \oplus_{i=1}^{q-1} n_i^{-1} + n_q^{-1}\mathbf{J}_{q-1} \end{pmatrix}.$$

D. L. Zimmerman, *Linear Model Theory*,
https://doi.org/10.1007/978-3-030-52074-8_10

The Schur complement of that same block of the constrained normal equations coefficient matrix is, as noted in the example, nonsingular and equal to $\begin{pmatrix} 0 & 1 \\ 1 & 0 \end{pmatrix}$, and its inverse is equal to itself (as is easily verified). Therefore, using Theorem 2.9.5a the inverse of the constrained normal equations coefficient matrix has lower right 2×2 block $\begin{pmatrix} 0 & 1 \\ 1 & 0 \end{pmatrix}$, upper right $q \times 2$ block

$$-\begin{pmatrix} n_q^{-1} & -n_q^{-1}\mathbf{1}_{q-1}^T \\ -n_q^{-1}\mathbf{1}_{q-1} & \oplus_{i=1}^{q-1} n_i^{-1} + n_q^{-1}\mathbf{J}_{q-1} \end{pmatrix} \begin{pmatrix} n_q & 0 \\ \mathbf{0}_{q-1} & \mathbf{0}_{q-1} \end{pmatrix} \begin{pmatrix} 0 & 1 \\ 1 & 0 \end{pmatrix} = \begin{pmatrix} 0 & -1 \\ \mathbf{0}_{q-1} & \mathbf{1}_{q-1} \end{pmatrix},$$

lower left $2 \times q$ block $\begin{pmatrix} 0 & \mathbf{0}_{q-1}^T \\ -1 & \mathbf{1}_{q-1}^T \end{pmatrix}$ (by the symmetry of the coefficient matrix), and upper left $q \times q$ block

$$\begin{aligned} &\begin{pmatrix} n_q^{-1} & -n_q^{-1}\mathbf{1}_{q-1}^T \\ -n_q^{-1}\mathbf{1}_{q-1} & \oplus_{i=1}^{q-1} n_i^{-1} + n_q^{-1}\mathbf{J}_{q-1} \end{pmatrix} + \begin{pmatrix} 0 & 1 \\ 0 & -\mathbf{1}_{q-1} \end{pmatrix} \begin{pmatrix} n_q & \mathbf{0}_{q-1}^T \\ 0 & \mathbf{0}_{q-1}^T \end{pmatrix} \\ &\times \begin{pmatrix} n_q^{-1} & -n_q^{-1}\mathbf{1}_{q-1}^T \\ -n_q^{-1}\mathbf{1}_{q-1} - n_q^{-1}\mathbf{1}_{q-1} & \oplus_{i=1}^{q-1} n_i^{-1} + n_q^{-1}\mathbf{J}_{q-1} \end{pmatrix} \\ &= \begin{pmatrix} n_q^{-1} & -n_q^{-1}\mathbf{1}_{q-1}^T \\ -n_q^{-1}\mathbf{1}_{q-1} & \oplus_{i=1}^{q-1} n_i^{-1} + n_q^{-1}\mathbf{J}_{q-1} \end{pmatrix}. \end{aligned}$$

Putting these blocks together, the inverse of the constrained normal equations coefficient matrix is

$$\begin{pmatrix} n_q^{-1} & -n_q^{-1}\mathbf{1}_{q-1}^T & 0 & -1 \\ -n_q^{-1}\mathbf{1}_{q-1} & \oplus_{i=1}^{q-1} n_i^{-1} + n_q^{-1}\mathbf{J}_{q-1} & \mathbf{0}_{q-1} & \mathbf{1}_{q-1} \\ 0 & \mathbf{0}_{q-1}^T & 0 & 1 \\ -1 & \mathbf{1}_{q-1}^T & 1 & 0 \end{pmatrix}.$$

Extracting $\mathbf{G}_{11}$ and $\mathbf{G}_{12}$ as the upper left $(q+1) \times (q+1)$ and upper right $(q+1) \times 1$ blocks, respectively, of this last matrix, we may use Theorem 10.1.4

to write the constrained least squares estimator of an estimable (under the model with constraint $\alpha_q = h$) function $\mathbf{c}^T\boldsymbol{\beta}$ as

$$
\begin{aligned}
\mathbf{c}^T\breve{\boldsymbol{\beta}} &= \mathbf{c}^T(\mathbf{G}_{11}\mathbf{X}^T\mathbf{y} + \mathbf{G}_{12}h) \\
&= \mathbf{c}^T\left[\begin{pmatrix} n_q^{-1} & -n_q^{-1}\mathbf{1}_{q-1}^T & 0 \\ -n_q^{-1}\mathbf{1}_{q-1} & \oplus_{i=1}^{q-1} n_i^{-1} + n_q^{-1}\mathbf{J}_{q-1} & \mathbf{0}_{q-1} \\ 0 & \mathbf{0}_{q-1}^T & 0 \end{pmatrix}\begin{pmatrix} y_{\cdot\cdot} \\ y_{1\cdot} \\ \vdots \\ y_{q\cdot} \end{pmatrix} + \begin{pmatrix} -1 \\ \mathbf{1}_{q-1} \\ 1 \end{pmatrix} h\right] \\
&= \mathbf{c}^T\left[\begin{pmatrix} n_q^{-1}(y_{\cdot\cdot} - \sum_{i=1}^{q-1} y_{i\cdot}) \\ -n_q^{-1}y_{\cdot\cdot}\mathbf{1}_{q-1} + \begin{pmatrix} \bar{y}_{1\cdot} \\ \vdots \\ \bar{y}_{q-1\cdot} \end{pmatrix} + n_q^{-1}(\sum_{i=1}^{q-1} y_{i\cdot})\mathbf{1}_{q-1} \\ 0 \end{pmatrix} + \begin{pmatrix} -1 \\ \mathbf{1}_{q-1} \\ 1 \end{pmatrix} h\right] \\
&= \mathbf{c}^T\begin{pmatrix} \bar{y}_{q\cdot} - h \\ \bar{y}_{1\cdot} - \bar{y}_{q\cdot} + h \\ \bar{y}_{2\cdot} - \bar{y}_{q\cdot} + h \\ \vdots \\ \bar{y}_{q-1\cdot} - \bar{y}_{q\cdot} + h \\ h \end{pmatrix}.
\end{aligned}
$$

(b) In this case the coefficient matrix of the constrained normal equations is

$$
\begin{pmatrix} n & \mathbf{n}^T & 0 \\ \mathbf{n} & \oplus_{i=1}^{q} n_i & \mathbf{1}_q \\ 0 & \mathbf{1}_q^T & 0 \end{pmatrix},
$$

which is nonsingular because $\sum_{i=1}^{n} \alpha_i$ is nonestimable under the unconstrained model. The upper left scalar element and the lower right $(q+1) \times (q+1)$ submatrix of this matrix are nonsingular also. In particular, by Theorem 2.9.5a the inverse of that submatrix is

$$
\begin{aligned}
&\begin{pmatrix} \oplus_{i=1}^{q} n_i & \mathbf{1}_q \\ \mathbf{1}_q^T & 0 \end{pmatrix}^{-1} \\
&= \begin{pmatrix} \oplus_{i=1}^{q} n_i^{-1} - m^{-1}\oplus_{i=1}^{q} n_i^{-1}\mathbf{1}_q\mathbf{1}_q^T\oplus_{i=1}^{q} n_i^{-1} & m^{-1}\oplus_{i=1}^{q} n_i^{-1}\mathbf{1}_q \\ m^{-1}\mathbf{1}_q^T\oplus_{i=1}^{q} n_i^{-1} & -m^{-1} \end{pmatrix}
\end{aligned}
$$

where $m = \sum_{i=1}^{q} n_i^{-1}$. Applying Theorem 2.9.5b with $\mathbf{P}$ in that theorem taken here to be

$$n - \begin{pmatrix} \mathbf{n}^T & 0 \end{pmatrix} \begin{pmatrix} \oplus_{i=1}^{q} n_i & \mathbf{1}_q \\ \mathbf{1}_q^T & 0 \end{pmatrix}^{-1} \begin{pmatrix} \mathbf{n} \\ 0 \end{pmatrix} = q^2/m,$$

we find, after tedious calculations, that the inverse of the constrained normal equations coefficient matrix is

$$\begin{pmatrix} n & \mathbf{n}^T & 0 \\ \mathbf{n} & \oplus_{i=1}^{q} n_i & \mathbf{1}_q \\ 0 & \mathbf{1}_q^T & 0 \end{pmatrix}^{-1} = \begin{pmatrix} m_{11} & \mathbf{m}_{12}^T \\ \mathbf{m}_{12} & \mathbf{M}_{22} \end{pmatrix}$$

where

$$m_{11} = m/q^2,$$

$$\mathbf{m}_{12}^T = \left(-(m/q^2)\mathbf{1}_q^T + (1/q)\mathbf{1}_q^T \oplus_{i=1}^{q} n_i^{-1} \right),$$

$$\mathbf{M}_{22} = \begin{pmatrix} \oplus_{i=1}^{q} n_i^{-1} + (m/q^2)\mathbf{1}_q\mathbf{1}_q^T - (1/q)\mathbf{1}_q^T \oplus_{i=1}^{q} n_i^{-1} - (1/q) \oplus_{i=1}^{q} n_i^{-1}\mathbf{1}_q\mathbf{1}_q^T & (1/q)\mathbf{1}_q \\ (1/q)\mathbf{1}_q^T & 0 \end{pmatrix}.$$

Then,

$$\begin{aligned} &\mathbf{G}_{11}\mathbf{X}^T\mathbf{y} \\ &= \begin{pmatrix} m/q^2 - (m/q^2)\mathbf{1}_q^T + (1/q)\mathbf{1}_q^T \oplus_{i=1}^{q} n_i^{-1} \\ -(m/q^2)\mathbf{1}_q + (1/q) \oplus_{i=1}^{q} n_i^{-1}\mathbf{1}_q \oplus_{i=1}^{q} n_i^{-1} + (m/q^2)\mathbf{1}_q\mathbf{1}_q^T \\ -(1/q)\mathbf{1}_q\mathbf{1}_q^T \oplus_{i=1}^{q} n_i^{-1} - (1/q) \oplus_{i=1}^{q} n_i^{-1}\mathbf{1}_q\mathbf{1}_q^T \end{pmatrix} \\ &\quad \times \begin{pmatrix} y_{\cdot\cdot} \\ y_{1\cdot} \\ \vdots \\ y_{q\cdot} \end{pmatrix} \\ &= \begin{pmatrix} (1/q)\sum_{j=1}^{q} \bar{y}_{j\cdot} \\ \bar{y}_{1\cdot} - (1/q)\sum_{j=1}^{q} \bar{y}_{j\cdot} \\ \vdots \\ \bar{y}_{q\cdot} - (1/q)\sum_{j=1}^{q} \bar{y}_{j\cdot} \end{pmatrix} \end{aligned}$$

and

$$\mathbf{G}_{12}h = \begin{pmatrix} -(1/q) \\ (1/q)\mathbf{1}_q \end{pmatrix} h.$$

Therefore, a general expression for the constrained least squares estimator of an estimable (under the model with constraint $\sum_{i=1}^q \alpha_i = h$) function $\mathbf{c}^T\boldsymbol{\beta}$ is

$$\mathbf{c}^T \begin{pmatrix} (1/q)\sum_{j=1}^q \bar{y}_{j\cdot} - (h/q) \\ \bar{y}_{1\cdot} - (1/q)\sum_{j=1}^q \bar{y}_{j\cdot} + (h/q) \\ \vdots \\ \bar{y}_{q\cdot} - (1/q)\sum_{j=1}^q \bar{y}_{j\cdot} + (h/q) \end{pmatrix}.$$

Constrained least squares estimators of μ and α_i are $(1/q)\sum_{j=1}^q \bar{y}_{j\cdot} - (1/q)h$ and $\bar{y}_{i\cdot} - (1/q)\sum_{j=1}^q \bar{y}_{j\cdot} + (1/q)h$, respectively. Constrained least squares estimators of $\mu + \alpha_i$ and $\alpha_i - \alpha_{i'}$ are $\bar{y}_{i\cdot}$ and $\bar{y}_{i\cdot} - \bar{y}_{i'\cdot}$, respectively, which coincide with their counterparts under the unconstrained model.

(c) By Theorem 3.3.7, one generalized inverse of the constrained normal equations coefficient matrix has lower right scalar element $-n_1$, upper right $(q+1)\times 1$ block

$$-\begin{pmatrix} 0 & \mathbf{0}_q^T \\ \mathbf{0}_q & \oplus_{i=1}^q n_i^{-1} \end{pmatrix} \begin{pmatrix} 1 \\ \mathbf{u}_1 \end{pmatrix} (-n_1) = \begin{pmatrix} 0 \\ 1 \\ \mathbf{0}_{q-1} \end{pmatrix},$$

lower left $1 \times (q+1)$ block

$$(-n_1)(1, \mathbf{u}_1^T) \begin{pmatrix} 0 & \mathbf{0}_q^T \\ \mathbf{0}_q & \oplus_{i=1}^q n_i^{-1} \end{pmatrix} = \begin{pmatrix} 0 \ 1 \ \mathbf{0}_{q-1}^T \end{pmatrix},$$

and upper left $(q+1)\times(q+1)$ block

$$\begin{pmatrix} 0 & \mathbf{0}_q^T \\ \mathbf{0}_q & \oplus_{i=1}^q n_i^{-1} \end{pmatrix} + \begin{pmatrix} 0 & \mathbf{0}_q^T \\ \mathbf{0}_q & \oplus_{i=1}^q n_i^{-1} \end{pmatrix} \begin{pmatrix} 1 \\ \mathbf{u}_1 \end{pmatrix} (-n_1)(1, \mathbf{u}_1^T) \begin{pmatrix} 0 & \mathbf{0}_q^T \\ \mathbf{0}_q & \oplus_{i=1}^q n_i^{-1} \end{pmatrix}$$
$$= \begin{pmatrix} \mathbf{0}_{2\times 2} & \mathbf{0}_{2\times(q-1)} \\ \mathbf{0}_{(q-1)\times 2} & \oplus_{i=2}^q n_i^{-1} \end{pmatrix}.$$

Therefore, one solution to the constrained normal equations is

$$\begin{pmatrix} 0 & 0 & \mathbf{0}_{q-1}^T & 0 \\ 0 & 0 & \mathbf{0}_{q-1}^T & 1 \\ \mathbf{0}_{q-1} & \mathbf{0}_{q-1} & \oplus_{i=2}^{q} n_i^{-1} & \mathbf{0}_{q-1} \\ 0 & 1 & \mathbf{0}_{q-1}^T & -n_1 \end{pmatrix} \begin{pmatrix} y_{\cdot\cdot} \\ y_{1\cdot} \\ y_{2\cdot} \\ \vdots \\ y_{q\cdot} \\ h \end{pmatrix} = \begin{pmatrix} 0 \\ h \\ \bar{y}_{2\cdot} \\ \vdots \\ \bar{y}_{q\cdot} \\ y_{1\cdot} - n_1 h \end{pmatrix},$$

so the constrained least squares estimator of an estimable (under the model with constraint $\mu + \alpha_1 = h$) function $\mathbf{c}^T \boldsymbol{\beta}$ is

$$\mathbf{c}^T \begin{pmatrix} 0 \\ h \\ \bar{y}_{2\cdot} \\ \vdots \\ \bar{y}_{q\cdot} \end{pmatrix}.$$

Thus, the constrained least squares estimators of $\mu + \alpha_1$, $\mu + \alpha_i$ $(i = 2, \ldots, q)$, $\alpha_i - \alpha_1$ $(i = 2, \ldots, q)$, and $\alpha_i - \alpha_{i'}$ $(i > i' = 2, \ldots, q)$ are, respectively, h, $\bar{y}_{i\cdot}$, $\bar{y}_{i\cdot} - h$, and $\bar{y}_{i\cdot} - \bar{y}_{i'\cdot}$. Finally, for $\mathbf{C}^T = (\mathbf{1}_q, \mathbf{I}_q)$,

$$\text{var}(\mathbf{C}^T \breve{\boldsymbol{\beta}}) = \sigma^2 (\mathbf{1}_q, \mathbf{I}_q) \begin{pmatrix} 0 & 0 & \mathbf{0}_{q-1}^T \\ 0 & 0 & \mathbf{0}_{q-1}^T \\ \mathbf{0}_{q-1} & \mathbf{0}_{q-1} & \oplus_{i=2}^{q} n_i^{-1} \end{pmatrix} \begin{pmatrix} \mathbf{1}_q^T \\ \mathbf{I}_q \end{pmatrix} = \sigma^2 \begin{pmatrix} 0 & \mathbf{0}_{q-1}^T \\ \mathbf{0}_{q-1} & \oplus_{i=2}^{q} n_i^{-1} \end{pmatrix}.$$

▶ **Exercise 2** Consider the one-factor model. For this model, the normal equations may be written without using matrix and vector notation as

$$n\mu + \sum_{i=1}^{q} n_i \alpha_i = y_{\cdot\cdot},$$

$$n_i \mu + n_i \alpha_i = y_{i\cdot} \quad (i = 1, \ldots, q).$$

Using the approach described in Example 7.1-2, it may be shown that one solution to these equations is $\hat{\boldsymbol{\beta}} = (0, \bar{y}_{1\cdot}, \bar{y}_{2\cdot}, \ldots, \bar{y}_{q\cdot})^T$. The following five functions might be of interest: μ, α_1, $\alpha_1 - \alpha_2$, $\mu + \frac{1}{q}\sum \alpha_i$, and $\mu + \frac{1}{n}\sum n_i \alpha_i$.

(a) For a model without constraints, which of these five functions are estimable and which are not? Give the BLUEs for those that are estimable.

(b) For each function that is nonestimable under a model without constraints, show that it is estimable under the model with the constraint $\alpha_q = 0$ and give its BLUE under this model.
(c) For each function that is estimable under the model without constraints, show that its least squares estimator in terms of the elements of $\hat{\boldsymbol{\beta}}$, and its constrained least squares estimator under the model with constraint $\alpha_q = 0$, are identical.

Solution

(a) $\alpha_1 - \alpha_2$, $\mu + \frac{1}{q}\sum_{i=1}^q \alpha_i$, and $\mu + \frac{1}{n}\sum_{i=1}^q n_i\alpha_i$ are estimable; μ and α_1 are not. Using the given solution, BLUEs for the estimable functions are, respectively, $\bar{y}_{1\cdot} - \bar{y}_{2\cdot}$, $(1/q)\sum_{i=1}^q \bar{y}_{i\cdot}$, and $(1/n)\sum_{i=1}^q n_i\bar{y}_{i\cdot} = (1/n)\sum_{i=1}^q \sum_{j=1}^{n_i} y_{ij} = \bar{y}_{\cdot\cdot}$.
(b) The constraint $\alpha_q = 0$ may be written as $\mathbf{A}\boldsymbol{\beta} = h$, where $\mathbf{A} = (\mathbf{0}_q^T, 1)$ (and $h = 0$). Then,

$$\mu = (1, \mathbf{0}_{q-1}^T, 1)\boldsymbol{\beta} - (\mathbf{0}_q^T, 1)\boldsymbol{\beta}.$$

In this expression, the first row vector multiplying $\boldsymbol{\beta}$ is an element of $\mathcal{R}(\mathbf{X})$, and the second such row vector is an element of $\mathcal{R}(\mathbf{A})$. Thus $\mu = \mathbf{c}^T\boldsymbol{\beta}$ where $\mathbf{c}^T \in \mathcal{R}\begin{pmatrix}\mathbf{X}\\ \mathbf{A}\end{pmatrix}$, so it is estimable under the constrained model. From the equation

$$n_q\mu + n_q\alpha_q = y_{q\cdot}$$

(the next-to-last equation in the constrained normal equations), we obtain $\breve{\mu} = \bar{y}_{q\cdot}$, and from the $q-1$ equations

$$n_i\mu + n_i\alpha_i = y_{i\cdot} \quad (i = 1, \ldots, q-1)$$

we obtain $\bar{y}_{q\cdot} + \alpha_i = \bar{y}_{i\cdot}$, i.e., $\breve{\alpha}_i = \bar{y}_{i\cdot} - \bar{y}_{q\cdot}$ $(i = 1, \ldots, q-1)$. Next consider α_1, which may be expressed as

$$\alpha_1 = (1, 1, \mathbf{0}_{q-1}^T)\boldsymbol{\beta} - (1, \mathbf{0}_{q-1}^T, 1)\boldsymbol{\beta} + (\mathbf{0}_q^T, 1)\boldsymbol{\beta};$$

the first two row vectors multiplying $\boldsymbol{\beta}$ are elements of $\mathcal{C}(\mathbf{X})$, and the third is an element of $\mathcal{R}(\mathbf{A})$. Thus α_1 is estimable under the constrained model, and its BLUE, using the previously stated solution to the constrained normal equations, is $\breve{\alpha}_1 = \bar{y}_{1\cdot} - \bar{y}_{q\cdot}$.
(c) The least squares estimator of $\alpha_1 - \alpha_2$ in terms of $\hat{\boldsymbol{\beta}}$ is

$$(0, 1, -1, \mathbf{0}_{q-2}^T)(0, \bar{y}_{1\cdot}, \bar{y}_{2\cdot}, \ldots, \bar{y}_{q\cdot})^T = \bar{y}_{1\cdot} - \bar{y}_{2\cdot}.$$

Its constrained least squares estimator in terms of the solution to the constrained normal equations obtained in part (b) is

$$\begin{aligned}&(0, 1, -1, \mathbf{0}_{q-2}^T)(\bar{y}_{q\cdot}, \bar{y}_{1\cdot} - \bar{y}_{q\cdot}, \bar{y}_{2\cdot} - \bar{y}_{q\cdot}, \ldots, \bar{y}_{q-1\cdot} - \bar{y}_{q\cdot}, 0)^T \\ &= (\bar{y}_{1\cdot} - \bar{y}_{q\cdot}) - (\bar{y}_{2\cdot} - \bar{y}_{q\cdot}) \\ &= \bar{y}_{1\cdot} - \bar{y}_{2\cdot}.\end{aligned}$$

The least squares estimator of $\mu + (1/q)\sum_{i=1}^{q} \alpha_i$ in terms of $\hat{\boldsymbol{\beta}}$ is

$$(1, (1/q)\mathbf{1}_q^T)\hat{\boldsymbol{\beta}} = (1/q)\sum_{i=1}^{q} \bar{y}_{i\cdot}.$$

Its constrained least squares estimator in terms of the solution to the constrained normal equations obtained in part (b) is

$$(1, (1/q)\mathbf{1}_q^T)(\bar{y}_{q\cdot}, \bar{y}_{1\cdot} - \bar{y}_{q\cdot}, \bar{y}_{2\cdot} - \bar{y}_{q\cdot}, \ldots, \bar{y}_{q-1\cdot} - \bar{y}_{q\cdot}, 0)^T = (1/q)\sum_{i=1}^{q} \bar{y}_{i\cdot}.$$

The least squares estimator of $\mu + (1/n)\sum_{i=1}^{q} n_i\alpha_i$ in terms of $\hat{\boldsymbol{\beta}}$ is

$$(1, (n_1/n), (n_2/n), \ldots, (n_q/n))\hat{\boldsymbol{\beta}} = (1/n)\sum_{i=1}^{q} n_i\bar{y}_{i\cdot} = \bar{y}_{\cdot\cdot}.$$

Its constrained least squares estimator in terms of the solution to the constrained normal equations obtained in part (b) is

$$\begin{aligned}&(1, (n_1/n), (n_2/n), \ldots, (n_q/n))(\bar{y}_{q\cdot}, \bar{y}_{1\cdot} - \bar{y}_{q\cdot}, \bar{y}_{2\cdot} - \bar{y}_{q\cdot}, \ldots, \bar{y}_{q-1\cdot} - \bar{y}_{q\cdot}, 0)^T \\ &= \bar{y}_{q\cdot} + \sum_{i=1}^{q-1}(n_i/n)(\bar{y}_{i\cdot} - \bar{y}_{q\cdot}) \\ &= \bar{y}_{q\cdot} + (1/n)\sum_{i=1}^{q-1} n_i\bar{y}_{i\cdot} - (1/n)\bar{y}_{q\cdot}\sum_{i=1}^{q-1} n_i \\ &= \bar{y}_{q\cdot} + (1/n)\sum_{i=1}^{q-1} n_i\bar{y}_{i\cdot} - [(n - n_q)/n]\bar{y}_{q\cdot} \\ &= \bar{y}_{q\cdot} + (1/n)\sum_{i=1}^{q-1}\sum_{j=1}^{n_i} y_{ij} - \bar{y}_{q\cdot} + (1/n)\sum_{j=1}^{n_q} y_{qj} \\ &= \bar{y}_{\cdot\cdot}\end{aligned}$$

Exercise 3 Consider the two-way main effects model, with exactly one observation per cell.

(a) Write out the normal equations for this model without using matrix and vector notation.
(b) Show that one solution to the normal equations is given by $\hat{\mu} = \bar{y}_{\cdot\cdot}$, $\hat{\alpha}_i = \bar{y}_{i\cdot} - \bar{y}_{\cdot\cdot}$ $(i = 1, \ldots, q)$, $\hat{\gamma}_j = \bar{y}_{\cdot j} - \bar{y}_{\cdot\cdot}$ $(j = 1, \ldots, m)$.
(c) Solve the constrained normal equations obtained by augmenting the model with an appropriate number of jointly nonestimable pseudo-constraints. List your pseudo-constraints explicitly and explain why they are jointly nonestimable.
(d) The following five functions might be of interest: μ, α_1, $\gamma_1 - \gamma_2$, $\alpha_1 - \gamma_2$, and $\mu + \frac{1}{q}\sum_{i=1}^{q} \alpha_i + \frac{1}{m}\sum_{j=1}^{m} \gamma_j$.
  (i) For a model without constraints, which of these five functions are estimable and which are not? Give the BLUEs for those that are estimable.
  (ii) For each function that is nonestimable under a model without constraints, show that it is estimable under the model with the pseudo-constraints you gave in part (c) and give its BLUE under this constrained model.
  (iii) For each function that is estimable under a model without constraints, show that its least squares estimator in terms of the elements of the solution to the normal equations given in part (b) is identical to its constrained least squares estimator under the model with the constraints you gave in part (c).

Solution

(a) The model matrix is

$$\mathbf{X} = (\mathbf{1}_{qm}, \mathbf{I}_q \otimes \mathbf{1}_m, \mathbf{1}_q \otimes \mathbf{I}_m),$$

so the normal equations are

$$(\mathbf{1}_{qm}, \mathbf{I}_q \otimes \mathbf{1}_m, \mathbf{1}_q \otimes \mathbf{I}_m)^T (\mathbf{1}_{qm}, \mathbf{1}_m, \mathbf{1}_q \otimes \mathbf{I}_m)\boldsymbol{\beta} = (\mathbf{1}_{qm}, \mathbf{I}_q \otimes \mathbf{1}_m, \mathbf{1}_q \otimes \mathbf{I}_m)^T \mathbf{y},$$

i.e.,

$$qm\mu + m\sum_{i=1}^{q} \alpha_i + q\sum_{j=1}^{m} \gamma_j = y_{\cdot\cdot}$$

$$m\mu + m\alpha_i + \sum_{j=1}^{m} \gamma_j = y_{i\cdot} \quad (i = 1, \ldots, q)$$

$$q\mu + \sum_{i=1}^{q} \alpha_i + q\gamma_j = y_{\cdot j} \quad (j = 1, \ldots, m).$$

(b) Substituting the purported solution for the parameters on the left-hand side of the normal equations results in

$$qm\bar{y}_{\cdot\cdot} + m\sum_{i=1}^{q}(\bar{y}_{i\cdot} - \bar{y}_{\cdot\cdot}) + q\sum_{j=1}^{m}(\bar{y}_{\cdot j} - \bar{y}_{\cdot\cdot}) = y_{\cdot\cdot}$$

$$m\bar{y}_{\cdot\cdot} + m(\bar{y}_{i\cdot} - \bar{y}_{\cdot\cdot}) + \sum_{j=1}^{m}(\bar{y}_{\cdot j} - \bar{y}_{\cdot\cdot}) = y_{i\cdot} \quad (i = 1, \ldots, q)$$

$$q\bar{y}_{\cdot\cdot} + \sum_{i=1}^{q}(\bar{y}_{i\cdot} - \bar{y}_{\cdot\cdot}) + q(\bar{y}_{\cdot j} - \bar{y}_{\cdot\cdot}) = y_{\cdot j} \quad (j = 1, \ldots, m),$$

demonstrating that the purported solution does indeed solve those equations.

(c) Recall from Example 6.1-3 that rank($\mathbf{X}$) $= q + m - 1$ while the number of parameters is $q+m+1$. Therefore, two jointly nonestimable pseudo-constraints are needed. Constraints $\mu = 0$ and $\alpha_1 = 0$ are suitable; they are nonestimable because the coefficients on the α_i's do not sum to the coefficient on μ (as illustrated in Example 6.1-3). The corresponding unique solution is as follows:

$$\breve{\mu} = 0, \quad \breve{\alpha}_1 = 0, \quad \breve{\alpha}_i = \bar{y}_{i\cdot} - \bar{y}_{1\cdot} \ (i = 2, \ldots, q),$$
$$\breve{\gamma}_j = \bar{y}_{\cdot j} + \bar{y}_{1\cdot} - \bar{y}_{\cdot\cdot} \ (j = 1, \ldots, m).$$

(d) (i) $\gamma_1 - \gamma_2$ and $\mu + (1/q)\sum_{i=1}^{q}\alpha_i + (1/m)\sum_{j=1}^{m}\gamma_j$ are estimable; μ, α_1, and $\alpha_1 - \gamma_2$ are nonestimable. Using the solution from part (b), BLUEs of the estimable functions are $[(\bar{y}_{\cdot 1} - \bar{y}_{\cdot\cdot}) - (\bar{y}_{\cdot 2} - \bar{y}_{\cdot\cdot})] = \bar{y}_{\cdot 1} - \bar{y}_{\cdot 2}$ and $\bar{y}_{\cdot\cdot} + (1/q)\sum_{i=1}^{q}(\bar{y}_{i\cdot} - \bar{y}_{\cdot\cdot}) + (1/m)\sum_{j=1}^{m}(\bar{y}_{\cdot j} - \bar{y}_{\cdot\cdot}) = \bar{y}_{\cdot\cdot}$, respectively.

(ii) $\mathbf{A} = \begin{pmatrix} 1 & \mathbf{0}_q^T & \mathbf{0}_m^T \\ 0 & \mathbf{u}_1^{(q)T} & \mathbf{0}_m^T \end{pmatrix}$, so $\mu = (1, \mathbf{0}_q^T, \mathbf{0}_m^T)\boldsymbol{\beta}$ and $\alpha_1 = (0, \mathbf{u}_1^{(q)T}, \mathbf{0}_m^T)\boldsymbol{\beta}$ plainly are estimable under the constrained model. So too is $\alpha_1 - \gamma_2 = \mu + 2\alpha_1 - (\mu + \alpha_1 + \gamma_2)$. Using the solution from part (c), their BLUEs are 0, 0, and $-(\bar{y}_{\cdot 2} + \bar{y}_{1\cdot} - \bar{y}_{\cdot\cdot})$, respectively.

(iii) Using the solution to the constrained normal equations from part (c), the BLUE of $\gamma_1 - \gamma_2$ is $[(\bar{y}_{\cdot 1} + \bar{y}_{1\cdot} - \bar{y}_{\cdot\cdot}) - (\bar{y}_{\cdot 2} + \bar{y}_{1\cdot} - \bar{y}_{\cdot\cdot})] = \bar{y}_{\cdot 1} - \bar{y}_{\cdot 2}$; and the BLUE of $\mu + (1/q)\sum_{i=1}^{q}\alpha_i + (1/m)\sum_{j=1}^{m}\gamma_j$ is $0 + (1/q)\sum_{i=2}^{q}(\bar{y}_{i\cdot} - \bar{y}_{1\cdot}) + (1/m)\sum_{j=1}^{m}(\bar{y}_{\cdot j} + \bar{y}_{1\cdot} - \bar{y}_{\cdot\cdot}) = \bar{y}_{\cdot\cdot}$.

▶ **Exercise 4** Consider the two-factor nested model.

(a) Write out the normal equations for this model without using matrix and vector notation.

(b) Show that one solution to the normal equations is given by $\hat{\mu} = \bar{y}_{\cdots}$, $\hat{\alpha}_i = \bar{y}_{i\cdot\cdot} - \bar{y}_{\cdots}$ $(i = 1, \ldots, q)$, $\hat{\gamma}_{ij} = \bar{y}_{ij\cdot} - \bar{y}_{i\cdot\cdot}$ $(i = 1, \ldots, q;\ j = 1, \ldots, m_i)$.

(c) Solve the constrained normal equations obtained by augmenting the model with an appropriate number of jointly nonestimable pseudo-constraints. List your pseudo-constraints explicitly and explain why they are jointly nonestimable.

(d) The following six functions might be of interest: μ, α_1, $\mu + \alpha_1$, $\gamma_{11} - \gamma_{12}$, $\mu + \alpha_1 + (1/m_1)\sum_{j=1}^{m_1} \gamma_{1j}$, and $\mu + (1/m.)\sum_{i=1}^{q} m_i\alpha_i + (1/m.)\sum_{i=1}^{q}\sum_{j=1}^{m_i} \gamma_{ij}$.
   (i) For a model without constraints, which of these six functions are estimable and which are not? Give the BLUEs for those that are estimable.
   (ii) For each function that is nonestimable under a model without constraints, show that it is estimable under the model with the pseudo-constraints you gave in part (c) and give its BLUE under this constrained model.
   (iii) For each function that is estimable under a model without constraints, show that its least squares estimator in terms of the elements of the solution to the normal equations given in part (b) is identical to its constrained least squares estimator under the model with the constraints you gave in part (c).

Solution

(a) The normal equations are

$$n_{\cdot\cdot}\mu + \sum_{i=1}^{q} n_{i\cdot}\alpha_i + \sum_{i=1}^{q}\sum_{j=1}^{m_i} n_{ij}\gamma_{ij} = y_{\cdots}$$

$$n_{i\cdot}\mu + n_{i\cdot}\alpha_i + \sum_{j=1}^{m_i} n_{ij}\gamma_{ij} = y_{i\cdot\cdot} \quad (i = 1, \ldots, q)$$

$$n_{ij}\mu + n_{ij}\alpha_i + n_{ij}\gamma_{ij} = y_{ij\cdot} \quad (i = 1, \ldots, q;\ j = 1, \ldots, m_i).$$

(b) Substituting the purported solution for the parameters in the left-hand side of the normal equations results in

$$n_{\cdot\cdot}\bar{y}_{\cdots} + \sum_{i=1}^{q} n_{i\cdot}(\bar{y}_{i\cdot\cdot} - \bar{y}_{\cdots}) + \sum_{i=1}^{q}\sum_{j=1}^{m_i} n_{ij}(\bar{y}_{ij\cdot} - \bar{y}_{\cdots}) = y_{\cdots}$$

$$n_{i\cdot}\bar{y}_{\cdots} + n_{i\cdot}(\bar{y}_{i\cdot\cdot} - \bar{y}_{\cdots}) + \sum_{j=1}^{m_i} n_{ij}(\bar{y}_{ij\cdot} - \bar{y}_{\cdots}) = y_{i\cdot\cdot} \quad (i = 1, \ldots, q)$$

$$n_{ij}\bar{y}_{\cdots} + n_{ij}(\bar{y}_{i\cdot\cdot} - \bar{y}_{\cdots}) + n_{ij}(\bar{y}_{ij\cdot} - \bar{y}_{i\cdot\cdot}) = y_{ij\cdot} \quad (i = 1, \ldots, q;\ j = 1, \ldots, m_i),$$

demonstrating that the purported solution does indeed solve those equations.

(c) Recall from Example 6.1-6 that rank($\mathbf{X}$) $= \sum_{i=1}^{q} m_i$, while the number of parameters is $q+1+\sum_{i=1}^{q} m_i$. So $q+1$ jointly nonestimable pseudo-constraints are needed. Constraints $\mu = 0$ and $\alpha_i = 0$ $(i = 1, \ldots, q)$ are suitable, yielding the unique solution $\breve{\mu} = 0$, $\breve{\alpha}_i = 0$, $\breve{\gamma}_{ij} = \bar{y}_{ij\cdot}$ for all i and j.

(d) (i) $\gamma_{11} - \gamma_{12}$, $\mu + \alpha_1 + (1/m_1)\sum_{j=1}^{m_1} \gamma_{1j}$, and $\mu + (1/m_\cdot)\sum_{i=1}^{q} m_i\alpha_i + (1/m_\cdot)\sum_{i=1}^{q}\sum_{j=1}^{m_i} \gamma_{ij}$ are estimable; μ, α_1, and $\mu + \alpha_1$ are not estimable. Using the solution from part (b), BLUEs of the estimable functions are $[(\bar{y}_{11\cdot} - \bar{y}_{\cdots}) - (\bar{y}_{12\cdot} - \bar{y}_{\cdots})] = \bar{y}_{11\cdot} - \bar{y}_{12\cdot}$, $\bar{y}_{\cdots} + (\bar{y}_{1\cdot\cdot} - \bar{y}_{\cdots}) + (1/m_1)\sum_{j=1}^{m_1}(\bar{y}_{1j\cdot} - \bar{y}_{1\cdot\cdot}) = (1/m_1)\sum_{j=1}^{m_1}\bar{y}_{1j\cdot}$, and $\bar{y}_{\cdots} + (1/m_\cdot)\sum_{i=1}^{q} m_i(\bar{y}_{i\cdot\cdot} - \bar{y}_{\cdots}) + (1/m_\cdot)\sum_{i=1}^{q}\sum_{j=1}^{m_i}(\bar{y}_{ij\cdot} - \bar{y}_{i\cdot\cdot}) = (1/m_\cdot)\sum_{i=1}^{q}\sum_{j=1}^{m_i}\bar{y}_{ij\cdot}$, respectively.

(ii) $\mathbf{A} = \begin{pmatrix} 1 & \mathbf{0}_q^T & \mathbf{0}_{m\cdot}^T \\ \mathbf{0}_q & \mathbf{I}_q & \mathbf{0}_{q\times m\cdot} \end{pmatrix}$, so $\mu = (1, \mathbf{0}_q^T, \mathbf{0}_{m\cdot}^T)\boldsymbol{\beta}$ and $\alpha_1 = (0, \mathbf{u}_1^{(q)T}, \mathbf{0}_{m\cdot}^T)\boldsymbol{\beta}$ plainly are estimable under the constrained model. So too is $\mu + \alpha_1 = (1, \mathbf{u}_1^{(q)T}, \mathbf{0}_{m\cdot}^T)\boldsymbol{\beta}$. Using the solution from (c), their BLUEs are all 0.

(iii) Using the solution to the constrained normal equations from part (c), the BLUE of $\gamma_{11} - \gamma_{12}$ is $\bar{y}_{11\cdot} - \bar{y}_{12\cdot}$; the BLUE of $\mu + \alpha_1 + (1/m)\sum_{j=1}^{m_1}\gamma_{1j}$ is $0 + 0 + (1/m_1)\sum_{j=1}^{m_1}\bar{y}_{1j\cdot} = (1/m_1)\sum_{j=1}^{m_1}\bar{y}_{1j\cdot}$; and the BLUE of $\mu + (1/m_\cdot)\sum_{i=1}^{q} m_i\alpha_i + (1/m_\cdot)\sum_{i=1}^{q}\sum_{j=1}^{m_i}\gamma_{ij}$ is $0 + (1/m_\cdot)\sum_{i=1}^{q} m_i \cdot 0 + (1/m_\cdot)\sum_{i=1}^{q}\sum_{j=1}^{m_i}\bar{y}_{ij\cdot} = (1/m_\cdot)\sum_{i=1}^{q}\sum_{j=1}^{m_i}\bar{y}_{ij\cdot}$.

▶ **Exercise 5** Consider the one-factor model

$$y_{ij} = \mu + \alpha_i + e_{ij} \quad (i = 1, 2; \quad j = 1, \ldots, r)$$

with the constraint

$$\alpha_1 + \alpha_2 = 24.$$

(a) Write out the constrained normal equations without using matrix and vector notation.
(b) Obtain the unique solution $\breve{\mu}$, $\breve{\alpha}_1$, and $\breve{\alpha}_2$ to the constrained normal equations.
(c) Is $\breve{\alpha}_1 - \breve{\alpha}_2$ identical to the least squares estimator of $\alpha_1 - \alpha_2$ under the corresponding unconstrained model? If you answer yes, verify your answer. If you answer no, give a reason why not.

Solution

(a)

$$\begin{aligned} 2r\mu + r\alpha_1 + r\alpha_2 &= y_{\cdot\cdot} \\ r\mu + r\alpha_1 &= y_{1\cdot} \\ r\mu + r\alpha_2 &= y_{2\cdot} \\ \alpha_1 + \alpha_2 &= 24 \end{aligned}$$

(b) The second and third equations imply that $\alpha_1 - \alpha_2 = (y_{1\cdot} - y_{2\cdot})/r = \bar{y}_{1\cdot} - \bar{y}_{2\cdot}$. This derived equation and the constraint equation yield $2\alpha_1 = \bar{y}_{1\cdot} - \bar{y}_{2\cdot} + 24$, which finally yields the solution

$$\breve{\alpha}_1 = (1/2)(\bar{y}_{1\cdot} - \bar{y}_{2\cdot}) + 12, \quad \breve{\alpha}_2 = (1/2)(\bar{y}_{2\cdot} - \bar{y}_{1\cdot}) + 12,$$
$$\breve{\mu} = \bar{y}_{1\cdot} - \breve{\alpha}_1 = (1/2)(\bar{y}_{1\cdot} + \bar{y}_{2\cdot}) - 12 = \bar{y}_{\cdot\cdot} - 12.$$

(c) Yes, because the constraint $\alpha_1 + \alpha_2 = 24$ is nonestimable under the unconstrained model. The constrained least squares estimator of $\alpha_1 - \alpha_2$ is

$$\breve{\alpha}_1 - \breve{\alpha}_2 = (1/2)(\bar{y}_{1\cdot} - \bar{y}_{2\cdot}) + 12 - [(1/2)(\bar{y}_{2\cdot} - \bar{y}_{1\cdot}) + 12] = \bar{y}_{1\cdot} - \bar{y}_{2\cdot},$$

which agrees with the least squares estimator of this function associated with the unconstrained model.

▶ **Exercise 6** Prove Theorem 10.1.6: Under the constrained Gauss–Markov model $\{\mathbf{y}, \mathbf{X}\boldsymbol{\beta}, \sigma^2\mathbf{I} : \mathbf{A}\boldsymbol{\beta} = \mathbf{h}\}$, the variance of the constrained least squares estimator of an estimable function $\mathbf{c}^T\boldsymbol{\beta}$ associated with the model $\{\mathbf{y}, \mathbf{X}\boldsymbol{\beta} : \mathbf{A}\boldsymbol{\beta} = \mathbf{h}\}$ is uniformly (in $\boldsymbol{\beta}$ and σ^2) less than that of any other linear unbiased estimator of $\mathbf{c}^T\boldsymbol{\beta}$.

Solution Because $\mathbf{c}^T\boldsymbol{\beta}$ is estimable under the constrained model, by Theorem 6.2.2 an n-vector $\mathbf{a}_1$ and a q-vector $\mathbf{a}_2$ exist such that $\mathbf{c}^T = \mathbf{a}_1^T\mathbf{X}^T\mathbf{X} + \mathbf{a}_2^T\mathbf{A}$. Let $\mathbf{c}^T\breve{\boldsymbol{\beta}}$ and $t_0 + \mathbf{t}^T\mathbf{y}$ be the constrained least squares estimator and any linear unbiased estimator, respectively, of $\mathbf{c}^T\boldsymbol{\beta}$ under the constrained model. Then (by Theorem 6.2.1) a q-vector $\mathbf{g}$ exists such that $t_0 = \mathbf{g}^T\mathbf{h}$ and $\mathbf{t}^T\mathbf{X} + \mathbf{g}^T\mathbf{A} = \mathbf{c}^T$. Consequently, by Theorem 4.2.2b,

$$\begin{aligned}
\mathrm{var}(t_0 + \mathbf{t}^T\mathbf{y}) &= \mathrm{var}(\mathbf{t}^T\mathbf{y}) \\
&= \mathrm{var}[\mathbf{c}^T\breve{\boldsymbol{\beta}} + (\mathbf{t}^T\mathbf{y} - \mathbf{c}^T\breve{\boldsymbol{\beta}})] \\
&= \mathrm{var}(\mathbf{c}^T\breve{\boldsymbol{\beta}}) + \mathrm{var}(\mathbf{t}^T\mathbf{y} - \mathbf{c}^T\breve{\boldsymbol{\beta}}) + 2\,\mathrm{cov}(\mathbf{c}^T\breve{\boldsymbol{\beta}}, \mathbf{t}^T\mathbf{y} - \mathbf{c}^T\breve{\boldsymbol{\beta}}),
\end{aligned}$$

where, by Corollary 4.2.3.1 and various parts of Theorem 3.3.8 (with $\mathbf{X}^T\mathbf{X}$ and $\mathbf{A}^T$ here playing the roles of $\mathbf{A}$ and $\mathbf{B}$ in the theorem),

$$\begin{aligned}
&\mathrm{cov}(\mathbf{c}^T\breve{\boldsymbol{\beta}}, \mathbf{t}^T\mathbf{y} - \mathbf{c}^T\breve{\boldsymbol{\beta}}) \\
&= \mathrm{cov}\{\mathbf{c}^T(\mathbf{G}_{11}\mathbf{X}^T\mathbf{y} + \mathbf{G}_{12}\mathbf{h}),\ [\mathbf{t}^T - \mathbf{c}^T(\mathbf{G}_{11}\mathbf{X}^T\mathbf{y} + \mathbf{G}_{12}\mathbf{h})]\} \\
&= \mathbf{c}^T\mathbf{G}_{11}\mathbf{X}^T(\sigma^2\mathbf{I})(\mathbf{t} - \mathbf{X}\mathbf{G}_{11}^T\mathbf{c}) \\
&= \sigma^2\mathbf{c}^T\mathbf{G}_{11}\mathbf{X}^T(\mathbf{t} - \mathbf{X}\mathbf{G}_{11}^T\mathbf{c}) \\
&= \sigma^2(\mathbf{a}_1\mathbf{X}^T\mathbf{X} + \mathbf{a}_2^T\mathbf{A})\mathbf{G}_{11}[\mathbf{X}^T\mathbf{t} - \mathbf{X}^T\mathbf{X}\mathbf{G}_{11}^T(\mathbf{X}^T\mathbf{X}\mathbf{a}_1 + \mathbf{A}^T\mathbf{a}_2)] \\
&= \sigma^2(\mathbf{a}_1\mathbf{X}^T\mathbf{X} + \mathbf{a}_2^T\mathbf{A})\mathbf{G}_{11}[(\mathbf{c} - \mathbf{A}^T\mathbf{g}) - \mathbf{X}^T\mathbf{X}\mathbf{G}_{11}\mathbf{X}^T\mathbf{X}\mathbf{a}_1 - \mathbf{X}^T\mathbf{X}\mathbf{G}_{11}^T\mathbf{A}^T\mathbf{a}_2]
\end{aligned}$$

$$= \sigma^2(\mathbf{a}_1\mathbf{X}^T\mathbf{X} + \mathbf{a}_2^T\mathbf{A})\mathbf{G}_{11}[\mathbf{c} - \mathbf{A}^T\mathbf{g} - (\mathbf{X}^T\mathbf{X}\mathbf{a}_1 + \mathbf{A}^T\mathbf{G}_{22}\mathbf{A}\mathbf{a}_1)]$$
$$= \sigma^2(\mathbf{a}_1\mathbf{X}^T\mathbf{X} + \mathbf{a}_2^T\mathbf{A})\mathbf{G}_{11}[\mathbf{c} - \mathbf{A}^T\mathbf{g} - (\mathbf{c}^T - \mathbf{A}^T\mathbf{a}_2 + \mathbf{A}^T\mathbf{G}_{22}\mathbf{A}\mathbf{a}_1)]$$
$$= 0.$$

Thus,

$$\begin{aligned}\text{var}(t_0 + \mathbf{t}^T\mathbf{y}) &= \text{var}(\mathbf{c}^T\breve{\boldsymbol{\beta}}) + \text{var}(\mathbf{t}^T\mathbf{y} - \mathbf{c}^T\breve{\boldsymbol{\beta}}) \\ &\geq \text{var}(\mathbf{c}^T\breve{\boldsymbol{\beta}}),\end{aligned}$$

and equality holds if and only if $\text{var}(\mathbf{t}^T\mathbf{y} - \mathbf{c}^T\breve{\boldsymbol{\beta}}) = 0$, i.e., if and only if $\sigma^2(\mathbf{t}^T - \mathbf{c}^T\mathbf{G}_{11}\mathbf{X}^T)(\mathbf{t}^T - \mathbf{c}^T\mathbf{G}_{11}\mathbf{X}^T)^T = 0$, i.e., if and only if $\mathbf{t}^T = \mathbf{c}^T\mathbf{G}_{11}\mathbf{X}^T$, i.e., if and only if $\mathbf{t}^T\mathbf{y} = \mathbf{c}^T\mathbf{G}_{11}\mathbf{X}^T\mathbf{y}$ for all $\mathbf{y}$, i.e., if and only if $t_0 + \mathbf{t}^T\mathbf{y} = \mathbf{g}^T\mathbf{h} + \mathbf{c}^T\mathbf{G}_{11}\mathbf{X}^T\mathbf{y}$ for all $\mathbf{y}$, i.e., if and only if $t_0 + \mathbf{t}^T\mathbf{y} = \mathbf{c}^T\mathbf{G}_{12}\mathbf{h} + \mathbf{c}^T\mathbf{G}_{11}\mathbf{X}^T\mathbf{y} = \mathbf{c}^T\breve{\boldsymbol{\beta}}$ for all $\mathbf{y}$ because, by various parts of Theorem 3.3.8,

$$\begin{aligned}\mathbf{g}^T\mathbf{A} &= \mathbf{c}^T - \mathbf{t}^T\mathbf{X} = \mathbf{c}^T - \mathbf{c}^T\mathbf{G}_{11}\mathbf{X}^T\mathbf{X} \\ &= \mathbf{c}^T - (\mathbf{a}_1^T\mathbf{X}^T\mathbf{X} + \mathbf{a}_2^T\mathbf{A})\mathbf{G}_{11}\mathbf{X}^T\mathbf{X} = \mathbf{c}^T - \mathbf{a}_1^T\mathbf{X}^T\mathbf{X}\mathbf{G}_{11}\mathbf{X}^T\mathbf{X} \\ &= \mathbf{c}^T - \mathbf{a}_1^T(\mathbf{X}^T\mathbf{X} + \mathbf{A}^T\mathbf{G}_{22}\mathbf{A}) = \mathbf{c}^T - \mathbf{a}_1^T\mathbf{X}^T\mathbf{X}(\mathbf{I} - \mathbf{G}_{12}\mathbf{A}) \\ &= \mathbf{c}^T - (\mathbf{c}^T - \mathbf{a}_2^T\mathbf{A})(\mathbf{I} - \mathbf{G}_{12}\mathbf{A}) = \mathbf{c}^T - (\mathbf{c}^T - \mathbf{a}_2^T\mathbf{A} - \mathbf{c}^T\mathbf{G}_{12}\mathbf{A} + \mathbf{a}_2^T\mathbf{A}\mathbf{G}_{12}\mathbf{A}) \\ &= \mathbf{c}^T\mathbf{G}_{12}\mathbf{A},\end{aligned}$$

implying that $\mathbf{g}^T\mathbf{h} = \mathbf{c}^T\mathbf{G}_{12}\mathbf{h}$.

▶ **Exercise 7** Prove Theorem 10.1.10: Define $\breve{\mathbf{y}} = \mathbf{X}\breve{\boldsymbol{\beta}}$ and $\breve{\mathbf{e}} = \mathbf{y} - \mathbf{X}\breve{\boldsymbol{\beta}}$, where $\mathbf{X}\breve{\boldsymbol{\beta}}$ is given by (10.6). Then, under the constrained Gauss–Markov model $\{\mathbf{y}, \mathbf{X}\boldsymbol{\beta}, \sigma^2\mathbf{I} : \mathbf{A}\boldsymbol{\beta} = \mathbf{h}\}$:

(a) $\text{E}(\breve{\mathbf{y}}) = \mathbf{X}\boldsymbol{\beta}$;
(b) $\text{E}(\breve{\mathbf{e}}) = \mathbf{0}$;
(c) $\text{var}(\breve{\mathbf{y}}) = \sigma^2\mathbf{P}_{\mathbf{X}(\mathbf{I}-\mathbf{A}^-\mathbf{A})}$;
(d) $\text{var}(\breve{\mathbf{e}}) = \sigma^2(\mathbf{I} - \mathbf{P}_{\mathbf{X}(\mathbf{I}-\mathbf{A}^-\mathbf{A})})$; and
(e) $\text{cov}(\breve{\mathbf{y}}, \breve{\mathbf{e}}) = \mathbf{0}$.

Solution Because $\mathbf{X}\boldsymbol{\beta}$ is estimable under the constrained model, $\mathbf{X}\breve{\boldsymbol{\beta}}$ is its constrained least squares estimator. By Theorem 10.1.4, $\mathbf{X}\breve{\boldsymbol{\beta}}$ is unbiased for $\mathbf{X}\boldsymbol{\beta}$. Thus $\text{E}(\breve{\mathbf{y}}) = \text{E}(\mathbf{X}\breve{\boldsymbol{\beta}}) = \mathbf{X}\boldsymbol{\beta}$, which proves part (a). Then $\text{E}(\breve{\mathbf{e}}) = \text{E}(\mathbf{y} - \mathbf{X}\breve{\boldsymbol{\beta}}) = \mathbf{X}\beta - \mathbf{X}\beta = 0$. This establishes part (b). Finally, by the symmetry and idempotency of $\mathbf{P}_{\mathbf{X}(\mathbf{I}-\mathbf{A}^-\mathbf{A})}$

(Theorem 10.1.8).

$$\text{var}(\breve{\mathbf{y}}) = \text{var}(\mathbf{P}_{\mathbf{X}(\mathbf{I}-\mathbf{A}^-\mathbf{A})}\mathbf{y}) = \mathbf{P}_{\mathbf{X}(\mathbf{I}-\mathbf{A}^-\mathbf{A})}(\sigma^2\mathbf{I})\mathbf{P}^T_{\mathbf{X}(\mathbf{I}-\mathbf{A}^-\mathbf{A}^-\mathbf{A})} = \sigma^2\mathbf{P}_{\mathbf{X}(\mathbf{I}-\mathbf{A}^-\mathbf{A})},$$

$$\begin{aligned}\text{var}(\breve{\mathbf{e}}) &= \text{var}[(\mathbf{I}-\mathbf{P}_{\mathbf{X}(\mathbf{I}-\mathbf{A}^-\mathbf{A})})\mathbf{y}] = (\mathbf{I}-\mathbf{P}_{\mathbf{X}(\mathbf{I}-\mathbf{A}^-\mathbf{A})})(\sigma^2\mathbf{I})(\mathbf{I}-\mathbf{P}_{\mathbf{X}(\mathbf{I}-\mathbf{A}^-\mathbf{A})})^T \\ &= \sigma^2(\mathbf{I}-\mathbf{P}_{\mathbf{X}(\mathbf{I}-\mathbf{A}^-\mathbf{A})}),\end{aligned}$$

and

$$\begin{aligned}\text{cov}(\breve{\mathbf{y}}, \breve{\mathbf{e}}) &= \text{cov}[\mathbf{P}_{\mathbf{X}(\mathbf{I}-\mathbf{A}^-\mathbf{A})}\mathbf{y}, (\mathbf{I}-\mathbf{P}_{\mathbf{X}(\mathbf{I}-\mathbf{A}^-\mathbf{A})})\mathbf{y}] = \mathbf{P}_{\mathbf{X}(\mathbf{I}-\mathbf{A}^-\mathbf{A})}(\sigma^2\mathbf{I})(\mathbf{I}-\mathbf{P}_{\mathbf{X}(\mathbf{I}-\mathbf{A}^-\mathbf{A})}) \\ &= \sigma^2(\mathbf{P}_{\mathbf{X}(\mathbf{I}-\mathbf{A}^-\mathbf{A})} - \mathbf{P}_{\mathbf{X}(\mathbf{I}-\mathbf{A}^-\mathbf{A})}) = \mathbf{0}.\end{aligned}$$

This establishes the remaining parts of the theorem.

▶ **Exercise 8** Under the constrained Gauss–Markov linear model

$$\mathbf{y} = \mathbf{X}\boldsymbol{\beta} + \mathbf{e}, \quad \mathbf{A}\boldsymbol{\beta} = \mathbf{h},$$

consider the estimation of an estimable function $\mathbf{c}^T\boldsymbol{\beta}$.

(a) Show that the least squares estimator of $\mathbf{c}^T\boldsymbol{\beta}$ associated with the corresponding unconstrained model is unbiased under the constrained model.
(b) Theorem 10.1.6, in tandem with the unbiasedness of the least squares estimator shown in part (a), reveals that the variance of the least squares estimator of $\mathbf{c}^T\boldsymbol{\beta}$ under the constrained Gauss–Markov model is at least as large as the variance of the constrained least squares estimator of $\mathbf{c}^T\boldsymbol{\beta}$ under that model. For the case in which the constraints are estimable under the corresponding unconstrained model, obtain an expression for the amount by which the former exceeds the latter and verify that this exceedance is nonnegative.
(c) Suppose that $n > \text{rank}(\mathbf{X})$, and let $\hat{\sigma}^2$ denote the residual mean square associated with the corresponding unconstrained model, i.e., $\hat{\sigma}^2 = \mathbf{y}^T(\mathbf{I} - \mathbf{P_X})\mathbf{y}/[n - \text{rank}(\mathbf{X})]$. Is $\hat{\sigma}^2$ unbiased under the constrained model?

Solution

(a) (a) As a special case of Theorem 6.2.1, the least squares estimator $\mathbf{c}^T(\mathbf{X}^T\mathbf{X})^-\mathbf{X}^T\mathbf{y}$ of $\mathbf{c}^T\boldsymbol{\beta}$ is unbiased if and only if a vector $\mathbf{g}$ exists such that $\mathbf{g}^T\mathbf{h} = 0$ and $\mathbf{c}^T(\mathbf{X}^T\mathbf{X})^-\mathbf{X}^T\mathbf{X} + \mathbf{g}^T\mathbf{A} = \mathbf{c}^T$, i.e., because $\mathbf{c}^T \in \mathcal{R}(\mathbf{X})$, if and only if a vector $\mathbf{g}$ exists such that $\mathbf{g}^T\mathbf{h} = 0$ and $\mathbf{c}^T + \mathbf{g}^T\mathbf{A} = \mathbf{c}^T$. Because $\mathbf{g} = \mathbf{0}$ is such a vector, the least squares estimator is unbiased.

(b) By Theorems 7.2.2 and 10.1.5, $\text{var}(\mathbf{c}^T\hat{\boldsymbol{\beta}}) = \sigma^2\mathbf{c}^T(\mathbf{X}^T\mathbf{X})^-\mathbf{c}$ and $\text{var}(\mathbf{c}^T\breve{\boldsymbol{\beta}}) = \sigma^2\mathbf{c}^T\mathbf{G}_{11}\mathbf{c}$, where $\mathbf{G}_{11}$ is the upper left $p \times p$ block of any generalized inverse $\begin{pmatrix} \mathbf{G}_{11} & \mathbf{G}_{12} \\ \mathbf{G}_{21} & \mathbf{G}_{22} \end{pmatrix}$ of $\begin{pmatrix} \mathbf{X}^T\mathbf{X} & \mathbf{A}^T \\ \mathbf{A} & \mathbf{0} \end{pmatrix}$. If the constraints are estimable under the unconstrained model, then as shown in Sect. 10.3, one generalized inverse of $\begin{pmatrix} \mathbf{X}^T\mathbf{X} & \mathbf{A}^T \\ \mathbf{A} & \mathbf{0} \end{pmatrix}$ has as its upper left $p \times p$ block the matrix

$$(\mathbf{X}^T\mathbf{X})^- - (\mathbf{X}^T\mathbf{X})^-\mathbf{A}^T[\mathbf{A}(\mathbf{X}^T\mathbf{X})^-\mathbf{A}^T]^-\mathbf{A}(\mathbf{X}^T\mathbf{X})^-.$$

So, in this case

$$\begin{aligned}\text{var}(\mathbf{c}^T\breve{\boldsymbol{\beta}}) &= \sigma^2\mathbf{c}^T(\mathbf{X}^T\mathbf{X})^-\mathbf{c} - \sigma^2\mathbf{c}^T(\mathbf{X}^T\mathbf{X})^-\mathbf{A}^T[\mathbf{A}(\mathbf{X}^T\mathbf{X})^-\mathbf{A}^T]^-\mathbf{A}(\mathbf{X}^T\mathbf{X})^-\mathbf{c} \\ &= \text{var}(\mathbf{c}^T\hat{\boldsymbol{\beta}}) - \sigma^2\mathbf{c}^T(\mathbf{X}^T\mathbf{X})^-\mathbf{A}^T[\mathbf{A}(\mathbf{X}^T\mathbf{X})^-\mathbf{A}^T]^-\mathbf{A}(\mathbf{X}^T\mathbf{X})^-\mathbf{c}.\end{aligned}$$

Let $\mathbf{a}$ represent an n-vector such that $\mathbf{a}^T\mathbf{X} = \mathbf{c}^T$, and let $\mathbf{M}$ represent a matrix such that $\mathbf{MX} = \mathbf{A}$. Then, $\text{var}(\mathbf{c}^T\hat{\boldsymbol{\beta}})$ exceeds $\text{var}(\mathbf{c}^T\tilde{\boldsymbol{\beta}})$ by the amount

$$\begin{aligned}&\sigma^2\mathbf{c}^T(\mathbf{X}^T\mathbf{X})^-\mathbf{A}^T[\mathbf{A}(\mathbf{X}^T\mathbf{X})^-\mathbf{A}^T]^-\mathbf{A}(\mathbf{X}^T\mathbf{X})^-\mathbf{c} \\ &= \sigma^2\mathbf{a}^T\mathbf{X}(\mathbf{X}^T\mathbf{X})^-\mathbf{X}^T\mathbf{M}^T[\mathbf{MX}(\mathbf{X}^T\mathbf{X})^-\mathbf{X}^T\mathbf{M}^T]^-\mathbf{MX}(\mathbf{X}^T\mathbf{X})^-\mathbf{X}^T\mathbf{a} \\ &= \sigma^2\mathbf{a}^T(\mathbf{MP_X})^T[(\mathbf{MP_X})(\mathbf{MP_X})^T]^-(\mathbf{MP_X})\mathbf{a} \\ &= \sigma^2\mathbf{a}^T\mathbf{P}_{(\mathbf{MP_X})^T}\mathbf{a},\end{aligned}$$

which is nonnegative because $\mathbf{P}_{(\mathbf{MP_X})^T}$ is nonnegative definite by Corollary 2.15.9.1.

(c) Yes, $\hat{\sigma}^2$ is unbiased under the constrained model. This may be shown in exactly the same manner that it was shown in the proof of Theorem 8.2.1 for an unconstrained model.

▶ **Exercise 9** Prove Theorem 10.4.2b: Consider a constrained linear model $\{\mathbf{y}, \mathbf{X}\boldsymbol{\beta} : \mathbf{A}\boldsymbol{\beta} = \mathbf{h}\}$ for which the constraints are jointly nonestimable under the corresponding unconstrained model, and let $(\breve{\boldsymbol{\beta}}^T, \breve{\boldsymbol{\lambda}}^T)^T$ be any solution to the constrained normal equations for the constrained model. Then $\mathbf{XG}_{11}\mathbf{X}^T = \mathbf{P_X}$, $\mathbf{XG}_{12}\mathbf{h} = \mathbf{0}$, and $\mathbf{h}^T\mathbf{G}_{22}\mathbf{h} = 0$. (Hint: Use Theorems 3.3.11 and 3.3.8b.)

Solution By Theorem 3.3.11, with $\begin{pmatrix} \mathbf{X}^T\mathbf{X} & \mathbf{A}^T \\ \mathbf{A} & \mathbf{0} \end{pmatrix}$ playing the role of $\begin{pmatrix} \mathbf{A}_{11} & \mathbf{A}_{12} \\ \mathbf{A}_{21} & \mathbf{A}_{22} \end{pmatrix}$ in that theorem, $\mathbf{G}_{11}$ is a generalized inverse of $\mathbf{X}^T\mathbf{X}$. It follows immediately that $\mathbf{XG}_{11}\mathbf{X}^T = \mathbf{P_X}$.

Next, by Theorem 3.3.8b, $\mathbf{X}^T\mathbf{XG}_{12}\mathbf{A} = \mathbf{A}^T\mathbf{G}_{21}\mathbf{X}^T\mathbf{X}$. The rows of the matrix on the left-hand side of this matrix equation are elements of $\mathcal{R}(\mathbf{A})$, and the rows of the matrix on the right-hand side are elements of $\mathcal{R}(\mathbf{X})$. Because $\mathcal{R}(\mathbf{X}) \cap \mathcal{R}(\mathbf{A}) = \{\mathbf{0}\}$, all elements of these two matrices are zero. In particular $\mathbf{X}^T\mathbf{X}\mathbf{G}_{12}\mathbf{A} = \mathbf{0}$, implying (by Theorem 3.3.2) that $\mathbf{X}\mathbf{G}_{12}\mathbf{A} = \mathbf{0}$, implying finally (upon post-multiplication of both sides of this last matrix equation by $\boldsymbol{\beta}$) that $\mathbf{X}\mathbf{G}_{12}\mathbf{h} = \mathbf{0}$.

Finally, because (as just shown) $\mathbf{X}^T\mathbf{X}\mathbf{G}_{12}\mathbf{A} = \mathbf{0}$, by Theorem 3.3.8b $\mathbf{A}^T\mathbf{G}_{22}\mathbf{A} = \mathbf{0}$. Pre- and post-multiplication of this last matrix equation by $\boldsymbol{\beta}^T$ and $\boldsymbol{\beta}$, respectively, yields $\mathbf{h}^T\mathbf{G}_{22}\mathbf{h} = 0$.

▶ **Exercise 10** For the case of estimable constraints:

(a) Obtain expression (10.14) for the constrained least squares estimator of an estimable function $\mathbf{c}^T\boldsymbol{\beta}$.
(b) Show that $\mathbf{P}_{\mathbf{X}(\mathbf{I}-\mathbf{A}^-\mathbf{A})} = \mathbf{P}_{\mathbf{X}} - \mathbf{P}_{\mathbf{X}(\mathbf{X}^T\mathbf{X})^-\mathbf{A}^T}$.
(c) Show that $(\mathbf{P}_{\mathbf{X}} - \mathbf{P}_{\mathbf{X}(\mathbf{X}^T\mathbf{X})^-\mathbf{A}^T})\mathbf{P}_{\mathbf{X}(\mathbf{X}^T\mathbf{X})^-\mathbf{A}^T} = \mathbf{0}$, $(\mathbf{P}_{\mathbf{X}} - \mathbf{P}_{\mathbf{X}(\mathbf{X}^T\mathbf{X})^-\mathbf{A}^T})(\mathbf{I} - \mathbf{P}_{\mathbf{X}}) = \mathbf{0}$, and $\mathbf{P}_{\mathbf{X}(\mathbf{X}^T\mathbf{X})^-\mathbf{A}^T}(\mathbf{I} - \mathbf{P}_{\mathbf{X}}) = \mathbf{0}$.
(d) Obtain expressions (10.16) and (10.17) for the Constrained Model and Constrained Residual second-order polynomial functions, respectively.

Solution

(a)

$$\begin{aligned}
\mathbf{c}^T\breve{\boldsymbol{\beta}} &= \mathbf{c}^T(\mathbf{G}_{11}\mathbf{X}^T\mathbf{y} + \mathbf{G}_{12}\mathbf{h}) \\
&= \mathbf{c}^T\{(\mathbf{X}^T\mathbf{X})^- - (\mathbf{X}^T\mathbf{X})^-\mathbf{A}^T[\mathbf{A}(\mathbf{X}^T\mathbf{X})^-\mathbf{A}^T]^-\mathbf{A}(\mathbf{X}^T\mathbf{X})^-\}\mathbf{X}^T\mathbf{y} \\
&\quad + \mathbf{c}^T(\mathbf{X}^T\mathbf{X})^-\mathbf{A}^T[\mathbf{A}(\mathbf{X}^T\mathbf{X})^-\mathbf{A}^T]^-\mathbf{h} \\
&= \mathbf{c}^T(\mathbf{X}^T\mathbf{X})^-\mathbf{X}^T\mathbf{y} - \mathbf{c}^T(\mathbf{X}^T\mathbf{X})^-\mathbf{A}^T[\mathbf{A}(\mathbf{X}^T\mathbf{X})^-\mathbf{A}^T]^-\mathbf{A}(\mathbf{X}^T\mathbf{X})^-\mathbf{X}^T\mathbf{y} \\
&\quad + \mathbf{c}^T(\mathbf{X}^T\mathbf{X})^-\mathbf{A}^T[\mathbf{A}(\mathbf{X}^T\mathbf{X})^-\mathbf{A}^T]^-\mathbf{h} \\
&= \mathbf{c}^T\hat{\boldsymbol{\beta}} - \mathbf{c}^T(\mathbf{X}^T\mathbf{X})^-\mathbf{A}^T[\mathbf{A}(\mathbf{X}^T\mathbf{X})^-\mathbf{A}^T]^-(\mathbf{A}\hat{\boldsymbol{\beta}} - \mathbf{h}).
\end{aligned}$$

(b)

$$\begin{aligned}
\mathbf{P}_{\mathbf{X}(\mathbf{I}-\mathbf{A}^-\mathbf{A})} &= \mathbf{X}\mathbf{G}_{11}\mathbf{X}^T \\
&= \mathbf{X}\{(\mathbf{X}^T\mathbf{X})^- - (\mathbf{X}^T\mathbf{X})^-\mathbf{A}^T[\mathbf{A}(\mathbf{X}^T\mathbf{A}^T]^-\mathbf{A}(\mathbf{X}^T\mathbf{X})^-\}\mathbf{X}^T \\
&= \mathbf{P}_{\mathbf{X}} - \mathbf{X}(\mathbf{X}^T\mathbf{X})^-\mathbf{A}^T[\mathbf{M}\mathbf{X}(\mathbf{X}^T\mathbf{X})^-\mathbf{A}^T]^-\mathbf{A}(\mathbf{X}^T\mathbf{X})^-\mathbf{X}^T \\
&= \mathbf{P}_{\mathbf{X}} - \mathbf{X}(\mathbf{X}^T\mathbf{X})^-\mathbf{A}^T[\mathbf{M}\mathbf{X}((\mathbf{X}^T\mathbf{X})^-)^T\mathbf{X}^T\mathbf{X}(\mathbf{X}^T\mathbf{X})^-\mathbf{A}^T]^-\mathbf{A}(\mathbf{X}^T\mathbf{X})^-\mathbf{X}^T \\
&= \mathbf{P}_{\mathbf{X}} - \mathbf{X}(\mathbf{X}^T\mathbf{X})^-\mathbf{A}^T[(\mathbf{X}(\mathbf{X}^T\mathbf{X})^-\mathbf{A}^T)^T\mathbf{X}(\mathbf{X}^T\mathbf{X})^-\mathbf{A}^T]^-[\mathbf{X}(\mathbf{X}^T\mathbf{X})^-\mathbf{A}]^T \\
&= \mathbf{P}_{\mathbf{X}} - \mathbf{P}_{\mathbf{X}(\mathbf{X}^T\mathbf{X})^-\mathbf{A}^T}
\end{aligned}$$

where we used Theorem 3.3.3b to obtain the fourth equality.

(c) Observe that $\mathbf{P_X} - \mathbf{P}_{\mathbf{X}(\mathbf{X}^T\mathbf{X})^-\mathbf{A}^T}$, $\mathbf{P}_{\mathbf{X}(\mathbf{X}^T\mathbf{X})^-\mathbf{A}^T}$, and $\mathbf{I} - \mathbf{P_X}$ are symmetric. Also, because $\mathcal{C}(\mathbf{X}(\mathbf{X}^T\mathbf{X})^-\mathbf{A}^T) \subseteq \mathcal{C}(\mathbf{X})$, we find that

$$\mathbf{P}_{\mathbf{X}(\mathbf{X}^T\mathbf{X})^-\mathbf{A}^T}\mathbf{P_X} = \mathbf{P}_{\mathbf{X}(\mathbf{X}^T\mathbf{X})^-\mathbf{A}^T}$$

and

$$\mathbf{P}_{\mathbf{X}(\mathbf{X}^T\mathbf{X})^-\mathbf{A}^T}(\mathbf{I} - \mathbf{P_X}) = \mathbf{0}.$$

Furthermore, using these results we obtain

$$(\mathbf{P_X} - \mathbf{P}_{\mathbf{X}(\mathbf{X}^T\mathbf{X})^-\mathbf{A}^T})\mathbf{P}_{\mathbf{X}(\mathbf{X}^T\mathbf{X})^-\mathbf{A}^T} = \mathbf{P}_{\mathbf{X}(\mathbf{X}^T\mathbf{X})^-\mathbf{A}^T} - \mathbf{P}_{\mathbf{X}(\mathbf{X}^T\mathbf{X})^-\mathbf{A}^T} = \mathbf{0}$$

and

$$(\mathbf{P_X} - \mathbf{P}_{\mathbf{X}(\mathbf{X}^T\mathbf{X})^-\mathbf{A}^T})(\mathbf{I} - \mathbf{P_X}) = \mathbf{P_X}(\mathbf{I} - \mathbf{P_X}) - \mathbf{P}_{\mathbf{X}(\mathbf{X}^T\mathbf{X})^-\mathbf{A}^T}(\mathbf{I} - \mathbf{P_X}) = \mathbf{0}.$$

(d) As noted in the derivation of (10.7), $\mathbf{h}^T\mathbf{G}_{21}\mathbf{X}^T\mathbf{y} = \mathbf{h}^T\mathbf{G}_{12}^T\mathbf{X}^T\mathbf{y} = \mathbf{y}^T\mathbf{X}\mathbf{G}_{12}\mathbf{h}$. Thus, the Constrained Model second-order polynomial may be written as

$$\begin{aligned}
&\mathbf{y}^T\mathbf{X}\mathbf{G}_{11}\mathbf{X}^T\mathbf{y} + \mathbf{h}^T\mathbf{G}_{21}\mathbf{X}^T\mathbf{y} + \mathbf{y}^T\mathbf{X}\mathbf{G}_{12}\mathbf{h} + \mathbf{h}^T\mathbf{G}_{22}\mathbf{h} \\
&= \mathbf{y}^T\mathbf{X}(\mathbf{X}^T\mathbf{X})^-\mathbf{X}^T\mathbf{y} - \mathbf{y}^T\mathbf{X}(\mathbf{X}^T\mathbf{X})^-\mathbf{X}^T\mathbf{A}^T[\mathbf{A}(\mathbf{X}^T\mathbf{X})^-\mathbf{A}^T]^-\mathbf{A}(\mathbf{X}^T\mathbf{X})^-\mathbf{X}^T\mathbf{y} \\
&\quad + \mathbf{h}^T[\mathbf{A}(\mathbf{X}^T\mathbf{X})^-\mathbf{A}^T]^-\mathbf{A}(\mathbf{X}^T\mathbf{X})^-\mathbf{X}^T\mathbf{y} \\
&\quad + \mathbf{y}^T\mathbf{X}(\mathbf{X}^T\mathbf{X})^-\mathbf{X}^T\mathbf{A}^T[\mathbf{A}(\mathbf{X}^T\mathbf{X})^-\mathbf{A}^T]^-\mathbf{h} - \mathbf{h}^T[\mathbf{A}(\mathbf{X}^T\mathbf{X})^-\mathbf{A}^T]^-\mathbf{h} \\
&= \mathbf{y}^T\mathbf{P_X}\mathbf{y} - \hat{\boldsymbol{\beta}}^T\mathbf{A}^T[\mathbf{A}(\mathbf{X}^T\mathbf{X})^-\mathbf{A}^T]^-\mathbf{A}\hat{\boldsymbol{\beta}} + \mathbf{h}^T[\mathbf{A}(\mathbf{X}^T\mathbf{X})^-\mathbf{A}^T]^-\mathbf{A}\hat{\boldsymbol{\beta}} \\
&\quad + \hat{\boldsymbol{\beta}}^T\mathbf{A}^T[\mathbf{A}(\mathbf{X}^T\mathbf{X})^-\mathbf{A}^T]^-\mathbf{h} - \mathbf{h}^T[\mathbf{A}(\mathbf{X}^T\mathbf{X})^-\mathbf{A}^T]^-\mathbf{h} \\
&= \mathbf{y}^T\mathbf{P_X}\mathbf{y} - (\mathbf{A}\hat{\boldsymbol{\beta}} - \mathbf{h})^T[\mathbf{A}(\mathbf{X}^T\mathbf{X})^-\mathbf{A}^T]^-(\mathbf{A}\hat{\boldsymbol{\beta}} - \mathbf{h}).
\end{aligned}$$

Then,

$$\begin{aligned}
Q(\breve{\boldsymbol{\beta}}) &= \mathbf{y}^T\mathbf{y} - \{\mathbf{y}^T\mathbf{P_X}\mathbf{y} - (\mathbf{A}\hat{\boldsymbol{\beta}} - \mathbf{h})^T[\mathbf{A}(\mathbf{X}^T\mathbf{X})^-\mathbf{A}^T]^-(\mathbf{A}\hat{\boldsymbol{\beta}} - \mathbf{h})\} \\
&= \mathbf{y}^T(\mathbf{I} - \mathbf{P_X})\mathbf{y} + (\mathbf{A}\hat{\boldsymbol{\beta}} - \mathbf{h})^T[\mathbf{A}(\mathbf{X}^T\mathbf{X})^-\mathbf{A}^T]^-(\mathbf{A}\hat{\boldsymbol{\beta}} - \mathbf{h}).
\end{aligned}$$

▶ **Exercise 11** Consider a constrained version of the simple linear regression model

$$y_i = \beta_1 + \beta_2 x_i + e_i \quad (i = 1, \ldots, n)$$

where the constraint is that the line $\beta_1 + \beta_2 x$ passes through a known point (a, b).

(a) Give the constrained normal equations for this model in matrix form, with numbers given, where possible, for the elements of the matrices and vectors involved.

(b) Give simplified expressions for the constrained least squares estimators of β_1 and β_2.

(c) Give an expression for the constrained residual mean square in terms of a solution to the constrained normal equations (and other quantities).

Solution

(a) The constrained normal equations are

$$\begin{pmatrix} n & \sum_{i=1}^n x_i & 1 \\ \sum_{i=1}^n x_i & \sum_{i=1}^n x_i^2 & a \\ 1 & a & 0 \end{pmatrix} \begin{pmatrix} \beta_1 \\ \beta_2 \\ \lambda \end{pmatrix} = \begin{pmatrix} \sum_{i=1}^n y_i \\ \sum_{i=1}^n x_i y_i \\ b \end{pmatrix}.$$

(b) A solution to the constrained normal equations in part (a) may be obtained by inverting the coefficient matrix of those equations, but a less messy approach is to multiply the first equation by a and subtract it from the second equation, yielding the equation

$$\beta_1 \sum_{i=1}^n (x_i - a) + \beta_2 \sum_{i=1}^n x_i(x_i - a) = \sum_{i=1}^n (x_i - a)y_i. \tag{10.1}$$

Then multiplying the third of the constrained normal equations by $\sum_{i=1}^n (x_i - a)$ and subtracting it from (10.1) yields the solution

$$\breve{\beta}_2 = \frac{\sum_{i=1}^n (x_i - a)(y_i - b)}{\sum_{i=1}^n (x_i - a)^2},$$

$$\breve{\beta}_1 = b - a\breve{\beta}_2,$$

$$\breve{\lambda} = \sum_{i=1}^n y_i - n\breve{\beta}_1 - \breve{\beta}_2 \sum_{i=1}^n x_i.$$

(c) $p_{\mathbf{A}}^* = \text{rank}\begin{pmatrix} \mathbf{X} \\ \mathbf{A} \end{pmatrix} - \text{rank}(\mathbf{A}) = 2 - 1 = 1$. Hence

$$\breve{\sigma}^2 = \frac{\mathbf{y}^T\mathbf{y} - \breve{\boldsymbol{\beta}}^T\mathbf{X}^T\mathbf{y} - \breve{\lambda}b}{n - p_{\mathbf{A}}^*} = \frac{\sum_{i=1}^n y_i^2 - \breve{\beta}_1 \sum_{i=1}^n y_i - \breve{\beta}_2 \sum_{i=1}^n x_i y_i - \breve{\lambda}b}{n-1}.$$

Exercise 12 Consider a constrained version of the mean-centered second-order polynomial regression model

$$y_i = \beta_1 + \beta_2(x_i - \bar{x}) + \beta_3(x_i - \bar{x})^2 + e_i \quad (i = 1, \ldots, 5)$$

where $x_i = i$, and the constraint is that the first derivative of the second-order polynomial is equal to 0 at $x = a$ where a is known.

(a) Give the constrained normal equations for this model in matrix form, with numbers given, where possible, for the elements of the matrices and vectors involved.
(b) Give nonmatrix expressions for the constrained least squares estimators of β_1, β_2, and β_3.
(c) Give a nonmatrix expression for the constrained residual mean square in terms of a solution to the constrained normal equations (and other quantities).

Solution

(a) The constrained normal equations are

$$\begin{pmatrix} n & 0 & \sum_{i=1}^n (x_i - \bar{x})^2 & 0 \\ 0 & \sum_{i=1}^n (x_i - \bar{x})^2 & \sum_{i=1}^n (x_i - \bar{x})^3 & 1 \\ \sum_{i=1}^n (x_i - \bar{x})^2 & \sum_{i=1}^n (x_i - \bar{x})^3 & \sum_{i=1}^n (x_i - \bar{x})^4 & 2(a - \bar{x}) \\ 0 & 1 & 2(a - \bar{x}) & 0 \end{pmatrix} \begin{pmatrix} \beta_1 \\ \beta_2 \\ \beta_3 \\ \lambda \end{pmatrix}$$

$$= \begin{pmatrix} \sum_{i=1}^n y_i \\ \sum_{i=1}^n (x_i - \bar{x}) y_i \\ \sum_{i=1}^n (x_i - \bar{x})^2 y_i \\ 0 \end{pmatrix}.$$

Using the specified values of x_i, $i = 1, \ldots, 5)$, these equations can be written as

$$\begin{pmatrix} 5 & 0 & 10 & 0 \\ 0 & 10 & 0 & 1 \\ 10 & 0 & 34 & 2a - 6 \\ 0 & 1 & 2a - 6 & 0 \end{pmatrix} \begin{pmatrix} \beta_1 \\ \beta_2 \\ \beta_3 \\ \lambda \end{pmatrix} = \begin{pmatrix} \sum_{i=1}^5 y_i \\ 2y_5 + y_4 - y_2 - 2y_1 \\ 4y_5 + y_4 + y_2 + 4y_1 \\ 0 \end{pmatrix}.$$

(b) Multiplying the second equation by $2a - 6$ and subtracting the result from the third equation yield

$$10\beta_1 - 10(2a-6)\beta_2 + 34\beta_3 = (4y_5 + y_4 + y_2 + 4y_1) - (2y_5 + y_4 - y_2 - 2y_1)(2a-6).$$

Subtracting twice the first equation from the equation just derived yields

$$-10(2a-6)\beta_2+14\beta_3 = (2y_5-y_4-y_2+2y_1)-(2y_5+y_4-y_2-2y_1)(2a-6).$$

Adding $10(2a-6)$ times the fourth equation to the equation just derived yields

$$[14+10(2a-6)^2]\beta_3 = (2y_5-y_4-y_2+2y_1)-(2y_5+y_4-y_2-2y_1)(2a-6).$$

Therefore,

$$\breve{\beta}_3 = \frac{(2y_5-y_4-y_2+2y_1)-(2y_5+y_4-y_2-2y_1)(2a-6)}{14+10(2a-6)^2},$$

$$\breve{\beta}_2 = -(2a-6)\breve{\beta}_3,$$

$$\breve{\beta}_1 = \bar{y}-2\breve{\beta}_3.$$

(c)

$$p^*_{\mathbf{A}} = \text{rank}\begin{pmatrix}\mathbf{X}\\ \mathbf{A}\end{pmatrix} - \text{rank}(\mathbf{A}) = 3-1=2, \quad n-p^*_{\mathbf{A}} = 5-2=3,$$

$$\breve{\sigma}^2 = \frac{\sum_{i=1}^5 y_i^2 - \breve{\beta}_1\sum_{i=1}^5 y_i - \breve{\beta}_2(2y_5+y_4-y_2-2y_1) - \breve{\beta}_3(4y_5+y_4+y_2+4y_1)}{3}.$$

▶ **Exercise 13** Recall from Sect. 10.4 that one can augment an unconstrained model of rank p^* with $p-p^*$ jointly nonestimable "pseudo-constraints" so as to obtain a unique solution to the constrained normal equations, from which the least squares estimators of all functions that are estimable under an unconstrained model may be obtained. Extend this idea to a constrained linear model $\mathbf{y} = \mathbf{X}\boldsymbol{\beta}+\mathbf{e}$, where $\boldsymbol{\beta}$ satisfies q_1 real, linearly independent, consistent constraints $\mathbf{A}_1\boldsymbol{\beta} = \mathbf{h}_1$, so as to obtain a unique solution to constrained normal equations corresponding to this constrained model augmented by pseudo-constraints $\mathbf{A}_2\boldsymbol{\beta}_2 = \mathbf{h}_2$, from which constrained least squares estimators of all functions that are estimable under the original constrained model may be obtained.

Solution Suppose that we impose q_2 linearly independent pseudo-constraints $\mathbf{A}_2\boldsymbol{\beta}_2 = \mathbf{h}_2$ upon the model with q_1 linearly independent real constraints $\mathbf{A}_1\boldsymbol{\beta}_1 = \mathbf{h}_1$ and that the pseudo-constraints are jointly nonestimable under the model with the true constraints, or equivalently

$$\mathcal{R}\begin{pmatrix}\mathbf{X}\\ \mathbf{A}_1\end{pmatrix} \cap \mathcal{R}(\mathbf{A}_2) = \{\mathbf{0}\}.$$

By Corollary 10.1.7.1,

$$\begin{aligned}
\text{rank}\begin{pmatrix} \mathbf{X}^T\mathbf{X} & \mathbf{A}_1^T & \mathbf{A}_2^T \\ \mathbf{A}_1 & \mathbf{0} & \mathbf{0} \\ \mathbf{A}_2 & \mathbf{0} & \mathbf{0} \end{pmatrix} &= \text{rank}\begin{pmatrix} \mathbf{X} \\ \mathbf{A}_1 \\ \mathbf{A}_2 \end{pmatrix} + \text{rank}\begin{pmatrix} \mathbf{A}_1 \\ \mathbf{A}_2 \end{pmatrix} \\
&= \text{rank}\begin{pmatrix} \mathbf{X} \\ \mathbf{A}_1 \end{pmatrix} + \text{rank}(\mathbf{A}_2) + \text{rank}\begin{pmatrix} \mathbf{A}_1 \\ \mathbf{A}_2 \end{pmatrix} \\
&= \text{rank}\begin{pmatrix} \mathbf{X} \\ \mathbf{A}_1 \end{pmatrix} + q_2 + (q_1 + q_2).
\end{aligned}$$

Thus, if we choose the number of pseudo-constraints to be $q_2 = p - \text{rank}\begin{pmatrix} \mathbf{X} \\ \mathbf{A}_1 \end{pmatrix}$, then

$$\text{rank}\begin{pmatrix} \mathbf{X}^T\mathbf{X} & \mathbf{A}_1^T & \mathbf{A}_2^T \\ \mathbf{A}_1 & \mathbf{0} & \mathbf{0} \\ \mathbf{A}_2 & \mathbf{0} & \mathbf{0} \end{pmatrix} = p + q_1 + q_2,$$

i.e., $\begin{pmatrix} \mathbf{X}^T\mathbf{X} & \mathbf{A}_1^T & \mathbf{A}_2^T \\ \mathbf{A}_1 & \mathbf{0} & \mathbf{0} \\ \mathbf{A}_2 & \mathbf{0} & \mathbf{0} \end{pmatrix}$ has full rank and is therefore invertible. Thus, if we augment the model with real constraints by $p-\text{rank}\begin{pmatrix} \mathbf{X} \\ \mathbf{A}_1 \end{pmatrix}$ pseudo-constraints that are jointly nonestimable under the model with the true constraints, then a unique solution to the constrained normal equations corresponding to the augmented model is given by

$$\begin{pmatrix} \mathbf{X}^T\mathbf{X} & \mathbf{A}_1^T & \mathbf{A}_2^T \\ \mathbf{A}_1 & \mathbf{0} & \mathbf{0} \\ \mathbf{A}_2 & \mathbf{0} & \mathbf{0} \end{pmatrix}^{-1} \begin{pmatrix} \mathbf{X}^T\mathbf{y} \\ \mathbf{h}_1 \\ \mathbf{h}_2 \end{pmatrix}.$$

Constrained least squares estimators of all functions that are estimable under the original constrained model may be obtained using this solution.

Best Linear Unbiased Estimation for the Aitken Model

11

This chapter presents exercises on best linear unbiased estimation for the Aitken model and provides solutions to those exercises.

▶ **Exercise 1** Prove Theorem 11.1.8: Under the positive definite Aitken model $\{\mathbf{y}, \mathbf{X}\boldsymbol{\beta}, \sigma^2\mathbf{W}\}$:

(a) $\mathrm{E}(\tilde{\mathbf{y}}) = \mathbf{X}\boldsymbol{\beta}$;
(b) $\mathrm{E}(\tilde{\mathbf{e}}) = \mathbf{0}$;
(c) $\mathrm{var}(\tilde{\mathbf{y}}) = \sigma^2\mathbf{X}(\mathbf{X}^T\mathbf{W}^{-1}\mathbf{X})^-\mathbf{X}^T$;
(d) $\mathrm{var}(\tilde{\mathbf{e}}) = \sigma^2[\mathbf{W} - \mathbf{X}(\mathbf{X}^T\mathbf{W}^{-1}\mathbf{X})^-\mathbf{X}^T]$;
(e) $\mathrm{cov}(\tilde{\mathbf{y}}, \tilde{\mathbf{e}}) = \mathbf{0}$.

The expressions in parts (c) and (d) are invariant to the choice of generalized inverse of $\mathbf{X}^T\mathbf{W}^{-1}\mathbf{X}$.

Solution

(a) $\mathrm{E}(\tilde{\mathbf{y}}) = \mathrm{E}(\tilde{\mathbf{P}}_\mathbf{X}\mathbf{y}) = \tilde{\mathbf{P}}_\mathbf{X}\mathbf{X}\boldsymbol{\beta} = \mathbf{X}\boldsymbol{\beta}$ by Theorem 11.1.7b.
(b) $\mathrm{E}(\tilde{\mathbf{e}}) = \mathrm{E}[(\mathbf{I} - \tilde{\mathbf{P}}_\mathbf{X})\mathbf{y}] = (\mathbf{I} - \tilde{\mathbf{P}}_\mathbf{X})\mathbf{X}\boldsymbol{\beta} = \mathbf{X}\boldsymbol{\beta} - \tilde{\mathbf{P}}_\mathbf{X}\mathbf{X}\boldsymbol{\beta} = \mathbf{0}$, again by Theorem 11.1.7b.
(c)

$$\begin{aligned}
\mathrm{var}(\tilde{\mathbf{y}}) &= \mathrm{var}(\tilde{\mathbf{P}}_\mathbf{X}\mathbf{y}) \\
&= \tilde{\mathbf{P}}_\mathbf{X}(\sigma^2\mathbf{W})\tilde{\mathbf{P}}_\mathbf{X}^T = \sigma^2\mathbf{X}(\mathbf{X}^T\mathbf{W}^{-1}\mathbf{X})^-\mathbf{X}^T\mathbf{W}^{-1}\mathbf{W}\mathbf{W}^{-1}\mathbf{X}[(\mathbf{X}^T\mathbf{W}^{-1}\mathbf{X})^-]^T\mathbf{X}^T \\
&= \sigma^2\mathbf{X}[(\mathbf{X}^T\mathbf{W}^{-1}\mathbf{X})^-]^T\mathbf{X}^T \\
&= \sigma^2\mathbf{X}(\mathbf{X}^T\mathbf{W}^{-1}\mathbf{X})^-\mathbf{X}^T
\end{aligned}$$

by Theorem 11.1.6a, b.

D. L. Zimmerman, *Linear Model Theory*,
https://doi.org/10.1007/978-3-030-52074-8_11

(d)

$$
\begin{aligned}
\text{var}(\tilde{\mathbf{e}}) &= \text{var}[(\mathbf{I} - \tilde{\mathbf{P}}_{\mathbf{X}})\mathbf{y}] \\
&= (\mathbf{I} - \tilde{\mathbf{P}}_{\mathbf{X}})(\sigma^2\mathbf{W})(\mathbf{I} - \tilde{\mathbf{P}}_{\mathbf{X}})^T \\
&= \sigma^2(\mathbf{W} - \tilde{\mathbf{P}}_{\mathbf{X}}\mathbf{W} - \mathbf{W}\tilde{\mathbf{P}}_{\mathbf{X}}^T + \tilde{\mathbf{P}}_{\mathbf{X}}\mathbf{W}\tilde{\mathbf{P}}_{\mathbf{X}}^T) \\
&= \sigma^2\{\mathbf{W} - \mathbf{X}(\mathbf{X}^T\mathbf{W}^{-1}\mathbf{X})^-\mathbf{X}^T - \mathbf{X}[(\mathbf{X}^T\mathbf{W}^{-1}\mathbf{X})^-]^T\mathbf{X}^T \\
&\quad +\mathbf{X}(\mathbf{X}^T\mathbf{W}^{-1}\mathbf{X})^-\mathbf{X}^T\mathbf{W}^{-1}\mathbf{X}[(\mathbf{X}^T\mathbf{W}^{-1}\mathbf{X})^-]^T\mathbf{X}^T\} \\
&= \sigma^2[\mathbf{W} - \mathbf{X}(\mathbf{X}^T\mathbf{W}^{-1}\mathbf{X})^-\mathbf{X}^T]
\end{aligned}
$$

by Theorem 11.1.6a, b.

(e)

$$
\begin{aligned}
\text{cov}(\tilde{\mathbf{y}}, \tilde{\mathbf{e}}) &= \text{cov}[\tilde{\mathbf{P}}_{\mathbf{X}}\mathbf{y}, (\mathbf{I} - \tilde{\mathbf{P}}_{\mathbf{X}})\mathbf{y}] \\
&= \tilde{\mathbf{P}}_{\mathbf{X}}(\sigma^2\mathbf{W})(\mathbf{I} - \tilde{\mathbf{P}}_{\mathbf{X}})^T \\
&= \sigma^2\mathbf{X}(\mathbf{X}^T\mathbf{W}^{-1}\mathbf{X})^-\mathbf{X}^T\{\mathbf{I} - \mathbf{W}^{-1}\mathbf{X}[(\mathbf{X}^T\mathbf{W}^{-1}\mathbf{X})^-]^T\mathbf{X}^T\} \\
&= \sigma^2\mathbf{X}(\mathbf{X}^T\mathbf{W}^{-1}\mathbf{X})^-\{\mathbf{X}^T - \mathbf{X}^T\mathbf{W}^{-1}\mathbf{X}[(\mathbf{X}^T\mathbf{W}^{-1}\mathbf{X})^-]^T\mathbf{X}^T\} \\
&= \mathbf{0},
\end{aligned}
$$

where for the last equality we used Theorem 11.1.6a.

▶ **Exercise 2** Prove Theorem 11.1.9b: Under the positive definite Aitken model $\{\mathbf{y}, \mathbf{X}\boldsymbol{\beta}, \sigma^2\mathbf{W}\}$, $\text{E}[\mathbf{y}^T(\mathbf{W}^{-1} - \tilde{\mathbf{P}}_{\mathbf{X}}^T\mathbf{W}^{-1}\tilde{\mathbf{P}}_{\mathbf{X}})\mathbf{y}] = (n - p^*)\sigma^2$.

Solution By Theorems 4.2.4, 1.1.7b, 3.3.5, 2.8.8, and 2.8.9,

$$
\begin{aligned}
\text{E}[\mathbf{y}^T(\mathbf{W}^{-1} - \tilde{\mathbf{P}}_{\mathbf{X}}^T\mathbf{W}^{-1}\tilde{\mathbf{P}}_{\mathbf{X}})\mathbf{y}] &= (\mathbf{X}\boldsymbol{\beta})^T(\mathbf{W}^{-1} - \tilde{\mathbf{P}}_{\mathbf{X}}^T\mathbf{W}^{-1}\tilde{\mathbf{P}}_{\mathbf{X}})\mathbf{X}\boldsymbol{\beta} \\
&\quad +\text{tr}[(\mathbf{W}^{-1} - \tilde{\mathbf{P}}_{\mathbf{X}}^T\mathbf{W}^{-1}\tilde{\mathbf{P}}_{\mathbf{X}})(\sigma^2\mathbf{W})] \\
&= 0 + \sigma^2\text{tr}[\mathbf{I}_n - \mathbf{X}^T\mathbf{W}^{-1}\mathbf{X}(\mathbf{X}^T\mathbf{W}^{-1}\mathbf{X})^-] \\
&= \sigma^2[n - \text{rank}(\mathbf{X}^T\mathbf{W}^{-1}\mathbf{X})] \\
&= \sigma^2[n - \text{rank}(\mathbf{W}^{-\frac{1}{2}}\mathbf{X})] \\
&= \sigma^2(n - p^*).
\end{aligned}
$$

▶ **Exercise 3** Prove Theorem 11.1.10: Let $\mathbf{C}^T\boldsymbol{\beta}$ be an estimable vector, and let $\mathbf{C}^T\tilde{\boldsymbol{\beta}}$ be its generalized least squares estimator associated with the positive definite Aitken model $\{\mathbf{y}, \mathbf{X}\boldsymbol{\beta}, \sigma^2\mathbf{W}\}$. Suppose that $n > p^*$, and let $\tilde{\sigma}^2$ be the generalized residual mean square for the same model.

(a) If the skewness matrix of $\mathbf{y}$ is null, then $\text{cov}(\mathbf{C}^T\tilde{\boldsymbol{\beta}}, \tilde{\sigma}^2) = \mathbf{0}$ under the specified model;
(b) If the skewness and excess kurtosis matrices of $\mathbf{y}$ are null, then $\text{var}(\tilde{\sigma}^2) = 2\sigma^4/(n - p^*)$ under the specified model.

Solution By Corollary 4.2.5.1,

$$\begin{aligned}\text{cov}(\mathbf{C}^T\tilde{\boldsymbol{\beta}}, \tilde{\sigma}^2) &= 2\mathbf{C}^T(\mathbf{X}^T\mathbf{W}^{-1}\mathbf{X})^-\mathbf{X}^T\mathbf{W}^{-1}(\sigma^2\mathbf{W}) \\ &\quad\times[\mathbf{W}^{-1} - \mathbf{W}^{-1}\mathbf{X}(\mathbf{W}^T\mathbf{W}^{-1}\mathbf{X})^-\mathbf{X}^T\mathbf{W}^{-1}]\mathbf{X}\boldsymbol{\beta}/(n - p^*) \\ &= 2\mathbf{C}^T(\mathbf{X}^T\mathbf{W}^{-1}\mathbf{X})^-\mathbf{X}^T\mathbf{W}^{-1}(\sigma^2\mathbf{W}) \\ &\quad\times[\mathbf{W}^{-1}\mathbf{X} - \mathbf{W}^{-1}\mathbf{X}(\mathbf{W}^T\mathbf{W}^{-1}\mathbf{X})^-\mathbf{X}^T\mathbf{W}^{-1}\mathbf{X}]\boldsymbol{\beta}/(n - p^*) \\ &= \mathbf{0},\end{aligned}$$

where the last equality follows from Theorem 11.1.6a. This establishes part (a). For part (b), by Corollary 4.2.6.3,

$$\begin{aligned}\text{var}(\tilde{\sigma}^2) &= 2\text{tr}\{[\mathbf{W}^{-1} - \mathbf{W}^{-1}\mathbf{X}(\mathbf{X}^T\mathbf{W}^{-1}\mathbf{X})^-\mathbf{X}^T\mathbf{W}^{-1}] \\ &\quad(\sigma^2\mathbf{W})[\mathbf{W}^{-1} - \mathbf{W}^{-1}\mathbf{X}(\mathbf{X}^T\mathbf{W}^{-1}\mathbf{X})^-\mathbf{X}^T\mathbf{W}^{-1}](\sigma^2\mathbf{W})/(n - p^*)^2\} \\ &\quad+4(\mathbf{X}\boldsymbol{\beta})^T[\mathbf{W}^{-1} - \mathbf{W}^{-1}\mathbf{X}(\mathbf{X}^T\mathbf{W}^{-1}\mathbf{X})^-\mathbf{X}^T\mathbf{W}^{-1}] \\ &\quad(\sigma^2\mathbf{W})[\mathbf{W}^{-1} - \mathbf{W}^{-1}\mathbf{X}(\mathbf{X}^T\mathbf{W}^{-1}\mathbf{X})^-\mathbf{X}^T\mathbf{W}^{-1}]\mathbf{X}\boldsymbol{\beta} \\ &= [2\sigma^4/(n - p^*)^2]\text{tr}\{[\mathbf{W}^{-1} - \mathbf{W}^{-1}\mathbf{X}(\mathbf{X}^T\mathbf{W}^{-1}\mathbf{X})^-\mathbf{X}^T\mathbf{W}^{-1}]\mathbf{W}\} \\ &= [2\sigma^4/(n - p^*)^2]\text{tr}\{[\mathbf{I} - \mathbf{W}^{-1}\mathbf{X}(\mathbf{X}^T\mathbf{W}^{-1}\mathbf{X})^-\mathbf{X}^T]\} \\ &= [2\sigma^4/(n - p^*)^2]\{\text{tr}(\mathbf{I}_n) - \text{tr}[(\mathbf{X}^T\mathbf{W}^{-1}\mathbf{X})(\mathbf{X}^T\mathbf{W}^{-1}\mathbf{X})^-]\} \\ &= 2\sigma^4/(n - p^*),\end{aligned}$$

where we used Theorems 11.1.6d and 3.3.5 for the second and fifth equalities, respectively.

▶ **Exercise 4** Obtain $\tilde{\beta}$ (the generalized least squares estimator of β) and its variance under an Aitken model for which $\mathbf{X} = \mathbf{1}_n$ and $\mathbf{W} = (1 - \rho)\mathbf{I}_n + \rho\mathbf{J}_n$, where ρ is a known scalar satisfying $-\frac{1}{n-1} < \rho < 1$. How does $\text{var}(\tilde{\beta})$ compare to σ^2/n?

Solution Using expression (2.1) for the inverse of a nonsingular compound symmetric matrix, we obtain

$$\tilde{\beta} = (\mathbf{X}^T\mathbf{W}^{-1}\mathbf{X})^{-1}\mathbf{X}^T\mathbf{W}^{-1}\mathbf{y} = \left[\mathbf{1}_n^T\left(\frac{1}{1-\rho}\mathbf{I}_n - \frac{\rho}{(1-\rho)(1-\rho+n\rho)}\mathbf{J}_n\right)\mathbf{1}_n\right]^{-1}$$
$$\times\mathbf{1}_n^T\left(\frac{1}{1-\rho}\mathbf{I}_n - \frac{\rho}{(1-\rho)(1-\rho+n\rho)}\mathbf{J}_n\right)\mathbf{y}$$
$$= \left(\frac{n}{1-\rho} - \frac{n^2\rho}{(1-\rho)(1-\rho+n\rho)}\right)^{-1}\left(\frac{y.}{1-\rho} - \frac{n\rho y.}{(1-\rho)(1-\rho+n\rho)}\right)$$
$$= \left(\frac{n(1-\rho+n\rho)-n^2\rho}{(1-\rho)(1-\rho+n\rho)}\right)^{-1}\left(\frac{y.(1-\rho+n\rho)-n\rho y.}{(1-\rho)(1-\rho+n\rho)}\right)$$
$$= \bar{y}.$$

Also, using an expression obtained above,

$$\begin{aligned}\text{var}(\tilde{\beta}) &= \sigma^2(\mathbf{X}^T\mathbf{W}^{-1}\mathbf{X})^{-1}\\ &= \sigma^2\left(\frac{n(1-\rho+n\rho)-n^2\rho}{(1-\rho)(1-\rho+n\rho)}\right)^{-1}\\ &= \sigma^2\left(\frac{1-\rho+n\rho}{n}\right)\\ &= \sigma^2\left(\frac{1}{n}+\frac{\rho(n-1)}{n}\right),\end{aligned}$$

which differs from σ^2/n by an amount $\sigma^2\rho(n-1)/n$.

▶ **Exercise 5** Obtain $\tilde{\beta}$ (the generalized least squares estimator of β) and its variance under an Aitken model for which $\mathbf{X} = \mathbf{1}_n$ and $\mathbf{W} = \mathbf{I} + \mathbf{a}\mathbf{a}^T$, where $\mathbf{a}$ is a known n-vector whose elements sum to 0. How does $\text{var}(\tilde{\beta})$ compare to σ^2/n?

Solution Using Corollary 2.9.7.1 and the fact that $\mathbf{1}_n^T\mathbf{a} = 0$, we obtain

$$\begin{aligned}\tilde{\beta} &= (\mathbf{X}^T\mathbf{W}^{-1}\mathbf{X})^{-1}\mathbf{X}^T\mathbf{W}^{-1}\mathbf{y}\\ &= \{\mathbf{1}_n^T[\mathbf{I}_n - (1+\mathbf{a}^T\mathbf{a})^{-1}\mathbf{a}\mathbf{a}^T]\mathbf{1}_n\}^{-1}\mathbf{1}_n^T[\mathbf{I}_n - (1+\mathbf{a}^T\mathbf{a})^{-1}\mathbf{a}\mathbf{a}^T]\mathbf{y}\\ &= (n-0)^{-1}(y.-0)\\ &= \bar{y}.\end{aligned}$$

Furthermore, $\text{var}(\tilde{\beta}) = \sigma^2(\mathbf{X}^T\mathbf{W}^{-1}\mathbf{X})^{-1} = \sigma^2\{\mathbf{1}_n^T[\mathbf{I}_n - (1+\mathbf{a}^T\mathbf{a})^{-1}\mathbf{a}\mathbf{a}^T]\mathbf{1}_n\}^{-1} = \sigma^2/n$.

▶ **Exercise 6** Consider the model

$$y_i = \beta x_i + e_i \quad (i = 1, \ldots, n)$$

where $x_i = i$ for all i and $(e_1, \ldots, e_n)^T$ has mean $\mathbf{0}$ and positive definite variance–covariance matrix

$$\sigma^2 \mathbf{W} = \sigma^2 \begin{pmatrix} 1 & 1 & 1 & \cdots & 1 \\ 1 & 2 & 2 & \cdots & 2 \\ 1 & 2 & 3 & \cdots & 3 \\ \vdots & \vdots & \vdots & \ddots & \vdots \\ 1 & 2 & 3 & \cdots & n \end{pmatrix}.$$

Obtain specialized expressions for the BLUE of β and its variance. (Hint: Observe that the model matrix is one of the columns of $\mathbf{W}$, which makes it possible to compute the BLUE without obtaining an expression for $\mathbf{W}^{-1}$.)

Solution Letting $\mathbf{w}_n$ denote the last column of $\mathbf{W}$, we find that

$$\begin{aligned} \tilde{\beta} &= (\mathbf{x}^T \mathbf{W}^{-1} \mathbf{x})^{-1} \mathbf{x}^T \mathbf{W}^{-1} \mathbf{y} \\ &= (\mathbf{w}_n^T \mathbf{W}^{-1} \mathbf{w}_n)^{-1} \mathbf{w}_n^T \mathbf{W}^{-1} \mathbf{y} \\ &= \left((1, 2, \ldots, n) \mathbf{u}_n^{(n)}\right)^{-1} \mathbf{u}_n^{(n)T} \mathbf{y} \\ &= y_n / n. \end{aligned}$$

Furthermore,

$$\text{var}(\tilde{\beta}) = \text{var}(y_n/n) = (1/n)^2 (\sigma^2 n) = \sigma^2/n.$$

▶ **Exercise 7** Consider the problem of best linear unbiased estimation of estimable linear functions of $\boldsymbol{\beta}$ under the constrained positive definite Aitken model $\{\mathbf{y}, \mathbf{X}\boldsymbol{\beta}, \sigma^2 \mathbf{W} : \mathbf{A}\boldsymbol{\beta} = \mathbf{h}\}$.

(a) Derive the system of equations that must be solved to obtain the BLUE (called the *constrained generalized least squares estimator*) under this model.
(b) Generalize Theorems 10.1.4, 10.1.5, and 10.1.12 to give explicit expressions for the constrained generalized least squares estimator of a vector $\mathbf{C}^T \boldsymbol{\beta}$ of estimable functions and its variance–covariance matrix, and for an unbiased estimator of σ^2, in terms of a generalized inverse of the coefficient matrix for the system of equations derived in part (a).
(c) If the constraints are jointly nonestimable, will the constrained generalized least squares estimators of all functions that are estimable under the constrained

model coincide with the (unconstrained) generalized least squares estimators of those functions? Explain.

Solution

(a) The Lagrangian function for the problem of minimizing $Q_{\mathbf{W}}(\boldsymbol{\beta}) \equiv (\mathbf{y} - \mathbf{X}\boldsymbol{\beta})^T\mathbf{W}^{-1}(\mathbf{y} - \mathbf{X}\boldsymbol{\beta})$ with respect to $\boldsymbol{\beta}$, subject to constraints $\mathbf{A}\boldsymbol{\beta} = \mathbf{h}$, is

$$L(\boldsymbol{\beta}, \boldsymbol{\lambda}) = (\mathbf{y} - \mathbf{X}\boldsymbol{\beta})^T\mathbf{W}^{-1}(\mathbf{y} - \mathbf{X}\boldsymbol{\beta}) + 2\boldsymbol{\lambda}^T(\mathbf{A}\boldsymbol{\beta} - \mathbf{h})$$

where $2\boldsymbol{\lambda}$ is a vector of Lagrange multipliers. Then,

$$\frac{\partial L}{\partial \boldsymbol{\beta}} = -2\mathbf{X}^T\mathbf{W}^{-1}\mathbf{y} + 2\mathbf{X}^T\mathbf{W}^{-1}\mathbf{X}\boldsymbol{\beta} + 2\mathbf{A}^T\boldsymbol{\lambda},$$

$$\frac{\partial L}{\partial \boldsymbol{\lambda}} = 2(\mathbf{A}\boldsymbol{\beta} - \mathbf{h}).$$

Equating these partial derivatives to $\mathbf{0}$ yields the system of equations

$$\begin{pmatrix} \mathbf{X}^T\mathbf{W}^{-1}\mathbf{X} & \mathbf{A}^T \\ \mathbf{A} & \mathbf{0} \end{pmatrix} \begin{pmatrix} \boldsymbol{\beta} \\ \boldsymbol{\lambda} \end{pmatrix} = \begin{pmatrix} \mathbf{X}^T\mathbf{W}^{-1}\mathbf{y} \\ \mathbf{h} \end{pmatrix}.$$

(b) Let $\begin{pmatrix} \mathbf{G}_{11} & \mathbf{G}_{12} \\ \mathbf{G}_{21} & \mathbf{G}_{22} \end{pmatrix}$ represent any generalized inverse of the coefficient matrix of the system of equations derived in part (a), and let $\begin{pmatrix} \breve{\boldsymbol{\beta}} \\ \breve{\boldsymbol{\lambda}} \end{pmatrix}$ be any solution to that system. Then, proceeding as in the proof of Theorem 10.1.4, we obtain

$$\mathbf{c}^T\breve{\boldsymbol{\beta}} = \begin{pmatrix} \mathbf{c}^T & \mathbf{0}^T \end{pmatrix} \begin{pmatrix} \mathbf{G}_{11} & \mathbf{G}_{12} \\ \mathbf{G}_{21} & \mathbf{G}_{22} \end{pmatrix} \begin{pmatrix} \mathbf{X}^T\mathbf{W}^{-1}\mathbf{y} \\ \mathbf{h} \end{pmatrix} = \mathbf{c}^T(\mathbf{G}_{11}\mathbf{X}^T\mathbf{W}^{-1}\mathbf{y} + \mathbf{G}_{12}\mathbf{h}).$$

Very similar arguments to those used in the proofs of Theorem 10.1.5 and 10.1.12 then establish that $\text{var}(\mathbf{C}^T\breve{\boldsymbol{\beta}}) = \sigma^2\mathbf{C}^T\mathbf{G}_{11}\mathbf{C}$ and that $\breve{\sigma}^2 = (\mathbf{y}^T\mathbf{W}^{-1}\mathbf{y} - \breve{\boldsymbol{\beta}}^T\mathbf{X}^T\mathbf{W}^{-1}\mathbf{y} - \breve{\boldsymbol{\lambda}}^T\mathbf{h})/(n - p_{\mathbf{A}}^*)$ is an unbiased estimator of σ^2, where $p_{\mathbf{A}}^*$ is defined exactly as it was in Theorem 10.1.8.

(c) Yes. Observe that the "top" subset of the system of equations derived in part (a) can be rearranged as

$$\mathbf{X}^T\mathbf{W}^{-1}(\mathbf{X}\breve{\boldsymbol{\beta}} - \mathbf{y}) = -\mathbf{A}^T\breve{\boldsymbol{\lambda}},$$

or equivalently as

$$(\mathbf{X}\breve{\boldsymbol{\beta}} - \mathbf{y})^T\mathbf{W}^{-1}\mathbf{X} = -\breve{\boldsymbol{\lambda}}^T\mathbf{A}.$$

If the constraints are jointly nonestimable, then $\mathcal{R}(\mathbf{X}) \cap \mathcal{R}(\mathbf{A}) = \{\mathbf{0}\}$, so both sides of the system of equations equal $\mathbf{0}^T$, yielding the system $\mathbf{X}^T\mathbf{W}^{-1}\mathbf{X}\breve{\boldsymbol{\beta}} = \mathbf{X}^T\mathbf{W}^{-1}\mathbf{y}$. Thus $\breve{\boldsymbol{\beta}}$ satisfies the (unconstrained) Aitken equations, implying that the constrained generalized least squares estimators of all functions that are estimable under a model with jointly nonestimable constraints will coincide with the (unconstrained) generalized least squares estimators of those functions.

▶ **Exercise 8** Consider an Aitken model in which

$$\mathbf{X} = \begin{pmatrix} 1 \\ 1 \\ -2 \end{pmatrix} \quad \text{and} \quad \mathbf{W} = \begin{pmatrix} 1 & -0.5 & -0.5 \\ -0.5 & 1 & -0.5 \\ -0.5 & -0.5 & 1 \end{pmatrix}.$$

Observe that the columns of $\mathbf{W}$ sum to $\mathbf{0}$; thus $\mathbf{W}$ is singular.

(a) Show that $\mathbf{t}^T\mathbf{y}$ and $\boldsymbol{\ell}^T\mathbf{y}$, where $\mathbf{t} = (\frac{1}{6}, \frac{1}{6}, -\frac{1}{3})^T$ and $\boldsymbol{\ell} = (\frac{1}{2}, \frac{1}{2}, 0)^T$, are BLUEs of β, thus establishing that there is not a unique BLUE of β under this model.

(b) Using Theorem 3.3.7, find a generalized inverse of $\begin{pmatrix} \mathbf{W} & \mathbf{X} \\ \mathbf{X}^T & \mathbf{0} \end{pmatrix}$ and use Theorem 11.2.3a to characterize the collection of all BLUEs of β.

Solution

(a) $\mathbf{t}^T\mathbf{X} = (\frac{1}{6}, \frac{1}{6}, -\frac{1}{3})\begin{pmatrix} 1 \\ 1 \\ -2 \end{pmatrix} = 1$ and $\mathbf{W}\mathbf{t} = \begin{pmatrix} 1 & -0.5 & -0.5 \\ -0.5 & 1 & -0.5 \\ -0.5 & -0.5 & 1 \end{pmatrix}\begin{pmatrix} \frac{1}{6} \\ \frac{1}{6} \\ -\frac{1}{3} \end{pmatrix} = (3/2)\begin{pmatrix} 1 \\ 1 \\ -2 \end{pmatrix}$, so by Corollary 11.2.2.1, $\mathbf{t}^T\mathbf{y}$ is a BLUE of β. Similarly, $\boldsymbol{\ell}^T\mathbf{X} = (\frac{1}{2}, \frac{1}{2}, 0)\begin{pmatrix} 1 \\ 1 \\ -2 \end{pmatrix} = 1$ and $\mathbf{W}\boldsymbol{\ell} = \begin{pmatrix} 1 & -0.5 & -0.5 \\ -0.5 & 1 & -0.5 \\ -0.5 & -0.5 & 1 \end{pmatrix}\begin{pmatrix} \frac{1}{2} \\ \frac{1}{2} \\ 0 \end{pmatrix} = (1/4)\begin{pmatrix} 1 \\ 1 \\ -2 \end{pmatrix}$, so $\boldsymbol{\ell}^T\mathbf{y}$ is also a BLUE of β.

(b) Observe that

$$\begin{pmatrix} 1 \\ 1 \\ -2 \end{pmatrix} = \begin{pmatrix} 1 \\ -0.5 \\ -0.5 \end{pmatrix} + \begin{pmatrix} -0.5 \\ 1 \\ -0.5 \end{pmatrix} - \begin{pmatrix} -0.5 \\ -0.5 \\ 1 \end{pmatrix},$$

implying that $\mathcal{C}(\mathbf{X}) \subseteq \mathcal{C}(\mathbf{W})$ and, by the symmetry of $\mathbf{W}$, that $\mathcal{R}(\mathbf{X}^T) \subseteq \mathcal{R}(\mathbf{W})$. Thus, by Theorem 3.3.7, one generalized inverse of $\begin{pmatrix} \mathbf{W} & \mathbf{X} \\ \mathbf{X}^T & \mathbf{0} \end{pmatrix}$ is $\begin{pmatrix} \frac{5}{9} & -\frac{1}{9} & \frac{2}{9} & \frac{1}{6} \\ -\frac{1}{9} & \frac{5}{9} & \frac{2}{9} & \frac{1}{6} \\ \frac{2}{9} & \frac{2}{9} & \frac{2}{9} & -\frac{1}{3} \\ \frac{1}{6} & \frac{1}{6} & -\frac{1}{3} & -\frac{1}{4} \end{pmatrix} \equiv \begin{pmatrix} \mathbf{G}_{11} & \mathbf{G}_{12} \\ \mathbf{G}_{21} & \mathbf{G}_{22} \end{pmatrix}$ where $\mathbf{G}_{11}$ is 3×3. It may be verified that

$$\mathbf{G}_{12} = \begin{pmatrix} \frac{1}{6} \\ \frac{1}{6} \\ -\frac{1}{3} \end{pmatrix}, \quad \mathbf{I}_3 - \mathbf{G}_{11}\mathbf{W} - \mathbf{G}_{12}\mathbf{X}^T = (1/3)\mathbf{J}_3, \quad \text{and} \quad \mathbf{G}_{11}\mathbf{X} = \mathbf{0}_3.$$

Then by Theorem 11.2.3a, the set of all BLUEs of β is

$$\left\{ \tilde{\mathbf{t}}^T\mathbf{y} : \tilde{\mathbf{t}} = \begin{pmatrix} \frac{1}{6} \\ \frac{1}{6} \\ -\frac{1}{3} \end{pmatrix} + ((1/3)\mathbf{J}_3, \mathbf{0}_3)\mathbf{z},\ \mathbf{z} \in \mathbb{R}^4 \right\} = \left\{ \tilde{\mathbf{t}}^T\mathbf{y} : \tilde{\mathbf{t}} = \begin{pmatrix} \frac{1}{6} \\ \frac{1}{6} \\ -\frac{1}{3} \end{pmatrix} + z\mathbf{1}_3,\ z \in \mathbb{R} \right\}.$$

▶ **Exercise 9** Prove Theorem 11.2.4: Let $\mathbf{C}^T\boldsymbol{\beta}$ be an estimable vector, and let $\tilde{\mathbf{t}}^T\mathbf{y}$ be a BLUE of it under the Aitken model $\{\mathbf{y}, \mathbf{X}\boldsymbol{\beta}, \sigma^2\mathbf{W}\}$. Suppose that $\text{rank}(\mathbf{W}, \mathbf{X}) > \text{rank}(\mathbf{X})$.

(a) If the skewness matrix of $\mathbf{y}$ is null, then $\text{cov}(\tilde{\mathbf{t}}^T\mathbf{y}, \tilde{\sigma}^2) = \mathbf{0}$;
(b) If the skewness and excess kurtosis matrices of $\mathbf{y}$ are null, then

$$\text{var}(\tilde{\sigma}^2) = \frac{2\sigma^4}{\text{rank}(\mathbf{W}, \mathbf{X}) - \text{rank}(\mathbf{X})}.$$

Solution

(a) By Corollary 4.2.5.1,

$$\begin{aligned} \text{cov}(\tilde{\mathbf{t}}^T\mathbf{y}, \tilde{\sigma}^2) &= \text{cov}\left\{ \left[\mathbf{c}^T\mathbf{G}_{12}^T + \mathbf{z}^T \begin{pmatrix} \mathbf{I}_n - \mathbf{W}\mathbf{G}_{11}^T - \mathbf{X}\mathbf{G}_{12}^T \\ -\mathbf{X}^T\mathbf{G}_{11} \end{pmatrix} \right] \mathbf{y}, \right. \\ &\qquad \left. \mathbf{y}^T\mathbf{G}_{11}\mathbf{y}/[\text{rank}(\mathbf{W}, \mathbf{X}) - \text{rank}(\mathbf{X})] \right\} \\ &= 2\left[\mathbf{c}^T\mathbf{G}_{12}^T + \mathbf{z}^T \begin{pmatrix} \mathbf{I}_n - \mathbf{W}\mathbf{G}_{11}^T - \mathbf{X}\mathbf{G}_{12}^T \\ -\mathbf{X}^T\mathbf{G}_{11} \end{pmatrix} \right] \\ &\qquad \times(\sigma^2\mathbf{W})\mathbf{G}_{11}\mathbf{X}\boldsymbol{\beta}/[\text{rank}(\mathbf{W}, \mathbf{X}) - \text{rank}(\mathbf{X})] = 0, \end{aligned}$$

where for the last equality we used Theorem 3.3.8c (with $\mathbf{W}$ and $\mathbf{X}$ here playing the roles of $\mathbf{A}$ and $\mathbf{B}$ in the theorem).

(b) For ease of writing, put $s = [\text{rank}(\mathbf{W}, \mathbf{X}) - \text{rank}(\mathbf{X})]$. By Corollary 4.2.6.3,

$$\begin{aligned}
\text{var}(\tilde{\sigma}^2) &= 2\text{tr}[(1/s)\mathbf{G}_{11}(\sigma^2\mathbf{W})(1/s)\mathbf{G}_{11}(\sigma^2\mathbf{W})] + 4(\mathbf{X}\boldsymbol{\beta})^T(1/s)\mathbf{G}_{11}(\sigma^2\mathbf{W})(1/s)\mathbf{G}_{11}\mathbf{X}\boldsymbol{\beta} \\
&= (2\sigma^4/s^2)\text{tr}(\mathbf{G}_{11}\mathbf{W}\mathbf{G}_{11}\mathbf{W}) \\
&= (2\sigma^4/s^2)\text{tr}[\mathbf{G}_{11}(\mathbf{W} + \mathbf{X}\mathbf{G}_{22}\mathbf{X}^T)] \\
&= (2\sigma^4/s^2)\text{tr}(\mathbf{G}_{11}\mathbf{W} - \mathbf{G}_{11}\mathbf{W}\mathbf{G}_{12}\mathbf{X}^T) \\
&= (2\sigma^4/s^2)[\text{tr}(\mathbf{W}\mathbf{G}_{11}) - \text{tr}(\mathbf{G}_{12}\mathbf{X}^T\mathbf{G}_{11}\mathbf{W})] \\
&= (2\sigma^4/s^2)[\text{rank}(\mathbf{W}, \mathbf{X}) - \text{rank}(\mathbf{X})] \\
&= 2\sigma^4/[\text{rank}(\mathbf{W}, \mathbf{X}) - \text{rank}(\mathbf{X})],
\end{aligned}$$

where for the second, third, fourth, and sixth equalities we used parts (c), (d), (b), (c), and (e) (in that order) of Theorem 3.3.8 (with $\mathbf{W}$ and $\mathbf{X}$ here playing the roles of $\mathbf{A}$ and $\mathbf{B}$ in the theorem).

▶ **Exercise 10** Results on constrained least squares estimation under the Gauss–Markov model $\{\mathbf{y}, \mathbf{X}\boldsymbol{\beta}, \sigma^2\mathbf{I} : \mathbf{A}\boldsymbol{\beta} = \mathbf{h}\}$ that were established in Chap. 10 may be used to obtain results for best linear unbiased estimation under the unconstrained Aitken model $\{\mathbf{y}, \mathbf{X}\boldsymbol{\beta}, \sigma^2\mathbf{W}\}$, as follows. Consider the constrained Gauss–Markov model $\{\mathbf{y}, \mathbf{X}\boldsymbol{\beta}, \sigma^2\mathbf{I} : \mathbf{A}\boldsymbol{\beta} = \mathbf{h}\}$. Regard the vector $\mathbf{h}$ on the right-hand side of the system of constraint equations as a vector of q "pseudo-observations" and append it to the vector of actual observations $\mathbf{y}$, and then consider the unconstrained Aitken model

$$\begin{pmatrix} \mathbf{y} \\ \mathbf{h} \end{pmatrix} = \begin{pmatrix} \mathbf{X} \\ \mathbf{A} \end{pmatrix}\boldsymbol{\beta} + \begin{pmatrix} \mathbf{e} \\ \mathbf{0} \end{pmatrix}, \qquad \text{var}\begin{pmatrix} \mathbf{e} \\ \mathbf{0} \end{pmatrix} = \sigma^2\begin{pmatrix} \mathbf{I} & \mathbf{0} \\ \mathbf{0} & \mathbf{0} \end{pmatrix}.$$

Prove that the constrained least squares estimator of a function $\mathbf{c}^T\boldsymbol{\beta}$ that is estimable under the constrained Gauss–Markov model is a BLUE of $\mathbf{c}^T\boldsymbol{\beta}$ under the corresponding unconstrained Aitken model.

Solution First observe (by Theorems 6.1.2 and 6.2.2) that the necessary and sufficient condition for $\mathbf{c}^T\boldsymbol{\beta}$ to be estimable under either model is $\mathbf{c}^T \in \mathcal{R}\begin{pmatrix} \mathbf{X} \\ \mathbf{A} \end{pmatrix}$ so if $\mathbf{c}^T\boldsymbol{\beta}$ is estimable under either model then it is estimable under the other.

Let $\begin{pmatrix} \mathbf{G}_{11}^* & \mathbf{G}_{12}^* \\ \mathbf{G}_{21}^* & \mathbf{G}_{22}^* \end{pmatrix}$, where $\mathbf{G}_{11}^*$ is $p \times p$, represent an arbitrary generalized inverse of $\begin{pmatrix} \mathbf{X}^T\mathbf{X} & \mathbf{A}^T \\ \mathbf{A} & \mathbf{0} \end{pmatrix}$, and let $\begin{pmatrix} \mathbf{G}_{11} & \mathbf{G}_{12} \\ \mathbf{G}_{21} & \mathbf{G}_{22} \end{pmatrix}$, where $\mathbf{G}_{11}$ is $(n+q) \times (n+q)$, represent an arbitrary generalized inverse of $\mathbf{S} = \begin{pmatrix} \mathbf{I}_n & \mathbf{0} & \mathbf{X} \\ \mathbf{0} & \mathbf{0} & \mathbf{A} \\ \mathbf{X}^T & \mathbf{A}^T & \mathbf{0} \end{pmatrix}$. Then form the matrix

$$\mathbf{R} = \begin{pmatrix} \mathbf{I} - \mathbf{X}\mathbf{G}_{11}^{*T}\mathbf{X}^T & -\mathbf{X}\mathbf{G}_{12}^{*} & \mathbf{X}\mathbf{G}_{11}^{*T} \\ -\mathbf{G}_{12}^{*T}\mathbf{X}^T & -\mathbf{G}_{22}^{*} & \mathbf{G}_{12}^{*T} \\ \mathbf{G}_{211} & \mathbf{G}_{212} & \mathbf{G}_{22} \end{pmatrix}$$

where $(\mathbf{G}_{211}, \mathbf{G}_{212}) = \mathbf{G}_{21}$. It suffices to show that $\mathbf{R}$ is a generalized inverse of $\mathbf{S}$, for if that is so, then by Theorem 11.2.3a, $\mathbf{c}^T \begin{pmatrix} \mathbf{X}\mathbf{G}_{11}^{*T} \\ \mathbf{G}_{12}^{*T} \end{pmatrix}^T \begin{pmatrix} \mathbf{y} \\ \mathbf{h} \end{pmatrix}$, or equivalently $\mathbf{c}^T(\mathbf{G}_{11}^{*}\mathbf{X}^T\mathbf{y} + \mathbf{G}_{12}^{*}\mathbf{h})$, is a BLUE of $\mathbf{c}^T\boldsymbol{\beta}$. By Theorem 10.1.4, this would establish that the constrained least squares estimator is a BLUE under the Aitken model.

To show that $\mathbf{R}$ is a generalized inverse of $\mathbf{S}$, it is helpful to first document a few results. Observe that

$$\begin{aligned} \mathbf{0} &= \mathbf{X}^T\mathbf{X} - \mathbf{X}^T\mathbf{X}\mathbf{G}_{11}^{*}\mathbf{X}^T\mathbf{X} + \mathbf{A}\mathbf{G}_{22}^{*}\mathbf{A}^T \\ &= \mathbf{X}^T\mathbf{X} - \mathbf{X}^T\mathbf{X}\mathbf{G}_{11}^{*}\mathbf{X}^T\mathbf{X} - \mathbf{X}^T\mathbf{X}\mathbf{G}_{12}^{*}\mathbf{A} \\ &= \mathbf{X}^T\mathbf{X} - \mathbf{X}^T\mathbf{X}\mathbf{G}_{11}^{*T}\mathbf{X}^T\mathbf{X} - \mathbf{A}^T\mathbf{G}_{12}^{*T}\mathbf{X}^T\mathbf{X}, \end{aligned}$$

where we used Theorem 3.3.8d, b for the first two equalities and ordinary matrix transposition for the third. The third equality implies that

$$\mathbf{X}^T - \mathbf{X}^T\mathbf{X}\mathbf{G}_{11}^{*T}\mathbf{X}^T - \mathbf{A}^T\mathbf{G}_{12}^{*T}\mathbf{X}^T = \mathbf{0} \tag{11.1}$$

by Theorem 3.3.2. Similarly, by Theorem 3.3.8f,

$$\mathbf{0} = \mathbf{X}^T\mathbf{X} - \mathbf{X}^T\mathbf{X}\mathbf{G}_{11}^{*T}\mathbf{X}^T\mathbf{X} + \mathbf{A}\mathbf{G}_{22}^{*}\mathbf{A}^T$$

and via similar manipulations we obtain

$$\mathbf{X}^T - \mathbf{X}^T\mathbf{X}\mathbf{G}_{11}^{*}\mathbf{X}^T - \mathbf{A}^T\mathbf{G}_{12}^{*T}\mathbf{X}^T = \mathbf{0},$$

or equivalently

$$\mathbf{X} - \mathbf{X}\mathbf{G}_{11}^{*T}\mathbf{X}^T\mathbf{X} - \mathbf{X}\mathbf{G}_{12}^{*}\mathbf{A} = \mathbf{0}. \tag{11.2}$$

Also, note that $\mathbf{A}\mathbf{G}_{11}^{*}\mathbf{X}^T\mathbf{X} = \mathbf{0}$ by Theorem 3.3.8c, implying (by Theorem 3.3.2 and matrix transposition) that

$$\mathbf{X}\mathbf{G}_{11}^{*T}\mathbf{A}^T = \mathbf{0}. \tag{11.3}$$

Let

$$\mathbf{SRS} = \begin{pmatrix} \mathbf{M}_{11} & \mathbf{M}_{12} \\ \mathbf{M}_{21} & \mathbf{M}_{22} \end{pmatrix}.$$

Then,

$$\begin{aligned} \mathbf{M}_{11} &= \begin{pmatrix} \mathbf{I} - \mathbf{X}\mathbf{G}_{11}^{*T}\mathbf{X}^T & \mathbf{0} \\ \mathbf{0} & \mathbf{0} \end{pmatrix} + \begin{pmatrix} \mathbf{X} \\ \mathbf{A} \end{pmatrix} \mathbf{G}_{21} \begin{pmatrix} \mathbf{I} & \mathbf{0} \\ \mathbf{0} & \mathbf{0} \end{pmatrix} + \begin{pmatrix} \mathbf{X}\mathbf{G}_{11}^{*T}\mathbf{X}^T & \mathbf{X}\mathbf{G}_{11}^{*T}\mathbf{A}^T \\ \mathbf{0} & \mathbf{0} \end{pmatrix} \\ &\quad + \begin{pmatrix} \mathbf{X} \\ \mathbf{A} \end{pmatrix} \mathbf{G}_{22}(\mathbf{X}^T, \mathbf{A}^T) \\ &= \begin{pmatrix} \mathbf{I} & \mathbf{0} \\ \mathbf{0} & \mathbf{0} \end{pmatrix} \end{aligned}$$

where we used Theorem 3.3.8b and (11.3). Furthermore, by (11.2) and Theorem 3.3.8a,

$$\mathbf{M}_{12} = \begin{pmatrix} \mathbf{X} - \mathbf{X}\mathbf{G}_{11}^{*T}\mathbf{X}^T\mathbf{X} - \mathbf{X}\mathbf{G}_{12}^{*}\mathbf{A} \\ \mathbf{0} \end{pmatrix} + \begin{pmatrix} \mathbf{X} \\ \mathbf{A} \end{pmatrix} \mathbf{G}_{21} \begin{pmatrix} \mathbf{X} \\ \mathbf{A} \end{pmatrix} = \begin{pmatrix} \mathbf{X} \\ \mathbf{A} \end{pmatrix},$$

and by (11.1) and Theorem 3.3.8c, a,

$$\mathbf{M}_{21} = (\mathbf{X}^T\mathbf{X}\mathbf{G}_{11}^{*T}\mathbf{X}^T + \mathbf{A}^T\mathbf{G}_{12}^{*T}\mathbf{X}^T,\ \mathbf{X}^T\mathbf{X}\mathbf{G}_{11}^{*T}\mathbf{A}^T + \mathbf{A}^T\mathbf{G}_{12}^{*T}\mathbf{A}^T) = (\mathbf{X}^T,\ \mathbf{A}^T).$$

Finally, by Theorem 3.3.8b,

$$\mathbf{M}_{22} = -\mathbf{X}^T\mathbf{X}\mathbf{G}_{12}^{*}\mathbf{A} - \mathbf{A}^T\mathbf{G}_{22}^{*}\mathbf{A} = -(-\mathbf{A}^T\mathbf{G}_{22}^{*}\mathbf{A}) - \mathbf{A}^T\mathbf{G}_{22}^{*}\mathbf{A} = \mathbf{0}.$$

Thus $\mathbf{SRS} = \mathbf{S}$.

▶ **Exercise 11** Consider the same simple linear regression setting described in Example 11.3-1.

(a) Assuming that the elements of $\mathbf{W}$ are such that $\mathbf{W}$ is nonnegative definite, determine additional conditions on those elements that are necessary and sufficient for the ordinary least squares estimator of the slope to be a BLUE of that parameter. (Express your conditions as a set of equations that the w_{ij}'s must satisfy, for example, $w_{11} + w_{12} = 3w_{22}$, etc.)
(b) Assuming that the elements of $\mathbf{W}$ are such that $\mathbf{W}$ is nonnegative definite, determine additional conditions on those elements that are necessary and sufficient for the ordinary least squares estimator of the intercept to be a BLUE of that parameter.

(c) Assuming that the elements of $\mathbf{W}$ are such that $\mathbf{W}$ is nonnegative definite, determine additional conditions on those elements that are necessary and sufficient for the ordinary least squares estimator of every estimable function to be a BLUE of that function. (Hint: Combine your results from parts (a) and (b).)

(d) Give numerical entries of a positive definite matrix $\mathbf{W} \neq \mathbf{I}$ for which:
   (i) the ordinary least squares estimator of the slope, but not of the intercept, is a BLUE;
   (ii) the ordinary least squares estimator of the intercept, but not of the slope, is a BLUE;
   (iii) the ordinary least squares estimators of the slope and intercept are BLUEs.

Solution

(a) By Theorem 11.3.1, the necessary and sufficient condition is that

$$\begin{pmatrix} w_{11} & w_{12} & w_{13} \\ w_{12} & w_{22} & w_{23} \\ w_{13} & w_{23} & w_{33} \end{pmatrix} \begin{pmatrix} 1 & -1 \\ 1 & 0 \\ 1 & 1 \end{pmatrix} \begin{pmatrix} 3 & 0 \\ 0 & 2 \end{pmatrix}^{-1} \begin{pmatrix} 0 \\ 1 \end{pmatrix} = \begin{pmatrix} a-b \\ a \\ a+b \end{pmatrix}$$

for some $a, b \in \mathbb{R}$, or equivalently that

$$\begin{pmatrix} -w_{11} + w_{13} \\ -w_{12} + w_{23} \\ -w_{13} + w_{33} \end{pmatrix} = \begin{pmatrix} a-b \\ a \\ a+b \end{pmatrix}$$

for some $a, b \in \mathbb{R}$. This may be re-expressed as

$$(-w_{12} + w_{23}) - (-w_{11} + w_{13}) = (-w_{13} + w_{33}) - (-w_{12} + w_{23}),$$

or equivalently as $w_{11} - w_{33} = 2(w_{12} - w_{23})$.

(b) By Theorem 11.3.1, the necessary and sufficient condition is that

$$\begin{pmatrix} w_{11} & w_{12} & w_{13} \\ w_{12} & w_{22} & w_{23} \\ w_{13} & w_{23} & w_{33} \end{pmatrix} \begin{pmatrix} 1 & -1 \\ 1 & 0 \\ 1 & 1 \end{pmatrix} \begin{pmatrix} 3 & 0 \\ 0 & 2 \end{pmatrix}^{-1} \begin{pmatrix} 1 \\ 0 \end{pmatrix} = \begin{pmatrix} a-b \\ a \\ a+b \end{pmatrix}$$

for some $a, b \in \mathbb{R}$, or equivalently that

$$\begin{pmatrix} w_{11} + w_{12} + w_{13} \\ w_{12} + w_{22} + w_{23} \\ w_{13} + w_{23} + w_{33} \end{pmatrix} = \begin{pmatrix} a-b \\ a \\ a+b \end{pmatrix}$$

for some $a, b \in \mathbb{R}$. This may be re-expressed as

$$(w_{13}+w_{23}+w_{33})-(w_{12}+w_{22}+w_{23}) = (w_{12}+w_{22}+w_{23})-(w_{11}+w_{12}++w_{13}),$$

or equivalently as $w_{11} + w_{33} + 2w_{13} = w_{12} + w_{23} + 2w_{22}$.

(c) Every linear combination of the intercept and slope is estimable, and every estimable function is a linear combination of the intercept and slope. Thus, the least squares estimator of every estimable function will be a BLUE of that function if and only if the least squares estimators of the slope and intercept are BLUEs. By the results of parts (a) and (b), this occurs if and only if $w_{11} - w_{33} = 2(w_{12} - w_{23})$ and $w_{11} + w_{33} + 2w_{13} = w_{12} + w_{23} + 2w_{22}$, which cannot be simplified further.

(d)

(i) A suitable $\mathbf{W}$-matrix is the one featured in Example 11.4-1, i.e., $\mathbf{W} = \begin{pmatrix} 1 & 0.5 & 0.25 \\ 0.5 & 1 & 0.5 \\ 0.25 & 0.5 & 1 \end{pmatrix}$.

(ii) A suitable $\mathbf{W}$-matrix is $\mathbf{W} = \begin{pmatrix} 1.5 & 0 & 0.5 \\ 0 & 2 & 0 \\ 0.5 & 0 & 1.5 \end{pmatrix}$, which is diagonally dominant so it is positive definite by Theorem 2.15.10.

(iii) A suitable $\mathbf{W}$-matrix is $\mathbf{W} = \begin{pmatrix} 1 & 0.5 & 0.5 \\ 0.5 & 1 & 0.5 \\ 0.5 & 0.5 & 1 \end{pmatrix}$, which we know to be positive definite by the results of Example 4.1.

▶ **Exercise 12** Determine the most general form that the variance–covariance matrix of an Aitken one-factor model can have in order for the ordinary least squares estimator of every estimable function under the model to be a BLUE of that function.

Solution Corollary 11.3.1.2 may be used to address this issue. Recall from Example 8.1-3 that for this model, $\mathbf{P_X} = \oplus_{i=1}^{q}(1/n_i)\mathbf{J}_{n_i}$. Now write $\mathbf{W}$ in the following partitioned form:

$$\mathbf{W} = \begin{pmatrix} \mathbf{W}_{11} & \mathbf{W}_{12} & \cdots & \mathbf{W}_{1q} \\ \mathbf{W}_{21} & \mathbf{W}_{22} & \cdots & \mathbf{W}_{2q} \\ \vdots & \vdots & \ddots & \vdots \\ \mathbf{W}_{q1} & \mathbf{W}_{q2} & \cdots & \mathbf{W}_{qq} \end{pmatrix},$$

where $\mathbf{W}_{ij}$ has dimensions $n_i \times n_j$ for $i, j = 1, 2, \ldots, q$. Since $\mathbf{W}$ is nonnegative definite and hence symmetric, we have $\mathbf{W}_{ij} = \mathbf{W}_{ji}^T$.

By Corollary 11.3.1.2, ordinary least squares estimators of all estimable functions are BLUEs of those functions if and only if $\mathbf{WP_X}$ is symmetric. Because

$$\mathbf{WP_X} = \begin{pmatrix} \frac{1}{n_1}\mathbf{W}_{11}\mathbf{J}_{n_1} & \frac{1}{n_2}\mathbf{W}_{12}\mathbf{J}_{n_2} & \cdots & \frac{1}{n_q}\mathbf{W}_{1q}\mathbf{J}_{n_q} \\ \frac{1}{n_1}\mathbf{W}_{21}\mathbf{J}_{n_1} & \frac{1}{n_2}\mathbf{W}_{22}\mathbf{J}_{n_2} & \cdots & \frac{1}{n_q}\mathbf{W}_{2q}\mathbf{J}_{n_q} \\ \vdots & \vdots & \ddots & \vdots \\ \frac{1}{n_1}\mathbf{W}_{q1}\mathbf{J}_{n_1} & \frac{1}{n_2}\mathbf{W}_{q2}\mathbf{J}_{n_2} & \cdots & \frac{1}{n_q}\mathbf{W}_{qq}\mathbf{J}_{n_q} \end{pmatrix},$$

we conclude that $\mathbf{WP_X}$ is symmetric if and only if $\frac{1}{n_i}\mathbf{J}_{n_i}\mathbf{W}_{ij} = \frac{1}{n_j}\mathbf{W}_{ij}\mathbf{J}_{n_j}$ for $i, j = 1, 2, \ldots, q$.

Now fix one particular pair $(i, j) \in \{(s, t) : s, t \in \{1, 2, \ldots, q\}\}$, let

$$\mathbf{W}_{ij} = \begin{pmatrix} w_{11} & w_{12} & \cdots & w_{1n_j} \\ w_{21} & w_{22} & \cdots & w_{2n_j} \\ \vdots & \vdots & & \vdots \\ w_{n_i1} & w_{n_i2} & \cdots & w_{n_in_j} \end{pmatrix},$$

and define $\overline{w}_{k\cdot} = \frac{1}{n_j}\sum_{s=1}^{n_j} w_{ks}$ to be the kth row mean and $\overline{w}_{\cdot l} = \frac{1}{n_i}\sum_{s=1}^{n_i} w_{sl}$ to be the lth column mean, for $k = 1, \ldots, n_i$ and $l = 1, \ldots, n_j$. Then,

$$\frac{1}{n_i}\mathbf{J}_{n_i}\mathbf{W}_{ij} = \frac{1}{n_i}\mathbf{1}_{n_i}\mathbf{1}_{n_i}^T\mathbf{W}_{ij} = \begin{pmatrix} \overline{w}_{\cdot 1} & \overline{w}_{\cdot 2} & \cdots & \overline{w}_{\cdot n_j} \\ \overline{w}_{\cdot 1} & \overline{w}_{\cdot 2} & \cdots & \overline{w}_{\cdot n_j} \\ \vdots & \vdots & & \vdots \\ \overline{w}_{\cdot 1} & \overline{w}_{\cdot 2} & \cdots & \overline{w}_{\cdot n_j} \end{pmatrix},$$

$$\frac{1}{n_j}\mathbf{W}_{ij}\mathbf{J}_{n_j} = \frac{1}{n_j}\mathbf{W}_{ij}\mathbf{1}_{n_j}\mathbf{1}_{n_j}^T = \begin{pmatrix} \overline{w}_{1\cdot} & \overline{w}_{1\cdot} & \cdots & \overline{w}_{1\cdot} \\ \overline{w}_{2\cdot} & \overline{w}_{2\cdot} & \cdots & \overline{w}_{2\cdot} \\ \vdots & \vdots & & \vdots \\ \overline{w}_{n_i\cdot} & \overline{w}_{n_i\cdot} & \cdots & \overline{w}_{n_i\cdot} \end{pmatrix}.$$

By comparing the elements in the two matrices above, we conclude that $\frac{1}{n_i}\mathbf{J}_{n_i}\mathbf{W}_{ij} = \frac{1}{n_j}\mathbf{W}_{ij}\mathbf{J}_{n_j}$ if and only if

$$\overline{w}_{1\cdot} = \overline{w}_{2\cdot} = \cdots = \overline{w}_{n_i\cdot} = \overline{w}_{\cdot 1} = \overline{w}_{\cdot 2} = \cdots = \overline{w}_{\cdot n_j},$$

i.e., if and only if every row and every column of $\mathbf{W}_{ij}$ have the same mean. Since the pair (i, j) is arbitrary, we obtain the following necessary and sufficient conditions on $\mathbf{W}$ in order for the ordinary least squares estimators to be BLUEs:

1. $\mathbf{W}$ is nonnegative definite.
2. When $\mathbf{W}$ is partitioned as in this solution, for each $i, j = 1, 2, \ldots, q$ every row and every column of $\mathbf{W}_{ij}$ have the same mean.

Exercise 13 Consider a k-part general mixed linear model

$$\mathbf{y} = \mathbf{X}_1\boldsymbol{\beta}_1 + \mathbf{X}_2\boldsymbol{\beta}_2 + \cdots \mathbf{X}_k\boldsymbol{\beta}_k + \mathbf{e}$$

for which $\mathbf{V}(\boldsymbol{\theta}) = \theta_0\mathbf{I} + \theta_1\mathbf{X}_1\mathbf{X}_1^T + \theta_2\mathbf{X}_2\mathbf{X}_2^T + \cdots \theta_k\mathbf{X}_k\mathbf{X}_k^T$, where $\boldsymbol{\theta} = (\theta_0, \theta_1, \theta_2, \ldots, \theta_k)^T$ is an unknown parameter belonging to the subset of $\mathbb{R}^{k+1}$ within which $\mathbf{V}(\boldsymbol{\theta})$ is positive definite for every $\boldsymbol{\theta}$.

(a) Show that the ordinary least squares estimator of every estimable function under this model is a BLUE of that function.
(b) For the special case of a two-way main effects model with two levels of Factor A, three levels of Factor B, and one observation per cell, determine numerical entries of a $\mathbf{V}(\boldsymbol{\theta}) \neq \mathbf{I}$ for which the ordinary least squares estimator of every estimable function is a BLUE. (Hint: Use part (a).)
(c) For the special case of a two-factor nested model with two levels of Factor A, two levels of Factor B within each level of Factor A, and two observations per cell, determine numerical entries of a $\mathbf{V}(\boldsymbol{\theta}) \neq \mathbf{I}$ for which the ordinary least squares estimator of every estimable function is a BLUE.

Solution

(a) Using Theorem 9.1.3b (twice), we find that

$$\begin{aligned}\mathbf{V}(\boldsymbol{\theta})\mathbf{P_X} &= (\theta_0\mathbf{I} + \theta_1\mathbf{X}_1\mathbf{X}_1^T + \theta_2\mathbf{X}_2\mathbf{X}_2^T + \cdots \theta_k\mathbf{X}_k\mathbf{X}_k^T)\mathbf{P_X} \\ &= \theta_0\mathbf{P_X} + \theta_1\mathbf{X}_1\mathbf{X}_1^T + \theta_2\mathbf{X}_2\mathbf{X}_2^T + \cdots \theta_k\mathbf{X}_k\mathbf{X}_k^T \\ &= \mathbf{P_X}(\theta_0\mathbf{I} + \theta_1\mathbf{X}_1\mathbf{X}_1^T + \theta_2\mathbf{X}_2\mathbf{X}_2^T + \cdots \theta_k\mathbf{X}_k\mathbf{X}_k^T) \\ &= \mathbf{P_X}\mathbf{V}(\boldsymbol{\theta}).\end{aligned}$$

The desired result follows by Corollary 11.3.1.2.

(b) There are a variety of correct answers, but one of them is obtained using the three-part decomposition of $\mathbf{X}$ given in Example 6.1-3, i.e.,

$$\mathbf{X} = (\mathbf{X}_1, \mathbf{X}_2, \mathbf{X}_3) = (\mathbf{1}_6, \mathbf{I}_2 \otimes \mathbf{1}_3, \mathbf{1}_2 \otimes \mathbf{I}_3)$$

and setting $\theta_0 = 6, \theta_1 = \theta_2 = \theta_3 = 1$. This yields

$$\begin{aligned}
\mathbf{V} &= 6\mathbf{I}_6 + \mathbf{1}_6\mathbf{1}_6^T + (\mathbf{I}_2 \otimes \mathbf{1}_3)(\mathbf{I}_2 \otimes \mathbf{1}_3)^T + (\mathbf{1}_2 \otimes \mathbf{I}_3)(\mathbf{1}_2 \otimes \mathbf{I}_3)^T \\
&= 6\mathbf{I}_6 + \mathbf{J}_6 + (\mathbf{I}_2 \otimes \mathbf{J}_3) + (\mathbf{J}_2 \otimes \mathbf{I}_3) \\
&= \begin{pmatrix} 9&2&2&2&1&1 \\ 2&9&2&1&2&1 \\ 2&2&9&1&1&2 \\ 2&1&1&9&2&2 \\ 1&2&1&2&9&2 \\ 1&1&2&2&2&9 \end{pmatrix},
\end{aligned}$$

which is symmetric, has positive diagonal elements, and is diagonally dominant; thus by Theorem 2.15.10 it is positive definite.

(c) Again there are many correct answers, but one of them is obtained by the three-part decomposition of $\mathbf{X}$ obtained in Exercise 5.4, i.e.,

$$\mathbf{X} = (\mathbf{X}_1, \mathbf{X}_2, \mathbf{X}_3) = (\mathbf{1}_8, \mathbf{I}_2 \otimes \mathbf{1}_4, \mathbf{I}_4 \otimes \mathbf{1}_2)$$

and setting $\theta_0 = 9, \theta_1 = \theta_2 = \theta_3 = 1$. This yields

$$\begin{aligned}
\mathbf{V} &= 9\mathbf{I}_8 + \mathbf{1}_8\mathbf{1}_8^T + (\mathbf{I}_2 \otimes \mathbf{1}_4)(\mathbf{I}_2 \otimes \mathbf{1}_4)^T + (\mathbf{I}_4 \otimes \mathbf{1}_2)(\mathbf{I}_4 \otimes \mathbf{1}_2)^T \\
&= 9\mathbf{I}_8 + \mathbf{J}_8 + (\mathbf{I}_2 \otimes \mathbf{J}_4) + (\mathbf{I}_4 \otimes \mathbf{J}_2) \\
&= \begin{pmatrix} 12&3&2&2&1&1&1&1 \\ 3&12&2&2&1&1&1&1 \\ 2&2&12&3&1&1&1&1 \\ 2&2&3&12&1&1&1&1 \\ 1&1&1&1&12&3&2&2 \\ 1&1&1&1&3&12&2&2 \\ 1&1&1&1&2&2&12&3 \\ 1&1&1&1&2&2&3&12 \end{pmatrix},
\end{aligned}$$

which is symmetric, has positive diagonal elements, and is diagonally dominant; thus by Theorem 2.15.10 it is positive definite.

▶ **Exercise 14** Consider the positive definite Aitken model $\{\mathbf{y}, \mathbf{X}\boldsymbol{\beta}, \sigma^2\mathbf{W}\}$, and suppose that $n > p^*$.

(a) Prove that the ordinary least squares estimator and generalized least squares estimator of every estimable function are equal if and only if $\mathbf{P_X} = \tilde{\mathbf{P}}_\mathbf{X}$.
(b) Define $\hat{\sigma}^2 = \mathbf{y}^T(\mathbf{I} - \mathbf{P_X})\mathbf{y}/(n - p^*)$; note that this would be the residual mean square under a Gauss–Markov model. Also define

$$\tilde{\sigma}^2 = \mathbf{y}^T[\mathbf{W}^{-1} - \mathbf{W}^{-1}\mathbf{X}(\mathbf{X}^T\mathbf{W}^{-1}\mathbf{X})^-\mathbf{X}^T\mathbf{W}^{-1}]\mathbf{y}/(n - p^*),$$

which is the generalized residual mean square. Suppose that the ordinary least squares estimator and generalized least squares estimator of every estimable function are equal. Prove that $\hat{\sigma}^2 = \tilde{\sigma}^2$ (for all $\mathbf{y}$) if and only if $\mathbf{W}(\mathbf{I} - \mathbf{P_X}) = \mathbf{I} - \mathbf{P_X}$.

(c) Suppose that $\mathbf{W} = \mathbf{I} + \mathbf{P_X A P_X}$ for some $n \times n$ matrix $\mathbf{A}$ such that $\mathbf{W}$ is positive definite. Prove that in this case the ordinary least squares estimator and generalized least squares estimator of every estimable function are equal and $\hat{\sigma}^2 = \tilde{\sigma}^2$ (for all $\mathbf{y}$).

Solution

(a) By Theorems 7.1.4 and 11.1.1, the ordinary least squares estimator and generalized least squares estimator of every estimable function are equal if and only if

$$\mathbf{c}^T(\mathbf{X}^T\mathbf{X})^-\mathbf{X}^T\mathbf{y} = \mathbf{c}^T(\mathbf{X}^T\mathbf{W}^{-1}\mathbf{X})^-\mathbf{X}^T\mathbf{W}^{-1}\mathbf{y}$$

for all $\mathbf{y}$ and all $\mathbf{c}^T \in \mathcal{R}(\mathbf{X})$, i.e., by Theorems 2.1.1 and 6.1.2, if and only if

$$\mathbf{a}^T\mathbf{X}(\mathbf{X}^T\mathbf{X})^-\mathbf{X}^T = \mathbf{a}^T\mathbf{X}(\mathbf{X}^T\mathbf{W}^{-1}\mathbf{X})^-\mathbf{X}^T\mathbf{W}^{-1}$$

for all $\mathbf{a}$, i.e., by another application of Theorem 2.1.1, if and only if

$$\mathbf{X}(\mathbf{X}^T\mathbf{X})^-\mathbf{X}^T = \mathbf{X}(\mathbf{X}^T\mathbf{W}^{-1}\mathbf{X})^-\mathbf{X}^T\mathbf{W}^{-1},$$

i.e., if and only if $\mathbf{P_X} = \tilde{\mathbf{P}}_\mathbf{X}$.

(b) $\hat{\sigma}^2 = \tilde{\sigma}^2$ (for all $\mathbf{y}$) if and only if

$$\mathbf{y}^T(\mathbf{I} - \mathbf{P_X})\mathbf{y} = \mathbf{y}^T[\mathbf{W}^{-1} - \mathbf{W}^{-1}\mathbf{X}(\mathbf{X}^T\mathbf{W}^{-1}\mathbf{X})^-\mathbf{X}^T\mathbf{W}^{-1}]\mathbf{y}$$

for all $\mathbf{y}$, i.e. (by the uniqueness assertion in Theorem 2.14.1), if and only if

$$\mathbf{I} - \mathbf{P_X} = \mathbf{W}^{-1} - \mathbf{W}^{-1}\mathbf{X}(\mathbf{X}^T\mathbf{W}^{-1}\mathbf{X})^-\mathbf{X}^T\mathbf{W}^{-1},$$

i.e., if and only if $\mathbf{I} - \mathbf{P_X} = \mathbf{W}^{-1} - \mathbf{W}^{-1}\tilde{\mathbf{P}}_\mathbf{X}$. By the given assumption, this last condition may be re-expressed, using part (a), as $\mathbf{I} - \mathbf{P_X} = \mathbf{W}^{-1} - \mathbf{W}^{-1}\mathbf{P_X}$ or equivalently (by pre-multiplication of both sides by $\mathbf{W}$) as $\mathbf{W}(\mathbf{I} - \mathbf{P_X}) = \mathbf{I} - \mathbf{P_X}$.

(c)

$$\mathbf{W}\mathbf{P_X} = (\mathbf{I} + \mathbf{P_X A P_X})\mathbf{P_X} = \mathbf{P_X} + \mathbf{P_X A P_X} = \mathbf{P_X}(\mathbf{I} + \mathbf{P_X A P_X}) = \mathbf{P_X}\mathbf{W},$$

implying (by Corollary 11.3.1.2) that the least squares estimator and generalized least squares estimator of every estimable function are equal. Furthermore, by Theorem 8.1.4,

$$\mathbf{W}(\mathbf{I} - \mathbf{P_X}) = (\mathbf{I} + \mathbf{P_X A P_X})(\mathbf{I} - \mathbf{P_X}) = \mathbf{I} - \mathbf{P_X} + \mathbf{P_X A P_X}(\mathbf{I} - \mathbf{P_X}) = \mathbf{I} - \mathbf{P_X}.$$

Thus by part (b), $\hat{\sigma}^2 = \tilde{\sigma}^2$ for all $\mathbf{y}$.

Model Misspecification

12

This chapter presents exercises on the effects of misspecifying the linear model and provides solutions to those exercises.

▶ **Exercise 1** Consider the underspecified mean structure scenario, and suppose that $\mathbf{c}_1^T\boldsymbol{\beta}_1$ is estimable under the correct model and $\text{rank}[(\mathbf{I}-\mathbf{P}_1)\mathbf{X}_2] = p_2$. Prove that $\mathbf{c}_1^T\hat{\boldsymbol{\beta}}_1^{(U)}$ and $\mathbf{c}_1^T\hat{\boldsymbol{\beta}}_1$ coincide if and only if $\mathbf{c}_1^T(\mathbf{X}_1^T\mathbf{X}_1)^-\mathbf{X}_1^T\mathbf{X}_2 = \mathbf{0}^T$.

Solution First note that $\mathbf{c}_1^T\hat{\boldsymbol{\beta}}_1^{(U)} = \mathbf{c}_1^T(\mathbf{X}_1^T\mathbf{X}_1)^-\mathbf{X}_1^T\mathbf{y}$. By Theorem 3.3.7a, the upper left $p_1 \times p_1$ and upper right $p_1 \times p_2$ blocks of one generalized inverse of the coefficient matrix of the two-part normal equations corresponding to the correct model are

$$(\mathbf{X}_1^T\mathbf{X}_1)^- + (\mathbf{X}_1^T\mathbf{X}_1)^-\mathbf{X}_1^T\mathbf{X}_2[\mathbf{X}_2^T(\mathbf{I}-\mathbf{P}_1)\mathbf{X}_2]^-\mathbf{X}_2^T\mathbf{X}_1(\mathbf{X}_1^T\mathbf{X}_1)^-$$

and

$$-(\mathbf{X}_1^T\mathbf{X}_1)^-\mathbf{X}_1^T\mathbf{X}_2[\mathbf{X}_2^T(\mathbf{I}-\mathbf{P}_1)\mathbf{X}_2]^-,$$

respectively. Thus,

$$\begin{aligned}\mathbf{c}_1^T\hat{\boldsymbol{\beta}}_1 &= \mathbf{c}_1^T(\mathbf{X}_1^T\mathbf{X}_1)^-\mathbf{X}_1^T\mathbf{y} + \mathbf{c}_1^T(\mathbf{X}_1^T\mathbf{X}_1)^-\mathbf{X}_1^T\mathbf{X}_2[\mathbf{X}_2^T(\mathbf{I}-\mathbf{P}_1)\mathbf{X}_2]^-\mathbf{X}_2^T\mathbf{X}_1(\mathbf{X}_1^T\mathbf{X}_1)^-\mathbf{X}_1^T\mathbf{y}\\ &\quad -\mathbf{c}_1^T(\mathbf{X}_1^T\mathbf{X}_1)^-\mathbf{X}_1^T\mathbf{X}_2[\mathbf{X}_2^T(\mathbf{I}-\mathbf{P}_1)\mathbf{X}_2]^-\mathbf{X}_2^T\mathbf{y}.\end{aligned}$$

D. L. Zimmerman, *Linear Model Theory*,
https://doi.org/10.1007/978-3-030-52074-8_12

So if $\mathbf{c}_1^T(\mathbf{X}_1^T\mathbf{X}_1)^-\mathbf{X}_1^T\mathbf{X}_2 = \mathbf{0}^T$, then $\mathbf{c}_1^T\hat{\boldsymbol{\beta}}_1 = \mathbf{c}_1^T\hat{\boldsymbol{\beta}}_1^{(U)}$. Conversely, if $\mathbf{c}_1^T\hat{\boldsymbol{\beta}}_1 = \mathbf{c}_1^T\hat{\boldsymbol{\beta}}_1^{(U)}$, then

$$\mathbf{c}_1^T(\mathbf{X}_1^T\mathbf{X}_1)^-\mathbf{X}_1^T\mathbf{X}_2[\mathbf{X}_2^T(\mathbf{I}-\mathbf{P}_1)\mathbf{X}_2]^-\mathbf{X}_2^T\mathbf{X}_1(\mathbf{X}_1^T\mathbf{X}_1)^-\mathbf{X}_1^T\mathbf{y}$$
$$-\mathbf{c}_1^T(\mathbf{X}_1^T\mathbf{X}_1)^-\mathbf{X}_1^T\mathbf{X}_2[\mathbf{X}_2^T(\mathbf{I}-\mathbf{P}_1)\mathbf{X}_2]^-\mathbf{X}_2^T\mathbf{y} = 0$$

for all $\mathbf{y}$, implying that

$$\mathbf{c}_1^T(\mathbf{X}_1^T\mathbf{X}_1)^-\mathbf{X}_1^T\mathbf{X}_2[\mathbf{X}_2^T(\mathbf{I}-\mathbf{P}_1)\mathbf{X}_2]^-\mathbf{X}_2^T(\mathbf{I}-\mathbf{P}_1) = \mathbf{0}^T.$$

Post-multiplying by $\mathbf{X}_2$ and letting $\mathbf{a}$ represent any vector such that $\mathbf{c}_1^T = \mathbf{a}^T(\mathbf{I}-\mathbf{P}_2)\mathbf{X}_1$, we obtain

$$\begin{aligned}
\mathbf{0}^T &= \mathbf{a}^T(\mathbf{I}-\mathbf{P}_2)\mathbf{P}_1\mathbf{X}_2[\mathbf{X}_2^T(\mathbf{I}-\mathbf{P}_1)\mathbf{X}_2]^-\mathbf{X}_2^T(\mathbf{I}-\mathbf{P}_1)\mathbf{X}_2 \\
&= \mathbf{a}^T(\mathbf{I}-\mathbf{P}_2)[(\mathbf{I}-\mathbf{P}_1)\mathbf{X}_2][\mathbf{X}_2^T(\mathbf{I}-\mathbf{P}_1)\mathbf{X}_2]^-\mathbf{X}_2^T(\mathbf{I}-\mathbf{P}_1)\mathbf{X}_2 \\
&= \mathbf{a}^T(\mathbf{I}-\mathbf{P}_2)[(\mathbf{I}-\mathbf{P}_1)\mathbf{X}_2] \\
&= \mathbf{a}^T(\mathbf{I}-\mathbf{P}_2)\mathbf{P}_1\mathbf{X}_2 \\
&= \mathbf{a}^T(\mathbf{I}-\mathbf{P}_2)\mathbf{X}_1(\mathbf{X}_1^T\mathbf{X}_1)^-\mathbf{X}_1^T\mathbf{X}_2 \\
&= \mathbf{c}_1^T(\mathbf{X}_1^T\mathbf{X}_1)^-\mathbf{X}_1^T\mathbf{X}_2
\end{aligned}$$

where we used Theorem 3.3.3a [with $(\mathbf{I}-\mathbf{P}_1)\mathbf{X}_2$ playing the role of $\mathbf{A}$ in the theorem] for the third equality.

▶ **Exercise 2** Consider an underspecified mean structure scenario, where both models satisfy Gauss–Markov assumptions. Suppose that $\mathbf{c}_1^T\boldsymbol{\beta}_1$ is estimable under the underspecified model and that $\mathbf{c}_2^T\boldsymbol{\beta}_2$ is estimable under the correct model. Prove that $\mathbf{c}_1^T\hat{\boldsymbol{\beta}}_1^{(U)}$ and $\mathbf{c}_2^T\hat{\boldsymbol{\beta}}_2$ are uncorrelated under the correct model.

Solution By the given estimability conditions, $\mathbf{c}_1^T = \mathbf{a}_1^T\mathbf{X}_1$ for some $\mathbf{a}_1$, and $\mathbf{c}_2^T = \mathbf{a}_2^T(\mathbf{I}-\mathbf{P}_1)\mathbf{X}_2$ for some $\mathbf{a}_2$. Thus we may write $\mathbf{c}_1^T\hat{\boldsymbol{\beta}}_1^{(U)}$ as $\mathbf{a}_1^T\mathbf{X}_1(\mathbf{X}_1^T\mathbf{X}_1)^-\mathbf{X}_1^T\mathbf{y} = \mathbf{a}_1^T\mathbf{P}_1\mathbf{y}$, and using Theorem 3.3.7 we have

$$\begin{aligned}
\mathbf{c}_2^T\hat{\boldsymbol{\beta}}_2 &= \mathbf{a}_2^T(\mathbf{I}-\mathbf{P}_1)\mathbf{X}_2\left(-[\mathbf{X}_2^T(\mathbf{I}-\mathbf{P}_1)\mathbf{X}_2]^-\mathbf{X}_2^T\mathbf{X}_1(\mathbf{X}_1^T\mathbf{X}_1)^-\ \ [\mathbf{X}_2^T(\mathbf{I}-\mathbf{P}_1)\mathbf{X}_2]^-\right)\begin{pmatrix}\mathbf{X}_1^T\mathbf{y}\\ \mathbf{X}_2^T\mathbf{y}\end{pmatrix} \\
&= \mathbf{a}_2^T(\mathbf{I}-\mathbf{P}_1)\mathbf{X}_2\{-[\mathbf{X}_2^T(\mathbf{I}-\mathbf{P}_1)\mathbf{X}_2]^-\mathbf{X}_2^T\mathbf{P}_1\mathbf{y} + [\mathbf{X}_2^T(\mathbf{I}-\mathbf{P}_1)\mathbf{X}_2]^-\mathbf{X}_2^T\mathbf{y}\} \\
&= \mathbf{a}_2^T(\mathbf{I}-\mathbf{P}_1)\mathbf{X}_2[\mathbf{X}_2^T(\mathbf{I}-\mathbf{P}_1)\mathbf{X}_2]^-\mathbf{X}_2^T(\mathbf{I}-\mathbf{P}_1)\mathbf{y} \\
&= \mathbf{a}_2^T(\mathbf{P}_{12}-\mathbf{P}_1)\mathbf{y},
\end{aligned}$$

where we used Theorem 9.1.2 for the last equality. Then, by Theorems 4.2.3 and 9.1.1b,

$$\text{cov}[\mathbf{a}_1^T\mathbf{P}_1\mathbf{y},\ \mathbf{a}_2^T(\mathbf{P}_{12}-\mathbf{P}_1)\mathbf{y}] = \mathbf{a}_1^T\mathbf{P}_1(\sigma^2\mathbf{I})(\mathbf{P}_{12}-\mathbf{P}_1)\mathbf{a}_2 = 0.$$

▶ **Exercise 3** Suppose that the Gauss–Markov no-intercept simple linear regression model is fit to some data, but the correct model is the Gauss–Markov simple linear regression model.

(a) Determine the bias of $\hat{\beta}_2^{(U)}$, and determine necessary and sufficient conditions for the bias to equal 0.
(b) Determine necessary and sufficient conditions for the mean squared error of $\hat{\beta}_2^{(U)}$ to be less than or equal to the mean squared error of the least squares estimator of β_2 under the correct model.

Solution

(a) By Theorem 12.1.2a, the bias is $(n\bar{x}/\sum_{i=1}^n x_i^2)\beta_1$, which is equal to 0 (for all β_1) if and only if $\bar{x} = 0$.
(b) By Theorem 12.1.4b, $\text{MSE}(\hat{\beta}_2^{(U)}) \leq \text{MSE}(\hat{\beta}_2)$ if and only if

$$[(n\bar{x}/\sum_{i=1}^n x_i^2)\beta_1]^2 \leq \sigma^2(n\bar{x}/\sum_{i=1}^n x_i^2)\{\mathbf{1}^T[\mathbf{I}-\mathbf{x}(\mathbf{x}^T\mathbf{x})^{-1}\mathbf{x}^T]\mathbf{1}\}^{-1}(n\bar{x}/\sum_{i=1}^n x_i^2),$$

i.e., if and only if $\beta_1^2 \leq \sigma^2(n - n^2\bar{x}^2/\sum_{i=1}^n x_i^2)^{-1}$, i.e., if and only if $\beta_1^2 \leq \sigma^2\sum_{i=1}^n x_i^2/(nSXX)$, i.e., if and only if

$$\frac{|\beta_1|}{\sqrt{\text{var}(\hat{\beta}_1)}} \leq 1.$$

▶ **Exercise 4** Consider a situation in which the true model for three observations is the two-way main effects Gauss–Markov model

$$y_{ij} = \mu + \alpha_i + \gamma_j + e_{ij}, \quad (i,j) \in \{(1,1)\,(1,2),\ (2,1)\}$$

but the model fitted to the observations is the (underspecified) one-factor Gauss–Markov model

$$y_{ij} = \mu + \alpha_i + e_{ij}, \quad (i,j) \in \{(1,1)\,(1,2),\ (2,1)\}.$$

Recall that the true model for such observations was considered in Examples 7.1-3 and 7.2-3.

(a) Is $\mu + \alpha_1$ estimable under the true model? Regardless of the answer to that question, obtain its least squares estimator under the fitted model. Also obtain the expectation, variance, and mean squared error of that estimator under the true model.
(b) Is $\alpha_1 - \alpha_2$ estimable under the true model? Regardless of the answer to that question, obtain its least squares estimator under the fitted model. Also obtain the expectation, variance, and mean squared error of that estimator under the true model.
(c) Obtain the least squares estimator of $\alpha_1 - \alpha_2$ under the true model, and obtain its expectation, variance, and mean squared error under the true model.
(d) Compare the variances of the estimators of $\alpha_1 - \alpha_2$ obtained in parts (b) and (c). Which theorem does this exemplify?
(e) Obtain an explicit condition on γ_1, γ_2, and σ^2 for the mean squared error of the estimator of $\alpha_1 - \alpha_2$ defined in part (b) to be smaller than the mean squared error of the estimator of $\alpha_1 - \alpha_2$ defined in part (c).

Solution

(a) No, $\mu + \alpha_1$ is not estimable under the correct model. But from Example 6.1-2 it is estimable under the underspecified model, and from Example 7.1-2 the estimator obtained by applying least squares to the underspecified model is $\bar{y}_{1\cdot} = (y_{11} + y_{12})/2$. By Theorem 12.1.2 (or by first principles), this estimator has mean $\mu + \alpha_1 + (\gamma_1 + \gamma_2)/2$, variance $\sigma^2/2$, and mean squared error $(\sigma^2/2) + (\gamma_1 + \gamma_2)^2/4$.
(b) Yes, $\alpha_1 - \alpha_2$ is estimable under the correct model by Theorem 6.1.4 because the two-way layout is connected. From Example 7.1-2, the estimator of $\alpha_1 - \alpha_2$ obtained by applying least squares to the underspecified model is $\bar{y}_{1\cdot} - y_{21}$. By Theorem 12.1.2 (or by first principles), this estimator has expectation $\alpha_1 - \alpha_2 - (\gamma_1 - \gamma_2)/2$, variance $3\sigma^2/2$, and mean squared error $(3\sigma^2/2) + (\gamma_1 - \gamma_2)^2/4$.
(c) From Example 7.1-3, the least squares estimator of $\alpha_1 - \alpha_2$ under the correct model is $y_{11} - y_{21}$, which has expectation $\alpha_1 - \alpha_2$, variance $2\sigma^2$, and mean squared error $2\sigma^2$.
(d) The variance of the estimator obtained in part (b) is 25% smaller than the variance of the estimator obtained in part (c). This exemplifies Theorem 12.1.4a.
(e) The estimator obtained in part (b) has smaller MSE than the estimator obtained in part (c) if and only if $(3\sigma^2/2) + (\gamma_1 - \gamma_2)^2/4 \leq 2\sigma^2$, i.e., if and only if $(\gamma_1 - \gamma_2)^2 \leq 2\sigma^2$.

▶ **Exercise 5** Consider a situation in which the true model for three observations is the one-factor Gauss–Markov model

$$y_{ij} = \mu + \alpha_i + e_{ij}, \quad (i, j) \in \{(1, 1)\,(1, 2),\ (2, 1)\}$$

but the model fitted to the observations is the (overspecified) two-way main effects Gauss–Markov model

$$y_{ij} = \mu + \alpha_i + \gamma_j + e_{ij}, \quad (i, j) \in \{(1, 1)\,(1, 2),\ (2, 1)\}.$$

Recall that the fitted model for such observations was considered in Examples 7.1-3 and 7.2-3.

(a) Obtain the least squares estimator of $\alpha_1 - \alpha_2$ under the fitted model, and obtain the expectation, variance, and mean squared error of that estimator under the true model.
(b) Obtain the least squares estimator of $\alpha_1 - \alpha_2$ under the true model, and obtain the expectation, variance, and mean squared error of that estimator under the true model.
(c) Which estimator has larger mean squared error? Which theorem does this exemplify?

Solution

(a) The least squares estimator of $\alpha_1 - \alpha_2$ under the fitted (overspecified model) is $y_{11} - y_{21}$. Under the correct model, this estimator has expectation $\alpha_1 - \alpha_2$, variance $2\sigma^2$, and mean squared error $2\sigma^2$.
(b) The least squares estimator of $\alpha_1 - \alpha_2$ under the correct model is $\bar{y}_{1\cdot} - y_{21}$. Under the correct model, this estimator has expectation $\alpha_1 - \alpha_2$, variance $3\sigma^2/2$, and mean squared error $3\sigma^2/2$.
(c) The estimator obtained in part (a) has larger mean squared error, exemplifying Theorem 12.1.7b.

▶ **Exercise 6** Prove Theorem 12.1.5b, c: In the underspecified mean structure scenario, if $n > \text{rank}(\mathbf{X}_1)$ then:

(b) if the skewness matrix of $\mathbf{y}$ is null, then $\text{cov}(\mathbf{c}_1^T\hat{\boldsymbol{\beta}}_1^{(U)}, \hat{\sigma}_U^2) = 0$; and
(c) if the skewness and excess kurtosis matrices of $\mathbf{y}$ are null, then

$$\text{var}(\hat{\sigma}_U^2) = \frac{2\sigma^4}{n - \text{rank}(\mathbf{X}_1)} + \frac{4\sigma^2\boldsymbol{\beta}_2^T\mathbf{X}_2^T(\mathbf{I} - \mathbf{P}_1)\mathbf{X}_2\boldsymbol{\beta}_2}{[n - \text{rank}(\mathbf{X}_1)]^2}.$$

Solution By Corollary 4.2.5.1,

$$\begin{aligned}
\text{cov}(\mathbf{c}_1^T\hat{\boldsymbol{\beta}}_1^{(U)}, \hat{\sigma}_U^2) &= \text{cov}\left(\mathbf{c}_1^T(\mathbf{X}_1^T\mathbf{X}_1)^-\mathbf{X}_1^T\mathbf{y}, \frac{\mathbf{y}^T(\mathbf{I} - \mathbf{P}_1)\mathbf{y}}{n - \text{rank}(\mathbf{X}_1)}\right) \\
&= 2\mathbf{c}_1^T(\mathbf{X}_1^T\mathbf{X}_1)^-\mathbf{X}_1^T(\sigma^2\mathbf{I})(\mathbf{I} - \mathbf{P}_1)(\mathbf{X}_1\boldsymbol{\beta}_1 + \mathbf{X}_2\boldsymbol{\beta}_2)/[n - \text{rank}(\mathbf{X}_1)] \\
&= 2\sigma^2\mathbf{c}_1^T(\mathbf{X}_1^T\mathbf{X}_1)^-(\mathbf{X}_1^T - \mathbf{X}_1^T\mathbf{P}_1)(\mathbf{X}_1\boldsymbol{\beta}_1 + \mathbf{X}_2\boldsymbol{\beta}_2)/[n - \text{rank}(\mathbf{X}_1)] \\
&= 0
\end{aligned}$$

because $\mathbf{X}_1^T \mathbf{P}_1 = \mathbf{X}_1^T$. This proves part (b) of the theorem. Next, by Corollary 4.2.6.3,

$$\begin{aligned}\mathrm{var}(\hat{\sigma}_U^2) &= \frac{2\mathrm{tr}[(\mathbf{I}-\mathbf{P}_1)(\sigma^2\mathbf{I})(\mathbf{I}-\mathbf{P}_1)(\sigma^2\mathbf{I})]}{[n-\mathrm{rank}(\mathbf{X}_1)]^2} \\ &\quad + \frac{4(\mathbf{X}_1\boldsymbol{\beta}_1+\mathbf{X}_2\boldsymbol{\beta}_2)^T(\mathbf{I}-\mathbf{P}_1)(\sigma^2\mathbf{I})(\mathbf{I}-\mathbf{P}_1)(\mathbf{X}_1\boldsymbol{\beta}_1+\mathbf{X}_2\boldsymbol{\beta}_2)}{[n-\mathrm{rank}(\mathbf{X}_1)]^2} \\ &= \frac{2\sigma^4\mathrm{tr}(\mathbf{I}-\mathbf{P}_1)}{[n-\mathrm{rank}(\mathbf{X}_1)]^2} + \frac{4\sigma^2\boldsymbol{\beta}_2^T\mathbf{X}_2^T(\mathbf{I}-\mathbf{P}_1)\mathbf{X}_2\boldsymbol{\beta}_2}{[n-\mathrm{rank}(\mathbf{X}_1)]^2} \\ &= \frac{2\sigma^4}{n-\mathrm{rank}(\mathbf{X}_1)} + \frac{4\sigma^2\boldsymbol{\beta}_2^T\mathbf{X}_2^T(\mathbf{I}-\mathbf{P}_1)\mathbf{X}_2\boldsymbol{\beta}_2}{[n-\mathrm{rank}(\mathbf{X}_1)]^2}\end{aligned}$$

where we used Theorem 8.1.3b for the second equality and Theorems 2.12.2 and 8.1.3b, d for the third equality. This proves part (c) of the theorem.

▶ **Exercise 7** In the underspecified simple linear regression settings considered in Example 12.1.1-1 and Exercise 12.3, obtain the bias and variance of $\hat{\sigma}_U^2$ and compare the latter to the variance of $\hat{\sigma}^2$ (assuming that $n > 2$ and that the skewness and excess kurtosis matrices of $\mathbf{y}$ are null).

Solution First consider the setting of Example 12.1.1-1. By Theorem 12.1.5a, the bias of $\hat{\sigma}_U^2$ is $\beta_2\mathbf{x}^T(\mathbf{I}-\frac{1}{n}\mathbf{J}_n)\mathbf{x}\beta_2/(n-1) = \beta_2^2 SXX/(n-1)$. By Theorem 12.1.5c,

$$\mathrm{var}(\hat{\sigma}_U^2) = \frac{2\sigma^4}{n-1} + \frac{4\sigma^2\beta_2\mathbf{x}^T(\mathbf{I}-\frac{1}{n}\mathbf{J}_n)\mathbf{x}\beta_2}{(n-1)^2} = \frac{2\sigma^4}{n-1} + \frac{4\sigma^2\beta_2^2 SXX}{(n-1)^2}.$$

By Theorem 8.2.3, $\mathrm{var}(\hat{\sigma}^2) = 2\sigma^4/(n-2)$. Therefore, $\mathrm{var}(\hat{\sigma}_U^2) \geq \mathrm{var}(\hat{\sigma}^2)$ if and only if

$$\frac{2\sigma^4}{n-1} + \frac{4\sigma^2\beta_2^2 SXX}{(n-1)^2} \geq \frac{2\sigma^4}{n-2},$$

i.e., if and only if

$$2\beta_2^2 SXX \geq \left(\frac{n-1}{n-2}\right)\sigma^2.$$

Next consider the setting of Exercise 12.3. By Theorem 12.1.5a, the bias of $\hat{\sigma}_U^2$ is $\beta_1 \mathbf{1}^T[\mathbf{I} - \mathbf{x}(\mathbf{x}^T\mathbf{x})^{-1}\mathbf{x}^T]\beta_1/(n-1) = \beta_1^2 nSXX/[(n-1)\sum_{i=1}^n x_i^2]$. Therefore, $\text{var}(\hat{\sigma}_U^2) \geq \text{var}(\hat{\sigma}^2)$ if and only if

$$\frac{2\sigma^4}{n-1} + \frac{4\sigma^2\beta_1^2 nSXX}{(n-1)^2\sum_{i=1}^n x_i^2} \geq \frac{2\sigma^4}{n-2},$$

i.e., if and only if

$$2\beta_1^2 nSXX \geq \left(\frac{n-1}{n-2}\right)\sigma^2 \sum_{i=1}^n x_i^2.$$

► **Exercise 8** Prove Theorem 12.1.8: In the overspecified mean structure scenario, suppose that $\mathbf{X}_1^T\mathbf{X}_2 = \mathbf{0}$ and that $\mathbf{c}_1^T\boldsymbol{\beta}_1$ is estimable. Then $\mathbf{c}_1^T\hat{\boldsymbol{\beta}}_1^{(O)}$ and $\mathbf{c}_1^T\hat{\boldsymbol{\beta}}_1$ coincide.

Solution If $\mathbf{X}_1^T\mathbf{X}_2 = \mathbf{0}$, then

$$\begin{aligned}
\mathbf{c}_1^T\hat{\boldsymbol{\beta}}_1^{(O)} &= (\mathbf{c}_1^T, \mathbf{0}^T)\begin{pmatrix} \mathbf{X}_1^T\mathbf{X}_1 & \mathbf{0} \\ \mathbf{0} & \mathbf{X}_2^T\mathbf{X}_2 \end{pmatrix}^- \begin{pmatrix} \mathbf{X}_1^T\mathbf{y} \\ \mathbf{X}_2^T\mathbf{y} \end{pmatrix} \\
&= \mathbf{c}_1^T(\mathbf{X}_1^T\mathbf{X}_1)^-\mathbf{X}_1^T\mathbf{y} \\
&= \mathbf{c}_1^T\hat{\boldsymbol{\beta}}_1,
\end{aligned}$$

where Theorem 3.3.7a was used for the second equality.

► **Exercise 9** Suppose that the Gauss–Markov simple linear regression model is fit to some data, but the correct model is the no-intercept version of the same model. Determine the mean squared error of $\hat{\beta}_2^{(O)}$ and compare it to that of $\hat{\beta}_2$.

Solution By Theorem 12.1.6b,

$$\begin{aligned}
\text{MSE}(\hat{\beta}_2^{(O)}) &= \frac{\sigma^2}{\sum_{i=1}^n x_i^2} + \frac{\sigma^2}{(\sum_{i=1}^n x_i^2)^2}\left(\sum_{i=1}^n x_i\right)^2 \{\mathbf{1}^T[\mathbf{I} - \mathbf{x}(\mathbf{x}^T\mathbf{x})^{-1}\mathbf{x}^T]\mathbf{1}\}^{-1} \\
&= \frac{\sigma^2}{\sum_{i=1}^n x_i^2} + \frac{\sigma^2}{(\sum_{i=1}^n x_i^2)^2}\left(\sum_{i=1}^n x_i\right)^2 \left(n - \frac{(\sum_{i=1}^n x_i)^2}{\sum_{i=1}^n x_i^2}\right)^{-1} \\
&= \frac{\sigma^2}{\sum_{i=1}^n x_i^2}\left(1 + \frac{n\bar{x}^2}{SXX}\right).
\end{aligned}$$

Also, $\text{MSE}(\hat{\beta}_2) = \text{var}(\hat{\beta}_2) = \sigma^2/\sum_{i=1}^n x_i^2$. Thus, $\text{MSE}(\hat{\beta}_2^{(O)}) \geq \text{MSE}(\hat{\beta}_2)$, with equality if and only if $\bar{x} = 0$.

▶ **Exercise 10** Prove Theorem 12.1.9: In the overspecified mean structure scenario, if $n > \text{rank}(\mathbf{X}_1, \mathbf{X}_2)$ then:

(a) $\text{E}(\hat{\sigma}_O^2) = \sigma^2$;
(b) if the skewness matrix of $\mathbf{y}$ is null, then $\text{cov}(\mathbf{c}_1^T \hat{\boldsymbol{\beta}}_1^{(O)}, \hat{\sigma}_O^2) = 0$; and
(c) if the skewness and excess kurtosis matrices of $\mathbf{y}$ are null, then

$$\text{var}(\hat{\sigma}_O^2) = 2\sigma^4/[n - \text{rank}(\mathbf{X}_1, \mathbf{X}_2)].$$

Solution By Theorem 4.2.4,

$$\begin{aligned}
\text{E}(\hat{\sigma}_O^2) &= \text{E}\{\mathbf{y}^T(\mathbf{I} - \mathbf{P}_{12})\mathbf{y}/[n - \text{rank}(\mathbf{X}_1, \mathbf{X}_2)]\} \\
&= \{\boldsymbol{\beta}_1^T \mathbf{X}_1^T (\mathbf{I} - \mathbf{P}_{12})\mathbf{X}_1 \boldsymbol{\beta}_1 + \text{tr}[(\mathbf{I} - \mathbf{P}_{12})(\sigma^2 \mathbf{I})]\}/[n - \text{rank}(\mathbf{X}_1, \mathbf{X}_2)] \\
&= \{0 + \sigma^2[n - \text{rank}(\mathbf{X}_1, \mathbf{X}_2)]\}/[n - \text{rank}(\mathbf{X}_1, \mathbf{X}_2)] \\
&= \sigma^2.
\end{aligned}$$

Next, by Corollary 4.2.5.1,

$$\text{cov}(\mathbf{c}_1^T \hat{\boldsymbol{\beta}}_1^{(O)}, \hat{\sigma}_O^2) = 2\mathbf{b}^T(\sigma^2 \mathbf{I})\{(\mathbf{I} - \mathbf{P}_{12})/[n - \text{rank}(\mathbf{X}_1, \mathbf{X}_2)]\}\mathbf{X}_1 \boldsymbol{\beta}_1 = 0$$

where the exact form of $\mathbf{b}$ is unimportant because $(\mathbf{I} - \mathbf{P}_{12})\mathbf{X}_1 = \mathbf{0}$. Last, by Corollary 4.2.6.3,

$$\begin{aligned}
\text{var}(\hat{\sigma}_O^2) &= 2\text{tr}\left[\left(\frac{\mathbf{I} - \mathbf{P}_{12}}{n - \text{rank}(\mathbf{X}_1 \mathbf{X}_2)}\right)(\sigma^2 \mathbf{I})\left(\frac{\mathbf{I} - \mathbf{P}_{12}}{n - \text{rank}(\mathbf{X}_1, \mathbf{X}_2)}\right)(\sigma^2 \mathbf{I})\right] \\
&\quad + 4\boldsymbol{\beta}_1^T \mathbf{X}_1^T \left(\frac{\mathbf{I} - \mathbf{P}_{12}}{n - \text{rank}(\mathbf{X}_1, \mathbf{X}_2)}\right)(\sigma^2 \mathbf{I})\left(\frac{\mathbf{I} - \mathbf{P}_{12}}{n - \text{rank}(\mathbf{X}_1, \mathbf{X}_2)}\right)\mathbf{X}_1 \boldsymbol{\beta}_1 \\
&= \frac{2\sigma^4}{n - \text{rank}(\mathbf{X}_1, \mathbf{X}_2)}.
\end{aligned}$$

▶ **Exercise 11** In the overspecified simple linear regression settings considered in Example 12.1.2-1 and Exercise 12.9, obtain the variance of $\hat{\sigma}_O^2$ (assuming that $n > 2$ and that the skewness and excess kurtosis matrices of $\mathbf{y}$ are null) and compare it to the variance of $\hat{\sigma}^2$.

Solution By Theorem 12.1.9c, the variance of $\hat{\sigma}_O^2$ is $2\sigma^4/(n-2)$ in both Example 12.1.2-1 and Exercise 12.9. The variance of $\hat{\sigma}^2$ in both cases is $2\sigma^4/(n-1)$.

Thus, in both cases the variance of $\hat{\sigma}_O^2$ exceeds that of $\hat{\sigma}^2$ by an amount $2\sigma^4\left(\frac{1}{n-2}-\frac{1}{n-1}\right)=2\sigma^4\left(\frac{1}{(n-1)(n-2)}\right)$.

▶ **Exercise 12** Prove Theorem 12.2.1a: In the misspecified variance–covariance structure scenario, let $\mathbf{c}^T\boldsymbol{\beta}$ be an estimable function. Then $\mathbf{c}^T\tilde{\boldsymbol{\beta}}_I$ and $\tilde{\mathbf{t}}_C^T\mathbf{y}$ coincide if and only if $\mathbf{W}_2\mathbf{W}_1^{-1}\mathbf{X}(\mathbf{X}^T\mathbf{W}_1^{-1}\mathbf{X})^-\mathbf{c} \in \mathcal{C}(\mathbf{X})$.

Solution Note that $\mathbf{c}^T\tilde{\boldsymbol{\beta}}_I = \mathbf{c}^T(\mathbf{X}^T\mathbf{W}_1^{-1}\mathbf{X})^-\mathbf{X}^T\mathbf{W}_1^{-1}\mathbf{y}$, invariant to the choice of generalized inverse of $\mathbf{X}^T\mathbf{W}_1^{-1}\mathbf{X}$, and that $\mathbf{c}^T = \mathbf{a}^T\mathbf{X}$ for some $\mathbf{a}$. Therefore, by Corollary 11.2.2.2, $\mathbf{c}^T\tilde{\boldsymbol{\beta}}_I$ is the BLUE of $\mathbf{c}^T\boldsymbol{\beta}$ if and only if $\mathbf{W}_2\mathbf{W}_1^{-1}[(\mathbf{X}^T\mathbf{W}_1^{-1}\mathbf{X})^-]^T\mathbf{X}^T\mathbf{a} \in \mathcal{C}(\mathbf{X})$, or equivalently (by Theorem 11.1.6b and Corollary 3.3.1.1) if and only if $\mathbf{W}_2\mathbf{W}_1^{-1}(\mathbf{X}^T\mathbf{W}_1^{-1}\mathbf{X})^-\mathbf{c} \in \mathcal{C}(\mathbf{X})$.

▶ **Exercise 13** Consider a misspecified variance–covariance structure scenario with $\mathbf{W}_1 = \mathbf{I}$, in which case $\tilde{\sigma}_I^2$ is the ordinary residual mean square.

(a) Show that $0 \le \mathrm{E}(\tilde{\sigma}_I^2) \le \frac{n}{n-p^*}\bar{\sigma}^2$, where $\bar{\sigma}^2 = (1/n)\sum_{i=1}^n \sigma^2 w_{2,ii}$ and $w_{2,ii}$ is the ith main diagonal element of $\mathbf{W}_2$. (Hint: Use Theorem 2.15.13.)
(b) Prove the following lemma, and then use it to show that the lower bound in part (a) is attained if and only if $\mathbf{W}_2 = \mathbf{P_X}\mathbf{B}\mathbf{P_X}$ for some nonnegative definite matrix $\mathbf{B}$ and that the upper bound is attained if and only if $\mathbf{W}_2 = (\mathbf{I}-\mathbf{P_X})\mathbf{C}(\mathbf{I}-\mathbf{P_X})$ for some nonnegative definite matrix $\mathbf{C}$. (Hint: To prove the lemma, use Theorems 3.2.3 and 2.15.9.)

Lemma Let $\mathbf{A}$ and $\mathbf{Q}$ represent an $m \times n$ matrix and an $n \times n$ nonnegative definite matrix, respectively, and let $\mathbf{A}^-$ represent any generalized inverse of $\mathbf{A}$. Then $\mathbf{AQ} = \mathbf{0}$ if and only if $\mathbf{Q} = (\mathbf{I} - \mathbf{A}^-\mathbf{A})\mathbf{B}(\mathbf{I} - \mathbf{A}^-\mathbf{A})^T$ for some $n \times n$ nonnegative definite matrix $\mathbf{B}$.

(Note: This exercise was inspired by results from Dufour [1986].)

Solution

(a) By Theorem 4.2.4,

$$\mathrm{E}(\tilde{\sigma}_I^2) = \mathrm{E}[\mathbf{y}^T(\mathbf{I}-\mathbf{P_X})\mathbf{y}/(n-p^*)] = \frac{\sigma^2}{n-p^*}\mathrm{tr}[(\mathbf{I}-\mathbf{P_X})\mathbf{W}_2].$$

But $\mathrm{tr}[(\mathbf{I}-\mathbf{P_X})\mathbf{W}_2] \ge 0$ by Theorem 2.15.13 because both $\mathbf{I}-\mathbf{P_X}$ and $\mathbf{W}_2$ are nonnegative definite. This establishes the lower bound. Note further that, by the same theorem,

$$\mathrm{tr}[(\mathbf{I}-\mathbf{P_X})\mathbf{W}_2] = \mathrm{tr}(\mathbf{W}_2) - \mathrm{tr}(\mathbf{P_X}\mathbf{W}_2) \le \mathrm{tr}(\mathbf{W}_2)$$

because both $\mathbf{P_X}$ and $\mathbf{W}_2$ are nonnegative definite. Thus,

$$E(\tilde{\sigma}_I^2) \leq \frac{\sigma^2}{n-p^*}\text{tr}(\mathbf{W}_2) = \frac{n}{n-p^*}\bar{\sigma}^2.$$

(b) The lower bound is attained if and only if $\text{tr}[(\mathbf{I}-\mathbf{P_X})\mathbf{W}_2] = 0$ or equivalently, by Theorem 2.15.13, if and only if $(\mathbf{I}-\mathbf{P_X})\mathbf{W}_2 = \mathbf{0}$. Applying the lemma with $\mathbf{A} = \mathbf{I}-\mathbf{P_X}$ and $\mathbf{Q} = \mathbf{W}_2$, we find that $\mathbf{W}_2 = [\mathbf{I}-(\mathbf{I}-\mathbf{P_X})^-(\mathbf{I}-\mathbf{P_X})]\mathbf{B}[\mathbf{I}-(\mathbf{I}-\mathbf{P_X})^-(\mathbf{I}-\mathbf{P_X})]^T = \mathbf{P_X}\mathbf{B}\mathbf{P_X}$ for some $n \times n$ nonnegative definite matrix $\mathbf{B}$. Similarly, the upper bound is attained if and only if $\text{tr}(\mathbf{P_X}\mathbf{W}_2) = 0$ or equivalently, by Theorem 2.15.13, if and only if $\mathbf{P_X}\mathbf{W}_2 = \mathbf{0}$. Applying the lemma with $\mathbf{A} = \mathbf{P_X}$ and $\mathbf{Q} = \mathbf{W}_2$, we find that $\mathbf{W}_2 = (\mathbf{I}-\mathbf{P_X^-}\mathbf{P_X})\mathbf{C}(\mathbf{I}-\mathbf{P_X^-}\mathbf{P_X})^T = (\mathbf{I}-\mathbf{P_X})\mathbf{C}(\mathbf{I}-\mathbf{P_X})$ for some $n \times n$ nonnegative definite matrix $\mathbf{C}$.

Now we prove the lemma. Proving sufficiency is trivial since $\mathbf{A}(\mathbf{I}-\mathbf{A}^-\mathbf{A}) = \mathbf{0}$. To prove the necessity, suppose that $\mathbf{AQ} = \mathbf{0}$ and let $\mathbf{Q}^{\frac{1}{2}}$ be the nonnegative definite square root of $\mathbf{Q}$ given by Definition 2.15.2. Then $\mathbf{A}\mathbf{Q}^{\frac{1}{2}}\mathbf{Q}^{\frac{1}{2}} = \mathbf{0}$, implying by Theorem 3.3.2 that $\mathbf{A}\mathbf{Q}^{\frac{1}{2}} = \mathbf{0}$. This last matrix equation implies, by Theorem 3.2.3, that $\mathbf{Q}^{\frac{1}{2}} = (\mathbf{I}-\mathbf{A}^-\mathbf{A})\mathbf{Z}$ for some matrix $\mathbf{Z}$. Thus,

$$\mathbf{Q} = (\mathbf{I}-\mathbf{A}^-\mathbf{A})\mathbf{Z}\mathbf{Z}^T(\mathbf{I}-\mathbf{A}^-\mathbf{A})^T = (\mathbf{I}-\mathbf{A}^-\mathbf{A})\mathbf{B}(\mathbf{I}-\mathbf{A}^-\mathbf{A})^T$$

where $\mathbf{B}\ (= \mathbf{Z}\mathbf{Z}^T)$ is nonnegative definite by Theorem 2.15.9.

▶ **Exercise 14** Consider the Aitken model $\{\mathbf{y}, \mathbf{X}\boldsymbol{\beta}, \sigma^2[\mathbf{I}+\mathbf{P_X}\mathbf{A}\mathbf{P_X}+(\mathbf{I}-\mathbf{P_X})\mathbf{B}(\mathbf{I}-\mathbf{P_X})]\}$, where $\mathbf{B}$ is a specified $n \times n$ nonnegative definite matrix and $\mathbf{A}$ is a specified $n \times n$ matrix such that $\text{var}(\mathbf{e})$ is nonnegative definite. Suppose that $n > p^*$.

(a) Let $\tilde{\sigma}_I^2 = \mathbf{y}^T(\mathbf{I}-\mathbf{P_X})\mathbf{y}/(n-p^*)$. Obtain a simplified expression for the bias of $\tilde{\sigma}_I^2$ under this model, and determine necessary and sufficient conditions on $\mathbf{A}$ and $\mathbf{B}$ for this bias to equal 0.
(b) Suppose that the skewness and excess kurtosis matrices of $\mathbf{y}$ are null. Obtain a simplified expression for $\text{var}(\tilde{\sigma}_I^2)$. How does $\text{var}(\tilde{\sigma}_I^2)$ compare to $\text{var}(\tilde{\sigma}_C^2)$?

Solution

(a) By Theorem 4.2.4,

$$\begin{aligned}
E(\tilde{\sigma}_I^2) &= E[\mathbf{y}^T(\mathbf{I}-\mathbf{P_X})\mathbf{y}/(n-p^*)] \\
&= 0 + \text{tr}\{(\mathbf{I}-\mathbf{P_X})\sigma^2[\mathbf{I}+\mathbf{P_X}\mathbf{A}\mathbf{P_X}+(\mathbf{I}-\mathbf{P_X})\mathbf{B}(\mathbf{I}-\mathbf{P_X})]\}/(n-p^*) \\
&= \sigma^2\{1+\text{tr}[(\mathbf{I}-\mathbf{P_X})\mathbf{B}]/(n-p^*)\}.
\end{aligned}$$

Thus, $\tilde{\sigma}_I^2$ is unbiased if and only if $\mathbf{B}$ is such that $\text{tr}[(\mathbf{I} - \mathbf{P_X})\mathbf{B}] = 0$; there are no conditions on $\mathbf{A}$. Furthermore, by Theorem 2.15.13, $\text{tr}[(\mathbf{I} - \mathbf{P_X})\mathbf{B}] = 0$ if and only if $(\mathbf{I} - \mathbf{P_X})\mathbf{B} = \mathbf{0}$.

(b) By Corollary 4.2.6.3,

$$
\begin{aligned}
\text{var}(\tilde{\sigma}_I^2) &= 2\text{tr}\{(\mathbf{I} - \mathbf{P_X})\sigma^2[\mathbf{I} + \mathbf{P_X A P_X} + (\mathbf{I} - \mathbf{P_X})\mathbf{B}(\mathbf{I} - \mathbf{P_X})](\mathbf{I} - \mathbf{P_X}) \\
&\quad \times \sigma^2[\mathbf{I} + \mathbf{P_X A P_X} + (\mathbf{I} - \mathbf{P_X})\mathbf{B}(\mathbf{I} - \mathbf{P_X})]\}/(n - p^*)^2 + 0 \\
&= \frac{2\sigma^4}{(n - p^*)^2}\text{tr}[(\mathbf{I} - \mathbf{P_X}) + 2(\mathbf{I} - \mathbf{P_X})\mathbf{B}(\mathbf{I} - \mathbf{P_X}) + (\mathbf{I} - \mathbf{P_X})\mathbf{B}(\mathbf{I} - \mathbf{P_X})\mathbf{B}(\mathbf{I} - \mathbf{P_X})] \\
&= \frac{2\sigma^4}{n - p^*} + \frac{2\sigma^4\text{tr}[2(\mathbf{I} - \mathbf{P_X})\mathbf{B}(\mathbf{I} - \mathbf{P_X}) + (\mathbf{I} - \mathbf{P_X})\mathbf{B}(\mathbf{I} - \mathbf{P_X})\mathbf{B}(\mathbf{I} - \mathbf{P_X})]}{(n - p^*)^2}.
\end{aligned}
$$

Note that $\text{var}(\tilde{\sigma}_C^2) = 2\sigma^4/(n - p^*)$, and $(\mathbf{I} - \mathbf{P_X})\mathbf{B}(\mathbf{I} - \mathbf{P_X})$ and $(\mathbf{I} - \mathbf{P_X})\mathbf{B}(\mathbf{I} - \mathbf{P_X})\mathbf{B}(\mathbf{I} - \mathbf{P_X})$ are nonnegative definite because $\mathbf{B}$ and $\mathbf{I} - \mathbf{P_X}$ are nonnegative definite, so by Theorem 2.15.1, $\text{var}(\tilde{\sigma}_I^2) \geq \text{var}(\tilde{\sigma}_C^2)$.

► **Exercise 15** Consider a special case of the misspecified variance–covariance structure scenario in which $\mathbf{W}_1 = \mathbf{I}$ and $\mathbf{W}_2$ is a nonnegative definite matrix such that $\text{tr}(\mathbf{W}_2) = n$ but is otherwise arbitrary.

(a) Show that $\tilde{\sigma}_I^2 \equiv \mathbf{y}^T(\mathbf{I} - \mathbf{P_X})\mathbf{y}/(n - p^*)$ is an unbiased estimator of σ^2 under the correct model if and only if $\text{tr}(\mathbf{P_X W}_2) = p^*$.

(b) As an even more special case, suppose that

$$
\mathbf{X} = \begin{pmatrix} 1 & 1 \\ 1 & -1 \\ 1 & 1 \\ 1 & -1 \end{pmatrix} \quad \text{and} \quad \mathbf{W}_2 = \begin{pmatrix} 1 & \rho & 0 & 0 \\ \rho & 1 & \rho & 0 \\ 0 & \rho & 1 & \rho \\ 0 & 0 & \rho & 1 \end{pmatrix}.
$$

Here ρ is a specified real number for which $\mathbf{W}_2$ is nonnegative definite. Using part (a), determine whether $\tilde{\sigma}_I^2$ is an unbiased estimator of σ^2 in this case.

Solution

(a) By Theorem 12.2.1d,

$$
\text{E}(\tilde{\sigma}_I^2) = \sigma^2\left(\frac{\text{tr}[(\mathbf{I} - \mathbf{P_X})\mathbf{W}_2]}{n - p^*}\right) = \sigma^2\left(\frac{n - \text{tr}(\mathbf{P_X W}_2)}{n - p^*}\right),
$$

so $\tilde{\sigma}_I^2$ is unbiased if and only if $\text{tr}(\mathbf{P_X W}_2) = p^*$.

(b)

$$\mathbf{P_X} = \mathbf{X}(\mathbf{X}^T\mathbf{X})^{-}\mathbf{X}^T = \begin{pmatrix} 1 & 1 \\ 1 & -1 \\ 1 & 1 \\ 1 & -1 \end{pmatrix} \begin{pmatrix} 4 & 0 \\ 0 & 4 \end{pmatrix}^{-1} \begin{pmatrix} 1 & 1 & 1 & 1 \\ 1 & -1 & 1 & -1 \end{pmatrix} = (1/4) \begin{pmatrix} 2 & 0 & 2 & 0 \\ 0 & 2 & 0 & 2 \\ 2 & 0 & 2 & 0 \\ 0 & 2 & 0 & 2 \end{pmatrix},$$

so

$$\mathbf{P_X}\mathbf{W}_2 = (1/4) \begin{pmatrix} 2 & 0 & 2 & 0 \\ 0 & 2 & 0 & 2 \\ 3 & 0 & 2 & 0 \\ 0 & 2 & 0 & 2 \end{pmatrix} \begin{pmatrix} 1 & \gamma & 0 & 0 \\ \gamma & 1 & \gamma & 0 \\ 0 & \gamma & 1 & \gamma \\ 0 & 0 & \gamma & 1 \end{pmatrix} = (1/2)\mathbf{I}_4,$$

which has trace equal to 2. Because $p^* = 2$, $\tilde{\sigma}_I^2$ is unbiased for σ^2.

▶ **Exercise 16** Consider a misspecified variance–covariance structure scenario in which the correct model is a heteroscedastic no-intercept linear regression model $\{\mathbf{y}, \mathbf{x}\beta, \sigma^2\text{diag}(x_1^2, x_2^2, \ldots, x_n^2)\}$ and the incorrect model is its Gauss–Markov counterpart $\{\mathbf{y}, \mathbf{x}\beta, \sigma^2\mathbf{I}\}$.

(a) Obtain specialized expressions for $\text{var}(\tilde{\beta}_I)$ and $\text{var}(\tilde{\beta}_C)$.
(b) Suppose that $x_i = 100/(101 - i)$ $(i = 1, \ldots, 100)$. Evaluate the expression for $\text{var}(\tilde{\beta}_I)$ obtained in part (a) for each $n = 1, \ldots, 100$. (You should write a short computer program to do this.) Are you surprised by the behavior of $\text{var}(\tilde{\beta}_I)$ as the sample size increases? Explain.

(Note: This exercise was inspired by results from Meng and Xie [2014].)

Solution

(a) Specializing the expressions in Theorem 12.2.1c, we obtain

$$\text{var}(\tilde{\beta}_I) = \sigma^2 \left(\sum_{i=1}^{n} x_i^2\right)^{-1} \mathbf{x}^T[\text{diag}(x_1^2, x_2^2, \ldots, x_n^2)]\mathbf{x} \left(\sum_{i=1}^{n} x_i^2\right)^{-1} = \sigma^2 \frac{\sum_{i=1}^{n} x_i^4}{(\sum_{i=1}^{n} x_i^2)^2}$$

and

$$\text{var}(\tilde{\beta}_C) = \sigma^2\{\mathbf{x}^T[\text{diag}(1/x_1^2, 1/x_2^2, \ldots, 1/x_n^2)]\mathbf{x}\}^{-1} = \sigma^2/n.$$

(b) It turns out that $\text{var}(\tilde{\beta}_I)$ decreases monotonically as n increases from 1 to 64, but surprisingly, $\text{var}(\tilde{\beta}_I)$ *increases* monotonically as n increases from 65 to 100. Partial results are listed in the following table:

n	$\text{var}(\tilde{\beta}_I)$
1	1.000000
2	0.500051
3	0.333424
4	0.250129
5	0.200167
⋮	⋮
63	0.021441
64	0.021436
65	0.021453
66	0.021492
⋮	⋮
98	0.133746
99	0.204172
100	0.404883

▶ **Exercise 17** Consider a misspecified variance–covariance structure scenario in which the incorrect model is $\{\mathbf{y}, \mathbf{1}\mu, \sigma^2\mathbf{I}\}$ and the correct model has the same mean structure but a variance–covariance matrix given by (5.13). Obtain a general expression for the efficiency (the ratio of variances) of the (incorrect) least squares estimator of μ relative to the (correct) generalized least squares estimator, and evaluate it for $n = 10$ and $\rho = -0.9, -0.7, -0.5, \ldots, 0.9$.

Solution Specializing one of the expressions in Theorem 12.2.1c, we obtain

$$\text{var}(\tilde{\mu}_I) = \sigma^2(1/n)\mathbf{1}^T\mathbf{W}\mathbf{1}(1/n) = \frac{\sigma^2}{n^2(1-\rho^2)}[n+2(n-1)\rho+2(n-2)\rho^2+\cdots+2\rho^{n-1}],$$

and from Example 11.1-3 we have

$$\text{var}(\tilde{\mu}_C) = \frac{\sigma^2}{(1-\rho)[2+(n-2)(1-\rho)]}.$$

Thus the efficiency of $\tilde{\mu}_I$ relative to $\tilde{\mu}_C$ is

$$Eff = \frac{\text{var}(\tilde{\mu}_C)}{\text{var}(\tilde{\mu}_I)} = \frac{n^2(1+\rho)}{[n-(n-2)\rho][n+2(n-1)\rho+2(n-2)\rho^2+\cdots+2\rho^{n-1}]}.$$

When $n = 10$ and $\rho = -0.9, -0.7, \ldots, 0.7, 0.9$, we obtain the following values of *Eff*:

ρ	*Eff*
−0.9	0.6831
−0.7	0.8603
−0.5	0.9455
−0.3	0.9835
−0.1	0.9983
0.1	0.9984
0.3	0.9861
0.5	0.9614
0.7	0.9299
0.9	0.9326

References

Dufour, J. (1986). Bias of S^2 in linear regressions with dependent errors. *The American Statistician, 40*, 284–285.

Meng, X. & Xie, X. (2014). I got more data, my model is more refined, but my estimator is getting worse! Am I just dumb? *Econometric Reviews, 33*, 218–250.

Best Linear Unbiased Prediction 13

This chapter presents exercises on best linear unbiased prediction under a linear model, and provides solutions to those exercises.

▶ **Exercise 1** Prove Theorem 13.1.2: A function $\tau = \mathbf{c}^T\boldsymbol{\beta} + u$ is predictable under the model $\left\{\begin{pmatrix}\mathbf{y}\\ u\end{pmatrix}, \begin{pmatrix}\mathbf{X}\boldsymbol{\beta}\\ 0\end{pmatrix}\right\}$ if and only if $\mathbf{c}^T\boldsymbol{\beta}$ is estimable under the model $\{\mathbf{y}, \mathbf{X}\boldsymbol{\beta}\}$, i.e., if and only if $\mathbf{c}^T \in \mathcal{R}(\mathbf{X})$.

Solution If $\mathbf{c}^T \in \mathcal{R}(\mathbf{X})$, then an n-vector $\mathbf{a}$ exists such that $\mathbf{a}^T\mathbf{X} = \mathbf{c}^T$, implying further that $\mathrm{E}(\mathbf{a}^T\mathbf{y}) = \mathbf{a}^T\mathbf{X}\boldsymbol{\beta} = \mathbf{c}^T\boldsymbol{\beta} = \mathrm{E}(\mathbf{c}^T\boldsymbol{\beta} + u)$ for all $\boldsymbol{\beta}$. Thus, $\mathbf{c}^T\boldsymbol{\beta}$ is predictable. Conversely, if $\mathbf{c}^T\boldsymbol{\beta} + u$ is predictable, then by definition and Theorem 13.1.1 there exists an unbiased predictor of $\mathbf{c}^T\boldsymbol{\beta} + u$ of the form $\mathbf{t}^T\mathbf{y}$, where $\mathbf{t}^T\mathbf{X} = \mathbf{c}^T$. Thus, $\mathbf{c}^T \in \mathcal{R}(\mathbf{X})$.

▶ **Exercise 2** Suppose that $n > p^*$. Find a nonnegative definite matrix $\begin{pmatrix}\mathbf{W} & \mathbf{k}\\ \mathbf{k}^T & h\end{pmatrix}$ for which the BLUP equations for a predictable function $\mathbf{c}^T\boldsymbol{\beta} + u$ are not consistent. (Hint: Take $\mathbf{W} = \mathbf{X}\mathbf{X}^T$ and determine a suitable $\mathbf{k}$.)

Solution Taking $\mathbf{W} = \mathbf{X}\mathbf{X}^T$, the BLUP equations for $\mathbf{c}^T\boldsymbol{\beta} + u$ are

$$\begin{pmatrix}\mathbf{X}\mathbf{X}^T & \mathbf{X}\\ \mathbf{X}^T & \mathbf{0}\end{pmatrix}\begin{pmatrix}\mathbf{t}\\ \boldsymbol{\lambda}\end{pmatrix} = \begin{pmatrix}\mathbf{k}\\ \mathbf{c}\end{pmatrix}.$$

In particular, the "top" subset of BLUP equations is $\mathbf{X}(\mathbf{X}^T\mathbf{t} + \boldsymbol{\lambda}) = \mathbf{k}$, which has a solution if and only if $\mathbf{k} \in \mathcal{C}(\mathbf{X})$. Therefore, those equations will not have a solution if $\mathbf{k}$ is taken to be any nonnull vector in $\mathcal{C}(\mathbf{I} - \mathbf{P_X})$. Such a vector exists because $n > p^*$. The value of h is irrelevant, so it can be taken to equal 1.

D. L. Zimmerman, *Linear Model Theory*,
https://doi.org/10.1007/978-3-030-52074-8_13

▶ **Exercise 3** Prove Corollary 13.2.1.1: Under the prediction-extended positive definite Aitken model described in Theorem 13.2.1, if $\mathbf{C}^T\boldsymbol{\beta} + \mathbf{u}$ is a vector of predictable functions and $\mathbf{L}^T$ is a matrix of constants having the same number of columns as $\mathbf{C}^T$ has rows, then $\mathbf{L}^T(\mathbf{C}^T\boldsymbol{\beta} + \mathbf{u})$ is predictable and its BLUP is $\mathbf{L}^T(\mathbf{C}^T\tilde{\boldsymbol{\beta}} + \tilde{\mathbf{u}})$.

Solution Each row of $\mathbf{L}^T\mathbf{C}^T$ is an element of $\mathcal{R}(\mathbf{X})$, $\mathrm{E}(\mathbf{L}^T\mathbf{u}) = \mathbf{0}$, and

$$\begin{aligned}\mathrm{var}\begin{pmatrix}\mathbf{y}\\ \mathbf{L}^T\mathbf{u}\end{pmatrix} &= \mathrm{var}\left[\begin{pmatrix}\mathbf{I} & \mathbf{0}\\ \mathbf{0} & \mathbf{L}^T\end{pmatrix}\begin{pmatrix}\mathbf{y}\\ \mathbf{u}\end{pmatrix}\right]\\ &= \begin{pmatrix}\mathbf{I} & \mathbf{0}\\ \mathbf{0} & \mathbf{L}^T\end{pmatrix}\sigma^2\begin{pmatrix}\mathbf{W} & \mathbf{K}\\ \mathbf{K}^T & \mathbf{H}\end{pmatrix}\begin{pmatrix}\mathbf{I} & \mathbf{0}\\ \mathbf{0} & \mathbf{L}\end{pmatrix}\\ &= \sigma^2\begin{pmatrix}\mathbf{W} & \mathbf{KL}\\ (\mathbf{KL})^T & \mathbf{L}^T\mathbf{HL}\end{pmatrix},\end{aligned}$$

which is nonnegative definite by Corollary 2.15.12.1. Therefore, $\mathbf{L}^T(\mathbf{C}^T\boldsymbol{\beta} + \mathbf{u})$ is predictable by Theorem 13.1.2, and by Theorem 13.2.1 its BLUP is

$$\mathbf{L}^T\mathbf{C}^T\tilde{\boldsymbol{\beta}} + (\mathbf{KL})^T\mathbf{Ey} = \mathbf{L}^T(\mathbf{C}^T\tilde{\boldsymbol{\beta}} + \tilde{\mathbf{u}}).$$

▶ **Exercise 4** Prove Theorem 13.2.3: Under a prediction-extended positive definite Aitken model for which $n > p^*$ and the skewness matrix of the joint distribution of $\mathbf{y}$ and $\mathbf{u}$ is null, $\mathrm{cov}(\tilde{\boldsymbol{\tau}} - \boldsymbol{\tau}, \tilde{\sigma}^2) = \mathbf{0}$.

Solution By Corollary 4.2.5.1,

$$\begin{aligned}&\mathrm{cov}(\tilde{\boldsymbol{\tau}} - \boldsymbol{\tau}, \tilde{\sigma}^2)\\ &= \mathrm{cov}\left[\left(\mathbf{C}^T(\mathbf{X}^T\mathbf{W}^{-1}\mathbf{X})^-\mathbf{X}^T\mathbf{W}^{-1} + \mathbf{K}^T\mathbf{E},\ -\mathbf{I}_s\right)\begin{pmatrix}\mathbf{y}\\ \mathbf{u}\end{pmatrix}, \begin{pmatrix}\mathbf{y}\\ \mathbf{u}\end{pmatrix}^T\begin{pmatrix}\mathbf{E} & \mathbf{0}\\ \mathbf{0} & \mathbf{0}\end{pmatrix}\begin{pmatrix}\mathbf{y}\\ \mathbf{u}\end{pmatrix}\right]\Big/(n-p^*)\\ &= 2\left(\mathbf{C}^T(\mathbf{X}^T\mathbf{W}^{-1}\mathbf{X})^-\mathbf{X}^T\mathbf{W}^{-1} + \mathbf{K}^T\mathbf{E},\ -\mathbf{I}_s\right)\sigma^2\begin{pmatrix}\mathbf{W} & \mathbf{K}\\ \mathbf{K}^T & \mathbf{H}\end{pmatrix}\begin{pmatrix}\mathbf{E} & \mathbf{0}\\ \mathbf{0} & \mathbf{0}\end{pmatrix}\begin{pmatrix}\mathbf{X}\boldsymbol{\beta}\\ \mathbf{0}\end{pmatrix}\Big/(n-p^*)\\ &= \mathbf{0},\end{aligned}$$

where the last equality follows from Theorem 11.1.6d.

▶ **Exercise 5** Prove Theorem 13.3.1a: For the prediction-extended Aitken model

$$\left\{\begin{pmatrix}\mathbf{y}\\ \mathbf{u}\end{pmatrix}, \begin{pmatrix}\mathbf{X}\boldsymbol{\beta}\\ \mathbf{0}\end{pmatrix}, \sigma^2\begin{pmatrix}\mathbf{W} & \mathbf{K}\\ \mathbf{K}^T & \mathbf{H}\end{pmatrix}\right\},$$

let $\begin{pmatrix} \mathbf{G}_{11} & \mathbf{G}_{12} \\ \mathbf{G}_{21} & \mathbf{G}_{22} \end{pmatrix}$ represent a generalized inverse of $\begin{pmatrix} \mathbf{W} & \mathbf{X} \\ \mathbf{X}^T & \mathbf{0} \end{pmatrix}$, let $\boldsymbol{\tau} = \mathbf{C}^T\boldsymbol{\beta} + \mathbf{u}$ be an s-vector of predictable functions, and suppose that $\mathcal{C}(\mathbf{K}) \subseteq \mathcal{C}(\mathbf{W}, \mathbf{X})$. Then the collection of BLUPs of $\boldsymbol{\tau}$ consists of all quantities of the form $\tilde{\mathbf{T}}^T\mathbf{y}$, where

$$\tilde{\mathbf{T}} = \mathbf{G}_{12}\mathbf{C} + \mathbf{G}_{11}\mathbf{K} + (\mathbf{I}_n - \mathbf{G}_{11}\mathbf{W} - \mathbf{G}_{12}\mathbf{X}^T,\ -\mathbf{G}_{11}\mathbf{X})\mathbf{Z},$$

where $\mathbf{Z}$ is an arbitrary $(n + p) \times s$ matrix. In particular, $(\mathbf{C}^T\mathbf{G}_{12}^T + \mathbf{K}^T\mathbf{G}_{11}^T)\mathbf{y}$ is a BLUP of $\boldsymbol{\tau}$.

Solution Because $\mathbf{C}^T\boldsymbol{\beta} + \mathbf{u}$ is predictable and $\mathcal{C}(\mathbf{K}) \subseteq \mathcal{C}(\mathbf{W}, \mathbf{X})$, the BLUP equations for each element of $\mathbf{C}^T\boldsymbol{\beta} + \mathbf{u}$ are consistent (Theorem 13.1.3). Therefore, by Theorem 3.2.3, all solutions to the BLUP equations for all elements of $\mathbf{C}^T\boldsymbol{\beta} + \mathbf{u}$ are given by

$$\begin{pmatrix} \mathbf{G}_{11} & \mathbf{G}_{12} \\ \mathbf{G}_{21} & \mathbf{G}_{22} \end{pmatrix}\begin{pmatrix} \mathbf{K} \\ \mathbf{C} \end{pmatrix} + \left[\mathbf{I} - \begin{pmatrix} \mathbf{G}_{11} & \mathbf{G}_{12} \\ \mathbf{G}_{21} & \mathbf{G}_{22} \end{pmatrix}\begin{pmatrix} \mathbf{W} & \mathbf{X} \\ \mathbf{X}^T & \mathbf{0} \end{pmatrix}\right]\mathbf{Z},$$

where $\mathbf{Z}$ ranges throughout the space of $(n + p) \times s$ matrices. The matrix of first n-component subvectors of such a solution is

$$\tilde{\mathbf{T}} = \mathbf{G}_{11}\mathbf{K} + \mathbf{G}_{12}\mathbf{C} + (\mathbf{I}_n - \mathbf{G}_{11}\mathbf{W} - \mathbf{G}_{12}\mathbf{X}^T,\ -\mathbf{G}_{11}\mathbf{X})\mathbf{Z}.$$

Thus all BLUPs of $\mathbf{C}^T\boldsymbol{\beta} + \mathbf{u}$ may be written as $\tilde{\mathbf{T}}^T\mathbf{y}$ for matrices $\tilde{\mathbf{T}}$ of this form. One BLUP in particular, obtained by setting $\mathbf{Z} = \mathbf{0}$, is $(\mathbf{G}_{11}\mathbf{K} + \mathbf{G}_{12}\mathbf{C})^T\mathbf{y}$, i.e., $(\mathbf{C}^T\mathbf{G}_{12}^T + \mathbf{K}^T\mathbf{G}_{11}^T)\mathbf{y}$.

▶ **Exercise 6** Prove the invariance result in Theorem 13.3.1b: For the prediction-extended Aitken model $\left\{\begin{pmatrix} \mathbf{y} \\ \mathbf{u} \end{pmatrix}, \begin{pmatrix} \mathbf{X}\boldsymbol{\beta} \\ \mathbf{0} \end{pmatrix}, \sigma^2\begin{pmatrix} \mathbf{W} & \mathbf{K} \\ \mathbf{K}^T & \mathbf{H} \end{pmatrix}\right\}$, let $\begin{pmatrix} \mathbf{G}_{11} & \mathbf{G}_{12} \\ \mathbf{G}_{21} & \mathbf{G}_{22} \end{pmatrix}$ represent a generalized inverse of $\begin{pmatrix} \mathbf{W} & \mathbf{X} \\ \mathbf{X}^T & \mathbf{0} \end{pmatrix}$, let $\boldsymbol{\tau} = \mathbf{C}^T\boldsymbol{\beta} + \mathbf{u}$ be an s-vector of predictable functions, and suppose that $\mathcal{C}(\mathbf{K}) \subseteq \mathcal{C}(\mathbf{W}, \mathbf{X})$. Then:

(b) the variance–covariance matrix of prediction errors associated with a vector of BLUPs $\tilde{\boldsymbol{\tau}} = (\mathbf{C}^T\mathbf{G}_{12}^T + \mathbf{K}^T\mathbf{G}_{11}^T)\mathbf{y}$ of a vector of predictable functions $\boldsymbol{\tau} = \mathbf{C}^T\boldsymbol{\beta} + \mathbf{u}$ is

$$\text{var}(\tilde{\boldsymbol{\tau}} - \boldsymbol{\tau}) = \sigma^2\left[\mathbf{H} - (\mathbf{K}^T, \mathbf{C}^T)\begin{pmatrix} \mathbf{G}_{11} & \mathbf{G}_{12} \\ \mathbf{G}_{21} & \mathbf{G}_{22} \end{pmatrix}\begin{pmatrix} \mathbf{K} \\ \mathbf{C} \end{pmatrix}\right],$$

invariant to the choice of generalized inverse.

Solution It suffices to show that $(\mathbf{K}^T, \mathbf{C}^T)\begin{pmatrix} \mathbf{G}_{11} & \mathbf{G}_{12} \\ \mathbf{G}_{21} & \mathbf{G}_{22} \end{pmatrix}\begin{pmatrix} \mathbf{K} \\ \mathbf{C} \end{pmatrix}$, i.e., $\mathbf{K}^T\mathbf{G}_{11}\mathbf{K} + \mathbf{K}^T\mathbf{G}_{12}\mathbf{C} + \mathbf{C}^T\mathbf{G}_{21}\mathbf{K} + \mathbf{C}^T\mathbf{G}_{22}\mathbf{C}$, is invariant to the choice of generalized inverse. Now, by Theorem 3.3.8f (with $\mathbf{W}$ and $\mathbf{X}$ here playing the roles of $\mathbf{A}$ and $\mathbf{B}$ in the theorem), all four of the following matrices have this invariance: $\mathbf{W}\mathbf{G}_{11}\mathbf{W}, \mathbf{W}\mathbf{G}_{12}\mathbf{X}^T, \mathbf{X}\mathbf{G}_{21}\mathbf{W}$, and $\mathbf{X}\mathbf{G}_{22}\mathbf{X}^T$. We consider each of these four matrices in turn. Using Theorem 3.3.8c and defining $\mathbf{F}$ and $\mathbf{L}$ as in the proof of the earlier part of Theorem 13.2.1b, we obtain

$$\begin{aligned}
\mathbf{K}^T\mathbf{G}_{11}\mathbf{K} &= (\mathbf{F}^T\mathbf{W} + \mathbf{L}^T\mathbf{X}^T)\mathbf{G}_{11}(\mathbf{W}\mathbf{F} + \mathbf{X}\mathbf{L}) \\
&= \mathbf{F}^T(\mathbf{W}\mathbf{G}_{11}\mathbf{W})\mathbf{F} + \mathbf{L}^T(\mathbf{X}^T\mathbf{G}_{11}\mathbf{W})\mathbf{F} + \mathbf{F}^T(\mathbf{W}\mathbf{G}_{11}\mathbf{X})\mathbf{L} + \mathbf{L}^T(\mathbf{X}^T\mathbf{G}_{11}\mathbf{X})\mathbf{L} \\
&= \mathbf{F}^T(\mathbf{W}\mathbf{G}_{11}\mathbf{W})\mathbf{F},
\end{aligned}$$

which establishes that $\mathbf{K}^T\mathbf{G}_{11}\mathbf{K}$ has the desired invariance. Using Theorem 3.3.8a, we obtain

$$\begin{aligned}
\mathbf{K}^T\mathbf{G}_{12}\mathbf{C} &= (\mathbf{F}^T\mathbf{W} + \mathbf{L}^T\mathbf{X}^T)\mathbf{G}_{12}\mathbf{X}^T\mathbf{A} = \mathbf{F}^T\mathbf{W}\mathbf{G}_{12}\mathbf{X}^T\mathbf{A} + \mathbf{L}^T\mathbf{X}^T\mathbf{G}_{12}\mathbf{X}^T\mathbf{A} \\
&= \mathbf{F}^T(\mathbf{W}\mathbf{G}_{12}\mathbf{X}^T)\mathbf{A} + \mathbf{L}^T\mathbf{X}^T\mathbf{A},
\end{aligned}$$

likewise establishing that $\mathbf{K}^T\mathbf{G}_{12}\mathbf{C}$ has the desired invariance. Similarly,

$$\mathbf{C}^T\mathbf{G}_{21}\mathbf{K} = \mathbf{A}^T\mathbf{X}\mathbf{G}_{21}(\mathbf{W}\mathbf{F} + \mathbf{X}\mathbf{L}) = \mathbf{A}^T(\mathbf{X}\mathbf{G}_{21}\mathbf{W})\mathbf{F} + \mathbf{A}^T\mathbf{X}\mathbf{L}$$

and $\mathbf{C}^T\mathbf{G}_{22}\mathbf{C} = \mathbf{A}^T(\mathbf{X}\mathbf{G}_{22}\mathbf{X}^T)\mathbf{A}$ have the desired invariance.

► **Exercise 7** In this exercise, you are to consider using the BLUE $\mathbf{c}^T\tilde{\boldsymbol{\beta}}$ of $\mathbf{c}^T\boldsymbol{\beta}$ as a predictor of a predictable $\tau = \mathbf{c}^T\boldsymbol{\beta} + u$ associated with the prediction-extended positive definite Aitken model $\left\{\begin{pmatrix} \mathbf{y} \\ u \end{pmatrix}, \begin{pmatrix} \mathbf{X}\boldsymbol{\beta} \\ 0 \end{pmatrix}, \sigma^2\begin{pmatrix} \mathbf{W} & \mathbf{k} \\ \mathbf{k}^T & h \end{pmatrix}\right\}$. Let us call this predictor the "BLUE-predictor."

(a) Show that the BLUE-predictor is a linear unbiased predictor of τ.
(b) Obtain an expression for the variance of the prediction error $\mathbf{c}^T\tilde{\boldsymbol{\beta}} - \tau$ corresponding to the BLUE-predictor.
(c) Because [from part (a)] the BLUE-predictor is a linear unbiased predictor, its prediction error variance [obtained in part (b)] must be at least as large as the prediction error variance of the BLUP of τ. Give an expression for how much larger it is, i.e., give an expression for $\text{var}(\mathbf{c}^T\tilde{\boldsymbol{\beta}} - \tau) - \text{var}(\tilde{\tau} - \tau)$.
(d) Determine a necessary and sufficient condition for your answer to part (c) to equal 0. Your condition should take the form $\mathbf{k} \in S$, where S is a certain set of vectors. Thus you will have established a necessary and sufficient condition for the BLUE of $\mathbf{c}^T\boldsymbol{\beta}$ to also be the BLUP of $\mathbf{c}^T\boldsymbol{\beta} + u$.

Solution

(a) $\mathbf{c}^T\tilde{\boldsymbol{\beta}} = \mathbf{c}^T(\mathbf{X}^T\mathbf{W}^{-1}\mathbf{X})^-\mathbf{X}^T\mathbf{W}^{-1}\mathbf{y}$ so $\mathbf{c}^T\tilde{\boldsymbol{\beta}}$ is a linear predictor. It is also unbiased because $\mathrm{E}(\mathbf{c}^T\tilde{\boldsymbol{\beta}}) = \mathbf{c}^T\boldsymbol{\beta} = \mathbf{c}^T\boldsymbol{\beta} + \mathrm{E}(u) = \mathrm{E}(\tau)$.

(b)

$$\begin{aligned}\mathrm{var}(\mathbf{c}^T\tilde{\boldsymbol{\beta}} - \tau) &= \mathrm{var}(\mathbf{c}^T\tilde{\boldsymbol{\beta}}) + \mathrm{var}(\tau) - 2\mathrm{cov}(\mathbf{c}^T\tilde{\boldsymbol{\beta}}, \tau)\\ &= \sigma^2\mathbf{c}^T(\mathbf{X}^T\mathbf{W}^{-1}\mathbf{X})^-\mathbf{c} + \sigma^2 h - 2\sigma^2\mathbf{c}^T(\mathbf{X}^T\mathbf{W}^{-1}\mathbf{X})^-\mathbf{X}^T\mathbf{W}^{-1}\mathbf{k}.\end{aligned}$$

(c) Using Theorem 13.2.3a,

$$\begin{aligned}\mathrm{var}(\mathbf{c}^T\tilde{\boldsymbol{\beta}} - \tau) - \mathrm{var}(\tilde{\tau} - \tau) &= \sigma^2[\mathbf{c}^T(\mathbf{X}^T\mathbf{W}^{-1}\mathbf{X})^-\mathbf{c} + h - 2\mathbf{c}^T(\mathbf{X}^T\mathbf{W}^{-1}\mathbf{X})^-\mathbf{X}^T\mathbf{W}^{-1}\mathbf{k}]\\ &\quad -\sigma^2[h - \mathbf{k}^T\mathbf{W}^{-1}\mathbf{k}\\ &\quad +(\mathbf{c}^T - \mathbf{k}^T\mathbf{W}^{-1}\mathbf{X})(\mathbf{X}^T\mathbf{W}^{-1}\mathbf{X})^-(\mathbf{c} - \mathbf{X}^T\mathbf{W}^{-1}\mathbf{k})]\\ &= \sigma^2[\mathbf{k}^T\mathbf{W}^{-1}\mathbf{k} - \mathbf{k}^T\mathbf{W}^{-1}\mathbf{X}(\mathbf{X}^T\mathbf{W}^{-1}\mathbf{X})^-\mathbf{X}^T\mathbf{W}^{-1}\mathbf{k}].\end{aligned}$$

(d) The difference in the answer to part (c) equals 0 if and only if $\mathbf{k}^T[\mathbf{W}^{-1} - \mathbf{W}^{-1}\mathbf{X}(\mathbf{X}^T\mathbf{W}^{-1}\mathbf{X})^-\mathbf{X}^T\mathbf{W}^{-1}]\mathbf{k} = 0$, i.e., if and only if $(\mathbf{W}^{-\frac{1}{2}}\mathbf{k})^T\{\mathbf{I} - (\mathbf{W}^{-\frac{1}{2}}\mathbf{X})[(\mathbf{W}^{-\frac{1}{2}}\mathbf{X})^T(\mathbf{W}^{-\frac{1}{2}}\mathbf{X})]^-(\mathbf{W}^{-\frac{1}{2}}\mathbf{X})^T\}\mathbf{W}^{-\frac{1}{2}}\mathbf{k} = 0$, i.e., if and only if $\mathbf{W}^{-\frac{1}{2}}\mathbf{k} \in \mathcal{C}(\mathbf{W}^{-\frac{1}{2}}\mathbf{X})$, i.e., if and only if $\mathbf{k} \in \mathcal{C}(\mathbf{X})$.

▶ **Exercise 8** Verify all results for the spatial prediction problem considered in Example 13.2-3. Then, obtain the BLUP of y_7 and the corresponding prediction error variance after making each one of the following modifications to the spatial configuration or model. For each modification, compare the weights corresponding to the elements of $\mathbf{y}$ with the weights in Example 13.2-3, and use your intuition to explain the notable differences. (Note: It is expected that you will use a computer to do this exercise.)

(a) Suppose that y_7 is to be observed at the grid point in the second row (from the bottom) of the first column (from the left).
(b) Suppose that the (i, j)th element of $\mathbf{W}$ is equal to $\exp(-d_{ij}/4)$.
(c) Suppose that the (i, j)th element of $\mathbf{W}$ is equal to 1 if $i = j$, or $0.5\exp(-d_{ij}/2)$ otherwise.

Solution First we establish a coordinate system with the origin placed at the bottom left corner and 1 unit length being the distance between two closest grid points on each axis. In addition, we label the observations from left to right and bottom-up if two observations are in the same column. In this way, the locations for each observation (including y_7) are given in the following table:

Observation	y_1	y_2	y_3	y_4	y_5	y_6	y_7
Location	(0, 0)	(1, 1)	(2, 3)	(3, 0)	(3, 2)	(4, 1)	(2, 1)

By definition, $\mathbf{W}=(w_{ij})$ with $w_{ij}=\exp(-d_{ij}/2)$ and $d_{ij} = \sqrt{(s_i - s_j)^2 + (t_i - t_j)^2}$, where observation i is at (s_i, t_i) and observation j is at (s_j, t_j). Numerical entries of $\mathbf{W}$ are shown as follows:

$$\mathbf{W} = \begin{pmatrix} 1.000 & 0.493 & 0.165 & 0.223 & 0.165 & 0.127 & 0.327 \\ 0.493 & 1.000 & 0.327 & 0.327 & 0.327 & 0.223 & 0.607 \\ 0.165 & 0.327 & 1.000 & 0.206 & 0.493 & 0.243 & 0.368 \\ 0.223 & 0.327 & 0.206 & 1.000 & 0.368 & 0.493 & 0.493 \\ 0.165 & 0.327 & 0.493 & 0.368 & 1.000 & 0.493 & 0.493 \\ 0.127 & 0.223 & 0.243 & 0.493 & 0.493 & 1.000 & 0.368 \\ 0.327 & 0.607 & 0.368 & 0.493 & 0.493 & 0.368 & 1.000 \end{pmatrix}.$$

Now consider the model

$$\begin{pmatrix} \mathbf{y} \\ y_7 \end{pmatrix} = \mu \mathbf{1}_7 + \begin{pmatrix} \mathbf{e} \\ e_7 \end{pmatrix},$$

$$\text{var} \begin{pmatrix} \mathbf{e} \\ e_7 \end{pmatrix} = \sigma^2 \mathbf{W} = \sigma^2 \begin{pmatrix} \mathbf{W}_1 & \mathbf{k} \\ \mathbf{k}^T & 1 \end{pmatrix},$$

where

$$\mathbf{W}_1 = \begin{pmatrix} 1.000 & 0.493 & 0.165 & 0.223 & 0.165 & 0.127 \\ 0.493 & 1.000 & 0.327 & 0.327 & 0.327 & 0.223 \\ 0.165 & 0.327 & 1.000 & 0.206 & 0.493 & 0.243 \\ 0.223 & 0.327 & 0.206 & 1.000 & 0.368 & 0.493 \\ 0.165 & 0.327 & 0.493 & 0.368 & 1.000 & 0.493 \\ 0.127 & 0.223 & 0.243 & 0.493 & 0.493 & 1.000 \end{pmatrix},$$

$$\mathbf{k}^T = \begin{pmatrix} 0.327 & 0.607 & 0.368 & 0.493 & 0.493 & 0.368 \end{pmatrix}.$$

Then, the BLUP of $y_7 = 1\mu + e_7$ is

$$\begin{aligned} \tilde{y}_7 &= [\mathbf{c}^T(\mathbf{X}^T\mathbf{W}_1^{-1}\mathbf{X})^-\mathbf{X}^T\mathbf{W}_1^{-1} + \mathbf{k}^T\mathbf{W}_1^{-1} - \mathbf{k}^T\mathbf{W}_1^{-1}\mathbf{X}(\mathbf{X}^T\mathbf{W}_1^{-1}\mathbf{X})^-\mathbf{X}^T\mathbf{W}_1^{-1}]\mathbf{y} \\ &= \begin{pmatrix} 0.017, & 0.422, & 0.065, & 0.247, & 0.218, & 0.030 \end{pmatrix}\mathbf{y} \\ &= 0.017y_1 + 0.422y_2 + 0.065y_3 + 0.247y_4 + 0.218y_5 + 0.030y_6. \end{aligned}$$

The BLUP's prediction error variance is

$$\begin{aligned}\mathrm{var}(\tilde{y}_7 - y_7) &= \sigma^2[1 - \mathbf{k}^T\mathbf{W}_1^{-1}\mathbf{k} + (\mathbf{c}^T - \mathbf{k}^T\mathbf{W}_1^{-1}\mathbf{X})(\mathbf{X}^T\mathbf{W}_1^{-1}\mathbf{X})^-(\mathbf{c}^T - \mathbf{k}^T\mathbf{W}_1^{-1}\mathbf{X})^T] \\ &= 0.478\sigma^2.\end{aligned}$$

In general, the closer the spatial location corresponding to an element of $\mathbf{y}$ is to that of y_7, the higher the corresponding weight will be (for example, y_2 has the highest weight). This result matches the intuition that closer observations are more highly correlated so it is reasonable to put more weight on closer observations in order to have a good prediction. It can also be observed that, even though two observations (y_4 and y_5) have the same distance to y_7, the weights for them are close but not necessarily the same.

(a) We preserve the coordinate system and notations in the original example but the location of y_7 is now (0, 1). Repeating the calculations, we obtain the following quantities:

$$\mathbf{W} = \begin{pmatrix} 1.000 & 0.493 & 0.165 & 0.223 & 0.165 & 0.127 & 0.607 \\ 0.493 & 1.000 & 0.327 & 0.327 & 0.327 & 0.223 & 0.607 \\ 0.165 & 0.327 & 1.000 & 0.206 & 0.493 & 0.243 & 0.243 \\ 0.223 & 0.327 & 0.206 & 1.000 & 0.368 & 0.493 & 0.206 \\ 0.165 & 0.327 & 0.493 & 0.368 & 1.000 & 0.493 & 0.206 \\ 0.127 & 0.223 & 0.243 & 0.493 & 0.493 & 1.000 & 0.135 \\ 0.607 & 0.607 & 0.243 & 0.206 & 0.206 & 0.135 & 1.000 \end{pmatrix},$$

$$\mathbf{W}_1 = \begin{pmatrix} 1.000 & 0.493 & 0.165 & 0.223 & 0.165 & 0.127 \\ 0.493 & 1.000 & 0.327 & 0.327 & 0.327 & 0.223 \\ 0.165 & 0.327 & 1.000 & 0.206 & 0.493 & 0.243 \\ 0.223 & 0.327 & 0.206 & 1.000 & 0.368 & 0.493 \\ 0.165 & 0.327 & 0.493 & 0.368 & 1.000 & 0.493 \\ 0.127 & 0.223 & 0.243 & 0.493 & 0.493 & 1.000 \end{pmatrix},$$

$$\mathbf{k}^T = \begin{pmatrix} 0.607 & 0.607 & 0.243 & 0.206 & 0.206 & 0.135 \end{pmatrix}.$$

Therefore, the BLUP of $y_7 = 1\mu + e_7$ is

$$\begin{aligned}\tilde{y}_7 &= [\mathbf{c}^T(\mathbf{X}^T\mathbf{W}_1^{-1}\mathbf{X})^-\mathbf{X}^T\mathbf{W}_1^{-1} + \mathbf{k}^T\mathbf{W}_1^{-1} - \mathbf{k}^T\mathbf{W}_1^{-1}\mathbf{X}(\mathbf{X}^T\mathbf{W}_1^{-1}\mathbf{X})^-\mathbf{X}^T\mathbf{W}_1^{-1}]\mathbf{y} \\ &= \begin{pmatrix} 0.453, & 0.414, & 0.094, & 0.005, & 0.004, & 0.029 \end{pmatrix}\mathbf{y} \\ &= 0.453y_1 + 0.414y_2 + 0.094y_3 + 0.005y_4 + 0.004y_5 + 0.029y_6,\end{aligned}$$

and the BLUP's prediction error variance is

$$\begin{aligned}\text{var}(\tilde{y}_7 - y_7) &= \sigma^2[1 - \mathbf{k}^T\mathbf{W}_1^{-1}\mathbf{k} + (\mathbf{c}^T - \mathbf{k}^T\mathbf{W}_1^{-1}\mathbf{X})(\mathbf{X}^T\mathbf{W}_1^{-1}\mathbf{X})^-(\mathbf{c}^T - \mathbf{k}^T\mathbf{W}_1^{-1}\mathbf{X})^T] \\ &= 0.517\sigma^2.\end{aligned}$$

Now, the weights of y_1 and y_2 in the BLUP are significantly higher than the others. However, in the original example the weights of $y_2, y_4,$ and y_5 are significantly higher. This corresponds to the fact that now y_7's location is much closer to those of y_1 and y_2 than to the locations of other observations, while in (a) y_7's location is much closer to those of y_2, y_4, and y_5 than to the locations of other observations. Another interesting observation is that even though y_6's location is farthest from y_7's, the weight for y_6 is not the smallest one.

(b) We preserve the coordinate system and notations in the original example but now $w_{ij} = \exp(-d_{ij}/4)$. Repeating the calculations, we obtain the following quantities:

$$\mathbf{W} = \begin{pmatrix} 1.000 & 0.702 & 0.406 & 0.472 & 0.406 & 0.357 & 0.572 \\ 0.702 & 1.000 & 0.572 & 0.572 & 0.572 & 0.472 & 0.779 \\ 0.406 & 0.572 & 1.000 & 0.454 & 0.702 & 0.493 & 0.607 \\ 0.472 & 0.572 & 0.454 & 1.000 & 0.607 & 0.702 & 0.702 \\ 0.406 & 0.572 & 0.702 & 0.607 & 1.000 & 0.702 & 0.702 \\ 0.357 & 0.472 & 0.493 & 0.702 & 0.702 & 1.000 & 0.607 \\ 0.572 & 0.779 & 0.607 & 0.702 & 0.702 & 0.607 & 1.000 \end{pmatrix},$$

$$\mathbf{W}_1 = \begin{pmatrix} 1.000 & 0.702 & 0.406 & 0.472 & 0.406 & 0.357 \\ 0.702 & 1.000 & 0.572 & 0.572 & 0.572 & 0.472 \\ 0.406 & 0.572 & 1.000 & 0.454 & 0.702 & 0.493 \\ 0.472 & 0.572 & 0.454 & 1.000 & 0.607 & 0.702 \\ 0.406 & 0.572 & 0.702 & 0.607 & 1.000 & 0.702 \\ 0.357 & 0.472 & 0.493 & 0.702 & 0.702 & 1.000 \end{pmatrix},$$

$$\mathbf{k}^T = \begin{pmatrix} 0.572 & 0.779 & 0.607 & 0.702 & 0.702 & 0.607 \end{pmatrix}.$$

Therefore, the BLUP of $y_7 = 1\mu + e_7$ is

$$\begin{aligned}\tilde{y}_7 &= [\mathbf{c}^T(\mathbf{X}^T\mathbf{W}_1^{-1}\mathbf{X})^-\mathbf{X}^T\mathbf{W}_1^{-1} + \mathbf{k}^T\mathbf{W}_1^{-1} - \mathbf{k}^T\mathbf{W}_1^{-1}\mathbf{X}(\mathbf{X}^T\mathbf{W}_1^{-1}\mathbf{X})^-\mathbf{X}^T\mathbf{W}_1^{-1}]\mathbf{y} \\ &= \begin{pmatrix} 0.001, & 0.45, & 0.049, & 0.261, & 0.23, & 0.01 \end{pmatrix}\mathbf{y} \\ &= 0.001y_1 + 0.45y_2 + 0.049y_3 + 0.261y_4 + 0.23y_5 + 0.01y_6,\end{aligned}$$

and the BLUP's prediction error variance is

$$\begin{aligned}\mathrm{var}(\tilde{y}_7 - y_7) &= \sigma^2[1 - \mathbf{k}^T\mathbf{W}_1^{-1}\mathbf{k} + (\mathbf{c}^T - \mathbf{k}^T\mathbf{W}_1^{-1}\mathbf{X})(\mathbf{X}^T\mathbf{W}_1^{-1}\mathbf{X})^-(\mathbf{c}^T - \mathbf{k}^T\mathbf{W}_1^{-1}\mathbf{X})^T]\\ &= 0.254\sigma^2.\end{aligned}$$

The weights in the BLUP corresponding to the elements of $\mathbf{y}$ are very similar to those in the original example. This corresponds to the fact that the locations of the observations are unchanged. However, the BLUP's prediction error variance is much smaller than it was in the original example. An intuitive explanation is that for the new variance–covariance matrix of $\mathbf{y}$, the correlations between any two observations are larger than those in the original example. With higher correlation, the prediction is more precise and thus the prediction error variance is smaller.

(c) We preserve the coordinate system and notations in the original example but now $w_{ij} = 1$ if $i = j$, and $w_{ij} = 0.5\exp(-d_{ij}/2)$ otherwise. Repeating the calculations, we obtain the following quantities:

$$\mathbf{W} = \begin{pmatrix} 1.000 & 0.247 & 0.082 & 0.112 & 0.082 & 0.064 & 0.163 \\ 0.247 & 1.000 & 0.163 & 0.163 & 0.163 & 0.112 & 0.303 \\ 0.082 & 0.163 & 1.000 & 0.103 & 0.247 & 0.122 & 0.184 \\ 0.112 & 0.163 & 0.103 & 1.000 & 0.184 & 0.247 & 0.247 \\ 0.082 & 0.163 & 0.247 & 0.184 & 1.000 & 0.247 & 0.247 \\ 0.064 & 0.112 & 0.122 & 0.247 & 0.247 & 1.000 & 0.184 \\ 0.163 & 0.303 & 0.184 & 0.247 & 0.247 & 0.184 & 1.000 \end{pmatrix},$$

$$\mathbf{W}_1 = \begin{pmatrix} 1.000 & 0.247 & 0.082 & 0.112 & 0.082 & 0.064 \\ 0.247 & 1.000 & 0.163 & 0.163 & 0.163 & 0.112 \\ 0.082 & 0.163 & 1.000 & 0.103 & 0.247 & 0.122 \\ 0.112 & 0.163 & 0.103 & 1.000 & 0.184 & 0.247 \\ 0.082 & 0.163 & 0.247 & 0.184 & 1.000 & 0.247 \\ 0.064 & 0.112 & 0.122 & 0.247 & 0.247 & 1.000 \end{pmatrix},$$

$$\mathbf{k}^T = \begin{pmatrix} 0.163 & 0.303 & 0.184 & 0.247 & 0.247 & 0.184 \end{pmatrix}.$$

Therefore, the BLUP of $y_7 = 1\mu + e_7$ is

$$\begin{aligned}\tilde{y}_7 &= [\mathbf{c}^T(\mathbf{X}^T\mathbf{W}_1^{-1}\mathbf{X})^-\mathbf{X}^T\mathbf{W}_1^{-1} + \mathbf{k}^T\mathbf{W}_1^{-1} - \mathbf{k}^T\mathbf{W}_1^{-1}\mathbf{X}(\mathbf{X}^T\mathbf{W}_1^{-1}\mathbf{X})^-\mathbf{X}^T\mathbf{W}_1^{-1}]\mathbf{y}\\ &= \begin{pmatrix} 0.124, & 0.256, & 0.133, & 0.194, & 0.175, & 0.119 \end{pmatrix}\mathbf{y}\\ &= 0.124y_1 + 0.256y_2 + 0.133y_3 + 0.194y_4 + 0.175y_5 + 0.119y_6,\end{aligned}$$

and the BLUP's prediction error variance is

$$\begin{aligned}\text{var}(\tilde{y}_7 - y_7) &= \sigma^2[1 - \mathbf{k}^T \mathbf{W}_1^{-1}\mathbf{k} + (\mathbf{c}^T - \mathbf{k}^T \mathbf{W}_1^{-1}\mathbf{X})(\mathbf{X}^T \mathbf{W}_1^{-1}\mathbf{X})^{-}(\mathbf{c}^T - \mathbf{k}^T \mathbf{W}_1^{-1}\mathbf{X})^T] \\ &= 0.843\sigma^2.\end{aligned}$$

Now, the weights corresponding to the elements of $\mathbf{y}$ do not vary as much as they did in the original example, even though the relative locations are unchanged. This is because the new variance–covariance matrix results in all observations being relatively weakly correlated and the differences in correlations are relatively small. Furthermore, due to the weak correlation between observations, it is not surprising to see that the BLUP's prediction error variance is much higher than it was in the original example.

▶ **Exercise 9** Observe that in Example 13.2-3 and Exercise 13.8, the "weights" (the coefficients on the elements of $\mathbf{y}$ in the expression for the BLUP) sum to one (apart from roundoff error). Explain why this is so.

Solution Let $\mathbf{t}^T\mathbf{y}$ represent the BLUP. The BLUP is an unbiased predictor, so by Theorem 13.1.1 $\mathbf{t}^T\mathbf{X} = \mathbf{c}^T$, which in this problem is $\mathbf{t}^T\mathbf{1} = 1$ (i.e., the weights sum to 1) because $\mathbf{X} = \mathbf{1}$ and $\mathbf{c} = 1$.

▶ **Exercise 10** Suppose that observations (x_i, y_i) follow the simple linear regression model

$$y_i = \beta_1 + \beta_2 x_i + e_i \quad (i = 1, \ldots, n),$$

where $n \geq 3$. Consider the problem of predicting an unobserved y-value, y_{n+1}, corresponding to a specified x-value, x_{n+1}. Assume that y_{n+1} follows the same basic model as the observed responses; that is, the joint model for y_{n+1} and the observed responses can be written as

$$y_i = \beta_1 + \beta_2 x_i + e_i \qquad (i = 1, \ldots, n, n+1),$$

where the e_i's satisfy Gauss–Markov assumptions with common variance $\sigma^2 > 0$. Example 13.2-1 established that the best linear unbiased predictor of y_{n+1} under this model is

$$\tilde{y}_{n+1} = \hat{\beta}_1 + \hat{\beta}_2 x_{n+1}$$

and its mean squared prediction error (MSPE) is $\sigma^2[1 + (1/n) + (x_{n+1} - \bar{x})^2/SXX]$. Although $\tilde{y}_{n+1}$ has smallest MSPE among all *unbiased* linear predictors of y_{n+1}, *biased* linear predictors of y_{n+1} exist that may have smaller MSPE than $\tilde{y}_{n+1}$. One such predictor is $\bar{y}$, the sample mean of the y_i's. For $x_{n+1} \neq \bar{x}$, show that $\bar{y}$ has smaller MSPE than $\tilde{y}_{n+1}$ if and only if $\beta_2^2 SXX/\sigma^2 < 1$.

Solution The mean squared prediction error of $\bar{y}$ is

$$\begin{aligned}
\mathrm{E}[(\bar{y}-y_{n+1})^2] &= \mathrm{var}(\bar{y}-y_{n+1})+[\mathrm{E}(\bar{y})-\mathrm{E}(y_{n+1})]^2 \\
&= \mathrm{var}(\bar{y})+\mathrm{var}(y_{n+1})-2\mathrm{cov}(\bar{y},y_{n+1})+[\beta_1+\beta_2\bar{x}-(\beta_1+\beta_2 x_{n+1})]^2 \\
&= \sigma^2[1+(1/n)]+\beta_2^2(x_{n+1}-\bar{x})^2.
\end{aligned}$$

Thus the mean squared prediction error of $\bar{y}$ is less than that of $\tilde{y}_{n+1}$ if and only if

$$\sigma^2[1+(1/n)]+\beta_2^2(x_{n+1}-\bar{x})^2 < \sigma^2[1+(1/n)+(x_{n+1}-\bar{x})^2/SXX],$$

i.e., if and only if $\beta_2^2 SXX/\sigma^2 < 1$.

▶ **Exercise 11** Obtain the expressions for $\mathrm{var}(\tilde{b}-b)$ and its unbiased estimator given in Example 13.4.2-1.

Solution Specializing the expressions in Theorem 13.4.2b, c using $\mathbf{C}=0, \mathbf{F}=1, \mathbf{X}=0, \mathbf{Z}=\mathbf{z}, \mathbf{G}=\sigma_b^2/\sigma^2 \equiv \psi$, and $\mathbf{R}=\mathbf{I}$, we obtain

$$\mathrm{var}(\tilde{b}-b)=\sigma^2 1(\psi^{-1}+\mathbf{z}^T\mathbf{z})^{-1}1+0=\frac{\sigma^2}{\psi^{-1}+\sum_{i=1}^n z_i^2}.$$

An unbiased estimator of $\mathrm{var}(\tilde{b}-b)$ may be obtained by replacing σ^2 in the numerator of this expression with

$$\tilde{\sigma}^2=\mathbf{y}^T[\mathbf{I}-(\psi^{-1}+\mathbf{z}^T\mathbf{z})^{-1}\mathbf{z}\mathbf{z}^T]\mathbf{y}/(n-0)=\left(\sum_{i=1}^n y_i^2-\frac{\left(\sum_{i=1}^n z_i y_i\right)^2}{\psi^{-1}+\sum_{i=1}^n z_i^2}\right)\Big/ n.$$

▶ **Exercise 12** Consider the random no-intercept simple linear regression model

$$y_i = bz_i + d_i \quad (i=1,\ldots,n),$$

where $b, d_1, d_2, \ldots, d_n$ are uncorrelated zero-mean random variables such that $\mathrm{var}(b)=\sigma_b^2>0$ and $\mathrm{var}(d_i)=\sigma^2>0$ for all i. Let $\psi=\sigma_b^2/\sigma^2$ and suppose that ψ is known. Let z represent a specified real number.

(a) Write out the mixed-model equations in as simple a form as possible, and solve them.
(b) Give a nonmatrix expression, simplified as much as possible, for the BLUP of bz.
(c) Give a nonmatrix expression, simplified as much as possible, for the prediction error variance of the BLUP of bz.

Solution

(a) $(\psi^{-1} + \sum_{i=1}^{n} z_i^2)b = \sum_{i=1}^{n} z_i y_i$, with solution $\dot{b} = \frac{\sum_{i=1}^{n} z_i y_i}{\psi^{-1}+\sum_{i=1}^{n} z_i^2}$.

(b) $\dot{b}z$.

(c) $\frac{\sigma^2 z^2}{\psi^{-1}+\sum_{i=1}^{n} z_i^2}$.

▶ **Exercise 13** Consider the following random-slope simple linear regression model:

$$y_i = \beta + bz_i + d_i \quad (i = 1, \cdots, n),$$

where β is an unknown parameter, b is a zero-mean random variable with variance $\sigma_b^2 > 0$, and the d_i's are uncorrelated (with each other and with b) zero-mean random variables with common variance $\sigma^2 > 0$. Let $\psi = \sigma_b^2/\sigma^2$ and suppose that ψ is known. The model equation may be written in matrix form as

$$\mathbf{y} = \beta \mathbf{1} + b\mathbf{z} + \mathbf{d}.$$

(a) Determine $\sigma^2 \mathbf{W} \equiv \text{var}(\mathbf{y})$.
(b) Write out the mixed-model equations in as simple a form as possible and solve them.
(c) Consider predicting the predictable function $\tau \equiv \beta + bz$, where z is a specified real number. Give an expression for the BLUP $\tilde{\tau}$ of τ in terms of a solution $(\dot{\beta}, \dot{b})^T$ to the mixed-model equations.
(d) Give a nonmatrix expression for $\tilde{\sigma}^2$, the generalized residual mean square, in terms of a solution to the mixed-model equations mentioned in part (c).
(e) Obtain a nonmatrix expression for the prediction error variance associated with $\tilde{\tau}$.

Solution

(a) $\text{var}(\mathbf{y}) = \text{var}(\beta\mathbf{1}+b\mathbf{z}+\mathbf{d}) = \text{var}(b\mathbf{z})+\text{var}(\mathbf{d}) = \sigma_b^2\mathbf{z}\mathbf{z}^T + \sigma^2\mathbf{I} = \sigma^2(\psi\mathbf{z}\mathbf{z}^T + \mathbf{I})$.
(b) The mixed-model equations simplify to

$$\begin{pmatrix} n & \sum_{i=1}^{n} z_i \\ \sum_{i=1}^{n} z_i & \sum_{i=1}^{n} z_i^2 + \psi^{-1} \end{pmatrix} \begin{pmatrix} \beta \\ b \end{pmatrix} = \begin{pmatrix} \sum_{i=1}^{n} y_i \\ \sum_{i=1}^{n} z_i y_i \end{pmatrix},$$

from which we obtain the solution

$$\dot{b} = \frac{\sum_{i=1}^{n}(z_i - \bar{z})y_i}{\sum_{i=1}^{n}(z_i - \bar{z})^2 + \psi^{-1}}, \qquad \dot{\beta} = \bar{y} - \dot{b}\bar{z}.$$

(c) By Corollary 13.4.3.1, the BLUP of $\tau \equiv \beta + bz$ is $\tilde{\tau} = \dot{\beta} + \dot{b}z$.
(d) By Corollary 13.4.3.2, $\tilde{\sigma}^2 = \frac{\mathbf{y}^T\mathbf{y} - \dot{\beta}\mathbf{1}_n^T\mathbf{y} - \dot{b}\mathbf{z}^T\mathbf{y}}{n-1} = \frac{\sum_{i=1}^n y_i(y_i - \dot{\beta} - \dot{b}z_i)}{n-1}$.
(e) By Theorem 13.4.4,

$$\begin{aligned}\text{var}(\tilde{\tau} - \tau) &= \sigma^2(1, z)\left(\frac{1}{n(\sum_{i=1}^n z_i^2 + \psi^{-1}) - (\sum_{i=1}^n z_i)^2}\right)\\ &\quad \begin{pmatrix} \sum_{i=1}^n z_i^2 + \psi^{-1} & -\sum_{i=1}^n z_i \\ -\sum_{i=1}^n z_i & n \end{pmatrix}\begin{pmatrix} 1 \\ z \end{pmatrix}\\ &= \sigma^2\left(\frac{\sum_{i=1}^n z_i^2 + \psi^{-1} - 2z\sum_{i=1}^n z_i + z^2 n}{n(\sum_{i=1}^n z_i^2 + \psi^{-1}) - (\sum_{i=1}^n z_i)^2}\right).\end{aligned}$$

▶ **Exercise 14** Consider the following random-intercept simple linear regression model:

$$y_i = b + \beta x_i + d_i \quad (i = 1, \cdots, n),$$

where β is an unknown parameter, b is a zero-mean random variable with variance $\sigma_b^2 > 0$, and the d_i's are uncorrelated (with each other and with b) zero-mean random variables with common variance $\sigma^2 > 0$. Let $\psi = \sigma_b^2/\sigma^2$ and suppose that ψ is known. The model equation may be written in matrix form as

$$\mathbf{y} = b\mathbf{1} + \beta\mathbf{x} + \mathbf{d}.$$

(a) Determine $\sigma^2\mathbf{W} \equiv \text{var}(\mathbf{y})$.
(b) Write out the mixed-model equations in as simple a form as possible and solve them.
(c) Consider predicting the predictable function $\tau \equiv b + \beta x$, where x is a specified real number. Give an expression for the BLUP $\tilde{\tau}$ of τ in terms of a solution $(\dot{\beta}, \dot{b})^T$ to the mixed-model equations.
(d) Give a nonmatrix expression for $\tilde{\sigma}^2$, the generalized residual mean square, in terms of a solution to the mixed-model equations mentioned in part (c).
(e) Obtain a nonmatrix expression for the prediction error variance associated with $\tilde{\tau}$.

Solution

(a) $\text{var}(\mathbf{y}) = \text{var}(b\mathbf{1} + \beta\mathbf{x} + \mathbf{d}) = \text{var}(b\mathbf{1}) + \text{var}(\mathbf{d}) = \sigma_b^2\mathbf{1}\mathbf{1}^T + \sigma^2\mathbf{I} = \sigma^2(\psi\mathbf{1}\mathbf{1}^T + \mathbf{I})$.
(b) The mixed-model equations simplify to

$$\begin{pmatrix} \sum_{i=1}^n x_i^2 & \sum_{i=1}^n x_i \\ \sum_{i=1}^n x_i & n + \psi^{-1} \end{pmatrix}\begin{pmatrix} \beta \\ b \end{pmatrix} = \begin{pmatrix} \sum_{i=1}^n x_i y_i \\ \sum_{i=1}^n y_i \end{pmatrix},$$

from which we obtain the solution

$$\dot{\beta} = \frac{\sum_{i=1}^n x_i y_i - \frac{\sum_{i=1}^n x_i \sum_{i=1}^n y_i}{n+\psi^{-1}}}{\sum_{i=1}^n x_i^2 - \frac{(\sum_{i=1}^n x_i)^2}{n+\psi^{-1}}}, \qquad \dot{b} = \frac{\sum_{i=1}^n y_i - \dot{\beta}\sum_{i=1}^n x_i}{n+\psi^{-1}}.$$

(c) By Corollary 13.4.3.1, the BLUP of $\tau \equiv b + \beta x$ is $\tilde{\tau} = \dot{b} + \dot{\beta}x$.
(d) By Corollary 13.4.3.2, $\tilde{\sigma}^2 = \frac{\mathbf{y}^T\mathbf{y} - \dot{\beta}\mathbf{x}^T\mathbf{y} - \dot{b}\mathbf{1}_n^T\mathbf{y}}{n-1} = \frac{\sum_{i=1}^n y_i(y_i - \dot{b} - \dot{\beta}x_i)}{n-1}$.
(e) By Theorem 13.4.4,

$$\begin{aligned}\text{var}(\tilde{\tau} - \tau) &= \sigma^2(x, 1)\left(\frac{1}{(n+\psi^{-1})\sum_{i=1}^n x_i^2 - (\sum_{i=1}^n x_i)^2}\right) \\ &\quad \begin{pmatrix} n+\psi^{-1} & -\sum_{i=1}^n x_i \\ -\sum_{i=1}^n x_i & \sum_{i=1}^n x_i^2 \end{pmatrix}\begin{pmatrix} x \\ 1 \end{pmatrix} \\ &= \sigma^2\left(\frac{x^2(n+\psi^{-1}) - 2x\sum_{i=1}^n x_i + \sum_{i=1}^n x_i^2}{(n+\psi^{-1})\sum_{i=1}^n x_i^2 - (\sum_{i=1}^n x_i)^2}\right).\end{aligned}$$

▶ **Exercise 15** Consider the following random simple linear regression model:

$$y_i = b_1 + b_2 z_i + d_i \quad (i = 1, \cdots, n),$$

where b_1 and b_2 are uncorrelated zero-mean random variables with variances $\sigma_{b_1}^2 > 0$ and $\sigma_{b_2}^2 > 0$, respectively, and the d_i's are uncorrelated (with each other and with b_1 and b_2) zero-mean random variables with common variance $\sigma^2 > 0$. Let $\psi_1 = \sigma_{b_1}^2/\sigma^2$ and $\psi_2 = \sigma_{b_2}^2/\sigma^2$, and suppose that ψ_1 and ψ_2 are known. The model equation may be written in matrix form as

$$\mathbf{y} = b_1\mathbf{1} + b_2\mathbf{z} + \mathbf{d}.$$

(a) Determine $\sigma^2\mathbf{W} \equiv \text{var}(\mathbf{y})$.
(b) Write out the mixed-model equations in as simple a form as possible and solve them.
(c) Consider predicting the predictable function $\tau \equiv b_1 + b_2 z$, where z is a specified real number. Give an expression for the BLUP $\tilde{\tau}$ of τ in terms of a solution $(\dot{b}_1, \dot{b}_2)^T$ to the mixed-model equations.
(d) Give a nonmatrix expression for $\tilde{\sigma}^2$, the generalized residual mean square, in terms of a solution to the mixed-model equations mentioned in part (c).
(e) Obtain a nonmatrix expression for the prediction error variance associated with $\tilde{\tau}$.

Solution

(a) $\text{var}(\mathbf{y}) = \text{var}(b_1\mathbf{1} + b_2\mathbf{z} + \mathbf{d}) = \text{var}(b_1\mathbf{1}) + \text{var}(b_2\mathbf{z}) + \text{var}(\mathbf{d}) = \sigma_{b_1}^2\mathbf{1}\mathbf{1}^T + \sigma_{b_2}^2\mathbf{z}\mathbf{z}^T + \sigma^2\mathbf{I} = \sigma^2(\psi_1\mathbf{1}\mathbf{1}^T + \psi_2\mathbf{z}\mathbf{z}^T + \mathbf{I})$.

(b) The mixed-model equations simplify to

$$\begin{pmatrix} n+\psi_1^{-1} & \sum_{i=1}^n z_i \\ \sum_{i=1}^n z_i & \sum_{i=1}^n z_i^2 + \psi_2^{-1} \end{pmatrix} \begin{pmatrix} b_1 \\ b_2 \end{pmatrix} = \begin{pmatrix} \sum_{i=1}^n y_i \\ \sum_{i=1}^n z_i y_i \end{pmatrix},$$

from which we obtain the solution

$$\dot{b}_2 = \frac{\sum_{i=1}^n z_i y_i - \frac{\sum_{i=1}^n z_i \sum_{i=1}^n y_i}{n+\psi_1^{-1}}}{\left(\sum_{i=1}^n z_i^2 + \psi_2^{-1}\right) - \frac{(\sum_{i=1}^n z_i)^2}{n+\psi_1^{-1}}},$$

$$\dot{b}_1 = \frac{\sum_{i=1}^n y_i - \breve{b}_2 \sum_{i=1}^n z_i}{n+\psi_1^{-1}}.$$

(c) By Corollary 13.4.3.1, the BLUP of $\tau \equiv b_1 + b_2 z$ is $\tilde{\tau} = \dot{b}_1 + \dot{b}_2 z$.

(d) By Corollary 13.4.3.2,

$$\tilde{\sigma}^2 = \frac{\mathbf{y}^T\mathbf{y} - (\dot{b}_1\mathbf{1}^T + \dot{b}_2\mathbf{z}^T)\mathbf{y}}{n} = \frac{\sum_{i=1}^n y_i(y_i - \dot{b}_1 - \dot{b}_2 z_i)}{n}.$$

(e) By Theorem 13.4.4,

$$\text{var}(\tilde{\tau} - \tau) = \sigma^2(1, z)\left(\frac{1}{(n+\psi_1^{-1})(\sum_{i=1}^n z_i^2 + \psi_2^{-1}) - (\sum_{i=1}^n z_i)^2}\right.$$
$$\left.\begin{pmatrix} \sum_{i=1}^n z_i^2 + \psi_2^{-1} & -\sum_{i=1}^n z_i \\ -\sum_{i=1}^n z_i & n+\psi_1^{-1} \end{pmatrix}\right)\begin{pmatrix} 1 \\ z \end{pmatrix}$$
$$= \sigma^2\left(\frac{\sum_{i=1}^n z_i^2 + \psi_2^{-1} - 2z\sum_{i=1}^n z_i + (n+\psi_1^{-1})z^2}{(n+\psi_1^{-1})(\sum_{i=1}^n z_i^2 + \psi_2^{-1}) - (\sum_{i=1}^n z_i)^2}\right).$$

▶ **Exercise 16** For the balanced one-factor random effects model considered in Example 13.4.2-2:

(a) Obtain specialized expressions for the variances and covariances of the prediction errors, $\{\widetilde{b_i - b_{i'}} - (b_i - b_{i'}) : i' > i = 1, \ldots, q\}$.
(b) Obtain a specialized expression for the generalized residual mean square, $\tilde{\sigma}^2$.

Solution

(a) First observe that

$$\begin{aligned}\widetilde{b_i - b_{i'}} - (b_i - b_{i'}) &= \frac{r\psi}{r\psi + 1}(\bar{y}_{i\cdot} - \bar{y}_{i'\cdot}) - (b_i - b_{i'}) \\ &= \frac{r\psi}{r\psi + 1}[(b_i + \bar{d}_{i\cdot}) - (b_{i'} + \bar{d}_{i'\cdot})] - (b_i - b_{i'}) \\ &= \frac{-1}{r\psi + 1}(b_i - b_{i'}) + \frac{r\psi}{r\psi + 1}(\bar{d}_{i\cdot} - \bar{d}_{i'\cdot}).\end{aligned}$$

Hence for $i \leq j, i' > i = 1, \ldots, q$, and $j' > j = 1, \ldots, q$,

$$\begin{aligned}&\text{cov}[\widetilde{b_i - b_{i'}} - (b_i - b_{i'}), \widetilde{b_j - b_{j'}} - (b_j - b_{j'})] \\ &= \frac{1}{(r\psi + 1)^2}\text{cov}[(-1)(b_i - b_{i'}) + r\psi(\bar{d}_{i\cdot} - \bar{d}_{i'\cdot}), (-1)(b_j - b_{j'}) + r\psi(\bar{d}_{j\cdot} - \bar{d}_{j'\cdot})] \\ &= \frac{1}{(r\psi + 1)^2}\text{cov}[(b_i - b_{i'}), (b_j - b_{j'})] + \frac{r^2\psi^2}{(r\psi + 1)^2}\text{cov}[(\bar{d}_{i\cdot} - \bar{d}_{i'\cdot}), (\bar{d}_{j\cdot} - \bar{d}_{j'\cdot})] \\ &= \begin{cases} \frac{2\sigma_b^2}{(r\psi+1)^2} + \frac{r^2\psi^2}{(r\psi+1)^2} \cdot \frac{2\sigma^2}{r} = \frac{2\psi\sigma^2}{r\psi+1} & \text{if } i = j \text{ and } i' = j', \\ \frac{\sigma_b^2}{(r\psi+1)^2} + \frac{r^2\psi^2}{(r\psi+1)^2} \cdot \frac{\sigma^2}{r} = \frac{\psi\sigma^2}{r\psi+1} & \text{if } i = j \text{ and } i' \neq j', \text{ or } i \neq j \text{ and } i' = j', \\ \frac{-\sigma_b^2}{(r\psi+1)^2} + \frac{r^2\psi^2}{(r\psi+1)^2} \cdot \frac{(-\sigma^2)}{r} = -\frac{\psi\sigma^2}{r\psi+1} & \text{if } i' = j, \\ 0 & \text{otherwise.} \end{cases}\end{aligned}$$

(b) By Theorem 13.4.2c, an unbiased estimator of σ^2 is

$$\begin{aligned}\tilde{\sigma}^2 &= (\mathbf{y} - \bar{y}_{\cdot\cdot}\mathbf{1}_{qr})^T \left[\mathbf{I} - (\oplus_{i=1}^q \mathbf{1}_r)(\psi^{-1}\mathbf{I}_q + r\mathbf{I}_q)^{-1}\left(\oplus_{i=1}^q \mathbf{1}_r^T\right)\right](\mathbf{y} - \bar{y}_{\cdot\cdot}\mathbf{1}_{qr})/(n-1) \\ &= \left[\sum_{i=1}^q\sum_{j=1}^r (y_{ij} - \bar{y}_{\cdot\cdot})^2 - \frac{r\psi}{r\psi + 1}\sum_{i=1}^q r(\bar{y}_{i\cdot} - \bar{y}_{\cdot\cdot})^2\right] \Big/ (n-1).\end{aligned}$$

▶ **Exercise 17** Consider a two-way layout with two rows and two columns, and one observation in three of the four cells, labelled as y_{11}, y_{12}, and y_{21} as depicted in the sketch below:

y_{11}	y_{12}
y_{21}	

No response is observed in the other cell, but we would like to use the existing data to predict such a response, which we label as y_{22} (the response, not its predictor). Obtain simplified expressions for the BLUP of y_{22} and its mean squared prediction error, under each of the following three models. Note: The following suggestions may make your work easier in this exercise:

- For part (a), reparameterize the model in such a way that the parameter vector is given by

$$\begin{pmatrix} \mu + \alpha_2 + \gamma_2 \\ \gamma_1 - \gamma_2 \\ \alpha_1 - \alpha_2 \end{pmatrix}.$$

  For part (b), reparameterize the model in such a way that the parameter vector is given by

$$\begin{pmatrix} \mu + \alpha_2 \\ \alpha_1 - \alpha_2 \end{pmatrix}.$$

- The following matrix inverses may be useful:

$$\begin{pmatrix} 3 & 2 & 2 \\ 2 & 2 & 1 \\ 2 & 1 & 2 \end{pmatrix}^{-1} = \begin{pmatrix} 3 & -2 & -2 \\ -2 & 2 & 1 \\ -2 & 1 & 2 \end{pmatrix},$$

$$\begin{pmatrix} 2 & 0 & 1 \\ 0 & 2 & 0 \\ 1 & 0 & 2 \end{pmatrix}^{-1} = \begin{pmatrix} \frac{2}{3} & 0 & -\frac{1}{3} \\ 0 & \frac{1}{2} & 0 \\ -\frac{1}{3} & 0 & \frac{2}{3} \end{pmatrix},$$

$$\begin{pmatrix} 3 & 1 & 1 \\ 1 & 3 & 0 \\ 1 & 0 & 3 \end{pmatrix}^{-1} = \begin{pmatrix} \frac{3}{7} & -\frac{1}{7} & -\frac{1}{7} \\ -\frac{1}{7} & \frac{8}{21} & \frac{1}{21} \\ -\frac{1}{7} & \frac{1}{21} & \frac{8}{21} \end{pmatrix}.$$

(a) The Gauss–Markov two-way main effects model $y_{ij} = \mu + \alpha_i + \gamma_j + e_{ij}$,

$$\mathrm{E}\begin{pmatrix} e_{11} \\ e_{12} \\ e_{21} \\ e_{22} \end{pmatrix} = \mathbf{0}, \quad \mathrm{var}\begin{pmatrix} e_{11} \\ e_{12} \\ e_{21} \\ e_{22} \end{pmatrix} = \sigma^2\mathbf{I}.$$

(b) A mixed two-way main effects model $y_{ij} = \mu + \alpha_i + b_j + d_{ij}$,

$$\mathrm{E}\begin{pmatrix} b_1 \\ b_2 \\ d_{11} \\ d_{12} \\ d_{21} \\ d_{22} \end{pmatrix} = \mathbf{0}, \quad \mathrm{var}\begin{pmatrix} b_1 \\ b_2 \\ d_{11} \\ d_{12} \\ d_{21} \\ d_{22} \end{pmatrix} = \begin{pmatrix} \sigma_b^2 \mathbf{I}_2 & \mathbf{0} \\ \mathbf{0} & \sigma^2 \mathbf{I}_4 \end{pmatrix},$$

where $\sigma_b^2/\sigma^2 = 1$.

(c) A random two-way main effects model $y_{ij} = \mu + a_i + b_j + d_{ij}$,

$$\mathrm{E}\begin{pmatrix} a_1 \\ a_2 \\ b_1 \\ b_2 \\ d_{11} \\ d_{12} \\ d_{21} \\ d_{22} \end{pmatrix} = \mathbf{0}, \quad \mathrm{var}\begin{pmatrix} a_1 \\ a_2 \\ b_1 \\ b_2 \\ d_{11} \\ d_{12} \\ d_{21} \\ d_{22} \end{pmatrix} = \begin{pmatrix} \sigma_a^2 \mathbf{I}_2 & \mathbf{0} & \mathbf{0} \\ \mathbf{0} & \sigma_b^2 \mathbf{I}_2 & \\ \mathbf{0} & \mathbf{0} & \sigma^2 \mathbf{I}_4 \end{pmatrix},$$

where $\sigma_b^2/\sigma^2 = \sigma_a^2/\sigma^2 = 1$.

Solution

(a) After reparameterizing as suggested and putting $\mathbf{y} = (y_{11}, y_{12}, y_{21})^T$, we have

$$\mathbf{X} = \begin{pmatrix} 1 & 1 & 1 \\ 1 & 0 & 1 \\ 1 & 1 & 0 \end{pmatrix},$$

and $\mathbf{W} = \mathbf{I}_3$, $\mathbf{k} = \mathbf{0}_3$, and $h = 1$. The BLUE of $\mu + \alpha_2 + \gamma_2$ is the first element of

$$\tilde{\boldsymbol{\beta}} = (\mathbf{X}^T\mathbf{X})^{-1}\mathbf{X}^T\mathbf{y} = \begin{pmatrix} 3 & 2 & 2 \\ 2 & 2 & 1 \\ 2 & 1 & 2 \end{pmatrix}^{-1} \begin{pmatrix} 1 & 1 & 1 \\ 1 & 0 & 1 \\ 1 & 1 & 0 \end{pmatrix} \begin{pmatrix} y_{11} \\ y_{12} \\ y_{21} \end{pmatrix}$$

$$= \begin{pmatrix} 3 & -2 & -2 \\ -2 & 2 & 1 \\ -2 & 1 & 2 \end{pmatrix} \begin{pmatrix} 1 & 1 & 1 \\ 1 & 0 & 1 \\ 1 & 1 & 0 \end{pmatrix} \begin{pmatrix} y_{11} \\ y_{12} \\ y_{21} \end{pmatrix},$$

which is $y_{12} + y_{21} - y_{11}$, and this is also the BLUP of y_{22} in this case (because $\mathbf{k} = \mathbf{0}$). The mean squared prediction error of this BLUP is $\text{var}(y_{12} + y_{21} - y_{11} - y_{22}) = 4\sigma^2$.

(b) After reparameterizing as suggested, we have

$$\mathbf{X} = \begin{pmatrix} 1 & 1 \\ 1 & 1 \\ 1 & 0 \end{pmatrix}, \quad \mathbf{W} = \begin{pmatrix} 2 & 0 & 1 \\ 0 & 2 & 0 \\ 1 & 0 & 2 \end{pmatrix}, \quad \mathbf{k} = \begin{pmatrix} 0 \\ 1 \\ 0 \end{pmatrix}, \quad h = 2.$$

The BLUE of $\boldsymbol{\beta}$ is

$$\begin{aligned}
\tilde{\boldsymbol{\beta}} &= (\mathbf{X}^T\mathbf{W}^{-1}\mathbf{X})^{-1}\mathbf{X}^T\mathbf{W}^{-1}\mathbf{y} \\
&= \left[\begin{pmatrix} 1 & 1 & 1 \\ 1 & 1 & 0 \end{pmatrix}\begin{pmatrix} 2/3 & 0 & -1/3 \\ 0 & 1/2 & 0 \\ -1/3 & 0 & 2/3 \end{pmatrix}\begin{pmatrix} 1 & 1 \\ 1 & 1 \\ 1 & 0 \end{pmatrix}\right]^{-1}\begin{pmatrix} 1 & 1 & 1 \\ 1 & 1 & 0 \end{pmatrix}\begin{pmatrix} 2/3 & 0 & -1/3 \\ 0 & 1/2 & 0 \\ -1/3 & 0 & 2/3 \end{pmatrix}\begin{pmatrix} y_{11} \\ y_{12} \\ y_{21} \end{pmatrix} \\
&= \begin{pmatrix} 7/6 & 5/6 \\ 5/6 & 7/6 \end{pmatrix}^{-1}\begin{pmatrix} 1/3 & 1/2 & 1/3 \\ 2/3 & 1/2 & -1/3 \end{pmatrix}\begin{pmatrix} y_{11} \\ y_{12} \\ y_{21} \end{pmatrix} \\
&= \begin{pmatrix} (-y_{11} + y_{12} + 4y_{21})/4 \\ (3y_{11} + y_{12} - 4y_{21})/4 \end{pmatrix}.
\end{aligned}$$

Then, the BLUP of y_{22} is

$$\begin{aligned}
\tilde{y}_{22} &= \begin{pmatrix} 1 & 0 \end{pmatrix}\tilde{\boldsymbol{\beta}} + \mathbf{k}^T\mathbf{W}^{-1}(\mathbf{y} - \mathbf{X}\tilde{\boldsymbol{\beta}}) \\
&= [(-y_{11} + y_{12} + 4y_{21})/4] \\
&\quad + \begin{pmatrix} 0 & 1 & 0 \end{pmatrix}\begin{pmatrix} 2/3 & 0 & -1/3 \\ 0 & 1/2 & 0 \\ -1/3 & 0 & 2/3 \end{pmatrix}\left[\begin{pmatrix} y_{11} \\ y_{12} \\ y_{21} \end{pmatrix} - \begin{pmatrix} 1 & 1 \\ 1 & 1 \\ 1 & 0 \end{pmatrix}\begin{pmatrix} (-y_{11} + y_{12} + 4y_{21})/4 \\ (3y_{11} + y_{12} - 4y_{21})/4 \end{pmatrix}\right] \\
&= [(-y_{11} + y_{12} + 4y_{21})/4] + (y_{12}/2) - (1/2)[(-y_{11} + y_{12} + 4y_{21})/4] \\
&\quad -(1/2)[(3y_{11} + y_{12} - 4y_{21})/4] \\
&= (-1/2)y_{11} + (1/2)y_{12} + y_{21}.
\end{aligned}$$

The mean squared prediction error of $\tilde{y}_{22}$ is

$$\begin{aligned}
\text{var}[(-1/2)y_{11} + (1/2)y_{12} + y_{21} - y_{22}] &= \sigma^2\begin{pmatrix} -1/2 & 1/2 & 1 & -1 \end{pmatrix}\begin{pmatrix} 2 & 0 & 1 & 0 \\ 0 & 2 & 0 & 1 \\ 1 & 0 & 2 & 0 \\ 0 & 1 & 0 & 2 \end{pmatrix}\begin{pmatrix} -1/2 \\ 1/2 \\ 1 \\ -1 \end{pmatrix} \\
&= 3\sigma^2.
\end{aligned}$$

(c) We have

$$\mathbf{X} = \mathbf{1}_3, \quad \mathbf{W} = \begin{pmatrix} 3 & 1 & 1 \\ 1 & 3 & 0 \\ 1 & 0 & 3 \end{pmatrix}, \quad \mathbf{k} = \begin{pmatrix} 0 \\ 1 \\ 1 \end{pmatrix}, \quad h = 3.$$

The BLUE of $\boldsymbol{\beta}$ ($\equiv \mu$) is

$$\tilde{\boldsymbol{\beta}} = \left[\mathbf{1}_3^T \begin{pmatrix} 3/7 & -1/7 & -1/7 \\ -1/7 & 8/21 & 1/21 \\ -1/7 & 1/21 & 8/21 \end{pmatrix} \mathbf{1}_3 \right]^{-1} \mathbf{1}_3^T \begin{pmatrix} 3/7 & -1/7 & -1/7 \\ -1/7 & 8/21 & 1/21 \\ -1/7 & 1/21 & 8/21 \end{pmatrix} \begin{pmatrix} y_{11} \\ y_{12} \\ y_{21} \end{pmatrix}$$

$$= (7/5) \begin{pmatrix} 1/7 & 2/7 & 2/7 \end{pmatrix} \begin{pmatrix} y_{11} \\ y_{12} \\ y_{21} \end{pmatrix}$$

$$= (1/5)y_{11} + (2/5)y_{12} + (2/5)y_{21}.$$

Then, the BLUP of y_{22} is

$$\begin{aligned} \tilde{y}_{22} &= (1/5)y_{11} + (2/5)y_{12} + (2/5)y_{21} \\ &\quad + \begin{pmatrix} 0 & 1 & 1 \end{pmatrix} \begin{pmatrix} 3/7 & -1/7 & -1/7 \\ -1/7 & 8/21 & 1/21 \\ -1/7 & 1/21 & 8/21 \end{pmatrix} \left[\begin{pmatrix} y_{11} \\ y_{12} \\ y_{21} \end{pmatrix} - \mathbf{1}_3[(1/5)y_{11} + (2/5)y_{12} + (2/5)y_{21}] \right] \\ &= (1/5)y_{11} + (2/5)y_{12} + (2/5)y_{21} + (-2/7)y_{11} + (3/7)y_{12} + (3/7)y_{21} \\ &\quad -(4/7)[(1/5)y_{11} + (2/5)y_{12} + (2/5)y_{21}] \\ &= (-1/5)y_{11} + (3/5)y_{12} + (3/5)y_{21}. \end{aligned}$$

The mean squared prediction error of $\tilde{y}_{22}$ is

$$\begin{aligned} \mathrm{var}[(-1/5)y_{11} + (3/5)y_{12} + (3/5)y_{21} - y_{22}] &= \sigma^2 \begin{pmatrix} -1/5 & 3/5 & 3/5 & -1 \end{pmatrix} \\ &\quad \begin{pmatrix} 3 & 1 & 1 & 0 \\ 1 & 3 & 0 & 1 \\ 1 & 0 & 3 & 1 \\ 0 & 1 & 1 & 3 \end{pmatrix} \begin{pmatrix} -1/5 \\ 3/5 \\ 3/5 \\ -1 \end{pmatrix} \\ &= (12/5)\sigma^2. \end{aligned}$$

Exercise 18 Consider the following mixed linear model for two observations, y_1 and y_2:

$$y_1 = 2\beta + b + d_1$$
$$y_2 = \beta + 3b + d_2,$$

where β is a fixed unknown (and unrestricted) parameter, and b, d_1, d_2 are independent random variables with zero means and variances $\sigma_b^2 = \text{var}(b)$ and $\sigma^2 = \text{var}(d_1) = \text{var}(d_2)$. Suppose that $\sigma_b^2/\sigma^2 = 1/2$.

(a) Write this model in the matrix form $\mathbf{y} = \mathbf{X}\boldsymbol{\beta} + \mathbf{Z}\mathbf{b} + \mathbf{d}$ and give an expression for var($\mathbf{y}$) in which σ^2 is the only unknown parameter.
(b) Compute the BLUE of β and the BLUP of b.

Solution

(a) $\mathbf{y} = \begin{pmatrix} 2 \\ 1 \end{pmatrix} \beta + \begin{pmatrix} 1 \\ 3 \end{pmatrix} b + \mathbf{d}$, and

$$\begin{aligned} \text{var}(\mathbf{y}) &= \text{var}(\mathbf{Zb} + \mathbf{d}) \\ &= \sigma_b^2 \mathbf{Z}\mathbf{Z}^T + \sigma^2 \mathbf{I} \\ &= \sigma^2 \left[(1/2) \begin{pmatrix} 1 \\ 3 \end{pmatrix} (1\ 3) + \mathbf{I} \right] \\ &= \sigma^2 \begin{pmatrix} 3/2 & 3/2 \\ 3/2 & 11/2 \end{pmatrix}. \end{aligned}$$

(b) The mixed-model equations are

$$\begin{pmatrix} 5 & 5 \\ 5 & 12 \end{pmatrix} \begin{pmatrix} \beta \\ b \end{pmatrix} = \begin{pmatrix} 2y_1 + y_2 \\ y_1 + 3y_2 \end{pmatrix}.$$

Their unique solution, which yields the BLUE of β and the BLUP of b, is

$$\begin{pmatrix} \dot{\beta} \\ \dot{b} \end{pmatrix} = \begin{pmatrix} 12/35 & -5/35 \\ -5/35 & 5/35 \end{pmatrix} \begin{pmatrix} 2y_1 + y_2 \\ y_1 + 3y_2 \end{pmatrix} = \begin{pmatrix} (19/35)y_1 - (3/35)y_2 \\ (-1/7)y_1 + (2/7)y_2 \end{pmatrix}.$$

Exercise 19 Consider the balanced mixed two-way main effects model

$$y_{ijk} = \mu + \alpha_i + b_j + d_{ijk} \quad (i = 1, \ldots, q;\ j = 1, \ldots, m;\ k = 1, \ldots, r),$$

where $\mu, \alpha_1, \ldots, \alpha_q$ are unknown parameters, the b_j's are uncorrelated zero-mean random variables with common variance $\sigma_b^2 > 0$, and the d_{ijk}'s are uncorrelated (with each other and with the b_j's) zero-mean random variables with common variance $\sigma^2 > 0$. Let $\psi = \sigma_b^2/\sigma^2$ and suppose that ψ is known.

(a) One solution to the Aitken equations for this model is $\tilde{\boldsymbol{\beta}} = (0, \bar{y}_{1\cdot\cdot}, \ldots, \bar{y}_{q\cdot\cdot})^T$. Using this solution, obtain specialized expressions for the BLUEs of $\mu+\alpha_i$ and $\alpha_i - \alpha_{i'}$ $(i' > i = 1, \ldots, q)$.
(b) Obtain a specialized expression for the variance–covariance matrix of the BLUEs of $\mu+\alpha_i$, and do likewise for the BLUEs of $\alpha_i - \alpha_{i'}$ $(i' > i = 1, \ldots, q)$.
(c) Obtain specialized expressions for the BLUPs of $\mu+\alpha_i+b_j$ $(i = 1, \ldots, q;\ j = 1, \ldots, m)$ and $b_j - b_{j'}$ $(j' > j = 1, \ldots, m)$.
(d) Obtain a specialized expression for the variance–covariance matrix of the BLUPs of $\mu + \alpha_i + b_j$, and do likewise for the BLUPs of $b_j - b_{j'}$ $(i = 1, \ldots, q;\ j' > j = 1, \ldots, m)$.

Solution

(a) The BLUE of any estimable function $\mathbf{c}^T\boldsymbol{\beta}$ is $\mathbf{c}^T\tilde{\boldsymbol{\beta}}$. Thus, the BLUE of $\mu+\alpha_i$ is $\widetilde{\mu+\alpha_i} = \bar{y}_{i\cdot\cdot}$ $(i = 1, \ldots, q)$, and the BLUE of $\alpha_i - \alpha_{i'}$ is $\widetilde{\alpha_i - \alpha_{i'}} = \bar{y}_{i\cdot\cdot} - \bar{y}_{i'\cdot\cdot}$ $(i' > i = 1, \ldots, q)$.
(b) $\bar{y}_{i\cdot\cdot} = \frac{1}{mr}\sum_{j=1}^m \sum_{k=1}^r y_{ijk} = \frac{1}{mr}\sum_{j=1}^m \sum_{k=1}^r (\mu+\alpha_i+b_j+d_{ijk}) = \mu+\alpha_i+\bar{b}+\bar{d}_{i\cdot\cdot}$. Therefore,

$$\begin{aligned}
\mathrm{cov}(\widetilde{\mu+\alpha_i}, \widetilde{\mu+\alpha_{i'}}) &= \mathrm{cov}(\bar{y}_{i\cdot\cdot}, \bar{y}_{i'\cdot\cdot}) \\
&= \mathrm{cov}(\bar{b}+\bar{d}_{i\cdot\cdot}, \bar{b}+\bar{d}_{i'\cdot\cdot}) \\
&= \mathrm{var}(\bar{b}) + \mathrm{cov}(\bar{d}_{i\cdot\cdot}, \bar{d}_{i'\cdot\cdot}) \\
&= \begin{cases} \frac{\sigma_b^2}{m} + \frac{\sigma^2}{mr} & \text{if } i = i', \\ \frac{\sigma_b^2}{m} & \text{if } i \neq i'. \end{cases}
\end{aligned}$$

And, for $i \le s$, $i' > i = 1, \ldots, q$, and $s' > s = 1, \ldots, q$,

$$\begin{aligned}
\mathrm{cov}(\widetilde{\alpha_i - \alpha_{i'}}, \widetilde{\alpha_s - \alpha_{s'}}) &= \mathrm{cov}(\bar{y}_{i\cdot\cdot} - \bar{y}_{i'\cdot\cdot}, \bar{y}_{s\cdot\cdot} - \bar{y}_{s'\cdot\cdot}) \\
&= \mathrm{cov}(\bar{y}_{i\cdot\cdot}, \bar{y}_{s\cdot\cdot}) - \mathrm{cov}(\bar{y}_{i'\cdot\cdot}, \bar{y}_{s\cdot\cdot}) - \mathrm{cov}(\bar{y}_{i\cdot\cdot}, \bar{y}_{s'\cdot\cdot}) + \mathrm{cov}(\bar{y}_{i'\cdot\cdot}, \bar{y}_{s'\cdot\cdot}) \\
&= \begin{cases} \frac{2\sigma^2}{mr} & \text{if } i = s \text{ and } i' = s', \\ \frac{\sigma^2}{mr} & \text{if } i = s \text{ and } i' \neq s', \text{ or } i \neq s \text{ and } i' = s', \\ -\frac{\sigma^2}{mr} & \text{if } i' = s, \\ 0 & \text{otherwise.} \end{cases}
\end{aligned}$$

(c) By Theorem 13.4.2a,

$$\tilde{\mathbf{b}} = (\mathbf{G}^{-1} + \mathbf{Z}^T\mathbf{Z})^{-1}\mathbf{Z}^T(\mathbf{y} - \mathbf{X}\tilde{\boldsymbol{\beta}}) = \frac{qr\psi}{1+qr\psi}\begin{pmatrix} \bar{y}_{\cdot 1\cdot} - \bar{y}_{\cdots} \\ \vdots \\ \bar{y}_{\cdot m\cdot} - \bar{y}_{\cdots} \end{pmatrix}.$$

Therefore, the BLUP of $\mu + \alpha_i + b_j$ is

$$\widetilde{\mu + \alpha_i + b_j} = \bar{y}_{i\cdot\cdot} + \frac{qr\psi}{1+qr\psi}(\bar{y}_{\cdot j\cdot} - \bar{y}_{\cdots}) \quad (i = 1, \ldots, q;\ j = 1, \ldots, m)$$

and the BLUP of $b_j - b_{j'}$ is

$$\widetilde{b_j - b_{j'}} = \frac{qr\psi}{1+qr\psi}(\bar{y}_{\cdot j\cdot} - \bar{y}_{\cdot j'\cdot}) \quad (j' > j = 1, \ldots, m).$$

(d) Note that $\bar{y}_{\cdot j\cdot} = \mu + \bar{\alpha} + b_j + \bar{d}_{\cdot j\cdot}$ and $\bar{y}_{\cdots} = \mu + \bar{\alpha} + \bar{b} + \bar{d}_{\cdots}$, so

$$\text{cov}(\bar{y}_{\cdot j\cdot}, \bar{y}_{\cdot j'\cdot}) = \text{cov}(b_j, b_{j'}) + \text{cov}(\bar{d}_{\cdot j\cdot}, \bar{d}_{\cdot j'\cdot}) = \begin{cases} \sigma_b^2 + \frac{\sigma^2}{qr} & \text{if } j = j', \\ 0 & \text{if } j \neq j', \end{cases}$$

$$\text{var}(\bar{y}_{\cdots}) = \text{var}(\bar{b} + \bar{d}_{\cdots}) = \frac{\sigma_b^2}{m} + \frac{\sigma^2}{qmr},$$

$$\text{cov}(\bar{y}_{i\cdot\cdot}, \bar{y}_{\cdot j\cdot}) = \text{cov}(\bar{b}, b_j) + \text{cov}(\bar{d}_{i\cdot\cdot}, \bar{d}_{\cdot j\cdot}) = \frac{\sigma_b^2}{m} + \frac{\sigma^2}{qmr} \quad \text{for } i = 1, \ldots, q;\ j = 1, \ldots, m,$$

$$\text{cov}(\bar{y}_{i\cdot\cdot}, \bar{y}_{\cdots}) = \text{cov}(\bar{b}, \bar{b}) + \text{cov}(\bar{d}_{i\cdot\cdot}, \bar{d}_{\cdots}) = \frac{\sigma_b^2}{m} + \frac{\sigma^2}{qmr} \quad \text{for } i = 1, \ldots, q,$$

$$\text{cov}(\bar{y}_{\cdot j\cdot}, \bar{y}_{\cdots}) = \text{cov}(b_j, \bar{b}) + \text{cov}(\bar{d}_{\cdot j\cdot}, \bar{d}_{\cdots}) = \frac{\sigma_b^2}{m} + \frac{\sigma^2}{qmr} \quad \text{for } j = 1, \ldots, m.$$

Thus,

$$\text{cov}(\widetilde{\mu + \alpha_i + b_j}, \widetilde{\mu + \alpha_{i'} + b_{j'}}) = \text{cov}(\bar{y}_{i\cdot\cdot} + \frac{qr\psi}{1+qr\psi}\bar{y}_{\cdot j\cdot} - \frac{qr\psi}{1+qr\psi}\bar{y}_{\cdots}, \bar{y}_{i'\cdot\cdot}$$

$$+ \frac{qr\psi}{1+qr\psi}\bar{y}_{\cdot j'\cdot} - \frac{qr\psi}{1+qr\psi}\bar{y}_{\cdots})$$

$$= \begin{cases} \frac{\sigma_b^2}{m} + \frac{\sigma^2}{mr} + \left(\frac{qr\psi}{1+qr\psi}\right)^2 (m-1)\left(\frac{\sigma_b^2}{m} + \frac{\sigma^2}{mr}\right) & \text{if } i = i' \text{ and } j = j', \\ \frac{\sigma_b^2}{m} + \frac{\sigma^2}{mr} + \left(\frac{qr\psi}{1+qr\psi}\right)^2 \left(\frac{\sigma_b^2}{m} + \frac{\sigma^2}{mr}\right) & \text{if } i = i' \text{ and } j \neq j', \\ \frac{\sigma_b^2}{m} + \left(\frac{qr\psi}{1+qr\psi}\right)^2 (m-1)\left(\frac{\sigma_b^2}{m} + \frac{\sigma^2}{mr}\right) & \text{if } i \neq i' \text{ and } j = j', \\ \frac{\sigma_b^2}{m} - \left(\frac{qr\psi}{1+qr\psi}\right)^2 \left(\frac{\sigma_b^2}{m} + \frac{\sigma^2}{mr}\right) & \text{if } i \neq i' \text{ and } j \neq j', \end{cases}$$

and for $j \leq t$, $j' > j = 1, \ldots, m$, and $t' > t = 1, \ldots, m$,

$$\operatorname{cov}(\widetilde{b_j - b_{j'}}, \widetilde{b_t - b_{t'}}) = \left(\frac{qr\psi}{1+qr\psi}\right)^2 \operatorname{cov}(\bar{y}_{\cdot j\cdot} - \bar{y}_{\cdot j'\cdot}, \bar{y}_{\cdot t\cdot} - \bar{y}_{\cdot t'\cdot})$$

$$= \begin{cases} 2\left(\frac{qr\psi}{1+qr\psi}\right)^2\left(\sigma_b^2 + \frac{\sigma^2}{qr}\right) & \text{if } j = t \text{ and } j' = t', \\ \left(\frac{qr\psi}{1+qr\psi}\right)^2\left(\sigma_b^2 + \frac{\sigma^2}{qr}\right) & \text{if } j = t \text{ and } j' \neq t', \\ & \text{or } j \neq t \text{ and } j' = t', \\ -\left(\frac{qr\psi}{1+qr\psi}\right)^2\left(\sigma_b^2 + \frac{\sigma^2}{qr}\right) & \text{if } j' = t, \\ 0 & \text{otherwise.} \end{cases}$$

▶ **Exercise 20** Consider a random two-way partially crossed model, analogous to the fixed-effects model introduced in Example 5.1.4-1, with one observation per cell, i.e.,

$$y_{ij} = \mu + b_i - b_j + d_{ij} \qquad (i \neq j = 1, \ldots, q),$$

where μ is an unknown parameter, the b_i's are uncorrelated zero-mean random variables with common variance $\sigma_b^2 > 0$, and the d_{ij}'s are uncorrelated (with each other and with the b_i's) zero-mean random variables with common variance $\sigma^2 > 0$. Let $\psi = \sigma_b^2/\sigma^2$ and suppose that ψ is known.

(a) This model is a special case of the general mixed linear model

$$\mathbf{y} = \mathbf{X}\boldsymbol{\beta} + \mathbf{Z}\mathbf{b} + \mathbf{d},$$

where $\operatorname{var}(\mathbf{d}) = \sigma^2\mathbf{I}$ and $\operatorname{var}(\mathbf{b}) = \sigma^2\mathbf{G}$ for some positive definite matrix $\mathbf{G}$. Write out the elements of the matrices $\mathbf{X}$, $\mathbf{Z}$, and $\mathbf{G}$ for this case.
(b) Write out the mixed-model equations for this model and solve them.
(c) Determine $\mathbf{W} \equiv (1/\sigma^2)\operatorname{var}(\mathbf{y})$ and determine $\mathbf{W}^{-1}$.
(d) Give expressions, in as simple a form as possible, for the BLUPs of $b_i - b_j$ $(j > i = 1, \ldots, q)$ and for the variance–covariance matrix of the prediction errors associated with these BLUPs. Specialize that variance–covariance matrix for the case $q = 4$.

Solution

(a) $\mathbf{X} = \mathbf{1}_{q(q-1)}$, $\mathbf{Z} = (\mathbf{v}_{ij}^T)_{i\neq j=1,\ldots,q}$, $\mathbf{G} = \psi\mathbf{I}_q$ where $\mathbf{v}_{ij} = \mathbf{u}_i^{(q)} - \mathbf{u}_j^{(q)}$ and $\psi = \sigma_b^2/\sigma^2$.
(b) Here $\mathbf{X}^T\mathbf{X} = \mathbf{1}_{q(q-1)}^T\mathbf{1}_{q(q-1)} = q(q-1)$, $\mathbf{X}^T\mathbf{Z} = \mathbf{1}_{q(q-1)}(\mathbf{v}_{ij}^T)_{i\neq j=1,\ldots,q} = \mathbf{0}_q^T$ (because each column of $(\mathbf{v}_{ij}^T)_{i\neq j=1,\ldots,q}$ has $q-1$ ones and $q-1$ minus ones), $\mathbf{G} = \psi\mathbf{I}_q$, and $\mathbf{Z}^T\mathbf{Z} = 2q\mathbf{I}_q - 2\mathbf{J}_q$. Furthermore, $\mathbf{X}^T\mathbf{y} = \mathbf{1}_{q(q-1)}\mathbf{y} =$

$\sum_{i,j:i\neq j} y_{ij}$ and $\mathbf{Z}^T\mathbf{y} = \left(\sum_{i:i\neq j}(y_{ij} - y_{ji})\right)_{j=1,\ldots,q}$. So the mixed-model equations may be written as

$$\begin{pmatrix} q(q-1) & \mathbf{0}_q^T \\ \mathbf{0}_q & (\psi^{-1} + 2q)\mathbf{I}_q - 2\mathbf{J}_q \end{pmatrix} \begin{pmatrix} \mu \\ \mathbf{b} \end{pmatrix} = \begin{pmatrix} \sum_{i,j:i\neq j} y_{ij} \\ \left(\sum_{i\neq j}(y_{ij} - y_{ji})\right)_{j=1,\ldots,q} \end{pmatrix},$$

with solution

$$\dot{\mu} = \frac{\sum_{i,j:i\neq j} y_{ij}}{q(q-1)}$$

and

$$\begin{aligned} \dot{\mathbf{b}} &= [(\psi^{-1} + 2q)\mathbf{I}_q - 2\mathbf{J}_q]^{-1} \left(\sum_{i:i\neq j}(y_{ij} - y_{ji})\right)_{j=1,\ldots,q} \\ &= [(\psi^{-1} + 2q)^{-1}\mathbf{I}_q + \frac{2}{(\psi^{-1} + 2q)\psi^{-1}}\mathbf{J}_q] \left(\sum_{i:i\neq j}(y_{ij} - y_{ji})\right)_{j=1,\ldots,q} \\ &= (\psi^{-1} + 2q)^{-1} \left(\sum_{i:i\neq j}(y_{ij} - y_{ji})\right)_{j=1,\ldots,q}, \end{aligned}$$

where we used expression (2.1) for the inverse of a nonsingular compound symmetric matrix.

(c) $\text{var}(\mathbf{y}) = \sigma^2(\psi\mathbf{Z}\mathbf{Z}^T + \mathbf{I}_{q(q-1)})$. Thus $\mathbf{W} = \psi\mathbf{Z}\mathbf{Z}^T + \mathbf{I}_{q(q-1)}$ where $(\mathbf{Z}\mathbf{Z}^T)_{ij,st}$, for $i \leq s$, $j > i = 1, \ldots, q$, and $t > s = 1, \ldots, q$, is given by

$$\begin{aligned} (\mathbf{Z}\mathbf{Z}^T)_{ij,st} = \mathbf{v}_{ij}^T\mathbf{v}_{st} &= (\mathbf{u}_i^{(q)} - \mathbf{u}_j^{(q)})^T(\mathbf{u}_s^{(q)} - \mathbf{u}_t^{(q)}) \\ &= \mathbf{u}_i^{(q)T}\mathbf{u}_s^{(q)} - \mathbf{u}_j^{(q)T}\mathbf{u}_s^{(q)} - \mathbf{u}_i^{(q)T}\mathbf{u}_t^{(q)} + \mathbf{u}_j^{(q)T}\mathbf{u}_t^{(q)} \\ &= \begin{cases} 2 & \text{if } i = s \text{ and } j = t, \\ 1 & \text{if } i = s \text{ and } j \neq t, \text{ or } i \neq s \text{ and } j = t, \\ -1 & \text{if } j = s, \\ 0 & \text{otherwise.} \end{cases} \end{aligned}$$

(The remaining elements are determined by symmetry.) Using the computational formulae for $\mathbf{W}^{-1}$, we obtain

$$\mathbf{G}^{-1} + \mathbf{Z}^T\mathbf{Z} = (\psi^{-1} + 2q)\mathbf{I}_q - 2\mathbf{J}_q,$$

$$(\mathbf{G}^{-1} + \mathbf{Z}^T\mathbf{Z})^{-1} = (\psi^{-1} + 2q)^{-1}\mathbf{I}_q + \frac{2}{(\psi^{-1} + 2q)\psi^{-1}}\mathbf{J}_q,$$

$$\begin{aligned}\mathbf{W}^{-1} = (\mathbf{I} + \mathbf{Z}\mathbf{G}\mathbf{Z}^T)^{-1} &= \mathbf{I} - \mathbf{Z}(\mathbf{G}^{-1} + \mathbf{Z}^T\mathbf{Z})^{-1}\mathbf{Z}^T \\ &= \mathbf{I} - (\psi^{-1} + 2q)^{-1}\mathbf{Z}\mathbf{Z}^T\end{aligned}$$

because $\mathbf{ZJ} = \mathbf{0}$ and $\mathbf{JZ}^T = \mathbf{0}$.

(d) Ignoring terms involving the fixed effects,

$$\begin{aligned}\tilde{\mathbf{b}} - \mathbf{b} &= (\psi^{-1} + 2q)^{-1}\mathbf{Z}^T(\mathbf{Zb} + \mathbf{d}) - \mathbf{b} \\ &= (\psi^{-1} + 2q)^{-1}[(2q\mathbf{I}_q - 2\mathbf{J}_q)\mathbf{b} - (\psi^{-1} + 2q)\mathbf{b} + \mathbf{Z}^T\mathbf{d}] \\ &= (\psi^{-1} + 2q)^{-1}[(-\psi^{-1} - 2\mathbf{J}_q)\mathbf{b} + \mathbf{Z}^T\mathbf{d}].\end{aligned}$$

Thus

$$\begin{aligned}\mathrm{var}(\tilde{\mathbf{b}} - \mathbf{b}) &= (\psi^{-1} + 2q)^{-2}[\sigma^2\psi(\psi^{-1}\mathbf{I}_q + 2\mathbf{J}_q)(\psi^{-1}\mathbf{I}_q + 2\mathbf{J}_q) + \mathbf{Z}^T(\sigma^2\mathbf{I}_{q(q-1)})] \\ &= \sigma^2(\psi^{-1} + 2q)^{-2}[\psi^{-1}\mathbf{I}_q + 4(1 + q\psi)\mathbf{J}_q + (2q\mathbf{I}_q - 2\mathbf{J}_q)] \\ &= \sigma^2(\psi^{-1} + 2q)^{-2}[(\psi^{-1} + 2q)\mathbf{I}_q + (2 + 4q\psi)\mathbf{J}_q]\end{aligned}$$

and

$$\begin{aligned}\mathrm{var}(\mathbf{Z}\tilde{\mathbf{b}} - \mathbf{Zb}) &= \mathbf{Z}\mathrm{var}(\tilde{\mathbf{b}} - \mathbf{b})\mathbf{Z}^T \\ &= \sigma^2(\psi^{-1} + 2q)^{-2}(\psi^{-1} + 2q)\mathbf{Z}\mathbf{Z}^T \\ &= \frac{\sigma^2}{\psi^{-1} + 2q}\mathbf{Z}\mathbf{Z}^T.\end{aligned}$$

So, in the special case $q = 4$, with rows (and columns) ordered as 12, 13, 14, 23, 24, 34, we have

$$\mathrm{var}(\mathbf{Z}\tilde{\mathbf{b}} - \mathbf{Zb}) = \frac{\sigma^2}{\psi^{-1} + 8}\begin{pmatrix} 2 & 1 & 1 & -1 & -1 & 0 \\ 1 & 2 & 1 & 1 & 0 & -1 \\ 1 & 1 & 2 & 0 & 1 & 1 \\ -1 & 1 & 0 & 2 & 1 & -1 \\ -1 & 0 & 1 & 1 & 2 & 1 \\ 0 & -1 & 1 & -1 & 1 & 2 \end{pmatrix}.$$

Exercise 21 Consider the balanced mixed two-factor nested model

$$y_{ijk} = \mu + \alpha_i + b_{ij} + d_{ijk} \qquad (i = 1, \ldots, q;\ j = 1, \ldots, m;\ k = 1, \ldots, r),$$

where $\mu, \alpha_1, \ldots, \alpha_q$ are unknown parameters, the b_{ij}'s are uncorrelated zero-mean random variables with common variance $\sigma_b^2 > 0$, and the d_{ijk}'s are uncorrelated (with each other and with the b_{ij}'s) zero-mean random variables with common variance $\sigma^2 > 0$. Let $\psi = \sigma_b^2/\sigma^2$ and suppose that ψ is known.

(a) This model is a special case of the general mixed linear model

$$\mathbf{y} = \mathbf{X}\boldsymbol{\beta} + \mathbf{Z}\mathbf{b} + \mathbf{d},$$

where $\text{var}(\mathbf{d}) = \sigma^2\mathbf{I}$ and $\text{var}(\mathbf{b}) = \sigma^2\mathbf{G}$ for some positive definite matrix $\mathbf{G}$. Specialize the matrices $\mathbf{X}$, $\mathbf{Z}$, and $\mathbf{G}$ for this model.

(b) Is $\alpha_1 - \alpha_2 + b_{31}$ (assuming that $q \geq 3$) a predictable function? Explain why or why not.

(c) Specialize the mixed-model equations for this model.

(d) Obtain specialized expressions for $\mathbf{W} = (1/\sigma^2)\text{var}(\mathbf{y})$ and $\mathbf{W}^{-1}$.

(e) It can be shown that one solution to the Aitken equations corresponding to this model is $\tilde{\boldsymbol{\beta}} = (0, \bar{y}_{1\cdot\cdot}, \ldots, \bar{y}_{q\cdot\cdot})$. Using this fact, give a specialized expression for the BLUP of $\mu + \alpha_i + b_{ij}$.

(f) Obtain a specialized expression for the prediction error variance of the BLUP of $\mu + \alpha_i + b_{ij}$.

Solution

(a)

$$\mathbf{X} = \left(\mathbf{1}_{qmr}\ \mathbf{I}_q \otimes \mathbf{1}_{mr}\right),\ \boldsymbol{\beta} = \begin{pmatrix} \mu \\ \alpha_1 \\ \vdots \\ \alpha_q \end{pmatrix},\ \mathbf{Z} = \mathbf{I}_{qm} \otimes \mathbf{1}_r,\ \mathbf{b} = \begin{pmatrix} b_{11} \\ b_{12} \\ \vdots \\ b_{qm} \end{pmatrix},\ \mathbf{G} = (\sigma_b^2/\sigma^2)\mathbf{I}_{qm}.$$

(b) Yes, by Theorem 13.1.2 because $\alpha_1 - \alpha_2$ is estimable.

(c)
$$\begin{pmatrix} qmr & mr\mathbf{1}_q^T & r\mathbf{1}_{qm}^T \\ mr\mathbf{1}_q & mr\mathbf{I}_q & r\mathbf{I}_q \otimes \mathbf{1}_m^T \\ r\mathbf{1}_{qm} & r\mathbf{I}_q \otimes \mathbf{1}_m & (r + \frac{\sigma^2}{\sigma_b^2})\mathbf{I}_{qm} \end{pmatrix} \begin{pmatrix} \mu \\ \alpha_1 \\ \vdots \\ \alpha_q \\ b_{11} \\ \vdots \\ b_{qm} \end{pmatrix} = \begin{pmatrix} y_{\cdots} \\ y_{1\cdot\cdot} \\ \vdots \\ y_{q\cdot\cdot} \\ y_{11\cdot} \\ \vdots \\ y_{qm\cdot} \end{pmatrix}.$$

(d)

$$\begin{aligned}\text{var}(\mathbf{y}) &= \mathbf{Z}\text{var}(\mathbf{b})\mathbf{Z}^T + \text{var}(\mathbf{d}) \\ &= (\mathbf{I}_{qm} \otimes \mathbf{1}_r)(\sigma_b^2 \mathbf{I}_{qm})(\mathbf{I}_{qm} \otimes \mathbf{1}_r^T) + \sigma^2 \mathbf{I}_{qmr} \\ &= \sigma_b^2(\mathbf{I}_{qm} \otimes \mathbf{J}_r) + \sigma^2(\mathbf{I}_{qm} \otimes \mathbf{I}_r) \\ &= \sigma^2 \mathbf{I}_{qm} \otimes (\mathbf{I}_r + \psi \mathbf{J}_r),\end{aligned}$$

where $\psi = \sigma_b^2/\sigma^2$. Thus $\mathbf{W} = (1/\sigma^2)\text{var}(\mathbf{y}) = \mathbf{I}_{qm} \otimes (\mathbf{I}_r + \psi \mathbf{J}_r)$. From Example 2.9-1, we find that

$$\mathbf{W}^{-1} = \mathbf{I}_{qm}^{-1} \otimes (\mathbf{I}_r + \psi \mathbf{J}_r)^{-1} = \mathbf{I}_{qm} \otimes (\mathbf{I}_r - \frac{\psi}{1 + r\psi}\mathbf{J}_r).$$

(e) Using the given solution to the Aitken equations, $\widetilde{\mu + \alpha_i} = \bar{y}_{i\cdot\cdot}$. Furthermore,

$$\begin{aligned}\tilde{b}_{ij} &= \mathbf{u}_{(i-1)q+j}^{(qm)T} \mathbf{G}\mathbf{Z}^T(\mathbf{I} + \mathbf{Z}\mathbf{G}\mathbf{Z}^T)^{-1}(\mathbf{y} - \mathbf{X}\tilde{\boldsymbol{\beta}}) \\ &= \mathbf{u}_{(i-1)q+j}^{(qm)T}(\psi \mathbf{I}_{qm})(\mathbf{I}_{qm} \otimes \mathbf{1}_r^T)[\mathbf{I}_{qm} \otimes (\mathbf{I}_r - \frac{\psi}{1 + r\psi}\mathbf{J}_r)](\mathbf{y} - \mathbf{X}\tilde{\boldsymbol{\beta}}) \\ &= \psi \mathbf{u}_{(i-1)q+j}^{(qm)T}[\mathbf{I}_{qm} \otimes (\mathbf{1}_r^T - \frac{r\psi}{1 + r\psi}\mathbf{1}_r^T)](\mathbf{y} - \mathbf{X}\tilde{\boldsymbol{\beta}}) \\ &= \frac{\psi}{1 + r\psi}\mathbf{u}_{(i-1)q+j}^{(qm)T}(\mathbf{I}_{qm} \otimes \mathbf{1}_r^T)(\mathbf{y} - \mathbf{X}\tilde{\boldsymbol{\beta}}) \\ &= \frac{r\psi}{1 + r\psi}(\bar{y}_{ij\cdot} - \bar{y}_{i\cdot\cdot}).\end{aligned}$$

Thus, the BLUP of $\mu + \alpha_i + b_{ij}$ is $\bar{y}_{i\cdot\cdot} + \left(\frac{r\psi}{1+r\psi}\right)(\bar{y}_{ij\cdot} - \bar{y}_{i\cdot\cdot}) = \left(\frac{r\psi}{1+r\psi}\right)\bar{y}_{ij\cdot} + \left(\frac{1}{1+r\psi}\right)\bar{y}_{i\cdot\cdot}$.

(f) Because $\bar{y}_{i\cdot\cdot} = \mu + \alpha_i + \bar{b}_{i\cdot} + \bar{d}_{i\cdot\cdot}$ and $\bar{y}_{ij\cdot} = \mu + \alpha_i + b_{ij} + \bar{d}_{ij\cdot}$, we obtain

$$\begin{aligned}\text{var}(\bar{y}_{i\cdot\cdot}) &= \frac{\sigma_b^2}{m} + \frac{\sigma^2}{mr}, \\ \text{var}(\bar{y}_{ij\cdot}) &= \sigma_b^2 + \frac{\sigma^2}{r}, \\ \text{cov}(\bar{y}_{i\cdot\cdot}, \bar{y}_{ij\cdot}) &= \text{cov}(\bar{b}_{i\cdot}, b_{ij1}) + \text{cov}(\bar{d}_{i\cdot\cdot}, \bar{d}_{ij\cdot}) \\ &= \frac{\sigma_b^2}{m} + \frac{\sigma^2}{mr},\end{aligned}$$

$$\text{cov}(\bar{y}_{i\cdot\cdot}, b_{ij}) = \frac{\sigma_b^2}{m},$$
$$\text{cov}(\bar{y}_{ij\cdot}, b_{ij}) = \sigma_b^2.$$

Consequently, the prediction error variance associated with the BLUP of $\mu + \alpha_i + b_{ij}$ is

$$\begin{aligned}
&\text{var}\left(\frac{1}{1+r\psi}\bar{y}_{i\cdot\cdot} + \frac{r\psi}{1+r\psi}\bar{y}_{ij\cdot} - b_{ij}\right) \\
&= \left(\frac{1}{1+r\psi}\right)^2 \text{var}(\bar{y}_{i\cdot\cdot}) + \left(\frac{r\psi}{1+r\psi}\right)^2 \text{var}(\bar{y}_{ij\cdot}) + \text{var}(b_{ij}) + \frac{2r\psi}{(1+r\psi)^2}\text{cov}(\bar{y}_{i\cdot\cdot}, \bar{y}_{ij\cdot}) \\
&\quad - \frac{2}{1+r\psi}\text{cov}(\bar{y}_{i\cdot\cdot}, b_{ij}) - \frac{2r\psi}{1+r\psi}\text{cov}(\bar{y}_{ij\cdot}, b_{ij}) \\
&= \frac{1}{(1+r\psi)^2}\left(\frac{\sigma_b^2}{m} + \frac{\sigma^2}{mr}\right) + \frac{(r\psi)^2}{(1+r\psi)^2}\left(\sigma_b^2 + \frac{\sigma^2}{r}\right) + \sigma_b^2 + \frac{2r\psi}{(1+r\psi)^2}\left(\frac{\sigma_b^2}{m} + \frac{\sigma^2}{mr}\right) \\
&\quad - \frac{2}{1+r\psi}\frac{\sigma_b^2}{m} - \frac{2r\psi}{1+r\psi}\sigma_b^2 \\
&= \frac{m-1}{m(1+r\psi)^2}\sigma_b^2 + \frac{1+mr^2\psi^2+2r\psi}{mr(1+r\psi)^2}\sigma^2.
\end{aligned}$$

▶ **Exercise 22** Consider a mixed effects model with known $\boldsymbol{\psi}$, where $\mathbf{G} = \mathbf{G}(\boldsymbol{\psi})$ and $\mathbf{R} = \mathbf{R}(\boldsymbol{\psi})$ are positive definite. Let $\mathbf{B}$ be the unique $q \times q$ positive definite matrix such that $\mathbf{B}^T\mathbf{B} = \mathbf{G}^{-1}$ (such a matrix exists by Theorem 2.15.12). Suppose that we transform this model by pre-multiplying all of its terms by $\mathbf{R}^{-\frac{1}{2}}$, then augment the transformed model with the model $\mathbf{y}^* = \mathbf{B}\mathbf{b} + \mathbf{h}$, where $\mathbf{y}^* = \mathbf{0}_q$ and $\mathbf{h}$ is a random q-vector satisfying $\text{E}(\mathbf{h}) = \mathbf{0}$ and $\text{var}(\mathbf{h}) = \sigma^2\mathbf{I}$. Show that any solution to the normal equations for this transformed/augmented model is also a solution to the mixed-model equations associated with the original mixed effects model, thus establishing that any computer software that can perform ordinary least squares regression can be coerced into obtaining BLUPs under a mixed effects model.

Solution The transformed/augmented model is

$$\begin{pmatrix} \mathbf{R}^{-\frac{1}{2}}\mathbf{y} \\ \mathbf{y}^* \end{pmatrix} = \begin{pmatrix} \mathbf{R}^{-\frac{1}{2}}\mathbf{X} & \mathbf{R}^{-\frac{1}{2}}\mathbf{Z} \\ \mathbf{0} & \mathbf{B} \end{pmatrix}\begin{pmatrix} \boldsymbol{\beta} \\ \mathbf{b} \end{pmatrix} + \begin{pmatrix} \mathbf{R}^{-\frac{1}{2}}\mathbf{d} \\ \mathbf{h} \end{pmatrix},$$

where

$$\text{E}\begin{pmatrix} \mathbf{R}^{-\frac{1}{2}}\mathbf{d} \\ \mathbf{h} \end{pmatrix} = \mathbf{0}_{n+q}, \quad \text{var}\begin{pmatrix} \mathbf{R}^{-\frac{1}{2}}\mathbf{d} \\ \mathbf{h} \end{pmatrix} = \sigma^2\mathbf{I}_{n+q}.$$

The normal equations corresponding to this model are

$$\begin{pmatrix} \mathbf{R}^{-\frac{1}{2}}\mathbf{X} & \mathbf{R}^{-\frac{1}{2}}\mathbf{Z} \\ \mathbf{0} & \mathbf{B} \end{pmatrix}^T \begin{pmatrix} \mathbf{R}^{-\frac{1}{2}}\mathbf{X} & \mathbf{R}^{-\frac{1}{2}}\mathbf{Z} \\ \mathbf{0} & \mathbf{B} \end{pmatrix} \begin{pmatrix} \boldsymbol{\beta} \\ \mathbf{b} \end{pmatrix} = \begin{pmatrix} \mathbf{R}^{-\frac{1}{2}}\mathbf{X} & \mathbf{R}^{-\frac{1}{2}}\mathbf{Z} \\ \mathbf{0} & \mathbf{B} \end{pmatrix}^T \begin{pmatrix} \mathbf{R}^{-\frac{1}{2}}\mathbf{y} \\ \mathbf{y}^* \end{pmatrix},$$

i.e.,

$$\begin{pmatrix} \mathbf{X}^T\mathbf{R}^{-\frac{1}{2}} & \mathbf{0}^T \\ \mathbf{Z}^T\mathbf{R}^{-\frac{1}{2}} & \mathbf{B}^T \end{pmatrix} \begin{pmatrix} \mathbf{R}^{-\frac{1}{2}}\mathbf{X} & \mathbf{R}^{-\frac{1}{2}}\mathbf{Z} \\ \mathbf{0} & \mathbf{B} \end{pmatrix} \begin{pmatrix} \boldsymbol{\beta} \\ \mathbf{b} \end{pmatrix} = \begin{pmatrix} \mathbf{X}^T\mathbf{R}^{-\frac{1}{2}} & \mathbf{0}^T \\ \mathbf{Z}^T\mathbf{R}^{-\frac{1}{2}} & \mathbf{B}^T \end{pmatrix} \begin{pmatrix} \mathbf{R}^{-\frac{1}{2}}\mathbf{y} \\ \mathbf{0} \end{pmatrix},$$

i.e.,

$$\begin{pmatrix} \mathbf{X}^T\mathbf{R}^{-1}\mathbf{X} & \mathbf{X}^T\mathbf{R}^{-1}\mathbf{Z} \\ \mathbf{Z}^T\mathbf{R}^{-1}\mathbf{X} & \mathbf{Z}^T\mathbf{R}^{-1}\mathbf{Z}+\mathbf{G}^{-1} \end{pmatrix} \begin{pmatrix} \boldsymbol{\beta} \\ \mathbf{b} \end{pmatrix} = \begin{pmatrix} \mathbf{X}^T\mathbf{R}^{-1}\mathbf{y} \\ \mathbf{Z}^T\mathbf{R}^{-1}\mathbf{y} \end{pmatrix}.$$

This last system of equations is the mixed-model equations associated with the original mixed effects model; thus, any solution to the normal equations for the transformed/augmented model is also a solution to the mixed-model equations associated with the original mixed effects model.

▶ **Exercise 23** Prove Theorem 13.4.4: Under the positive definite mixed effects model with known $\boldsymbol{\psi}$, let $\dot{\boldsymbol{\beta}}$ and $\dot{\mathbf{b}}$ be the components of any solution to the mixed-model equations; let $\boldsymbol{\tau} = \mathbf{C}^T\boldsymbol{\beta} + \mathbf{F}^T\mathbf{b}$ represent a vector of predictable linear functions; and let $\tilde{\boldsymbol{\tau}} = \mathbf{C}^T\dot{\boldsymbol{\beta}} + \mathbf{F}^T\dot{\mathbf{b}}$ be the BLUP of $\boldsymbol{\tau}$. Then

$$\mathrm{var}(\tilde{\boldsymbol{\tau}} - \boldsymbol{\tau}) = \sigma^2 \begin{pmatrix} \mathbf{C} \\ \mathbf{F} \end{pmatrix}^T \begin{pmatrix} \mathbf{X}^T\mathbf{R}^{-1}\mathbf{X} & \mathbf{X}^T\mathbf{R}^{-1}\mathbf{Z} \\ \mathbf{Z}^T\mathbf{R}^{-1}\mathbf{X} & \mathbf{G}^{-1}+\mathbf{Z}^T\mathbf{R}^{-1}\mathbf{Z} \end{pmatrix}^{-} \begin{pmatrix} \mathbf{C} \\ \mathbf{F} \end{pmatrix},$$

invariant to the choice of generalized inverse of the coefficient matrix of the mixed-model equations. (Hint: For the invariance, use Theorem 3.2.1.)

Solution Observe that the lower right block $\mathbf{S} \equiv \mathbf{G}^{-1} + \mathbf{Z}^T\mathbf{R}^{-1}\mathbf{Z}$ of the coefficient matrix of the mixed-model equations is nonsingular, implying by Theorem 3.3.7b that

$$\begin{pmatrix} \mathbf{M}^- & -\mathbf{M}^-\mathbf{X}^T\mathbf{R}^{-1}\mathbf{Z}\mathbf{S}^{-1} \\ -\mathbf{S}^{-1}\mathbf{Z}^T\mathbf{R}^{-1}\mathbf{X}\mathbf{M}^- & \mathbf{S}^{-1}+\mathbf{S}^{-1}\mathbf{Z}^T\mathbf{R}^{-1}\mathbf{X}\mathbf{M}^-\mathbf{X}^T\mathbf{R}^{-1}\mathbf{Z}\mathbf{S}^{-1} \end{pmatrix},$$

where $\mathbf{M} = \mathbf{X}^T(\mathbf{R}^{-1} - \mathbf{R}^{-1}\mathbf{Z}\mathbf{S}^{-1}\mathbf{Z}^T\mathbf{R}^{-1})\mathbf{X}$, is a generalized inverse of the coefficient matrix. Denote this particular generalized inverse by $\mathbf{T}$. Pre-multiplication and post-multiplication of $\mathbf{T}$ by $(\mathbf{C}^T, \mathbf{F}^T)$ and its transpose, respectively, yields

$$\begin{aligned}
&\mathbf{C}^T\mathbf{M}^-\mathbf{C} - \mathbf{C}^T\mathbf{M}^-\mathbf{X}^T\mathbf{R}^{-1}\mathbf{Z}\mathbf{S}^{-1}\mathbf{F} - \mathbf{F}^T\mathbf{S}^{-1}\mathbf{Z}^T\mathbf{R}^{-1}\mathbf{X}\mathbf{M}^-\mathbf{C} + \mathbf{F}^T\mathbf{S}^{-1}\mathbf{F} \\
&+\mathbf{F}^T\mathbf{S}^{-1}\mathbf{Z}^T\mathbf{R}^{-1}\mathbf{X}\mathbf{M}^-\mathbf{X}^T\mathbf{R}^{-1}\mathbf{Z}\mathbf{S}^{-1}\mathbf{F} \\
= &\mathbf{F}^T\mathbf{S}^{-1}\mathbf{F} + (\mathbf{C}^T - \mathbf{F}^T\mathbf{S}^{-1}\mathbf{Z}^T\mathbf{R}^{-1}\mathbf{X}) \\
&\times[\mathbf{X}^T(\mathbf{R}^{-1} - \mathbf{R}^{-1}\mathbf{Z}\mathbf{S}^{-1}\mathbf{Z}^T\mathbf{R}^{-1})\mathbf{X}]^-(\mathbf{C}^T - \mathbf{F}^T\mathbf{S}^{-1}\mathbf{Z}^T\mathbf{R}^{-1}\mathbf{X})^T,
\end{aligned}$$

which matches the expression for $(1/\sigma^2)\text{var}(\tilde{\boldsymbol{\tau}} - \boldsymbol{\tau})$ given in Theorem 13.4.2b. Now let $\mathbf{A}$ denote the coefficient matrix of the mixed-model equations, and let $\mathbf{B}$ denote an arbitrary generalized inverse of $\mathbf{A}$. By Theorem 3.1.2, $\mathbf{B} = \mathbf{T} + \mathbf{U} - \mathbf{TAUAT}$ for some matrix $\mathbf{U}$ of appropriate dimensions. We obtain

$$\begin{aligned}
\mathbf{TA} &= \begin{pmatrix} \mathbf{M}^- & -\mathbf{M}^-\mathbf{X}^T\mathbf{R}^{-1}\mathbf{Z}\mathbf{S}^{-1} \\ -\mathbf{S}^{-1}\mathbf{Z}^T\mathbf{R}^{-1}\mathbf{X}\mathbf{M}^- & \mathbf{S}^{-1} + \mathbf{S}^{-1}\mathbf{Z}^T\mathbf{R}^{-1}\mathbf{X}\mathbf{M}^-\mathbf{X}^T\mathbf{R}^{-1}\mathbf{Z}\mathbf{S}^{-1} \end{pmatrix} \begin{pmatrix} \mathbf{X}^T\mathbf{R}^{-1}\mathbf{X} & \mathbf{X}^T\mathbf{R}^{-1}\mathbf{Z} \\ \mathbf{Z}^T\mathbf{R}^{-1}\mathbf{X} & \mathbf{G}^{-1} + \mathbf{Z}^T\mathbf{R}^{-1}\mathbf{X} \end{pmatrix} \\
&= \begin{pmatrix} \mathbf{M}^-\mathbf{M} & \mathbf{0} \\ \mathbf{0} & \mathbf{I} \end{pmatrix}
\end{aligned}$$

because $\mathbf{X}(\mathbf{I} - \mathbf{M}^-\mathbf{M}) = \mathbf{0}$ by Theorem 11.1.6a. Similarly, we obtain

$$\mathbf{AT} = \begin{pmatrix} \mathbf{M}\mathbf{M}^- & \mathbf{0} \\ \mathbf{0} & \mathbf{I} \end{pmatrix}$$

because $(\mathbf{I} - \mathbf{M}\mathbf{M}^-)\mathbf{X}^T = \mathbf{0}$. Therefore, upon partitioning $\mathbf{T}$ and $\mathbf{U}$ appropriately, we have

$$\begin{aligned}
\mathbf{B} &= \begin{pmatrix} \mathbf{T}_{11} & \mathbf{T}_{12} \\ \mathbf{T}_{21} & \mathbf{T}_{22} \end{pmatrix} + \begin{pmatrix} \mathbf{U}_{11} & \mathbf{U}_{12} \\ \mathbf{U}_{21} & \mathbf{U}_{22} \end{pmatrix} \\
&+ \begin{pmatrix} \mathbf{M}^-\mathbf{M} & \mathbf{0} \\ \mathbf{S}^{-1}\mathbf{Z}^T\mathbf{R}^{-1}\mathbf{X}(\mathbf{I} - \mathbf{M}^-\mathbf{M}) & \mathbf{I} \end{pmatrix} \begin{pmatrix} \mathbf{U}_{11} & \mathbf{U}_{12} \\ \mathbf{U}_{21} & \mathbf{U}_{22} \end{pmatrix} \begin{pmatrix} \mathbf{M}\mathbf{M}^- & (\mathbf{I} - \mathbf{M}^-\mathbf{M})\mathbf{X}^T\mathbf{R}^{-1}\mathbf{Z}\mathbf{S}^{-1} \\ \mathbf{0} & \mathbf{I} \end{pmatrix} \\
&= \begin{pmatrix} \mathbf{T}_{11} + \mathbf{U}_{11} - \mathbf{M}^-\mathbf{M}\mathbf{U}_{11}\mathbf{M}\mathbf{M}^- & \mathbf{T}_{12} + \mathbf{U}_{12} - \mathbf{M}^-\mathbf{M}\mathbf{U}_{12} \\ \mathbf{T}_{21} + \mathbf{U}_{21} - \mathbf{U}_{21}\mathbf{M}\mathbf{M}^- & \mathbf{T}_{22} \end{pmatrix}.
\end{aligned}$$

Finally, the predictability of $\mathbf{C}^T\boldsymbol{\beta} + \mathbf{F}^T\mathbf{b}$ implies that $\mathbf{C}^T = \mathbf{A}^T\mathbf{X}$ for some matrix $\mathbf{A}$, so

$$\begin{aligned}
&\begin{pmatrix} \mathbf{C}^T & \mathbf{F}^T \end{pmatrix} \mathbf{B} \begin{pmatrix} \mathbf{C} \\ \mathbf{F} \end{pmatrix} \\
&= \begin{pmatrix} \mathbf{A}^T\mathbf{X} & \mathbf{F}^T \end{pmatrix} \begin{pmatrix} \mathbf{T}_{11} + \mathbf{U}_{11} - \mathbf{M}^-\mathbf{M}\mathbf{U}_{11}\mathbf{M}\mathbf{M}^- & \mathbf{T}_{12} + \mathbf{U}_{12} - \mathbf{M}^-\mathbf{M}\mathbf{U}_{12} \\ \mathbf{T}_{21} + \mathbf{U}_{21} - \mathbf{U}_{21}\mathbf{M}\mathbf{M}^- & \mathbf{T}_{22} \end{pmatrix} \begin{pmatrix} \mathbf{X}^T\mathbf{A} \\ \mathbf{F} \end{pmatrix} \\
&= \mathbf{A}^T\mathbf{X}\mathbf{T}_{11}\mathbf{X}^T\mathbf{A} + \mathbf{A}^T\mathbf{X}(\mathbf{U}_{11} - \mathbf{M}^-\mathbf{M}\mathbf{U}_{11}\mathbf{M}\mathbf{M}^-)\mathbf{X}^T\mathbf{A} + \mathbf{A}^T\mathbf{X}\mathbf{T}_{12}\mathbf{F} + \mathbf{A}^T\mathbf{X}(\mathbf{U}_{12} - \mathbf{M}^-\mathbf{M}\mathbf{U}_{12})\mathbf{F} \\
&\quad + \mathbf{F}^T\mathbf{T}_{21}\mathbf{X}^T\mathbf{A} + \mathbf{F}^T(\mathbf{U}_{21} - \mathbf{U}_{21}\mathbf{M}\mathbf{M}^-)\mathbf{X}^T\mathbf{A} + \mathbf{F}^T\mathbf{T}_{22}\mathbf{F} \\
&= \mathbf{C}^T\mathbf{T}_{11}\mathbf{C} + \mathbf{C}^T\mathbf{T}_{12}\mathbf{F} + \mathbf{F}^T\mathbf{T}_{21}\mathbf{C} + \mathbf{F}^T\mathbf{T}_{22}\mathbf{F}.
\end{aligned}$$

This establishes the claimed invariance.

▶ **Exercise 24** For Example 13.4.5-1:

(a) Verify the given expressions for $\mathbf{P_1}$, $\mathbf{P}_\alpha$, $\mathbf{P_Z}$, $\mathbf{P}_\gamma$, and $\mathbf{P}_\xi$.
(b) Verify all the expressions for the pairwise products of $\mathbf{P_1}$, $\mathbf{P}_\alpha$, $\mathbf{P_Z}$, $\mathbf{P}_\gamma$, and $\mathbf{P}_\xi$.
(c) Verify that the matrices of the six quadratic forms in decomposition (13.12) are idempotent.
d) Verify the given expressions for the expected mean squares.

Solution

(a) Making extensive use of Theorems 2.17.4 and 2.17.5 and the fact that $\mathbf{A}^- \otimes \mathbf{B}^-$ is a generalized inverse of $\mathbf{A} \otimes \mathbf{B}$ (cf. Exercise 3.11), we obtain

$$\mathbf{P_1} = \mathbf{1}_{qrm}(\mathbf{1}_{qrm}^T\mathbf{1}_{qrm})^-\mathbf{1}_{qrm}^T = (1/qrm)\mathbf{J}_{qrm},$$

$$\begin{aligned}
\mathbf{P}_\alpha &= (\mathbf{I}_q \otimes \mathbf{1}_{rm})[(\mathbf{I}_q \otimes \mathbf{1}_{rm})^T(\mathbf{I}_q \otimes \mathbf{1}_{rm})]^-(\mathbf{I}_q \otimes \mathbf{1}_{rm})^T = (\mathbf{I}_q \otimes \mathbf{1}_{rm})(\mathbf{I}_q \otimes rm)^-(\mathbf{I}_q \otimes \mathbf{1}_{rm})^T \\
&= (1/rm)(\mathbf{I}_q \otimes \mathbf{J}_{rm}),
\end{aligned}$$

$$\begin{aligned}
\mathbf{P_Z} &= (\mathbf{I}_{qr} \otimes \mathbf{1}_m)[(\mathbf{I}_{qr} \otimes \mathbf{1}_m)^T(\mathbf{I}_{qr} \otimes \mathbf{1}_m)]^-(\mathbf{I}_{qr} \otimes \mathbf{1}_m)^T = (\mathbf{I}_{qr} \otimes \mathbf{1}_m)(\mathbf{I}_{qr} \otimes m)^-(\mathbf{I}_{qr} \otimes \mathbf{1}_m)^T \\
&= (1/m)(\mathbf{I}_{qr} \otimes \mathbf{J}_m),
\end{aligned}$$

$$\begin{aligned}
\mathbf{P}_\gamma &= (\mathbf{1}_{qr} \otimes \mathbf{I}_m)[(\mathbf{1}_{qr} \otimes \mathbf{I}_m)^T(\mathbf{1}_{qr} \otimes \mathbf{I}_m)]^-(\mathbf{1}_{qr} \otimes \mathbf{I}_m)^T = (\mathbf{1}_{qr} \otimes \mathbf{I}_m)(qr \otimes \mathbf{I}_m)^-(\mathbf{1}_{qr} \otimes \mathbf{I}_m)^T \\
&= (1/qr)(\mathbf{J}_{qr} \otimes \mathbf{I}_m),
\end{aligned}$$

$$\begin{aligned}
\mathbf{P}_\xi &= (\mathbf{I}_1 \otimes \mathbf{1}_r \otimes \mathbf{I}_m)[(\mathbf{I}_1 \otimes \mathbf{1}_r \otimes \mathbf{I}_m)^T(\mathbf{I}_1 \otimes \mathbf{1}_r \otimes \mathbf{I}_m)]^-(\mathbf{I}_1 \otimes \mathbf{1}_r \otimes \mathbf{I}_m)^T \\
&= (\mathbf{I}_1 \otimes \mathbf{1}_r \otimes \mathbf{I}_m)(\mathbf{I}_q \otimes r \otimes \mathbf{I}_m)^-(\mathbf{I}_1 \otimes \mathbf{1}_r \otimes \mathbf{I}_m)^T \\
&= (1/r)(\mathbf{I}_q \otimes \mathbf{J}_r \otimes \mathbf{I}_m).
\end{aligned}$$

(b) Using Theorem 2.17.5 repeatedly, we obtain

$$\begin{aligned}
\mathbf{P}_\alpha\mathbf{P}_1 &= (1/rm)(\mathbf{I}_q \otimes \mathbf{J}_{rm})(1/qrm)\mathbf{J}_{qrm} \\
&= (1/qr^2m^2)(\mathbf{I}_q \otimes \mathbf{J}_{rm})(\mathbf{J}_q \otimes \mathbf{J}_{rm}) \\
&= (1/qr^2m^2)(\mathbf{J}_q \otimes rm\mathbf{J}_{rm}) \\
&= (1/qrm)\mathbf{J}_{qrm} = \mathbf{P}_1, \\
\mathbf{P}_\mathbf{Z}\mathbf{P}_1 &= (1/m)(\mathbf{I}_{qr} \otimes \mathbf{J}_m)(1/qrm)(\mathbf{J}_{qr} \otimes \mathbf{J}_m) \\
&= (1/qrm^2)(\mathbf{J}_{qr} \otimes m\mathbf{J}_m) \\
&= (1/qrm)\mathbf{J}_{qrm} = \mathbf{P}_1, \\
\mathbf{P}_\gamma\mathbf{P}_1 &= (1/qr)(\mathbf{J}_{qr} \otimes \mathbf{I}_m)(1/qrm)(\mathbf{J}_{qr} \otimes \mathbf{J}_m) \\
&= (1/q^2r^2m)(qr\mathbf{J}_{qr} \otimes \mathbf{J}_m) \\
&= (1/qrm)\mathbf{J}_{qrm} = \mathbf{P}_1, \\
\mathbf{P}_\xi\mathbf{P}_1 &= (1/r)(\mathbf{I}_q \otimes \mathbf{J}_r \otimes \mathbf{I}_m)(1/qrm)(\mathbf{J}_q \otimes \mathbf{J}_r \otimes \mathbf{J}_m) \\
&= (1/qr^2m)(\mathbf{J}_q \otimes r\mathbf{J}_r \otimes \mathbf{J}_m) \\
&= (1/qrm)\mathbf{J}_{qrm} = \mathbf{P}_1, \\
\mathbf{P}_\alpha\mathbf{P}_\mathbf{Z} &= (1/rm)(\mathbf{I}_q \otimes \mathbf{J}_{rm})(1/m)(\mathbf{I}_{qr} \otimes \mathbf{J}_m) \\
&= (1/rm^2)(\mathbf{I}_q \otimes \mathbf{J}_r \otimes \mathbf{J}_m)(\mathbf{I}_q \otimes \mathbf{I}_r \otimes \mathbf{J}_m) \\
&= (1/rm)(\mathbf{I}_q \otimes \mathbf{J}_{rm}) = \mathbf{P}_\alpha, \\
\mathbf{P}_\alpha\mathbf{P}_\gamma &= (1/rm)(\mathbf{I}_q \otimes \mathbf{J}_{rm})(1/qr)(\mathbf{J}_{qr} \otimes \mathbf{I}_m) \\
&= (1/qr^2m)(\mathbf{I}_q \otimes \mathbf{J}_r \otimes \mathbf{J}_m)(\mathbf{J}_q \otimes \mathbf{J}_r \otimes \mathbf{I}_m) \\
&= (1/qrm)\mathbf{J}_{qrm} = \mathbf{P}_1, \\
\mathbf{P}_\alpha\mathbf{P}_\xi &= (1/rm)(\mathbf{I}_q \otimes \mathbf{J}_{rm})(1/r)(\mathbf{I}_q \otimes \mathbf{J}_r \otimes \mathbf{I}_m) \\
&= (1/r^2m)(\mathbf{I}_q \otimes \mathbf{J}_r \otimes \mathbf{J}_m)(\mathbf{I}_q \otimes \mathbf{J}_r \otimes \mathbf{I}_m) \\
&= (1/rm)(\mathbf{I}_q \otimes \mathbf{J}_{rm}) = \mathbf{P}_\alpha, \\
\mathbf{P}_\mathbf{Z}\mathbf{P}_\gamma &= (1/m)(\mathbf{I}_{qr} \otimes \mathbf{J}_m)(1/qr)(\mathbf{J}_{qr} \otimes \mathbf{I}_m) \\
&= (1/qrm)\mathbf{J}_{qrm} = \mathbf{P}_1, \\
\mathbf{P}_\mathbf{Z}\mathbf{P}_\xi &= (1/m)(\mathbf{I}_{qr} \otimes \mathbf{J}_m)(1/r)(\mathbf{I}_q \otimes \mathbf{J}_r \otimes \mathbf{I}_m) \\
&= (1/rm)(\mathbf{I}_q \otimes \mathbf{I}_r \otimes \mathbf{J}_m)(\mathbf{I}_q \otimes \mathbf{J}_r \otimes \mathbf{I}_m) \\
&= (1/rm)(\mathbf{I}_q \otimes \mathbf{J}_{rm}) = \mathbf{P}_\alpha, \\
\mathbf{P}_\gamma\mathbf{P}_\xi &= (1/qr)(\mathbf{J}_{qr} \otimes \mathbf{I}_m)(1/r)(\mathbf{I}_q \otimes \mathbf{J}_r \otimes \mathbf{I}_m)
\end{aligned}$$

$$= (1/qr^2)(\mathbf{J}_q \otimes \mathbf{J}_r \otimes \mathbf{I}_m)(\mathbf{I}_q \otimes \mathbf{J}_r \otimes \mathbf{I}_m)$$
$$= (1/m)(\mathbf{J}_{qr} \otimes \mathbf{I}_m) = \mathbf{P}_\gamma.$$

(c) We know already that $\mathbf{P_1P_1}$ is idempotent. Using the results from part (b), we obtain

$$\begin{aligned}
(\mathbf{P}_\alpha - \mathbf{P_1})(\mathbf{P}_\alpha - \mathbf{P_1}) &= \mathbf{P}_\alpha - \mathbf{P_1} - \mathbf{P_1} + \mathbf{P_1} \\
&= \mathbf{P}_\alpha - \mathbf{P_1}, \\
(\mathbf{P_Z} - \mathbf{P}_\alpha)(\mathbf{P_Z} - \mathbf{P}_\alpha) &= \mathbf{P_Z} - \mathbf{P}_\alpha - \mathbf{P}_\alpha + \mathbf{P}_\alpha \\
&= \mathbf{P_Z} - \mathbf{P}_\alpha, \\
(\mathbf{P}_\gamma - \mathbf{P_1})(\mathbf{P}_\gamma - \mathbf{P_1}) &= \mathbf{P}_\gamma - \mathbf{P_1} - \mathbf{P_1} + \mathbf{P_1} \\
&= \mathbf{P}_\gamma - \mathbf{P_1}, \\
(\mathbf{P}_\xi - \mathbf{P}_\alpha - \mathbf{P}_\gamma + \mathbf{P_1})(\mathbf{P}_\xi - \mathbf{P}_\alpha - \mathbf{P}_\gamma + \mathbf{P_1}) &= \mathbf{P}_\xi + \mathbf{P}_\alpha + \mathbf{P}_\gamma + \mathbf{P_1} - 2\mathbf{P}_\alpha - 2\mathbf{P}_\gamma + 2\mathbf{P_1} \\
&\quad +2\mathbf{P_1} - 2\mathbf{P_1} - 2\mathbf{P_1} \\
&= \mathbf{P}_\xi - \mathbf{P}_\alpha - \mathbf{P}_\gamma + \mathbf{P_1}, \\
(\mathbf{I} - \mathbf{P_Z} + \mathbf{P}_\alpha - \mathbf{P}_\xi)(\mathbf{I} - \mathbf{P_Z} + \mathbf{P}_\alpha - \mathbf{P}_\xi) &= \mathbf{I} + \mathbf{P_Z} + \mathbf{P}_\alpha + \mathbf{P}_\xi - 2\mathbf{P_Z} + 2\mathbf{P}_\alpha - 2\mathbf{P}_\xi \\
&\quad -2\mathbf{P}_\alpha + \mathbf{P}_\alpha - 2\mathbf{P}_\alpha \\
&= \mathbf{I} - \mathbf{P_Z} + \mathbf{P}_\alpha - \mathbf{P}_\xi.
\end{aligned}$$

(d) First observe that

$$\mathrm{E}(\bar{y}_{\cdot\cdot\cdot}) = \frac{1}{qrm}\sum_{i=1}^{q}\sum_{j=1}^{r}\sum_{k=1}^{m}(\mu + \alpha_i + \gamma_k + \xi_{ik}) = \mu + \bar{\alpha}_\cdot + \bar{\gamma}_\cdot + \bar{\xi}_{\cdot\cdot},$$

$$\mathrm{E}(\bar{y}_{i\cdot\cdot}) = \frac{1}{rm}\sum_{j=1}^{r}\sum_{k=1}^{m}(\mu + \alpha_i + \gamma_k + \xi_{ik}) = \mu + \alpha_i + \bar{\gamma}_\cdot + \bar{\xi}_{i\cdot},$$

$$\mathrm{E}(\bar{y}_{ij\cdot}) = \frac{1}{m}\sum_{k=1}^{m}(\mu + \alpha_i + \gamma_k + \xi_{ik}) = \mu + \alpha_i + \bar{\gamma}_\cdot + \bar{\xi}_{i\cdot},$$

$$\mathrm{E}(\bar{y}_{\cdot\cdot k}) = \frac{1}{qr}\sum_{i=1}^{q}\sum_{j=1}^{r}(\mu + \alpha_i + \gamma_k + \xi_{ik}) = \mu + \bar{\alpha}_\cdot + \gamma_k + \bar{\xi}_{\cdot k},$$

$$\mathrm{E}(\bar{y}_{i\cdot k}) = \frac{1}{r}\sum_{j=1}^{r}(\mu + \alpha_i + \gamma_k + \xi_{ik}) = \mu + \alpha_i + \gamma_k + \bar{\xi}_{ik}.$$

Then

$$\begin{aligned}
\mathrm{E}\left[rm\sum_{i=1}^{q}(\bar{y}_{i\cdot\cdot}-\bar{y}_{\cdot\cdot\cdot})^2\right] &= rm\sum_{i=1}^{q}[\mathrm{E}(\bar{y}_{i\cdot\cdot})-\mathrm{E}(\bar{y}_{\cdot\cdot\cdot})]^2 \\
&\quad +\mathrm{tr}\{[(1/rm)(\mathbf{I}_q\otimes\mathbf{J}_{rm})-(1/qrm)\mathbf{J}_{qrm}] \\
&\quad \sigma^2[\mathbf{I}_{qrm}+(\sigma_b^2/\sigma^2)(\mathbf{I}_{qr}\otimes\mathbf{J}_m)]\} \\
&= rm\sum_{i=1}^{q}[(\alpha_i-\bar{\alpha}_\cdot)+(\bar{\xi}_{i\cdot}-\bar{\xi}_{\cdot\cdot})]^2 \\
&\quad +\sigma^2\mathrm{tr}[(1/rm)(\mathbf{I}_q\otimes\mathbf{J}_{rm}-(1/qrm)\mathbf{J}_{qrm}] \\
&\quad +\sigma_b^2\mathrm{tr}[(1/rm)(\mathbf{I}_q\otimes\mathbf{J}_r\otimes m\mathbf{J}_m)-(1/qrm)(\mathbf{J}_{qr}\otimes m\mathbf{J}_m)] \\
&= rm\sum_{i=1}^{q}[(\alpha_i-\bar{\alpha}_\cdot)+(\bar{\xi}_{i\cdot}-\bar{\xi}_{\cdot\cdot})]^2+(q-1)(\sigma^2+m\sigma_b^2),
\end{aligned}$$

$$\begin{aligned}
\mathrm{E}\left[m\sum_{i=1}^{q}\sum_{j=1}^{r}(\bar{y}_{ij\cdot}-\bar{y}_{i\cdot\cdot})^2\right] &= m\sum_{i=1}^{q}\sum_{j=1}^{r}[\mathrm{E}(\bar{y}_{ij\cdot})-\mathrm{E}(\bar{y}_{i\cdot\cdot})]^2 \\
&\quad +\mathrm{tr}\{[(1/m)(\mathbf{I}_{qr}\otimes\mathbf{J}_m)-(1/rm)(\mathbf{I}_q\otimes\mathbf{J}_{rm})]\sigma^2[\mathbf{I}_{qrm} \\
&\quad +(\sigma_b^2/\sigma^2)(\mathbf{I}_{qr}\otimes\mathbf{J}_m)]\} \\
&= 0+\sigma^2\mathrm{tr}[(1/m)(\mathbf{I}_{qr}\otimes\mathbf{J}_m)-(1/rm)(\mathbf{I}_q\otimes\mathbf{J}_{rm}] \\
&\quad +\sigma_b^2\mathrm{tr}[(1/m)(\mathbf{I}_{qr}\otimes m\mathbf{J}_m)-(1/rm)(\mathbf{I}_q\otimes\mathbf{J}_r\otimes m\mathbf{J}_m)] \\
&= q(r-1)(\sigma^2+m\sigma_b^2),
\end{aligned}$$

$$\begin{aligned}
\mathrm{E}\left[qr\sum_{k=1}^{m}(\bar{y}_{\cdot\cdot k}-\bar{y}_{\cdot\cdot\cdot})^2\right] &= qr\sum_{k=1}^{m}[\mathrm{E}(\bar{y}_{\cdot\cdot k})-\mathrm{E}(\bar{y}_{\cdot\cdot\cdot})]^2 \\
&\quad +\mathrm{tr}\{[(1/qr)(\mathbf{J}_{qr}\otimes\mathbf{I}_m)-(1/qrm)\mathbf{J}_{qrm}]\sigma^2[\mathbf{I}_{qrm} \\
&\quad +(\sigma_b^2/\sigma^2)(\mathbf{I}_{qr}\otimes\mathbf{J}_m)]\} \\
&= qr\sum_{k=1}^{m}[(\gamma_k-\bar{\gamma}_\cdot)+(\bar{\xi}_{\cdot k}-\bar{\xi}_{\cdot\cdot})]^2 \\
&\quad +\sigma^2\mathrm{tr}[(1/qr)(\mathbf{J}_{qr}\otimes\mathbf{I}_m)-(1/qrm)(\mathbf{J}_{qrm}] \\
&\quad +\sigma_b^2\mathrm{tr}[(1/qr)(\mathbf{J}_{qr}\otimes\mathbf{J}_m)-(1/qrm)(\mathbf{J}_{qr}\otimes m\mathbf{J}_m)] \\
&= qr\sum_{k=1}^{m}[(\gamma_k-\bar{\gamma}_\cdot)+(\bar{\xi}_{\cdot k}-\bar{\xi}_{\cdot\cdot})]^2+(m-1)\sigma^2,
\end{aligned}$$

$$
\begin{aligned}
&\mathrm{E}\left[r\sum_{i=1}^{q}\sum_{k=1}^{m}(\bar{y}_{i\cdot k}-\bar{y}_{i\cdot\cdot}-\bar{y}_{\cdot\cdot k}+\bar{y}_{\cdot\cdot\cdot})^2\right]\\
&=r\sum_{i=1}^{q}\sum_{k=1}^{m}[\mathrm{E}(\bar{y}_{i\cdot k})-\mathrm{E}(\bar{y}_{i\cdot\cdot})-\mathrm{E}(\bar{y}_{\cdot\cdot k})\\
&\quad+\mathrm{E}(\bar{y}_{\cdot\cdot\cdot})]^2+\mathrm{tr}\{[(1/r)(\mathbf{I}_q\otimes\mathbf{J}_r\otimes\mathbf{I}_m)-(1/rm)(\mathbf{I}_q\otimes\mathbf{J}_{rm})\\
&\quad-(1/qr)(\mathbf{J}_{qr}\otimes\mathbf{I}_m)+(1/qrm)\mathbf{J}_{qrm}]\sigma^2[\mathbf{I}_{qrm}+(\sigma_b^2/\sigma^2)(\mathbf{I}_{qr}\otimes\mathbf{J}_m)]\}\\
&=r\sum_{i=1}^{q}\sum_{k=1}^{m}(\xi_{ik}-\bar{\xi}_{i\cdot}-\bar{\xi}_{\cdot k}+\bar{\xi}_{\cdot\cdot})^2\\
&\quad+\sigma^2\mathrm{tr}[(1/r)(\mathbf{I}_q\otimes\mathbf{J}_r\otimes\mathbf{I}_m)-(1/rm)(\mathbf{I}_q\otimes\mathbf{J}_{rm})\\
&\quad-(1/qr)(\mathbf{J}_{qr}\otimes\mathbf{I}_m)+(1/qrm)\mathbf{J}_{qrm}]\\
&\quad+\sigma_b^2\mathrm{tr}[(1/r)(\mathbf{I}_q\otimes\mathbf{J}_r\otimes\mathbf{J}_m)-(1/rm)(\mathbf{I}_q\otimes\mathbf{J}_r\otimes m\mathbf{J}_m)\\
&\quad-(1/qr)(\mathbf{J}_{qr}\otimes\mathbf{I}_m)+(1/qrm)(\mathbf{J}_{qr}\otimes m\mathbf{J}_m)]\\
&=r\sum_{i=1}^{q}\sum_{k=1}^{m}(\xi_{ik}-\bar{\xi}_{i\cdot}-\bar{\xi}_{\cdot k}+\bar{\xi}_{\cdot\cdot})^2+\sigma^2(qm-q-m+1),
\end{aligned}
$$

$$
\begin{aligned}
&\mathrm{E}\left[\sum_{i=1}^{q}\sum_{j=1}^{r}\sum_{k=1}^{m}(\bar{y}_{ijk}-\bar{y}_{ij\cdot}-\bar{y}_{i\cdot k}+\bar{y}_{i\cdot\cdot})^2\right]\\
&=\sum_{i=1}^{q}\sum_{j=1}^{r}\sum_{k=1}^{m}[\mathrm{E}(\bar{y}_{ijk})-\mathrm{E}(\bar{y}_{ij\cdot})-\mathrm{E}(\bar{y}_{i\cdot k})+\mathrm{E}(\bar{y}_{i\cdot\cdot})]^2\\
&\quad+\mathrm{tr}\{[\mathbf{I}-(1/m)(\mathbf{I}_{qr}\otimes\mathbf{J}_m)+(1/rm)(\mathbf{I}_q\otimes\mathbf{J}_{rm})\\
&\quad-(1/r)(\mathbf{I}_q\otimes\mathbf{J}_r\otimes\mathbf{I}_m)]\sigma^2[\mathbf{I}_{qrm}+(\sigma_b^2/\sigma^2)(\mathbf{I}_{qr}\otimes\mathbf{J}_m)]\}\\
&=0+\sigma^2\mathrm{tr}[\mathbf{I}-(1/m)(\mathbf{I}_{qr}\otimes\mathbf{J}_m)+(1/rm)(\mathbf{I}_q\otimes\mathbf{J}_{rm})-(1/r)(\mathbf{I}_q\otimes\mathbf{J}_r\otimes\mathbf{I}_m)]\\
&\quad+\sigma_b^2\mathrm{tr}[(\mathbf{I}_{qr}\otimes\mathbf{J}_m)-(1/m)(\mathbf{I}_{qr}\otimes m\mathbf{J}_m)\\
&\quad+(1/rm)(\mathbf{I}_q\otimes\mathbf{J}_r\otimes m\mathbf{J}_m)-(1/r)(\mathbf{I}_q\otimes\mathbf{J}_r\otimes\mathbf{J}_m)]\\
&=\sigma^2(qrm-qr+q-qm).
\end{aligned}
$$

The expected mean squares listed in Example 13.4.5-1 may be obtained by dividing each of the expected sums of squares given above by the corresponding rank.

▶ **Exercise 25** For the split-plot model considered in Example 13.4.5-1, obtain simplified expressions for:

(a) $\text{var}(\bar{y}_{i\cdot\cdot} - \bar{y}_{i'\cdot\cdot})$ $(i \neq i')$;
(b) $\text{var}(\bar{y}_{\cdot\cdot k} - \bar{y}_{\cdot\cdot k'})$ $(k \neq k')$;
(c) $\text{var}(\bar{y}_{i\cdot k} - \bar{y}_{i\cdot k'})$ $(k \neq k')$;
(d) $\text{var}(\bar{y}_{i\cdot k} - \bar{y}_{i'\cdot k'})$ $(i \neq i')$.

Solution

(a) Note that

$$\bar{y}_{i\cdot\cdot} = \frac{1}{rm}\sum_{j=1}^{r}\sum_{k=1}^{m}(\mu+\alpha_i+b_{ij}+\gamma_k+\xi_{ik}+d_{ijk}) = \mu+\alpha_i+\bar{b}_{i\cdot}+\bar{\gamma}_{\cdot}+\bar{\xi}_{i\cdot}+\bar{d}_{i\cdot\cdot}.$$

It follows that for $i \neq i'$,

$$\begin{aligned}\text{var}(\bar{y}_{i\cdot\cdot} - \bar{y}_{i'\cdot\cdot}) &= \text{var}[(\bar{b}_{i\cdot} + \bar{d}_{i\cdot\cdot}) - (\bar{b}_{i'\cdot} + \bar{d}_{i'\cdot\cdot})] \\ &= \text{var}(\bar{b}_{i\cdot}) + \text{var}(\bar{b}_{i'\cdot}) + \text{var}(\bar{d}_{i\cdot\cdot}) + \text{var}(\bar{d}_{i'\cdot\cdot}) \\ &= \frac{2}{rm}(m\sigma_b^2 + \sigma^2).\end{aligned}$$

(b) Note that

$$\bar{y}_{\cdot\cdot k} = \frac{1}{qr}\sum_{i=1}^{q}\sum_{j=1}^{r}(\mu+\alpha_i+b_{ij}+\gamma_k+\xi_{ik}+d_{ijk}) = \mu+\bar{\alpha}_{\cdot}+\bar{b}_{\cdot\cdot}+\gamma_k+\bar{\xi}_{\cdot k}+\bar{d}_{\cdot\cdot k}.$$

It follows that for $k \neq k'$,

$$\begin{aligned}\text{var}(\bar{y}_{\cdot\cdot k} - \bar{y}_{\cdot\cdot k'}) &= \text{var}[(\bar{d}_{\cdot\cdot k}) - (\bar{d}_{\cdot\cdot k'})] \\ &= \text{var}(\bar{d}_{\cdot\cdot k}) + \text{var}(\bar{d}_{\cdot\cdot k'}) \\ &= \frac{2\sigma^2}{qr}.\end{aligned}$$

(c) Note that

$$\bar{y}_{i\cdot k} = \frac{1}{r}\sum_{j=1}^{r}(\mu + \alpha_i + b_{ij} + \gamma_k + \xi_{ik} + d_{ijk}) = \mu + \alpha_i + \bar{b}_{i\cdot} + \gamma_k + \xi_{ik} + \bar{d}_{i\cdot k}.$$

It follows that for $k \neq k'$,

$$\begin{aligned}\text{var}(\bar{y}_{i\cdot k} - \bar{y}_{i\cdot k'}) &= \text{var}[(\bar{b}_{i\cdot} + \bar{d}_{i\cdot k}) - (\bar{b}_{i\cdot} + \bar{d}_{i\cdot k'})] \\ &= \text{var}(\bar{d}_{i\cdot k}) + \text{var}(\bar{d}_{i\cdot k'}) \\ &= \frac{2\sigma^2}{r}.\end{aligned}$$

(d) Note that

$$\bar{y}_{i\cdot k} = \frac{1}{r}\sum_{j=1}^{r}(\mu + \alpha_i + b_{ij} + \gamma_k + \xi_{ik} + d_{ijk}) = \mu + \alpha_i + \bar{b}_{i\cdot} + \gamma_k + \xi_{ik} + \bar{d}_{i\cdot k}.$$

It follows that for $i \neq i'$,

$$\begin{aligned}\text{var}(\bar{y}_{i\cdot k} - \bar{y}_{i'\cdot k'}) &= \text{var}[(\bar{b}_{i\cdot} + \bar{d}_{i\cdot k}) - (\bar{b}_{i'\cdot} + \bar{d}_{i'\cdot k'})] \\ &= \text{var}(\bar{b}_{i\cdot}) + \text{var}(\bar{b}_{i'\cdot}) + \text{var}(\bar{d}_{i\cdot k}) + \text{var}(\bar{d}_{i'\cdot k'}) \\ &= \frac{2(\sigma_b^2 + \sigma^2)}{r}.\end{aligned}$$

Distribution Theory

14

This chapter presents exercises on distribution theory relevant to linear models, and provides solutions to those exercises.

▶ **Exercise 1** Let $\mathbf{x}$ represent an n-dimensional random vector with mean vector $\boldsymbol{\mu}$ and variance–covariance matrix $\boldsymbol{\Sigma}$. Use Theorem 14.1.1 to show that $\mathbf{x}$ has a n-variate normal distribution if and only if, for every n-vector of constants $\mathbf{a}$, $\mathbf{a}^T\mathbf{x}$ has a (univariate) normal distribution.

Solution Suppose that $\mathbf{a}^T\mathbf{x}$ has a (univariate) normal distribution for every n-vector of constants $\mathbf{a}$. Note that $\text{E}(\mathbf{a}^T\mathbf{x}) = \mathbf{a}^T\boldsymbol{\mu}$ and $\text{var}(\mathbf{x}^T\mathbf{A}\mathbf{x}) = \mathbf{a}^T\boldsymbol{\Sigma}\mathbf{a}$. Then, denoting the mgf of $\mathbf{a}^T\mathbf{x}$ by $m_{\mathbf{a}^T\mathbf{x}}(t)$ and denoting the mgf of $\mathbf{x}$, if it exists, by $m^*_{\mathbf{x}}(\mathbf{t})$, we find that for all $\mathbf{a}$,

$$m^*_{\mathbf{x}}(\mathbf{a}) = \text{E}[\exp(\mathbf{a}^T\mathbf{x})] = m_{\mathbf{a}^T\mathbf{x}}(1) = \exp(\mathbf{a}^T\boldsymbol{\mu} + \frac{1}{2}\mathbf{a}^T\boldsymbol{\Sigma}\mathbf{a}).$$

It follows by Theorems 14.1.1 and 14.1.4 that $\mathbf{x} \sim \text{N}(\boldsymbol{\mu}, \boldsymbol{\Sigma})$. Conversely, suppose that $\mathbf{x} \sim \text{N}(\boldsymbol{\mu}, \boldsymbol{\Sigma})$. Then denoting the mgf of $\mathbf{x}$ by $m^*_{\mathbf{x}}(\mathbf{t})$ and denoting the mgf of $\mathbf{a}^T\mathbf{x}$, if it exists, by $m_{\mathbf{a}^T\mathbf{x}}(t)$, we find that for all t,

$$\begin{aligned} m_{\mathbf{a}^T\mathbf{x}}(t) &= \text{E}[\exp(t\mathbf{a}^T\mathbf{x})] = \text{E}\{\exp[(t\mathbf{a})^T\mathbf{x}]\} = m^*_{\mathbf{x}}(t\mathbf{a}) = \exp[(t\mathbf{a})^T\boldsymbol{\mu} + \frac{1}{2}(t\mathbf{a})^T\boldsymbol{\Sigma}(t\mathbf{a})] \\ &= \exp[t(\mathbf{a}^T\boldsymbol{\mu}) + \frac{1}{2}t^2(\mathbf{a}^T\boldsymbol{\Sigma}\mathbf{a})]. \end{aligned}$$

It follows from Theorems 14.1.1 and 14.1.4 that $\mathbf{a}^T\mathbf{x} \sim \text{N}(\mathbf{a}^T\boldsymbol{\mu}, \mathbf{a}^T\boldsymbol{\Sigma}\mathbf{a})$.

▶ **Exercise 2** Use Theorem 14.1.3 to construct an alternate proof of Theorem 4.2.4 for the special case in which $\mathbf{x} \sim \text{N}(\boldsymbol{\mu}, \boldsymbol{\Sigma})$ where $\boldsymbol{\Sigma}$ is positive definite.

D. L. Zimmerman, *Linear Model Theory*,
https://doi.org/10.1007/978-3-030-52074-8_14

Solution

$$
\begin{aligned}
\mathrm{E}(\mathbf{x}^T\mathbf{A}\mathbf{x}) &= \int_{-\infty}^{\infty}\cdots\int_{-\infty}^{\infty}\mathbf{x}^T\mathbf{A}\mathbf{x}(2\pi)^{-n/2}|\boldsymbol{\Sigma}|^{-\frac{1}{2}} \\
&\quad \exp\{-[\mathbf{x}^T\boldsymbol{\Sigma}^{-1}\mathbf{x}-2\boldsymbol{\mu}^T\boldsymbol{\Sigma}^{-1}\mathbf{x}+\boldsymbol{\mu}^T\boldsymbol{\Sigma}^{-1}\boldsymbol{\mu}]/2\}\,d\mathbf{x} \\
&= (2\pi)^{-n/2}|\boldsymbol{\Sigma}|^{-\frac{1}{2}}\left(\frac{1}{2}\right)\pi^{n/2}\left(\frac{1}{2}\right)^{-n/2}|\boldsymbol{\Sigma}|^{\frac{1}{2}}[\mathrm{tr}(2\mathbf{A}\boldsymbol{\Sigma})-0 \\
&\quad +\frac{1}{2}(-\boldsymbol{\mu}^T\boldsymbol{\Sigma}^{-1})(2\boldsymbol{\Sigma})\mathbf{A}(2\boldsymbol{\Sigma})(-\boldsymbol{\Sigma}^{-1}\boldsymbol{\mu})] \\
&\quad \times\exp\left[\frac{1}{4}(-\boldsymbol{\mu}^T\boldsymbol{\Sigma}^{-1})(2\boldsymbol{\Sigma})(-\boldsymbol{\Sigma}^{-1}\boldsymbol{\mu})-\frac{1}{2}\boldsymbol{\mu}^T\boldsymbol{\Sigma}^{-1}\boldsymbol{\mu}\right] \\
&= \mathrm{tr}(\mathbf{A}\boldsymbol{\Sigma})+\boldsymbol{\mu}^T\mathbf{A}\boldsymbol{\mu},
\end{aligned}
$$

where the next-to-last equality results from putting $a_0 = 0$, $\mathbf{a} = \mathbf{0}$, $\mathbf{A} = \mathbf{A}$, $b_0 = \frac{1}{2}\boldsymbol{\mu}^T\boldsymbol{\Sigma}^{-1}\boldsymbol{\mu}$, $\mathbf{b} = \boldsymbol{\Sigma}^{-1}\boldsymbol{\mu}$, and $\mathbf{B} = \frac{1}{2}\boldsymbol{\Sigma}^{-1}$ into Theorem 14.1.3.

▶ **Exercise 3** Suppose that $\mathbf{y}$ follows a normal positive definite Aitken model $\{\mathbf{y}, \mathbf{X}\boldsymbol{\beta}, \sigma^2\mathbf{W}\}$, and that $\mathbf{X}$ has full column rank. Use the "Factorization Theorem" [e.g., Theorem 6.2.6 of Casella and Berger (2002)] to show that the maximum likelihood estimators of $\boldsymbol{\beta}$ and σ^2 derived in Example 14.1-1 are sufficient statistics for those parameters.

Solution By Theorem 14.1.5, the likelihood function is

$$
\begin{aligned}
L(\boldsymbol{\beta},\sigma^2) &= (2\pi)^{-n/2}|\sigma^2\mathbf{W}|^{-\frac{1}{2}}\exp[-(\mathbf{y}-\mathbf{X}\boldsymbol{\beta})^T(\sigma^2\mathbf{W})^{-1}(\mathbf{y}-\mathbf{X}\boldsymbol{\beta})/2] \\
&= (2\pi)^{-n/2}(\sigma^2)^{-n/2}|\mathbf{W}|^{-\frac{1}{2}}\exp[-(\mathbf{y}-\mathbf{X}\boldsymbol{\beta})^T\mathbf{W}^{-1}(\mathbf{y}-\mathbf{X}\boldsymbol{\beta})/2\sigma^2],
\end{aligned}
$$

where we have used Theorem 2.11.4. By the same argument used in Theorem 7.1.2 for the ordinary residual sum of squares function, for the generalized residual sum of squares function we have

$$
(\mathbf{y}-\mathbf{X}\boldsymbol{\beta})^T\mathbf{W}^{-1}(\mathbf{y}-\mathbf{X}\boldsymbol{\beta}) = (\mathbf{y}-\mathbf{X}\tilde{\boldsymbol{\beta}})^T\mathbf{W}^{-1}(\mathbf{y}-\mathbf{X}\tilde{\boldsymbol{\beta}})+(\tilde{\boldsymbol{\beta}}-\boldsymbol{\beta})^T\mathbf{X}^T\mathbf{W}^{-1}\mathbf{X}(\tilde{\boldsymbol{\beta}}-\boldsymbol{\beta}).
$$

So we may rewrite the likelihood function as

$$
\begin{aligned}
L(\boldsymbol{\beta},\sigma^2) &= (2\pi)^{-n/2}|\mathbf{W}|^{-\frac{1}{2}}(\sigma^2)^{-n/2}\exp\{-[(\mathbf{y}-\mathbf{X}\tilde{\boldsymbol{\beta}})^T\mathbf{W}^{-1}(\mathbf{y}-\mathbf{X}\tilde{\boldsymbol{\beta}}) \\
&\quad +(\tilde{\boldsymbol{\beta}}-\boldsymbol{\beta})^T\mathbf{X}^T\mathbf{W}^{-1}\mathbf{X}(\tilde{\boldsymbol{\beta}}-\boldsymbol{\beta})]/2\sigma^2\} \\
&= (2\pi)^{-n/2}|\mathbf{W}|^{-\frac{1}{2}}(\sigma^2)^{-n/2}\exp\{-(n/2\sigma^2)[\tilde{\sigma}^2+(\tilde{\boldsymbol{\beta}}-\boldsymbol{\beta})^T\mathbf{X}^T\mathbf{W}^{-1}\mathbf{X}(\tilde{\boldsymbol{\beta}}-\boldsymbol{\beta})/n],\}
\end{aligned}
$$

where $\tilde{\sigma}^2 = (\mathbf{y} - \mathbf{X}\tilde{\boldsymbol{\beta}})^T\mathbf{W}^{-1}(\mathbf{y} - \mathbf{X}\tilde{\boldsymbol{\beta}})/n$. This shows that the likelihood function can be factored into a product of two functions, $(2\pi)^{-n/2}|\mathbf{W}|^{-\frac{1}{2}}$ and $(\sigma^2)^{-n/2}\exp\{-(n/2\sigma^2)[\tilde{\sigma}^2 + (\tilde{\boldsymbol{\beta}} - \boldsymbol{\beta})^T\mathbf{X}^T\mathbf{W}^{-1}\mathbf{X}(\tilde{\boldsymbol{\beta}} - \boldsymbol{\beta})/n]\}$, such that the first function does not depend on $(\boldsymbol{\beta}, \sigma^2)$ and the second function, which does depend on $(\boldsymbol{\beta}, \sigma^2)$, depends on $\mathbf{y}$ only through $(\tilde{\boldsymbol{\beta}}, \tilde{\sigma}^2)$. By the factorization theorem, this establishes that the maximum likelihood estimators comprise a sufficient statistic for $(\boldsymbol{\beta}, \sigma^2)$.

▶ **Exercise 4** Suppose that $\mathbf{y}$ follows the normal constrained Gauss–Markov model $\{\mathbf{y}, \mathbf{X}\boldsymbol{\beta}, \sigma^2\mathbf{I} : \mathbf{A}\boldsymbol{\beta} = \mathbf{h}\}$. Let $\mathbf{c}^T\boldsymbol{\beta}$ be an estimable function under this model. Obtain maximum likelihood estimators of $\mathbf{c}^T\boldsymbol{\beta}$ and σ^2.

Solution The log-likelihood function is

$$\log L(\boldsymbol{\beta}, \sigma^2) = -\frac{n}{2}\log 2\pi - \frac{n}{2}\log \sigma^2 - \frac{Q(\boldsymbol{\beta})}{2\sigma^2},$$

where $Q(\boldsymbol{\beta}) = (\mathbf{y} - \mathbf{X}\boldsymbol{\beta})^T(\mathbf{y} - \mathbf{X}\boldsymbol{\beta})$ is the residual sum of squares function. Recall from Definition 10.1.3 of the constrained least squares estimator of $\mathbf{c}^T\boldsymbol{\beta}$ that $Q(\boldsymbol{\beta})$ is minimized, over all $\boldsymbol{\beta}$ satisfying the constraints $\mathbf{A}\boldsymbol{\beta} = \mathbf{h}$, at $\breve{\boldsymbol{\beta}}$ where $\breve{\boldsymbol{\beta}}$ is the first p-component subvector of a solution to the constrained normal equations

$$\begin{pmatrix} \mathbf{X}^T\mathbf{X} & \mathbf{A}^T \\ \mathbf{A} & \mathbf{0} \end{pmatrix}\begin{pmatrix} \boldsymbol{\beta} \\ \boldsymbol{\lambda} \end{pmatrix} = \begin{pmatrix} \mathbf{X}^T\mathbf{y} \\ \mathbf{h} \end{pmatrix}.$$

Because these equations are free of σ^2, for each fixed value of σ^2 $\log L(\boldsymbol{\beta}, \sigma^2)$ is maximized with respect to $\boldsymbol{\beta}$, among those $\boldsymbol{\beta}$ that satisfy the constraints, at $\breve{\boldsymbol{\beta}}$. It then remains to maximize

$$\log L(\breve{\boldsymbol{\beta}}, \sigma^2) = -\frac{n}{2}\log 2\pi - \frac{n}{2}\log \sigma^2 - \frac{Q(\breve{\boldsymbol{\beta}})}{2\sigma^2}$$

with respect to σ^2. Taking the first derivative with respect to σ^2 yields

$$\frac{\partial \log L(\breve{\boldsymbol{\beta}}, \sigma^2)}{\partial \sigma^2} = -\frac{n}{2\sigma^2} + \frac{Q(\breve{\boldsymbol{\beta}})}{2\sigma^4} = \left(\frac{n}{2\sigma^4}\right)(\ddot{\sigma}^2 - \sigma^2),$$

where $\ddot{\sigma}^2 = \frac{Q(\breve{\boldsymbol{\beta}})}{n}$. Furthermore,

$$\frac{\partial \log L(\tilde{\boldsymbol{\beta}}, \sigma^2)}{\partial \sigma^2} \begin{cases} > 0 \text{ for } \sigma^2 < \ddot{\sigma}^2, \\ = 0 \text{ for } \sigma^2 = \ddot{\sigma}^2, \\ < 0 \text{ for } \sigma^2 > \ddot{\sigma}^2, \end{cases}$$

so that $L(\boldsymbol{\beta}, \sigma^2)$ attains a global maximum at $(\boldsymbol{\beta}, \sigma^2) = (\breve{\boldsymbol{\beta}}, \ddot{\sigma}^2)$ [unless $Q(\breve{\boldsymbol{\beta}}) = 0$, in which case $L(\boldsymbol{\beta}, \sigma^2)$ does not have a maximum; however, this is an event of probability 0.] Thus, the maximum likelihood estimators of $\mathbf{c}^T\boldsymbol{\beta}$ and σ^2 are $\mathbf{c}^T\breve{\boldsymbol{\beta}}$ and $Q(\breve{\boldsymbol{\beta}})/n$, respectively.

▶ **Exercise 5** Prove Theorem 14.1.8: Suppose that $\mathbf{x} \sim \mathrm{N}_n(\boldsymbol{\mu}, \boldsymbol{\Sigma})$, and partition $\mathbf{x}$, $\boldsymbol{\mu}$, and $\boldsymbol{\Sigma}$ conformably as

$$\mathbf{x} = \begin{pmatrix} \mathbf{x}_1 \\ \vdots \\ \mathbf{x}_m \end{pmatrix}, \quad \boldsymbol{\mu} = \begin{pmatrix} \boldsymbol{\mu}_1 \\ \vdots \\ \boldsymbol{\mu}_m \end{pmatrix}, \quad \boldsymbol{\Sigma} = \begin{pmatrix} \boldsymbol{\Sigma}_{11} & \cdots & \boldsymbol{\Sigma}_{1m} \\ \vdots & & \vdots \\ \boldsymbol{\Sigma}_{m1} & \cdots & \boldsymbol{\Sigma}_{mm} \end{pmatrix}.$$

Then, $\mathbf{x}_1, \ldots, \mathbf{x}_m$ are mutually independent if and only if $\boldsymbol{\Sigma}_{ij} = \mathbf{0}$ for $j \neq i = 1, \ldots, m$.

Solution Let $m_{\mathbf{x}}(\mathbf{t}), m_{\mathbf{x}_1}(\mathbf{t}_1), \ldots, m_{\mathbf{x}_m}(\mathbf{t}_m)$ denote the mgfs of $\mathbf{x}, \mathbf{x}_1, \ldots, \mathbf{x}_m$ where $\mathbf{t} = (\mathbf{t}_1^T, \ldots, \mathbf{t}_m^T)^T$. Suppose that $\boldsymbol{\Sigma}_{ij} = \mathbf{0}$ for all $j \neq i = 1, \ldots, m$. Then for all $\mathbf{t} \in \mathbb{R}^n$,

$$\begin{aligned} m_{\mathbf{x}}(\mathbf{t}) &= \exp(\mathbf{t}^T\boldsymbol{\mu} + \frac{1}{2}\mathbf{t}^T\boldsymbol{\Sigma}\mathbf{t}) \\ &= \exp\left(\sum_{i=1}^m \mathbf{t}_i^T\boldsymbol{\mu}_i + \frac{1}{2}\sum_{i=1}^m\sum_{j=1}^m \mathbf{t}_i^T\boldsymbol{\Sigma}_{ij}\mathbf{t}_j\right) \\ &= \exp(\mathbf{t}_1^T\boldsymbol{\mu}_1 + \frac{1}{2}\mathbf{t}_1^T\boldsymbol{\Sigma}_{11}\mathbf{t}_1)\cdots\exp(\mathbf{t}_m^T\boldsymbol{\mu}_m + \frac{1}{2}\mathbf{t}_m^T\boldsymbol{\Sigma}_{mm}\mathbf{t}_m) \\ &= \prod_{i=1}^m m_{\mathbf{x}_i}(\mathbf{t}_i), \end{aligned}$$

where we used Theorem 14.1.4 for the first and last equalities. Thus, by Theorem 14.1.2, $\mathbf{x}_1, \ldots, \mathbf{x}_m$ are mutually independent. Conversely, suppose that $\mathbf{x}_1, \ldots, \mathbf{x}_m$ are mutually independent. Then for $j \neq i = 1, \ldots, m$,

$$\boldsymbol{\Sigma}_{ij} = \mathrm{E}[(\mathbf{x}_i - \boldsymbol{\mu}_i)(\mathbf{x}_j - \boldsymbol{\mu}_j)^T] = \mathrm{E}(\mathbf{x}_i - \boldsymbol{\mu}_i)\mathrm{E}(\mathbf{x}_j - \boldsymbol{\mu}_j)^T = \mathbf{0}.$$

▶ **Exercise 6** Derive the distributions specified by (14.5) and (14.6) in Example 14.1-3.

Solution The BLUP is unbiased by definition so $E(\mathbf{C}^T\tilde{\boldsymbol{\beta}}+\tilde{\mathbf{u}}) = \mathbf{C}^T\boldsymbol{\beta}$. Furthermore, using results from the proof of Theorem 13.2.2,

$$\begin{aligned}
\text{var}(\mathbf{C}^T\tilde{\boldsymbol{\beta}}+\tilde{\mathbf{u}}) &= \text{var}(\mathbf{C}^T\tilde{\boldsymbol{\beta}}) + \text{var}(\tilde{\mathbf{u}}) + \text{cov}(\mathbf{C}^T\tilde{\boldsymbol{\beta}}, \tilde{\mathbf{u}}) + [\text{cov}(\mathbf{C}^T\tilde{\boldsymbol{\beta}}, \tilde{\mathbf{u}})]^T \\
&= \sigma^2\mathbf{C}^T(\mathbf{X}^T\mathbf{W}^{-1}\mathbf{X})^-\mathbf{C} + \sigma^2[\mathbf{K}^T\mathbf{W}^{-1}\mathbf{K} - \mathbf{K}^T\mathbf{W}^{-1}\mathbf{X}(\mathbf{X}^T\mathbf{W}^{-1}\mathbf{X})^-\mathbf{X}^T\mathbf{W}^{-1}\mathbf{K}] \\
&\quad +\mathbf{0} + \mathbf{0}^T,
\end{aligned}$$

which is easily seen to match (14.5). Furthermore, $E(\tilde{\boldsymbol{\tau}} - \boldsymbol{\tau}) = E(\tilde{\boldsymbol{\tau}}) - E(\boldsymbol{\tau}) = \mathbf{C}^T\boldsymbol{\beta} - \mathbf{C}^T\boldsymbol{\beta} = \mathbf{0}$; $E(\tilde{\mathbf{e}}) = \mathbf{0}$ (by Theorem 11.1.8b); $\text{var}(\tilde{\boldsymbol{\tau}} - \boldsymbol{\tau}) = \sigma^2\mathbf{Q}$ by Theorem 13.2.2; $\text{var}(\tilde{\mathbf{e}}) = \sigma^2[\mathbf{W} - \mathbf{X}(\mathbf{X}^T\mathbf{W}^{-1}\mathbf{X})^-\mathbf{X}^T]$ by Theorem 11.1.8d; and, letting $\mathbf{A}$ represent any matrix such that $\mathbf{A}^T\mathbf{X} = \mathbf{C}^T$,

$$\begin{aligned}
\text{cov}(\tilde{\boldsymbol{\tau}} - \boldsymbol{\tau}, \tilde{\mathbf{e}}) &= \text{cov}(\mathbf{C}^T\tilde{\boldsymbol{\beta}} + \mathbf{K}^T\mathbf{W}^{-1}\tilde{\mathbf{e}} - \mathbf{u}, \tilde{\mathbf{e}}) \\
&= \text{cov}(\mathbf{A}^T\tilde{\mathbf{y}}, \tilde{\mathbf{e}}) + \text{cov}(\mathbf{K}^T\mathbf{W}^{-1}\tilde{\mathbf{e}}, \tilde{\mathbf{e}}) - \text{cov}[\mathbf{u}, (\mathbf{I} - \tilde{\mathbf{P}}_\mathbf{X})\mathbf{y}] \\
&= \mathbf{0} + \sigma^2\mathbf{K}^T\mathbf{W}^{-1}[\mathbf{W} - \mathbf{X}(\mathbf{X}^T\mathbf{W}^{-1}\mathbf{X})^-\mathbf{X}^T] - \sigma^2\mathbf{K}^T(\mathbf{I} - \tilde{\mathbf{P}}_\mathbf{X})^T \\
&= \sigma^2[\mathbf{K}^T - \mathbf{K}^T\mathbf{W}^{-1}\mathbf{X}(\mathbf{X}^T\mathbf{W}^{-1}\mathbf{X})^-\mathbf{X}^T] \\
&\quad -\sigma^2[\mathbf{K}^T - \mathbf{K}^T\mathbf{W}^{-1}\mathbf{X}(\mathbf{X}^T\mathbf{W}^{-1}\mathbf{X})^-\mathbf{X}^T] \\
&= \mathbf{0},
\end{aligned}$$

where the third equality holds by Theorem 11.1.8d, e.

▶ **Exercise 7** Suppose that $\mathbf{x} \sim N_4(\boldsymbol{\mu}, \sigma^2\mathbf{W})$, where $\boldsymbol{\mu} = (1, 2, 3, 4)^T$ and $\mathbf{W} = \frac{1}{2}\mathbf{I}_4 + \frac{1}{2}\mathbf{J}_4$. Find the conditional distribution of $\mathbf{Ax}$ given that $\mathbf{Bx} = \mathbf{c}$, where

$$\mathbf{A} = \begin{pmatrix} 2 & 0 & -1 & -1 \end{pmatrix}, \quad \mathbf{B} = \begin{pmatrix} 0 & 0 & 2 & -2 \\ 1 & 1 & 1 & 1 \end{pmatrix}, \quad \mathbf{c} = \begin{pmatrix} 2 \\ 2 \end{pmatrix}.$$

Solution Because $\begin{pmatrix} \mathbf{Ax} \\ \mathbf{Bx} \end{pmatrix} = \begin{pmatrix} \mathbf{A} \\ \mathbf{B} \end{pmatrix}\mathbf{x}$, by Theorem 14.1.6 the joint distribution of $\mathbf{Ax}$ and $\mathbf{Bx}$ is normal with mean

$$\begin{pmatrix} \mathbf{A} \\ \mathbf{B} \end{pmatrix}\boldsymbol{\mu} = \begin{pmatrix} 2 & 0 & -1 & -1 \\ 0 & 0 & 2 & -2 \\ 1 & 1 & 1 & 1 \end{pmatrix}\begin{pmatrix} 1 \\ 2 \\ 3 \\ 4 \end{pmatrix} = \begin{pmatrix} -5 \\ -2 \\ 10 \end{pmatrix}$$

and variance–covariance matrix

$$\begin{pmatrix} \mathbf{A} \\ \mathbf{B} \end{pmatrix}\sigma^2\left(\frac{1}{2}\mathbf{I}_4 + \frac{1}{2}\mathbf{J}_4\right)\begin{pmatrix} \mathbf{A}^T & \mathbf{B}^T \end{pmatrix} = \frac{\sigma^2}{2}\begin{pmatrix} \mathbf{AA}^T & \mathbf{AB}^T \\ \mathbf{BA}^T & \mathbf{BB}^T + \mathbf{BJB}^T \end{pmatrix} = \sigma^2\begin{pmatrix} 3 & 0 & 0 \\ 0 & 4 & 0 \\ 0 & 0 & 10 \end{pmatrix}$$

because $\mathbf{AJ}_4 = \mathbf{0}_4^T$. By Theorem 14.1.8, $\mathbf{Ax}$ and $\mathbf{Bx}$ are independent; therefore, the conditional distribution of $\mathbf{Ax}$ given that $\mathbf{Bx} = \mathbf{c}$ is the same as the marginal distribution of $\mathbf{Ax}$, which by Theorem 14.1.7 is $\mathrm{N}(-5, 3\sigma^2)$.

▶ **Exercise 8** Suppose that $(y_1, y_2, y_3)^T$ satisfies the stationary autoregressive model of order one described in Example 5.2.4-1 with $n = 3$, and assume that the joint distribution of $(y_1, y_2, y_3)^T$ is trivariate normal. Determine the conditional distribution of $(y_1, y_3)^T$ given that $y_2 = 1$. What can you conclude about the dependence of y_1 and y_3, conditional on y_2?

Solution We have

$$\begin{pmatrix} y_1 \\ y_2 \\ y_3 \end{pmatrix} \sim \mathrm{N}\left(\mu\mathbf{1}_3, \frac{\sigma^2}{1-\rho^2}\begin{pmatrix} 1 & \rho & \rho^2 \\ \rho & 1 & \rho \\ \rho^2 & \rho & 1 \end{pmatrix}\right),$$

which upon rearrangement may be re-expressed as

$$\begin{pmatrix} y_1 \\ y_3 \\ y_2 \end{pmatrix} \sim \mathrm{N}\left(\mu\mathbf{1}_3, \frac{\sigma^2}{1-\rho^2}\begin{pmatrix} 1 & \rho^2 & \rho \\ \rho^2 & 1 & \rho \\ \rho & \rho & 1 \end{pmatrix}\right).$$

Thus, by Theorem 14.1.9, the conditional distribution of $(y_1, y_3)^T$ given that $y_2 = 1$ is bivariate normal with mean

$$\mu\mathbf{1}_2 + (1-\mu)\begin{pmatrix} \rho \\ \rho \end{pmatrix} = \begin{pmatrix} \rho + (1-\rho)\mu \\ \rho + (1-\rho)\mu \end{pmatrix}$$

and variance–covariance matrix

$$\frac{\sigma^2}{1-\rho^2}\begin{pmatrix} 1 & \rho^2 \\ \rho^2 & 1 \end{pmatrix} - \frac{\sigma^2}{1-\rho^2}\begin{pmatrix} \rho \\ \rho \end{pmatrix}\begin{pmatrix} \rho & \rho \end{pmatrix} = \sigma^2\mathbf{I}_2.$$

Because the off-diagonal elements of this variance–covariance matrix equal 0, we conclude, by Theorem 14.1.8, that y_1 and y_3 are conditionally independent given y_2.

▶ **Exercise 9** Let $\mathbf{x} = (x_i)$ be a random n-vector whose distribution is multivariate normal with mean vector $\boldsymbol{\mu} = (\mu_i)$ and variance–covariance matrix $\boldsymbol{\Sigma} = (\sigma_{ij})$. Show, by differentiating the moment generating function, that

$$E[(x_i - \mu_i)(x_j - \mu_j)(x_k - \mu_k)(x_\ell - \mu_\ell)] = \sigma_{ij}\sigma_{k\ell} + \sigma_{ik}\sigma_{j\ell} + \sigma_{i\ell}\sigma_{jk}$$

for $i, j, k, \ell = 1, \ldots, n$.

Solution By Theorem 14.1.6, $\mathbf{x} - \boldsymbol{\mu} \sim \text{N}(\mathbf{0}, \boldsymbol{\Sigma})$, hence by Theorem 14.1.4, $m_{\mathbf{x}-\boldsymbol{\mu}}(\mathbf{t}) = \exp(\mathbf{t}^T\boldsymbol{\Sigma}\mathbf{t}/2)$. Now first observe that

$$\frac{\partial}{\partial t_i}\frac{\mathbf{t}^T\boldsymbol{\Sigma}\mathbf{t}}{2} = \frac{1}{2}\frac{\partial}{\partial t_i}\left(\sum_{a=1}^{n}\sigma_{aa}t_a^2 + 2\sum_{a=1}^{n}\sum_{b>a}\sigma_{ab}t_a t_b\right) = \frac{1}{2}\left(2\sigma_{ii}t_i + 2\sum_{b\neq i}\sigma_{ib}t_b\right) = \sum_{b=1}^{n}\sigma_{ib}t_b.$$

Thus

$$\begin{aligned}
\frac{\partial m_{\mathbf{x}-\boldsymbol{\mu}}(\mathbf{t})}{\partial t_i} &= \exp(\mathbf{t}^T\boldsymbol{\Sigma}\mathbf{t}/2)\frac{\partial}{\partial t_i}\frac{\mathbf{t}^T\boldsymbol{\Sigma}\mathbf{t}}{2} \\
&= \exp(\mathbf{t}^T\boldsymbol{\Sigma}\mathbf{t}/2)\left(\sum_{a=1}^{n}\sigma_{ia}t_a\right), \\
\frac{\partial^2 m_{\mathbf{x}-\boldsymbol{\mu}}(\mathbf{t})}{\partial t_i t_j} &= \frac{\partial}{\partial t_j}\exp(\mathbf{t}^T\boldsymbol{\Sigma}\mathbf{t}/2)\left(\sum_{a=1}^{n}\sigma_{ia}t_a\right) \\
&= \exp(\mathbf{t}^T\boldsymbol{\Sigma}\mathbf{t}/2)\left(\sum_{a=1}^{n}\sigma_{ia}t_a\right)\left(\sum_{b=1}^{n}\sigma_{jb}t_b\right) + \sigma_{ij}\exp(\mathbf{t}^T\boldsymbol{\Sigma}\mathbf{t}/2), \\
\frac{\partial^3 m_{\mathbf{x}-\boldsymbol{\mu}}(\mathbf{t})}{\partial t_i t_j t_k} &= \exp(\mathbf{t}^T\boldsymbol{\Sigma}\mathbf{t}/2)\left(\sum_{a=1}^{n}\sigma_{ia}t_a\right)\left(\sum_{b=1}^{n}\sigma_{jb}t_b\right)\left(\sum_{c=1}^{n}\sigma_{kc}t_c\right) \\
&\quad + \exp(\mathbf{t}^T\boldsymbol{\Sigma}\mathbf{t}/2)\sigma_{ik}\left(\sum_{b=1}^{n}\sigma_{jb}t_b\right) + \exp(\mathbf{t}^T\boldsymbol{\Sigma}\mathbf{t}/2)\sigma_{jk}\left(\sum_{a=1}^{n}\sigma_{ia}t_a\right) \\
&\quad + \exp(\mathbf{t}^T\boldsymbol{\Sigma}\mathbf{t}/2)\sigma_{ij}\left(\sum_{c=1}^{n}\sigma_{kc}t_c\right), \\
\frac{\partial^4 m_{\mathbf{x}-\boldsymbol{\mu}}(\mathbf{t})}{\partial t_i t_j t_k t_\ell} &= \exp(\mathbf{t}^T\boldsymbol{\Sigma}\mathbf{t}/2)\left(\sum_{a=1}^{n}\sigma_{ia}t_a\right)\left(\sum_{b=1}^{n}\sigma_{jb}t_b\right)\left(\sum_{c=1}^{n}\sigma_{kc}t_c\right)\left(\sum_{d=1}^{n}\sigma_{\ell d}t_d\right) \\
&\quad + \exp(\mathbf{t}^T\boldsymbol{\Sigma}\mathbf{t}/2)\sigma_{i\ell}\left(\sum_{b=1}^{n}\sigma_{jb}t_b\right)\left(\sum_{c=1}^{n}\sigma_{kc}t_c\right) \\
&\quad + \exp(\mathbf{t}^T\boldsymbol{\Sigma}\mathbf{t}/2)\sigma_{j\ell}\left(\sum_{a=1}^{n}\sigma_{ia}t_a\right)\left(\sum_{c=1}^{n}\sigma_{kc}t_c\right) \\
&\quad + \exp(\mathbf{t}^T\boldsymbol{\Sigma}\mathbf{t}/2)\sigma_{k\ell}\left(\sum_{a=1}^{n}\sigma_{ia}t_a\right)\left(\sum_{b=1}^{n}\sigma_{jb}t_b\right)
\end{aligned}$$

$$+\exp(\mathbf{t}^T\boldsymbol{\Sigma}\mathbf{t}/2)\sigma_{ik}\left(\sum_{b=1}^{n}\sigma_{jb}t_b\right)\left(\sum_{d=1}^{n}\sigma_{\ell d}t_d\right)+\exp(\mathbf{t}^T\boldsymbol{\Sigma}\mathbf{t}/2)\sigma_{ik}\sigma_{j\ell}$$

$$+\exp(\mathbf{t}^T\boldsymbol{\Sigma}\mathbf{t}/2)\sigma_{jk}\left(\sum_{a=1}^{n}\sigma_{ia}t_a\right)\left(\sum_{d=1}^{n}\sigma_{\ell d}t_d\right)+\exp(\mathbf{t}^T\boldsymbol{\Sigma}\mathbf{t}/2)\sigma_{jk}\sigma_{i\ell}$$

$$+\exp(\mathbf{t}^T\boldsymbol{\Sigma}\mathbf{t}/2)\sigma_{ij}\left(\sum_{c=1}^{n}\sigma_{kc}t_c\right)\left(\sum_{d=1}^{n}\sigma_{\ell d}t_d\right)+\exp(\mathbf{t}^T\boldsymbol{\Sigma}\mathbf{t}/2)\sigma_{ij}\sigma_{k\ell}.$$

Finally,

$$\begin{aligned}\mathrm{E}[(x_i-\mu_i)(x_j-\mu_j)(x_k-\mu_k)(x_\ell-\mu_\ell)] &= \left.\frac{\partial^4 m_{\mathbf{x}-\boldsymbol{\mu}}(\mathbf{t})}{\partial t_i t_j t_k t_\ell}\right|_{\mathbf{t}=\mathbf{0}} \\ &= \sigma_{ij}\sigma_{k\ell}+\sigma_{ik}\sigma_{j\ell}+\sigma_{i\ell}\sigma_{jk}.\end{aligned}$$

▶ **Exercise 10** Let u and v be random variables whose joint distribution is $\mathrm{N}_2(\mathbf{0},\boldsymbol{\Sigma})$, where $\mathrm{rank}(\boldsymbol{\Sigma})=2$. Show that $\mathrm{corr}(u^2,v^2)=[\mathrm{corr}(u,v)]^2$.

Solution Write σ_{ij} $(i,j=1,2)$ for the elements of $\boldsymbol{\Sigma}$ and let $\mathbf{x}=\begin{pmatrix}u\\v\end{pmatrix}$. Now

$$u^2=\mathbf{x}^T\begin{pmatrix}1&0\\0&0\end{pmatrix}\mathbf{x}\quad\text{and}\quad v^2=\mathbf{x}^T\begin{pmatrix}0&0\\0&1\end{pmatrix}\mathbf{x}$$

so by Corollary 4.2.6.3,

$$\begin{aligned}\mathrm{cov}(u^2,v^2)&=2\mathrm{tr}\left[\begin{pmatrix}1&0\\0&0\end{pmatrix}\begin{pmatrix}\sigma_{11}&\sigma_{12}\\\sigma_{21}&\sigma_{22}\end{pmatrix}\begin{pmatrix}0&0\\0&1\end{pmatrix}\begin{pmatrix}\sigma_{11}&\sigma_{12}\\\sigma_{21}&\sigma_{22}\end{pmatrix}\right]\\&=\sigma_{12}^2.\end{aligned}$$

Similar calculations yield $\mathrm{var}(u^2)=\sigma_{11}^2$ and $\mathrm{var}(v^2)=\sigma_{22}^2$. Therefore,

$$\mathrm{corr}(u^2,v^2)=\frac{\mathrm{cov}(u^2,v^2)}{\sqrt{\mathrm{var}(u^2)\mathrm{var}(v^2)}}=\frac{\sigma_{12}^2}{\sigma_{11}\sigma_{22}}=[\mathrm{corr}(u,v)]^2.$$

▶ **Exercise 11** Suppose that $\mathbf{x}\sim\mathrm{N}_n(\mathbf{a},\sigma^2\mathbf{I})$ where $\mathbf{a}=(a_i)$ is an n-vector of constants such that $\mathbf{a}^T\mathbf{a}=1$. Let $\mathbf{b}$ represent another n-vector such that $\mathbf{b}^T\mathbf{b}=1$ and $\mathbf{a}^T\mathbf{b}=0$. Define $\mathbf{P_a}=\mathbf{aa}^T/(\mathbf{a}^T\mathbf{a})=\mathbf{aa}^T$ and $\mathbf{P_b}=\mathbf{bb}^T/(\mathbf{b}^T\mathbf{b})=\mathbf{bb}^T$.

(a) Find the distribution of $\mathbf{b}^T\mathbf{x}$.
(b) Find the distribution of $\mathbf{P_a}\mathbf{x}$.
(c) Find the distribution of $\mathbf{P_b}\mathbf{x}$.

(d) Find the conditional distribution of x_1 (the first element of $\mathbf{x}$) given that $\mathbf{a}^T\mathbf{x} = 0$. Assume that $n \geq 2$.

Solution

(a) By Theorem 14.1.6, $\mathbf{b}^T\mathbf{x} \sim \text{N}(\mathbf{b}^T\mathbf{a}, \mathbf{b}^T(\sigma^2\mathbf{I})\mathbf{b})$, i.e., $\text{N}(0, \sigma^2)$.
(b) By Theorem 14.1.6, $\mathbf{P_a x} \sim \text{N}_n(\mathbf{P_a a}, \mathbf{P_a}(\sigma^2\mathbf{I})\mathbf{P_a})$, i.e., $\text{N}_n(\mathbf{a}, \sigma^2\mathbf{P_a})$.
(c) By Theorem 14.1.6, $\mathbf{P_b x} \sim \text{N}_n(\mathbf{P_b a}, \mathbf{P_b}(\sigma^2\mathbf{I})\mathbf{P_b})$, i.e., $\text{N}_n(\mathbf{0}, \sigma^2\mathbf{P_b})$.
(d) By Theorem 14.1.6, $\begin{pmatrix} x_1 \\ \mathbf{a}^T\mathbf{x} \end{pmatrix} = \begin{pmatrix} \mathbf{u}_1^T \\ \mathbf{a}^T \end{pmatrix} \mathbf{x}$ (where $\mathbf{u}_1$ is the first unit vector) has a bivariate normal distribution with mean vector $\begin{pmatrix} \mathbf{u}_1^T \\ \mathbf{a}^T \end{pmatrix} \mathbf{a} = \begin{pmatrix} a_1 \\ 1 \end{pmatrix}$ and variance–covariance matrix $\begin{pmatrix} \mathbf{u}_1^T \\ \mathbf{a}^T \end{pmatrix} (\sigma^2\mathbf{I}) \begin{pmatrix} \mathbf{u}_1^T \\ \mathbf{a}^T \end{pmatrix}^T = \sigma^2 \begin{pmatrix} 1 & a_1 \\ a_1 & 1 \end{pmatrix}$. Then, by Theorem 14.1.9, the conditional distribution of x_1 given that $\mathbf{a}^T\mathbf{x} = 0$ is normal with mean

$$a_1 + a_1(1)^{-1}(0 - 1) = 0$$

and variance

$$\sigma^2[1 - a_1(1)^{-1}a_1] = \sigma^2(1 - a_1^2).$$

► **Exercise 12** Use the mgf of a $\chi^2(\nu, \lambda)$ distribution to obtain its mean and variance.

Solution $m_u(t) = (1 - 2t)^{-\frac{\nu}{2}} \exp\left(\frac{2t\lambda}{1-2t}\right)$ for $t < \frac{1}{2}$, so

$$\begin{aligned}
\frac{d}{dt} m_u(t) &= \frac{-\nu/2}{(1-2t)^{\frac{\nu}{2}+1}}(-2)\exp\left(\frac{2t\lambda}{1-2t}\right) + \frac{1}{(1-2t)^{\frac{\nu}{2}}}\left(\frac{2\lambda}{1-2t} - \frac{2t\lambda(-2)}{(1-2t)^2}\right)\exp\left(\frac{2t\lambda}{1-2t}\right) \\
&= \exp\left(\frac{2t\lambda}{1-2t}\right)\left[\frac{\nu+2\lambda}{(1-2t)^{\frac{\nu}{2}+1}} + \frac{4t\lambda}{(1-2t)^{\frac{\nu}{2}+2}}\right]
\end{aligned}$$

and

$$\begin{aligned}
\frac{d^2}{dt^2} m_u(t) &= \frac{1}{(1-2t)^{\frac{\nu}{2}}}\left(2\lambda + \frac{4t\lambda}{1-2t}\right)\exp\left(\frac{2t\lambda}{1-2t}\right)\left[\frac{\nu+2\lambda}{(1-2t)^{\frac{\nu}{2}+1}} + \frac{4t\lambda}{(1-2t)^{\frac{\nu}{2}+2}}\right] \\
&\quad + \exp\left(\frac{2t\lambda}{1-2t}\right)\left[\frac{(\nu+2\lambda)[-(\frac{\nu}{2}+2)](-2) + 4\lambda}{(1-2t)^{\frac{\nu}{2}+2}} + \frac{4t\lambda[-(\frac{\nu}{2}+2)](-2)}{(1-2t)^{\frac{\nu}{2}+3}}\right] \\
&= \frac{1}{(1-2t)^{\frac{\nu}{2}}}\left(2\lambda + \frac{4t\lambda}{1-2t}\right)\exp\left(\frac{2t\lambda}{1-2t}\right)\left[\frac{\nu+2\lambda}{(1-2t)^{\frac{\nu}{2}+1}} + \frac{4t\lambda}{(1-2t)^{\frac{\nu}{2}+2}}\right] \\
&\quad + \exp\left(\frac{2t\lambda}{1-2t}\right)\left[\frac{(\nu+2\lambda)(\nu+2) + 4\lambda}{(1-2t)^{\frac{\nu}{2}+2}} + \frac{4t\lambda(\nu+2)}{(1-2t)^{\frac{\nu}{2}+3}}\right]
\end{aligned}$$

Then

$$\begin{aligned}\mathrm{E}(u) &= \frac{d}{dt}m_u(t)\Big|_{t=0} = \nu + 2\lambda,\\ \mathrm{var}(u) &= \frac{d^2}{dt^2}m_u(t)\Big|_{t=0} - [E(u)]^2\\ &= (2\lambda)(\nu+2\lambda) + (\nu+2\lambda)(\nu+2) + 4\lambda - (\nu^2 + 4\lambda\nu + 4\lambda^2)\\ &= 2\nu + 8\lambda.\end{aligned}$$

▶ **Exercise 13** Suppose that $\mathbf{x} \sim \mathrm{N}_n(\boldsymbol{\mu}, \boldsymbol{\Sigma})$, where $\mathrm{rank}(\boldsymbol{\Sigma}) = n$. Let $\mathbf{A}$ represent a nonnull nonnegative definite matrix of constants such that $\mathbf{A\Sigma}$ is not idempotent. It has been proposed that the distribution of $\mathbf{x}^T\mathbf{A}\mathbf{x}$ may be approximated by a multiple of a central chi-square distribution, specifically by $c\chi^2(f)$ where c and f are chosen so that $\mathbf{x}^T\mathbf{A}\mathbf{x}$ and $c\chi^2(f)$ have the same mean and variance.

(a) Determine expressions for the appropriate c and f.
(b) Under the normal Gauss–Markov model $\{\mathbf{y}, \mathbf{X}\boldsymbol{\beta}, \sigma^2\mathbf{I}\}$, suppose that $\mathbf{L} \neq \mathbf{P_X}$ is a matrix such that $\mathrm{E}(\mathbf{Ly}) = \mathbf{X}\boldsymbol{\beta}$. Define $\breve{\mathbf{e}} = (\mathbf{I} - \mathbf{L})\mathbf{y}$. Using part (a), determine an approximation to the distribution of $\breve{\mathbf{e}}^T\breve{\mathbf{e}}$.

Solution

(a) Using the proposed approximation and expressions in Sect. 14.2 for the mean and variance of a central chi-square distribution, we obtain approximating equations

$$\mathrm{E}(\mathbf{y}^T\mathbf{A}\mathbf{y}) \doteq \mathrm{E}[c\chi^2(f)] = cf$$

and

$$\mathrm{Var}(\mathbf{y}^T\mathbf{A}\mathbf{y}) \doteq \mathrm{Var}[c\chi^2(f)] = 2c^2 f.$$

Using Theorem 4.2.4 and solving, we obtain

$$\begin{aligned}c &= \frac{\mathrm{Var}(\mathbf{y}^T\mathbf{A}\mathbf{y})}{2\mathrm{E}(\mathbf{y}^T\mathbf{A}\mathbf{y})} = \frac{2\boldsymbol{\mu}^T\mathbf{A\Sigma A}\boldsymbol{\mu} + \mathrm{tr}(\mathbf{A\Sigma A\Sigma})}{\boldsymbol{\mu}^T\mathbf{A}\boldsymbol{\mu} + \mathrm{tr}(\mathbf{A\Sigma})},\\ f &= \frac{\mathrm{E}(\mathbf{y}^T\mathbf{A}\mathbf{y})}{c} = \frac{[\boldsymbol{\mu}^T\mathbf{A}\boldsymbol{\mu} + \mathrm{tr}(\mathbf{A\Sigma})]^2}{2\boldsymbol{\mu}^T\mathbf{A\Sigma A}\boldsymbol{\mu} + \mathrm{tr}(\mathbf{A\Sigma A\Sigma})}.\end{aligned}$$

(b) Because $\mathbf{Ly}$ is unbiased for $\mathbf{X}\boldsymbol{\beta}$, $\mathrm{E}(\mathbf{Ly}) = \mathbf{X}\boldsymbol{\beta}$ for all $\boldsymbol{\beta}$, implying that $\mathbf{LX}\boldsymbol{\beta} = \mathbf{X}\boldsymbol{\beta}$ for all $\boldsymbol{\beta}$, or equivalently that $\mathbf{LX} = \mathbf{X}$. Note that $\breve{\mathbf{e}}^T\breve{\mathbf{e}} = \mathbf{y}^T(\mathbf{I}-\mathbf{L})^T(\mathbf{I}-\mathbf{L})\mathbf{y} \equiv \mathbf{y}^T\mathbf{A}\mathbf{y}$, say, where $\mathbf{A} = (\mathbf{I}-\mathbf{L})^T(\mathbf{I}-\mathbf{L})$. By part (a), $\mathbf{y}^T\mathbf{A}\mathbf{y}$ is approximately distributed as $c\chi^2(f)$ where

$$c = \frac{2(\mathbf{X}\boldsymbol{\beta})^T(\mathbf{I}-\mathbf{L})^T(\mathbf{I}-\mathbf{L})\sigma^2\mathbf{I}(\mathbf{I}-\mathbf{L})^T(\mathbf{I}-\mathbf{L})\mathbf{X}\boldsymbol{\beta} + \text{tr}[(\mathbf{I}-\mathbf{L})^T(\mathbf{I}-\mathbf{L})\sigma^2\mathbf{I}(\mathbf{I}-\mathbf{L})^T(\mathbf{I}-\mathbf{L})\sigma^2\mathbf{I}]}{(\mathbf{X}\boldsymbol{\beta})^T(\mathbf{I}-\mathbf{L})^T(\mathbf{I}-\mathbf{L})\mathbf{X}\boldsymbol{\beta} + \text{tr}[(\mathbf{I}-\mathbf{L})^T(\mathbf{I}-\mathbf{L})\sigma^2\mathbf{I}]}$$

$$= \frac{\sigma^2\text{tr}[(\mathbf{I}-\mathbf{L})^T(\mathbf{I}-\mathbf{L})(\mathbf{I}-\mathbf{L})^T(\mathbf{I}-\mathbf{L})]}{\text{tr}[(\mathbf{I}-\mathbf{L})^T(\mathbf{I}-\mathbf{L})]}$$

and

$$f = \frac{\{(\mathbf{X}\boldsymbol{\beta})^T(\mathbf{I}-\mathbf{L})^T(\mathbf{I}-\mathbf{L})\mathbf{X}\boldsymbol{\beta} + \text{tr}[(\mathbf{I}-\mathbf{L})^T(\mathbf{I}-\mathbf{L})\sigma^2\mathbf{I}]\}^2}{2(\mathbf{X}\boldsymbol{\beta})^T(\mathbf{I}-\mathbf{L})^T(\mathbf{I}-\mathbf{L})\sigma^2\mathbf{I}(\mathbf{I}-\mathbf{L})^T(\mathbf{I}-\mathbf{L})\mathbf{X}\boldsymbol{\beta} + \text{tr}[(\mathbf{I}-\mathbf{L})^T(\mathbf{I}-\mathbf{L})\sigma^2\mathbf{I}(\mathbf{I}-\mathbf{L})^T(\mathbf{I}-\mathbf{L})\sigma^2\mathbf{I}]}$$

$$= \frac{\{\text{tr}[(\mathbf{I}-\mathbf{L})^T(\mathbf{I}-\mathbf{L})]\}^2}{\text{tr}[(\mathbf{I}-\mathbf{L})^T(\mathbf{I}-\mathbf{L})(\mathbf{I}-\mathbf{L})^T(\mathbf{I}-\mathbf{L})]}.$$

► **Exercise 14** Suppose that the conditions of part (a) of Corollary 14.2.6.1 hold, i.e., suppose that $\mathbf{x} \sim \text{N}_n(\boldsymbol{\mu}, \boldsymbol{\Sigma})$ where rank$(\boldsymbol{\Sigma}) = n$ and that $\mathbf{A}$ is a nonnull symmetric $n \times n$ matrix of constants such that $\mathbf{A}\boldsymbol{\Sigma}$ is idempotent. Prove that $\mathbf{A}$ is nonnegative definite.

Solution Because $\mathbf{A}\boldsymbol{\Sigma}$ is idempotent, $\mathbf{A}\boldsymbol{\Sigma}\mathbf{A}\boldsymbol{\Sigma} = \mathbf{A}\boldsymbol{\Sigma}$. Post-multiplying both sides of this matrix equation by $\boldsymbol{\Sigma}^{-1}$ yields $\mathbf{A}\boldsymbol{\Sigma}\mathbf{A} = \mathbf{A}$. Therefore, for any n-vector $\mathbf{x}$, $\mathbf{x}^T\mathbf{A}\mathbf{x} = \mathbf{x}^T\mathbf{A}\boldsymbol{\Sigma}\mathbf{A}\mathbf{x} = \mathbf{x}^T\mathbf{A}^T\boldsymbol{\Sigma}\mathbf{A}\mathbf{x} = (\mathbf{A}\mathbf{x})^T\boldsymbol{\Sigma}(\mathbf{A}\mathbf{x}) \geq 0$ because $\boldsymbol{\Sigma}$ is positive definite.

► **Exercise 15** Consider the special case of Theorem 14.3.1 in which there are only two functions of interest and they are quadratic forms, i.e., $w_1 = \mathbf{x}^T\mathbf{A}_1\mathbf{x}$ and $w_2 = \mathbf{x}^T\mathbf{A}_2\mathbf{x}$. Show that, in the following cases, necessary and sufficient conditions for these two quadratic forms to be distributed independently are:

(a) $\boldsymbol{\Sigma}\mathbf{A}_1\boldsymbol{\Sigma}\mathbf{A}_2\boldsymbol{\Sigma} = \mathbf{0}$, $\boldsymbol{\Sigma}\mathbf{A}_1\boldsymbol{\Sigma}\mathbf{A}_2\boldsymbol{\mu} = \mathbf{0}$, $\boldsymbol{\Sigma}\mathbf{A}_2\boldsymbol{\Sigma}\mathbf{A}_1\boldsymbol{\mu} = \mathbf{0}$, and $\boldsymbol{\mu}^T\mathbf{A}_1\boldsymbol{\Sigma}\mathbf{A}_2\boldsymbol{\mu} = 0$, when there are no restrictions on $\mathbf{A}_1$ and $\mathbf{A}_2$ beyond those stated in the theorem.
(b) $\mathbf{A}_1\boldsymbol{\Sigma}\mathbf{A}_2\boldsymbol{\Sigma} = \mathbf{0}$ and $\mathbf{A}_1\boldsymbol{\Sigma}\mathbf{A}_2\boldsymbol{\mu} = \mathbf{0}$ when $\mathbf{A}_1$ is nonnegative definite.
(c) $\mathbf{A}_1\boldsymbol{\Sigma}\mathbf{A}_2 = \mathbf{0}$ when $\mathbf{A}_1$ and $\mathbf{A}_2$ are nonnegative definite.

Solution

(a) These conditions are trivial specializations of the conditions of Theorem 14.3.1 to the case in which $k = 2$ and $\boldsymbol{\ell}_1 = \boldsymbol{\ell}_2 = \mathbf{0}$.
(b) If $\mathbf{A}_1\boldsymbol{\Sigma}\mathbf{A}_2\boldsymbol{\Sigma} = \mathbf{0}$ and $\mathbf{A}_1\boldsymbol{\Sigma}\mathbf{A}_2\boldsymbol{\mu} = \mathbf{0}$, then $\boldsymbol{\Sigma}\mathbf{A}_1\boldsymbol{\Sigma}\mathbf{A}_2\boldsymbol{\Sigma} = \boldsymbol{\Sigma}(\mathbf{A}_1\boldsymbol{\Sigma}\mathbf{A}_2\boldsymbol{\Sigma}) = \boldsymbol{\Sigma}\mathbf{0} = \mathbf{0}$; $\boldsymbol{\Sigma}\mathbf{A}_1\boldsymbol{\Sigma}\mathbf{A}_2\boldsymbol{\mu} = \boldsymbol{\Sigma}(\mathbf{A}_1\boldsymbol{\Sigma}\mathbf{A}_2\boldsymbol{\mu}) = \boldsymbol{\Sigma}\mathbf{0} = \mathbf{0}$; $\boldsymbol{\Sigma}\mathbf{A}_2\boldsymbol{\Sigma}\mathbf{A}_1\boldsymbol{\mu} = (\mathbf{A}_1\boldsymbol{\Sigma}\mathbf{A}_2\boldsymbol{\Sigma})^T\boldsymbol{\mu} = \mathbf{0}\boldsymbol{\mu} = \mathbf{0}$; and $\boldsymbol{\mu}^T\mathbf{A}_1\boldsymbol{\Sigma}\mathbf{A}_2\boldsymbol{\mu} = \boldsymbol{\mu}^T(\mathbf{A}_1\boldsymbol{\Sigma}\mathbf{A}_2\boldsymbol{\mu}) = \boldsymbol{\mu}^T\mathbf{0} = 0$. Thus the conditions of part (a) are satisfied so w_1 and w_2 are independent. Conversely, if w_1 and w_2 are independent and $\mathbf{A}_1$ is nonnegative definite, then the conditions of part (a) are satisfied and they may be written as $\boldsymbol{\Sigma}\mathbf{A}_1\boldsymbol{\Sigma}\mathbf{A}_2\boldsymbol{\Sigma} = (\mathbf{A}_1^{\frac{1}{2}}\boldsymbol{\Sigma})^T(\mathbf{A}_1^{\frac{1}{2}}\boldsymbol{\Sigma})\mathbf{A}_2\boldsymbol{\Sigma} = \mathbf{0}$ and $\boldsymbol{\Sigma}\mathbf{A}_1\boldsymbol{\Sigma}\mathbf{A}_2\boldsymbol{\mu} = (\mathbf{A}_1^{\frac{1}{2}}\boldsymbol{\Sigma})^T(\mathbf{A}_1^{\frac{1}{2}}\boldsymbol{\Sigma})\mathbf{A}_2\boldsymbol{\mu} = \mathbf{0}$ where $\mathbf{A}_1^{\frac{1}{2}}$ is the unique

nonnegative definite square root of $\mathbf{A}_1$. Thus, by Theorem 3.3.2 $\mathbf{A}_1^{\frac{1}{2}}\boldsymbol{\Sigma}\mathbf{A}_2\boldsymbol{\Sigma} = \mathbf{0}$ and $\mathbf{A}_1^{\frac{1}{2}}\boldsymbol{\Sigma}\mathbf{A}_2\boldsymbol{\mu} = \mathbf{0}$. Pre-multiplying both equalities by $\mathbf{A}_1^{\frac{1}{2}}$ yields $\mathbf{A}_1\boldsymbol{\Sigma}\mathbf{A}_2\boldsymbol{\Sigma} = \mathbf{0}$ and $\mathbf{A}_1\boldsymbol{\Sigma}\mathbf{A}_2\boldsymbol{\mu} = \mathbf{0}$.

(c) If $\mathbf{A}_1\boldsymbol{\Sigma}\mathbf{A}_2 = \mathbf{0}$, then $\mathbf{A}_1\boldsymbol{\Sigma}\mathbf{A}_2\boldsymbol{\Sigma} = (\mathbf{A}_1\boldsymbol{\Sigma}\mathbf{A}_2)\boldsymbol{\Sigma} = \mathbf{0}\boldsymbol{\Sigma} = \mathbf{0}$ and $\mathbf{A}_1\boldsymbol{\Sigma}\mathbf{A}_2\boldsymbol{\mu} = (\mathbf{A}_1\boldsymbol{\Sigma}\mathbf{A}_2)\boldsymbol{\mu} = \mathbf{0}\boldsymbol{\mu} = \mathbf{0}$. Thus the conditions of part (b) are satisfied so w_1 and w_2 are independent. Conversely, if w_1 and w_2 are independent and $\mathbf{A}_1$ and $\mathbf{A}_2$ are nonnegative definite, then the conditions of part (b) are satisfied, the first of which may be written as $\mathbf{A}_1(\mathbf{A}_2^{\frac{1}{2}}\boldsymbol{\Sigma})^T(\mathbf{A}_2^{\frac{1}{2}}\boldsymbol{\Sigma}) = \mathbf{0}$ where $\mathbf{A}_2^{\frac{1}{2}}$ is the unique nonnegative definite square root of $\mathbf{A}_2$. This implies, by Theorem 3.3.2, that $\mathbf{A}_1\boldsymbol{\Sigma}\mathbf{A}_2^{\frac{1}{2}} = \mathbf{0}$. Post-multiplication of this equality by $\mathbf{A}_2^{\frac{1}{2}}$ yields $\mathbf{A}_1\boldsymbol{\Sigma}\mathbf{A}_2 = \mathbf{0}$.

▶ **Exercise 16** Suppose that $\mathbf{x}_1, \mathbf{x}_2$, and $\mathbf{x}_3$ are independent and identically distributed as $\mathrm{N}_n(\boldsymbol{\mu}, \boldsymbol{\Sigma})$, where $\mathrm{rank}(\boldsymbol{\Sigma}) = n$. Define

$$\mathbf{Q} = (\mathbf{x}_1, \mathbf{x}_2, \mathbf{x}_3)\mathbf{A}(\mathbf{x}_1, \mathbf{x}_2, \mathbf{x}_3)^T,$$

where $\mathbf{A}$ is a symmetric idempotent 3×3 matrix of constants. Let $r = \mathrm{rank}(\mathbf{A})$ and let $\mathbf{c}$ represent a nonzero n-vector of constants. Determine the distribution of $\mathbf{c}^T\mathbf{Q}\mathbf{c}/(\mathbf{c}^T\boldsymbol{\Sigma}\mathbf{c})$. [Hint: First obtain the distribution of $(\mathbf{x}_1, \mathbf{x}_2, \mathbf{x}_3)^T\mathbf{c}$.]

Solution By the given conditions, $\mathbf{c}^T\mathbf{x}_1, \mathbf{c}^T\mathbf{x}_2$, and $\mathbf{c}^T\mathbf{x}_3$ are independent and identically distributed as $\mathrm{N}(\mathbf{c}^T\boldsymbol{\mu}, \mathbf{c}^T\boldsymbol{\Sigma}\mathbf{c})$, implying that $(\mathbf{x}_1, \mathbf{x}_2, \mathbf{x}_3)^T\mathbf{c} \sim \mathrm{N}_3[(\mathbf{c}^T\boldsymbol{\mu})\mathbf{1}_3, \mathrm{diag}(\mathbf{c}^T\boldsymbol{\Sigma}\mathbf{c}, \mathbf{c}^T\boldsymbol{\Sigma}\mathbf{c}, \mathbf{c}^T\boldsymbol{\Sigma}\mathbf{c})]$. Partially standardizing, we have

$$(\mathbf{c}^T\boldsymbol{\Sigma}\mathbf{c})^{-\frac{1}{2}}(\mathbf{x}_1, \mathbf{x}_2, \mathbf{x}_3)^T\mathbf{c} \sim \mathrm{N}_3[(\mathbf{c}^T\boldsymbol{\Sigma}\mathbf{c})^{-\frac{1}{2}}(\mathbf{c}^T\boldsymbol{\mu})\mathbf{1}_3, \mathbf{I}_3).$$

Because $\mathbf{A}\mathbf{I}_3 = \mathbf{A}$ is idempotent, by Corollary 14.2.6.1 we have

$$\mathbf{c}^T\mathbf{Q}\mathbf{c}/(\mathbf{c}^T\boldsymbol{\Sigma}\mathbf{c}) \sim \chi^2[r, \frac{1}{2}(\mathbf{c}^T\boldsymbol{\Sigma}\mathbf{c})^{-1}(\mathbf{c}^T\boldsymbol{\mu})^2\mathbf{1}_3^T\mathbf{A}\mathbf{1}_3].$$

▶ **Exercise 17** Suppose that $\mathbf{x} \sim \mathrm{N}_n(\boldsymbol{\mu}, \boldsymbol{\Sigma})$, where $\mathrm{rank}(\boldsymbol{\Sigma}) = n$, and partition $\mathbf{x}$, $\boldsymbol{\mu}$, and $\boldsymbol{\Sigma}$ conformably as follows:

$$\mathbf{x} = \begin{pmatrix} \mathbf{x}_1 \\ \mathbf{x}_2 \end{pmatrix}, \quad \boldsymbol{\mu} = \begin{pmatrix} \boldsymbol{\mu}_1 \\ \boldsymbol{\mu}_2 \end{pmatrix}, \quad \boldsymbol{\Sigma} = \begin{pmatrix} \boldsymbol{\Sigma}_{11} & \boldsymbol{\Sigma}_{12} \\ \boldsymbol{\Sigma}_{21} & \boldsymbol{\Sigma}_{22} \end{pmatrix},$$

where $\mathbf{x}_1$ is $n_1 \times 1$ and $\mathbf{x}_2$ is $n_2 \times 1$.

(a) Obtain the distribution of $Q_1 = (\mathbf{x}_1 - \boldsymbol{\Sigma}_{12}\boldsymbol{\Sigma}_{22}^{-1}\mathbf{x}_2)^T(\boldsymbol{\Sigma}_{11} - \boldsymbol{\Sigma}_{12}\boldsymbol{\Sigma}_{22}^{-1}\boldsymbol{\Sigma}_{21})^{-1}(\mathbf{x}_1 - \boldsymbol{\Sigma}_{12}\boldsymbol{\Sigma}_{22}^{-1}\mathbf{x}_2)$.

(b) Obtain the distribution of $Q_2 = \mathbf{x}_2^T\boldsymbol{\Sigma}_{22}^{-1}\mathbf{x}_2$.

(c) Suppose that $\boldsymbol{\mu}_2 = \mathbf{0}$. Obtain the distribution of Q_1/Q_2 (suitably scaled).

Solution

(a) Observe that

$$\mathbf{x}_1 - \boldsymbol{\Sigma}_{12}\boldsymbol{\Sigma}_{22}^{-1}\mathbf{x}_2 = \begin{pmatrix} \mathbf{I}_{n_1} & -\boldsymbol{\Sigma}_{12}\boldsymbol{\Sigma}_{22}^{-1} \end{pmatrix} \begin{pmatrix} \mathbf{x}_1 \\ \mathbf{x}_2 \end{pmatrix}.$$

Thus, by Theorem 14.1.6, $\mathbf{x}_1 - \boldsymbol{\Sigma}_{12}\boldsymbol{\Sigma}_{22}^{-1}\mathbf{x}_2$ is normally distributed with mean vector

$$\begin{pmatrix} \mathbf{I}_{n_1} & -\boldsymbol{\Sigma}_{12}\boldsymbol{\Sigma}_{22}^{-1} \end{pmatrix} \begin{pmatrix} \boldsymbol{\mu}_1 \\ \boldsymbol{\mu}_2 \end{pmatrix} = \boldsymbol{\mu}_1 - \boldsymbol{\Sigma}_{12}\boldsymbol{\Sigma}_{22}^{-1}\boldsymbol{\mu}_2$$

and variance–covariance matrix

$$\begin{pmatrix} \mathbf{I}_{n_1} & -\boldsymbol{\Sigma}_{12}\boldsymbol{\Sigma}_{22}^{-1} \end{pmatrix} \begin{pmatrix} \boldsymbol{\Sigma}_{11} & \boldsymbol{\Sigma}_{12} \\ \boldsymbol{\Sigma}_{21} & \boldsymbol{\Sigma}_{22} \end{pmatrix} \begin{pmatrix} \mathbf{I}_{n_1} \\ -\boldsymbol{\Sigma}_{22}^{-1}\boldsymbol{\Sigma}_{21} \end{pmatrix} = \boldsymbol{\Sigma}_{11} - \boldsymbol{\Sigma}_{12}\boldsymbol{\Sigma}_{22}^{-1}\boldsymbol{\Sigma}_{21}.$$

Because $\boldsymbol{\Sigma}$ is positive definite, the $n_1 \times n_1$ matrix $\boldsymbol{\Sigma}_{11} - \boldsymbol{\Sigma}_{12}\boldsymbol{\Sigma}_{22}^{-1}\boldsymbol{\Sigma}_{21}$ is also positive definite by Theorem 2.15.7a. So by Theorem 14.2.4,

$$Q_1 \sim \chi^2\left(n_1, (\boldsymbol{\mu}_1 - \boldsymbol{\Sigma}_{12}\boldsymbol{\Sigma}_{22}^{-1}\boldsymbol{\mu}_2)^T(\boldsymbol{\Sigma}_{11} - \boldsymbol{\Sigma}_{12}\boldsymbol{\Sigma}_{22}^{-1}\boldsymbol{\Sigma}_{21})^{-1}(\boldsymbol{\mu}_1 - \boldsymbol{\Sigma}_{12}\boldsymbol{\Sigma}_{22}^{-1}\boldsymbol{\mu}_2)/2\right).$$

(b) $\mathbf{x}_2 \sim \mathrm{N}(\boldsymbol{\mu}_2, \boldsymbol{\Sigma}_{22})$ by Theorem 14.1.7. By the positive definiteness of $\boldsymbol{\Sigma}$ and Theorem 2.15.7a, $\boldsymbol{\Sigma}_{22}$ is positive definite, so its rank is n_2. By Theorem 14.2.4, $Q_2 \sim \chi^2(n_2, \boldsymbol{\mu}_2^T\boldsymbol{\Sigma}_{22}^{-1}\boldsymbol{\mu}_2/2)$.

(c) Because

$$\begin{pmatrix} \mathbf{x}_1 - \boldsymbol{\Sigma}_{12}\boldsymbol{\Sigma}_{22}^{-1}\mathbf{x}_2 \\ \mathbf{x}_2 \end{pmatrix} = \begin{pmatrix} \mathbf{I} & -\boldsymbol{\Sigma}_{12}\boldsymbol{\Sigma}_{22}^{-1} \\ \mathbf{0} & \mathbf{I} \end{pmatrix} \begin{pmatrix} \mathbf{x}_1 \\ \mathbf{x}_2 \end{pmatrix},$$

the joint distribution of $\mathbf{x}_1 - \boldsymbol{\Sigma}_{12}\boldsymbol{\Sigma}_{22}^{-1}\mathbf{x}_2$ and $\mathbf{x}_2$ is multivariate normal, and their matrix of covariances is

$$\begin{pmatrix} \mathbf{I} & -\boldsymbol{\Sigma}_{12}\boldsymbol{\Sigma}_{22}^{-1} \end{pmatrix} \begin{pmatrix} \boldsymbol{\Sigma}_{11} & \boldsymbol{\Sigma}_{12} \\ \boldsymbol{\Sigma}_{21} & \boldsymbol{\Sigma}_{22} \end{pmatrix} \begin{pmatrix} \mathbf{0} \\ \mathbf{I} \end{pmatrix} = \mathbf{0}.$$

Thus $\mathbf{x}_1 - \boldsymbol{\Sigma}_{12}\boldsymbol{\Sigma}_{22}^{-1}\mathbf{x}_2$ and $\mathbf{x}_2$ are independent by Theorem 14.1.8. Moreover, when $\boldsymbol{\mu}_2 = \mathbf{0}$, the results of parts (a) and (b) specialize as follows: $Q_1 \sim \chi^2(n_1, \boldsymbol{\mu}_1^T(\boldsymbol{\Sigma}_{11} - \boldsymbol{\Sigma}_{12}\boldsymbol{\Sigma}_{22}^{-1}\boldsymbol{\Sigma}_{21})^{-1}\boldsymbol{\mu}_1/2)$ and $Q_2 \sim \chi^2(n_2)$. Thus, by Definition 14.5.1,

$$\frac{Q_1/n_1}{Q_2/n_2} \sim \mathrm{F}(n_1, n_2, \boldsymbol{\mu}_1^T(\boldsymbol{\Sigma}_{11} - \boldsymbol{\Sigma}_{12}\boldsymbol{\Sigma}_{22}^{-1}\boldsymbol{\Sigma}_{21})^{-1}\boldsymbol{\mu}_1/2).$$

▶ **Exercise 18** Suppose that $\mathbf{x} \sim \mathrm{N}_n(\mathbf{0}, \mathbf{I})$ and that $\mathbf{A}$ and $\mathbf{B}$ are $n \times n$ symmetric idempotent matrices of ranks p and q, respectively, where $p > q$. Show, by completing the following three steps, that $\mathbf{x}^T(\mathbf{A} - \mathbf{B})\mathbf{x} \sim \chi^2(p - q)$ if $\mathrm{corr}(\mathbf{x}^T\mathbf{A}\mathbf{x}, \mathbf{x}^T\mathbf{B}\mathbf{x}) = \sqrt{q/p}$.

(a) Show that $\mathrm{tr}[(\mathbf{I} - \mathbf{A})\mathbf{B}] = 0$ if and only if $\mathrm{corr}(\mathbf{x}^T\mathbf{A}\mathbf{x}, \mathbf{x}^T\mathbf{B}\mathbf{x}) = \sqrt{q/p}$.
(b) Show that $(\mathbf{I} - \mathbf{A})\mathbf{B} = \mathbf{0}$ if and only if $\mathrm{tr}[(\mathbf{I} - \mathbf{A})\mathbf{B}] = 0$.
(c) Show that $\mathbf{x}^T(\mathbf{A} - \mathbf{B})\mathbf{x} \sim \chi^2(p - q)$ if $(\mathbf{I} - \mathbf{A})\mathbf{B} = \mathbf{0}$.

Solution

(a) By Corollary 4.2.6.3,

$$\begin{aligned} \mathrm{cov}(\mathbf{x}^T\mathbf{A}\mathbf{x}, \mathbf{x}^T\mathbf{B}\mathbf{x}) &= 2 \cdot \mathrm{tr}(\mathbf{AIBI}) + 4 \cdot \mathbf{0}^T\mathbf{AIB0} = 2\mathrm{tr}(\mathbf{AB}), \\ \mathrm{var}(\mathbf{x}^T\mathbf{A}\mathbf{x}) &= 2 \cdot \mathrm{tr}(\mathbf{AIAI}) + 4 \cdot \mathbf{0}^T\mathbf{AIA0} = 2\mathrm{tr}(\mathbf{A}) = 2p, \\ \mathrm{var}(\mathbf{x}^T\mathbf{B}\mathbf{x}) &= 2 \cdot \mathrm{tr}(\mathbf{BIBI}) + 4 \cdot \mathbf{0}^T\mathbf{BIB0} = 2\mathrm{tr}(\mathbf{B}) = 2q. \end{aligned}$$

So $\mathrm{corr}(\mathbf{x}^T\mathbf{A}\mathbf{x}, \mathbf{x}^T\mathbf{B}\mathbf{x}) = 2\mathrm{tr}(\mathbf{AB})/\sqrt{4pq} = \mathrm{tr}(\mathbf{AB})/\sqrt{pq}$, which is equal to $\sqrt{q/p}$ if and only if $\mathrm{tr}(\mathbf{AB}) = q = \mathrm{rank}(\mathbf{B})$, i.e., if and only if $\mathrm{tr}(\mathbf{AB}) = \mathrm{tr}(\mathbf{B})$, i.e., if and only if $\mathrm{tr}[(\mathbf{I} - \mathbf{A})\mathbf{B}] = 0$.

(b) If $(\mathbf{I}-\mathbf{A})\mathbf{B} = \mathbf{0}$, then $\mathrm{tr}[(\mathbf{I}-\mathbf{A})\mathbf{B}] = 0$ trivially. Conversely, if $\mathrm{tr}[(\mathbf{I}-\mathbf{A})\mathbf{B}] = 0$, then

$$\mathrm{tr}\{[(\mathbf{I}-\mathbf{A})\mathbf{B}]^T[(\mathbf{I}-\mathbf{A})\mathbf{B}]\} = \mathrm{tr}[\mathbf{B}(\mathbf{I}-\mathbf{A})\mathbf{B}] = \mathrm{tr}[(\mathbf{I}-\mathbf{A})\mathbf{BB}] = \mathrm{tr}[(\mathbf{I}-\mathbf{A})\mathbf{B}] = 0.$$

By Theorem 2.10.4, $(\mathbf{I} - \mathbf{A})\mathbf{B} = \mathbf{0}$.

(c) If $(\mathbf{I} - \mathbf{A})\mathbf{B} = \mathbf{0}$ then $\mathbf{B} = \mathbf{AB}$, so $\mathbf{BA} = \mathbf{B}^T\mathbf{A}^T = (\mathbf{AB})^T = \mathbf{B}^T = \mathbf{B}$. Then

$$(\mathbf{A} - \mathbf{B})(\mathbf{A} - \mathbf{B}) = \mathbf{AA} - \mathbf{BA} - \mathbf{AB} + \mathbf{BB} = \mathbf{A} - \mathbf{B} - \mathbf{B} + \mathbf{B} = \mathbf{A} - \mathbf{B},$$

i.e., $\mathbf{A} - \mathbf{B}$ is idempotent. Furthermore, $\mathrm{rank}(\mathbf{A} - \mathbf{B}) = \mathrm{tr}(\mathbf{A} - \mathbf{B}) = \mathrm{tr}(\mathbf{A}) - \mathrm{tr}(\mathbf{B}) = p - q$. Thus, by Corollary 14.2.6.1, $\mathbf{x}^T(\mathbf{A} - \mathbf{B})\mathbf{x} \sim \chi^2(p - q)$.

▶ **Exercise 19** Suppose that $\mathbf{x} \sim \mathrm{N}_n(\boldsymbol{\mu}, \boldsymbol{\Sigma})$, where $\mathrm{rank}(\boldsymbol{\Sigma}) = n$. Partition $\mathbf{x}$, $\boldsymbol{\mu}$, and $\boldsymbol{\Sigma}$ conformably as

$$\begin{pmatrix} \mathbf{x}_1 \\ \mathbf{x}_2 \end{pmatrix}, \quad \begin{pmatrix} \boldsymbol{\mu}_1 \\ \boldsymbol{\mu}_2 \end{pmatrix}, \quad \text{and} \quad \begin{pmatrix} \boldsymbol{\Sigma}_{11} & \boldsymbol{\Sigma}_{12} \\ \boldsymbol{\Sigma}_{21} & \boldsymbol{\Sigma}_{22} \end{pmatrix},$$

where $\mathbf{x}_1$ is $n_1 \times 1$ and $\mathbf{x}_2$ is $n_2 \times 1$. Let $\mathbf{A}$ and $\mathbf{B}$ represent symmetric $n_1 \times n_1$ and $n_2 \times n_2$ matrices of constants, respectively. Determine a necessary and sufficient condition for $\mathbf{x}_1^T\mathbf{A}\mathbf{x}_1$ and $\mathbf{x}_2^T\mathbf{B}\mathbf{x}_2$ to be independent.

Solution By Corollary 14.3.1.2a, $\mathbf{x}_1^T\mathbf{A}\mathbf{x}_1 = \begin{pmatrix}\mathbf{x}_1\\ \mathbf{x}_2\end{pmatrix}^T \begin{pmatrix}\mathbf{A} & \mathbf{0}\\ \mathbf{0} & \mathbf{0}\end{pmatrix}\begin{pmatrix}\mathbf{x}_1\\ \mathbf{x}_2\end{pmatrix}$ and $\mathbf{x}_2^T\mathbf{B}\mathbf{x}_2 = \begin{pmatrix}\mathbf{x}_1\\ \mathbf{x}_2\end{pmatrix}^T \begin{pmatrix}\mathbf{0} & \mathbf{0}\\ \mathbf{0} & \mathbf{B}\end{pmatrix}\begin{pmatrix}\mathbf{x}_1\\ \mathbf{x}_2\end{pmatrix}$ are independent if and only if

$$\begin{pmatrix}\mathbf{A} & \mathbf{0}\\ \mathbf{0} & \mathbf{0}\end{pmatrix}\begin{pmatrix}\boldsymbol{\Sigma}_{11} & \boldsymbol{\Sigma}_{12}\\ \boldsymbol{\Sigma}_{21} & \boldsymbol{\Sigma}_{22}\end{pmatrix}\begin{pmatrix}\mathbf{0} & \mathbf{0}\\ \mathbf{0} & \mathbf{B}\end{pmatrix} = \mathbf{0},$$

i.e., if and only if $\mathbf{A}\boldsymbol{\Sigma}_{12}\mathbf{B} = \mathbf{0}$.

▶ **Exercise 20** Suppose that $\mathbf{x} \sim \mathrm{N}_n(\mathbf{a}, \sigma^2\mathbf{I})$ where $\mathbf{a}$ is a nonzero n-vector of constants. Let $\mathbf{b}$ represent another nonzero n-vector. Define $\mathbf{P}_{\mathbf{a}} = \mathbf{a}\mathbf{a}^T/(\mathbf{a}^T\mathbf{a})$ and $\mathbf{P}_{\mathbf{b}} = \mathbf{b}\mathbf{b}^T/(\mathbf{b}^T\mathbf{b})$.

(a) Determine the distribution of $\mathbf{x}^T\mathbf{P}_{\mathbf{a}}\mathbf{x}/\sigma^2$. (Simplify as much as possible here and in all parts of this exercise.)
(b) Determine the distribution of $\mathbf{x}^T\mathbf{P}_{\mathbf{b}}\mathbf{x}/\sigma^2$.
(c) Determine, in as simple a form as possible, a necessary and sufficient condition for the two quadratic forms in parts (a) and (b) to be independent.
(d) Under the condition in part (c), can the expressions for the parameters in either or both of the distributions in parts (a) and (b) be simplified further? If so, how?

Solution

(a) Note that $\frac{1}{\sigma}\mathbf{x} \sim \mathrm{N}(\frac{1}{\sigma}\mathbf{a}, \mathbf{I})$ and that $\mathbf{P}_{\mathbf{a}}\mathbf{I} = \mathbf{P}_{\mathbf{a}}$ is idempotent with $\mathrm{rank}(\mathbf{P}_{\mathbf{a}}) = 1$. Then by part (a) of Corollary 14.2.6.1, $\mathbf{x}^T\mathbf{P}_{\mathbf{a}}\mathbf{x}/\sigma^2 = (\frac{1}{\sigma}\mathbf{x})^T\mathbf{P}_{\mathbf{a}}(\frac{1}{\sigma}\mathbf{x}) \sim \chi^2(1, \frac{1}{2}(\frac{1}{\sigma}\mathbf{a})^T\mathbf{P}_{\mathbf{a}}(\frac{1}{\sigma}\mathbf{a}))$, i.e., $\chi^2(1, \frac{1}{2\sigma^2}\mathbf{a}^T\mathbf{a})$.
(b) Note that $\mathbf{P}_{\mathbf{b}}\mathbf{I} = \mathbf{P}_{\mathbf{b}}$ is idempotent with $\mathrm{rank}(\mathbf{P}_{\mathbf{b}}) = 1$. Then by part (a) of Corollary 14.2.6.1, $\mathbf{x}^T\mathbf{P}_{\mathbf{b}}\mathbf{x}/\sigma^2 = (\frac{1}{\sigma}\mathbf{x})^T\mathbf{P}_{\mathbf{b}}(\frac{1}{\sigma}\mathbf{x}) \sim \chi^2(1, \frac{1}{2}(\frac{1}{\sigma}\mathbf{a})^T\mathbf{P}_{\mathbf{b}}(\frac{1}{\sigma}\mathbf{a}))$, i.e., $\chi^2(1, \frac{1}{2\sigma^2}\mathbf{a}^T\mathbf{P}_{\mathbf{b}}\mathbf{a})$.
(c) By Corollary 14.3.1.2a, the two quadratic forms in parts (a) and (b) are independent if and only if $\mathbf{P}_{\mathbf{a}}\mathbf{I}\mathbf{P}_{\mathbf{b}} = \mathbf{0}$, i.e., if and only if $\mathbf{a}\mathbf{a}^T\mathbf{b}\mathbf{b}^T/(\mathbf{a}^T\mathbf{a}\mathbf{b}^T\mathbf{b}) = \mathbf{0}$, i.e., if and only if $\mathbf{a}\mathbf{a}^T\mathbf{b}\mathbf{b}^T = \mathbf{0}$. This last condition holds if $\mathbf{a}^T\mathbf{b} = 0$, and it implies (by pre-multiplication and post-multiplication of both sides by $\mathbf{a}^T$ and $\mathbf{b}^T$, respectively, and then dividing both sides by the nonzero scalars $\mathbf{a}^T\mathbf{a}$ and $\mathbf{b}^T\mathbf{b}$) that $\mathbf{a}^T\mathbf{b} = 0$. Thus a necessary and sufficient condition is that $\mathbf{a}^T\mathbf{b} = 0$, i.e., that $\mathbf{a}$ and $\mathbf{b}$ are orthogonal.
(d) If $\mathbf{a}^T\mathbf{b} = 0$, the noncentrality parameter of the distribution of $\mathbf{x}^T\mathbf{P}_{\mathbf{b}}\mathbf{x}/\sigma^2$ simplifies to 0 because $\mathbf{a}^T\mathbf{P}_{\mathbf{b}}\mathbf{a} = \mathbf{a}^T(1/\mathbf{b}^T\mathbf{b})\mathbf{b}\mathbf{b}^T\mathbf{a} = 0$. The noncentrality parameter of the distribution of $\mathbf{x}^T\mathbf{P}_{\mathbf{a}}\mathbf{x}/\sigma^2$ does not simplify further.

▶ **Exercise 21** Suppose that $\mathbf{y}$ follows the normal Aitken model $\{\mathbf{y}, \mathbf{X}\boldsymbol{\beta}, \sigma^2\mathbf{W}\}$. Let $\mathbf{A}$ represent any $n \times n$ nonnegative definite matrix such that $\mathbf{y}^T\mathbf{A}\mathbf{y}$ is an unbiased estimator of σ^2.

(a) Show that $\mathbf{y}^T\mathbf{A}\mathbf{y}$ is uncorrelated with every linear function $\mathbf{a}^T\mathbf{y}$.
(b) Suppose further that $\mathbf{a}^T\mathbf{y}$ is a BLUE of its expectation. Show that $\mathbf{y}^T\mathbf{A}\mathbf{y}$ and $\mathbf{a}^T\mathbf{y}$ are distributed independently.
(c) The unbiasedness of $\mathbf{y}^T\mathbf{A}\mathbf{y}$ for σ^2 implies that the main diagonal elements of $\mathbf{AW}$ satisfy a certain property. Give this property. If $\mathbf{W} = k\mathbf{I} + (1 - k)\mathbf{J}$ for $k \in [0, 1]$, what must k equal (in terms of $\mathbf{A}$) in order for this property to be satisfied?

Solution

(a) By Theorem 4.2.4, the unbiasedness of $\mathbf{y}^T\mathbf{A}\mathbf{y}$ for σ^2 implies that

$$\boldsymbol{\beta}^T\mathbf{X}^T\mathbf{A}\mathbf{X}\boldsymbol{\beta} + \sigma^2\text{tr}(\mathbf{AW}) = \sigma^2 \quad \text{for all } \boldsymbol{\beta} \text{ and all } \sigma^2 > 0.$$

Putting $\boldsymbol{\beta} = \mathbf{0}_p$ into this equation yields $\text{tr}(\mathbf{AW}) = 1$, and back-substituting this into the same equation yields the result

$$\boldsymbol{\beta}^T\mathbf{X}^T\mathbf{A}\mathbf{X}\boldsymbol{\beta} = 0 \quad \text{for all } \boldsymbol{\beta}.$$

Writing $\mathbf{A}$ as $\mathbf{A}^{\frac{1}{2}}\mathbf{A}^{\frac{1}{2}}$, this last result can be rewritten as $\boldsymbol{\beta}^T\mathbf{X}^T\mathbf{A}^{\frac{1}{2}}\mathbf{A}^{\frac{1}{2}}\mathbf{X}\boldsymbol{\beta} = 0$ for all $\boldsymbol{\beta}$, implying that $\mathbf{A}^{\frac{1}{2}}\mathbf{X}\boldsymbol{\beta} = \mathbf{0}$ for all $\boldsymbol{\beta}$, implying further that $\mathbf{A}^{\frac{1}{2}}\mathbf{X} = \mathbf{0}$, implying still further (by pre-multiplying both sides of the equality by $\mathbf{A}^{\frac{1}{2}}$ that $\mathbf{AX} = \mathbf{0}$. Then, by Corollary 4.2.5.1,

$$\text{cov}(\mathbf{a}^T\mathbf{y}, \mathbf{y}^T\mathbf{A}\mathbf{y}) = 2\mathbf{a}^T(\sigma^2\mathbf{W})\mathbf{A}(\mathbf{X}\boldsymbol{\beta}) = 2\sigma^2\mathbf{a}^T\mathbf{W}\mathbf{0}\boldsymbol{\beta} = 0,$$

so $\mathbf{y}^T\mathbf{A}\mathbf{y}$ is uncorrelated with every linear function $\mathbf{a}^T\mathbf{y}$.

(b) If $\mathbf{a}^T\mathbf{y}$ is a BLUE of its expectation, then by Corollary 11.2.2.2 $\mathbf{Wa} = \mathbf{Xq}$ for some p-vector $\mathbf{q}$. Thus

$$\mathbf{A}(\sigma^2\mathbf{W})\mathbf{a} = \sigma^2\mathbf{AWa} = \sigma^2\mathbf{AXq} = \mathbf{0},$$

where we used the fact, shown in part (a), that $\mathbf{AX} = \mathbf{0}$. The independence of $\mathbf{y}^T\mathbf{A}\mathbf{y}$ and $\mathbf{a}^T\mathbf{y}$ follows immediately from Corollary 14.3.1.2b.

(c) The unbiasedness of $\mathbf{y}^T\mathbf{A}\mathbf{y}$ for σ^2 implies, as shown in part (a), that $\text{tr}(\mathbf{AW}) = 1$, i.e., that the main diagonal elements of $\mathbf{AW}$ sum to one. In the special case $\mathbf{W} = k\mathbf{I} + (1 - k)\mathbf{J}$, we have

$$1 = \text{tr}\{\mathbf{A}[k\mathbf{I} + (1 - k)\mathbf{J}]\} = k\text{tr}(\mathbf{A}) + (1 - k)\text{tr}(\mathbf{AJ}),$$

implying that

$$k = \frac{1 - \text{tr}(\mathbf{AJ})}{\text{tr}(\mathbf{A}) - \text{tr}(\mathbf{AJ})}.$$

► **Exercise 22** Suppose that $\mathbf{x} \sim \text{N}_n(\mathbf{1}\mu, \sigma^2[(1-\rho)\mathbf{I} + \rho\mathbf{J}])$ where $n \geq 2$, $-\infty < \mu < \infty$, $\sigma^2 > 0$, and $0 < \rho < 1$. Define $\mathbf{A} = \frac{1}{n}\mathbf{J}$ and $\mathbf{B} = \frac{1}{n-1}(\mathbf{I} - \frac{1}{n}\mathbf{J})$.

(a) Determine $\text{var}(\mathbf{x}^T\mathbf{A}\mathbf{x})$.
(b) Determine $\text{var}(\mathbf{x}^T\mathbf{B}\mathbf{x})$.
(c) Are $\mathbf{x}^T\mathbf{A}\mathbf{x}$ and $\mathbf{x}^T\mathbf{B}\mathbf{x}$ independent? Explain.
(d) Suppose that n is even. Let $\mathbf{x}_1$ represent the vector consisting of the first $n/2$ elements of $\mathbf{x}$ and let $\mathbf{x}_2$ represent the vector consisting of the remaining $n/2$ elements of $\mathbf{x}$. Determine the distribution of $\mathbf{x}_1 - \mathbf{x}_2$.
(e) Suppose that $n = 3$, in which case $\mathbf{x} = (x_1, x_2, x_3)^T$. Determine the conditional distribution of $(x_1, x_2)^T$ given that $x_3 = 1$.

Solution

(a) First observe that $\left(\frac{1}{n}\mathbf{J}\right)\sigma^2[(1-\rho)\mathbf{I} + \rho\mathbf{J}] = \frac{\sigma^2}{n}[(1-\rho)\mathbf{J} + \rho n\mathbf{J}] = \frac{\sigma^2}{n}(1 - \rho + \rho n)\mathbf{J}$. By Corollary 4.2.6.3,

$$\begin{aligned}
\text{var}(\mathbf{x}^T\mathbf{A}\mathbf{x}) &= 2\text{tr}\left\{\left(\frac{1}{n}\mathbf{J}\right)\sigma^2[(1-\rho)\mathbf{I} + \rho\mathbf{J}]\left(\frac{1}{n}\mathbf{J}\right)\sigma^2[(1-\rho)\mathbf{I} + \rho\mathbf{J}]\right\} \\
&\quad +4(\mu\mathbf{1})^T\left(\frac{1}{n}\mathbf{J}\right)\sigma^2[(1-\rho)\mathbf{I} + \rho\mathbf{J}]\left(\frac{1}{n}\mathbf{J}\right)(\mu\mathbf{1}) \\
&= 2\left(\frac{\sigma^4}{n^2}\right)(1-\rho+\rho n)^2\text{tr}(\mathbf{JJ}) + 4\mu^2\left(\frac{\sigma^2}{n^2}\right)(1-\rho+\rho n)\mathbf{1}^T\mathbf{JJ1} \\
&= 2\sigma^4(1-\rho+\rho n)^2 + 4n\mu^2\sigma^2(1-\rho+\rho n).
\end{aligned}$$

(b) First observe that $\frac{1}{n-1}(\mathbf{I} - \frac{1}{n}\mathbf{J})\sigma^2[(1-\rho)\mathbf{I} + \rho\mathbf{J}] = \frac{\sigma^2(1-\rho)}{n-1}(\mathbf{I} - \frac{1}{n}\mathbf{J})$. By Corollary 4.2.6.3 again,

$$\begin{aligned}
\text{var}(\mathbf{x}^T\mathbf{B}\mathbf{x}) &= 2\text{tr}\left\{\frac{1}{n-1}\left(\mathbf{I} - \frac{1}{n}\mathbf{J}\right)\sigma^2[(1-\rho)\mathbf{I} + \rho\mathbf{J}]\frac{1}{n-1}\left(\mathbf{I} - \frac{1}{n}\mathbf{J}\right)\sigma^2[(1-\rho)\mathbf{I} + \rho\mathbf{J}]\right\} \\
&\quad +4(\mu\mathbf{1})^T\frac{1}{n-1}\left(\mathbf{I} - \frac{1}{n}\mathbf{J}\right)\sigma^2[(1-\rho)\mathbf{I} + \rho\mathbf{J}]\frac{1}{n-1}\left(\mathbf{I} - \frac{1}{n}\mathbf{J}\right)(\mu\mathbf{1}) \\
&= \frac{2\sigma^4(1-\rho)^2}{(n-1)^2}\text{tr}\left[\left(\mathbf{I} - \frac{1}{n}\mathbf{J}\right)\left(\mathbf{I} - \frac{1}{n}\mathbf{J}\right)\right] + \frac{4\mu^2\sigma^2(1-\rho)}{(n-1)^2}\mathbf{1}^T\left(\mathbf{I} - \frac{1}{n}\mathbf{J}\right)\mathbf{1} \\
&= \frac{2\sigma^4(1-\rho)^2}{n-1}.
\end{aligned}$$

(c) Using the result established at the outset of the solution to part (a), we observe that

$$\left(\frac{1}{n}\mathbf{J}\right)\sigma^2[(1-\rho)\mathbf{I}+\rho\mathbf{J}]\frac{1}{n-1}\left(\mathbf{I}-\frac{1}{n}\mathbf{J}\right) = \frac{\sigma^2(1-\rho+\rho n)}{n(n-1)}\mathbf{J}\left(\mathbf{I}-\frac{1}{n}\mathbf{J}\right) = \mathbf{0}.$$

Therefore, by Corollary 14.3.1.2a, $\mathbf{x}^T\mathbf{A}\mathbf{x}$ and $\mathbf{x}^T\mathbf{B}\mathbf{x}$ are independent.

(d) Observe that $\mathbf{x}_1 - \mathbf{x}_2 = \begin{pmatrix} \mathbf{I}_{n/2} & -\mathbf{I}_{n/2} \end{pmatrix} \begin{pmatrix} \mathbf{x}_1 \\ \mathbf{x}_2 \end{pmatrix}$. Therefore, by Theorem 14.1.6, $\mathbf{x}_1 - \mathbf{x}_2$ has a normal distribution with mean vector

$$\begin{pmatrix} \mathbf{I}_{n/2} & -\mathbf{I}_{n/2} \end{pmatrix} \begin{pmatrix} \mu\mathbf{1}_{n/2} \\ \mu\mathbf{1}_{n/2} \end{pmatrix} = \mu\mathbf{1}_{n/2} - \mu\mathbf{1}_{n/2} = \mathbf{0}$$

and variance–covariance matrix

$$\begin{pmatrix} \mathbf{I}_{n/2} & -\mathbf{I}_{n/2} \end{pmatrix} \begin{pmatrix} \sigma^2[(1-\rho)\mathbf{I}_{n/2}+\rho\mathbf{J}_{n/2}] & \sigma^2\rho\mathbf{J}_{n/2} \\ \sigma^2\rho\mathbf{J}_{n/2} & \sigma^2[(1-\rho)\mathbf{I}_{n/2}+\rho\mathbf{J}_{n/2}] \end{pmatrix} \begin{pmatrix} \mathbf{I}_{n/2} \\ -\mathbf{I}_{n/2} \end{pmatrix}$$
$$= \sigma^2[(1-\rho)\mathbf{I}_{n/2}+\rho\mathbf{J}_{n/2}] + \sigma^2[(1-\rho)\mathbf{I}_{n/2}+\rho\mathbf{J}_{n/2}] - 2\sigma^2\rho\mathbf{J}_{n/2}$$
$$= 2\sigma^2(1-\rho)\mathbf{I}_{n/2}.$$

(e) By Theorem 14.1.9, the conditional distribution of $(x_1, x_2)^T$, given that $x_3 = 1$, is normal with mean vector

$$\mu\mathbf{1}_2 + \sigma^2\rho\mathbf{1}_2(\sigma^2)^{-1}(1-\mu) = [\mu + \rho(1-\mu)]\mathbf{1}_2$$

and variance–covariance matrix

$$\sigma^2[(1-\rho)\mathbf{I}_2 + \rho\mathbf{J}_2] - \sigma^2\rho\mathbf{1}_2(\sigma^2)^{-1}\sigma^2\rho\mathbf{1}_2^T = \sigma^2[(1-\rho)\mathbf{I}_2 + (\rho-\rho^2)\mathbf{J}_2].$$

▶ **Exercise 23** Suppose that $\mathbf{x} \sim \mathrm{N}_n(\boldsymbol{\mu}, c\mathbf{I}+\mathbf{b}\mathbf{1}_n^T+\mathbf{1}_n\mathbf{b}^T)$, where $c\mathbf{I}+\mathbf{b}\mathbf{1}_n^T+\mathbf{1}_n\mathbf{b}^T$ is positive definite.

(a) Determine the distribution of $\sum_{i=1}^n (x_i - \bar{x})^2$ (suitably scaled), and fully specify the parameters of that distribution.
(b) Determine, in as simple a form as possible, a necessary and sufficient condition on $\mathbf{b}$ for $\sum_{i=1}^n (x_i - \bar{x})^2$ and $\bar{x}^2$ to be independent.

Solution

(a) Let $\boldsymbol{\Sigma} \equiv c\mathbf{I} + \mathbf{b}\mathbf{1}_n^T + \mathbf{1}_n\mathbf{b}^T$. Then $\sum_{i=1}^n (x_i - \bar{x})^2 = \mathbf{x}^T(\mathbf{I} - \frac{1}{n}\mathbf{J})\mathbf{x} \equiv \mathbf{x}^T\mathbf{A}\mathbf{x}$, say, where

$$\mathbf{A}\boldsymbol{\Sigma} = (\mathbf{I} - \frac{1}{n}\mathbf{J})(c\mathbf{I} + \mathbf{b}\mathbf{1}_n^T + \mathbf{1}_n\mathbf{b}^T) = c(\mathbf{I} - \frac{1}{n}\mathbf{J})(\mathbf{I} + \frac{1}{c}\mathbf{b}\mathbf{1}_n^T).$$

Now,

$$\begin{aligned}\left(\frac{1}{c}\mathbf{A}\boldsymbol{\Sigma}\right)\left(\frac{1}{c}\mathbf{A}\boldsymbol{\Sigma}\right) &= (\mathbf{I} - \frac{1}{n}\mathbf{J})[(\mathbf{I} + \frac{1}{c}\mathbf{b}\mathbf{1}_n^T)(\mathbf{I} - \frac{1}{n}\mathbf{J})](\mathbf{I} + \frac{1}{c}\mathbf{b}\mathbf{1}_n^T)\\ &= (\mathbf{I} - \frac{1}{n}\mathbf{J})(\mathbf{I} - \frac{1}{n}\mathbf{J}_n)(\mathbf{I} + \frac{1}{c}\mathbf{b}\mathbf{1}^T)\\ &= (\mathbf{I} - \frac{1}{n}\mathbf{J})(\mathbf{I} + \frac{1}{c}\mathbf{b}\mathbf{1}_n^T).\end{aligned}$$

Thus, by Corollary 14.2.6.1, $\frac{1}{c}\sum_{i=1}^n (x_i - \bar{x})^2 \sim \chi^2(df, NCP)$ where

$$df = \text{rank}\left(\frac{1}{c}\mathbf{A}\right) = \text{rank}(\mathbf{I} - \frac{1}{n}\mathbf{J}) = n - 1,$$

$$NCP = \frac{1}{2c}\sum_{i=1}^n [\mathrm{E}(x_i) - \mathrm{E}(\bar{x})]^2 = \frac{1}{2c}\sum_{i=1}^n (\mu_i - \bar{\mu})^2,$$

where $\boldsymbol{\mu} = (\mu_i)$.

(b) $\bar{x}^2 = (\frac{1}{n}\mathbf{1}^T\mathbf{x})^2 = \frac{1}{n^2}\mathbf{x}^T\mathbf{J}\mathbf{x}$. Thus, by Corollary 14.3.1.2a, $\sum_{i=1}^n (x_i - \bar{x})^2$ and $\bar{x}^2$ are independent if and only if $(\mathbf{I} - \frac{1}{n}\mathbf{J})(\mathbf{I} + \frac{1}{c}\mathbf{b}\mathbf{1}_n^T)\mathbf{J} = \mathbf{0}$, i.e., if and only if $(\mathbf{I} - \frac{1}{n}\mathbf{J})\mathbf{b}\mathbf{1}_n^T\mathbf{J} = \mathbf{0}$, i.e., if and only if $n(\mathbf{I} - \frac{1}{n}\mathbf{1}\mathbf{1}^T)\mathbf{b}\mathbf{1}_n^T = \mathbf{0}$, i.e., if and only if $\mathbf{b} = a\mathbf{1}$ for some $a \in \mathbb{R}$.

▶ **Exercise 24** Suppose that

$$\begin{pmatrix}\mathbf{x}_1\\ \mathbf{x}_2\end{pmatrix} \sim \mathrm{N}\left(\begin{pmatrix}\boldsymbol{\mu}_1\\ \boldsymbol{\mu}_2\end{pmatrix}, \begin{pmatrix}\boldsymbol{\Sigma}_{11} & \boldsymbol{\Sigma}_{12}\\ \boldsymbol{\Sigma}_{21} & \boldsymbol{\Sigma}_{22}\end{pmatrix}\right),$$

where $\mathbf{x}_1$ and $\boldsymbol{\mu}_1$ are n_1-vectors, $\mathbf{x}_2$ and $\boldsymbol{\mu}_2$ are n_2-vectors, $\boldsymbol{\Sigma}_{11}$ is $n_1 \times n_1$, and $\boldsymbol{\Sigma}$ is nonsingular. Let $\mathbf{A}$ represent an $n_1 \times n_1$ symmetric matrix of constants and let $\mathbf{B}$ represent a matrix of constants having n_2 columns.

(a) Determine, in as simple a form as possible, a necessary and sufficient condition for $\mathbf{x}_1^T\mathbf{A}\mathbf{x}_1$ and $\mathbf{B}\mathbf{x}_2$ to be uncorrelated.
(b) Determine, in as simple a form as possible, a necessary and sufficient condition for $\mathbf{x}_1^T\mathbf{A}\mathbf{x}_1$ and $\mathbf{B}\mathbf{x}_2$ to be independent.

(c) Using your answer to part (a), give a necessary and sufficient condition for $\mathbf{x}_1^T\mathbf{x}_1$ and $\mathbf{x}_2$ to be uncorrelated.
(d) Using your answer to part (b), give a necessary and sufficient condition for $\mathbf{x}_1^T\mathbf{x}_1$ and $\mathbf{x}_2$ to be independent.
(e) Consider the special case in which $\mathbf{\Sigma} = (1-\rho)\mathbf{I}+\rho\mathbf{J}$, where $\frac{-1}{n-1} < \rho < 1$. Can $\mathbf{x}_1^T\mathbf{x}_1$ and $\mathbf{x}_2$ be uncorrelated? Can $\mathbf{x}_1^T\mathbf{x}_1$ and $\mathbf{x}_2$ be independent? If the answer to either or both of these questions is yes, give a necessary and sufficient condition for the result to hold.

Solution

(a) By Corollary 4.2.5.1, $\mathbf{x}_1^T\mathbf{A}\mathbf{x}_1$ and $\mathbf{B}\mathbf{x}_2$ are uncorrelated if and only if

$$2\begin{pmatrix}\mathbf{0} & \mathbf{B}\end{pmatrix}\begin{pmatrix}\mathbf{\Sigma}_{11} & \mathbf{\Sigma}_{12} \\ \mathbf{\Sigma}_{21} & \mathbf{\Sigma}_{22}\end{pmatrix}\begin{pmatrix}\mathbf{A} & \mathbf{0} \\ \mathbf{0} & \mathbf{0}\end{pmatrix}\begin{pmatrix}\boldsymbol{\mu}_1 \\ \boldsymbol{\mu}_2\end{pmatrix} = \mathbf{0},$$

i.e., if and only if $\mathbf{B}\mathbf{\Sigma}_{21}\mathbf{A}\boldsymbol{\mu}_1 = \mathbf{0}$.
(b) By Corollary 14.3.1.2b, $\mathbf{x}_1^T\mathbf{A}\mathbf{x}_1$ and $\mathbf{B}\mathbf{x}_2$ are independent if and only if

$$\begin{pmatrix}\mathbf{A} & \mathbf{0} \\ \mathbf{0} & \mathbf{0}\end{pmatrix}\begin{pmatrix}\mathbf{\Sigma}_{11} & \mathbf{\Sigma}_{12} \\ \mathbf{\Sigma}_{21} & \mathbf{\Sigma}_{22}\end{pmatrix}\begin{pmatrix}\mathbf{0} & \mathbf{B}\end{pmatrix}^T = \mathbf{0},$$

i.e., if and only if $\mathbf{A}\mathbf{\Sigma}_{12}\mathbf{B}^T = \mathbf{0}$, or equivalently, if and only if $\mathbf{B}\mathbf{\Sigma}_{21}\mathbf{A} = \mathbf{0}$.
(c) By part (a) with $\mathbf{A} = \mathbf{I}_{n_1}$ and $\mathbf{B} = \mathbf{I}_{n_2}$, $\mathbf{x}_1^T\mathbf{x}_1$ and $\mathbf{x}_2$ are uncorrelated if and only if $\mathbf{\Sigma}_{21}\boldsymbol{\mu}_1 = \mathbf{0}$.
(d) By part (b) with $\mathbf{A} = \mathbf{I}_{n_1}$ and $\mathbf{B} = \mathbf{I}_{n_2}$, $\mathbf{x}_1^T\mathbf{x}_1$ and $\mathbf{x}_2$ are independent if and only if $\mathbf{\Sigma}_{12} = \mathbf{0}$.
(e) If $\mathbf{\Sigma} = (1-\rho)\mathbf{I} + \rho\mathbf{J}$, where $0 < \rho < 1$, then the necessary and sufficient condition for uncorrelatedness in part (c) may be written as $\rho\mathbf{J}_{n_2 \times n_1}\boldsymbol{\mu}_1 = \mathbf{0}$, or equivalently as $\mathbf{1}_{n_1}^T\boldsymbol{\mu}_1 = 0$. However, the necessary and sufficient condition for independence in part (d) is not satisfied because $\rho\mathbf{J}_{n_1 \times n_2} \neq \mathbf{0}$.

► **Exercise 25** Suppose that $\mathbf{x} \sim \mathrm{N}_n(\boldsymbol{\mu}, \mathbf{\Sigma})$, where $\mathrm{rank}(\mathbf{\Sigma}) = n$. Let $\mathbf{A}$ represent a nonnull symmetric idempotent matrix such that $\mathbf{A}\mathbf{\Sigma} = k\mathbf{A}$ for some $k \neq 0$. Show that:

(a) $\mathbf{x}^T\mathbf{A}\mathbf{x}/k \sim \chi^2(\nu, \lambda)$ for some ν and λ, and determine ν and λ.
(b) $\mathbf{x}^T\mathbf{A}\mathbf{x}$ and $\mathbf{x}^T(\mathbf{I} - \mathbf{A})\mathbf{x}$ are independent.
(c) $\mathbf{A}\mathbf{x}$ and $(\mathbf{I} - \mathbf{A})\mathbf{x}$ are independent.
(d) $k = \mathrm{tr}(\mathbf{A}\mathbf{\Sigma}\mathbf{A})/\mathrm{rank}(\mathbf{A})$.

Solution

(a) Observe that $(\frac{1}{k}\mathbf{A})\boldsymbol{\Sigma} = (\frac{1}{k})(k\mathbf{A}) = \mathbf{A}$, which is idempotent. Thus, by Corollary 14.2.6.1, $\mathbf{x}^T\mathbf{A}\mathbf{x}/k = \mathbf{x}^T(\frac{1}{k}\mathbf{A})\mathbf{x} \sim \chi^2(df, NCP)$ where $df = \text{rank}[(\frac{1}{k}\mathbf{A})(k\mathbf{A})] = \text{rank}(\mathbf{A})$ and $NCP = \frac{1}{2}\boldsymbol{\mu}^T(\frac{1}{k}\mathbf{A})\boldsymbol{\mu} = \frac{1}{2k}\boldsymbol{\mu}^T\mathbf{A}\boldsymbol{\mu}$.
(b) $\mathbf{A}\boldsymbol{\Sigma}(\mathbf{I}-\mathbf{A}) = k\mathbf{A}(\mathbf{I}-\mathbf{A}) = k(\mathbf{A}-\mathbf{A}) = \mathbf{0}$ because $\mathbf{A}$ is idempotent. Thus, by Corollary 14.3.1.2a, $\mathbf{x}^T\mathbf{A}\mathbf{x}$ and $\mathbf{x}^T(\mathbf{I}-\mathbf{A})\mathbf{x}$ are independent.
(c) As shown in the solution to part (b), $\mathbf{A}\boldsymbol{\Sigma}(\mathbf{I}-\mathbf{A}) = \mathbf{0}$. Thus, by Corollary 14.3.1.2c, $\mathbf{A}\mathbf{x}$ and $(\mathbf{I}-\mathbf{A})\mathbf{x}$ are independent.
(d) $\text{tr}(\mathbf{A}\boldsymbol{\Sigma}\mathbf{A}) = \text{tr}[(k\mathbf{A})\mathbf{A}] = k\text{tr}(\mathbf{A}) = k\text{rank}(\mathbf{A})$ because $\mathbf{A}$ is idempotent.

▶ **Exercise 26** Suppose that $\mathbf{x} \sim \text{N}_n(\mu\mathbf{1}, \boldsymbol{\Sigma})$, where μ is an unknown parameter, $\boldsymbol{\Sigma} = \mathbf{I} + \mathbf{C}\mathbf{C}^T$ where $\mathbf{C}$ is known, $\mathbf{C} \neq \mathbf{0}$, and $\mathbf{C}^T\mathbf{1} = \mathbf{0}$. Let $u_1 = \mathbf{x}^T\boldsymbol{\Sigma}^{-1}\mathbf{x}$ and let $u_i = \mathbf{x}^T\mathbf{A}_i\mathbf{x}$ $(i = 2, 3)$ where $\mathbf{A}_2$ and $\mathbf{A}_3$ are nonnull symmetric matrices.

(a) Prove or disprove: There is no nonnull symmetric matrix $\mathbf{A}_2$ for which u_1 and u_2 are independent.
(b) Find two nonnull symmetric matrices $\mathbf{A}_2$ and $\mathbf{A}_3$ such that u_2 and u_3 are independent.
(c) Determine which, if either, of u_2 and u_3 [as you defined them in part (b)] have noncentral chi-square distributions. If either of them does have such a distribution, give the parameters of the distribution in as simple a form as possible.

Solution

(a) By Corollary 14.3.1.2a, u_1 and u_2 are independent if and only if $\boldsymbol{\Sigma}^{-1}\boldsymbol{\Sigma}\mathbf{A}_2 = \mathbf{0}$, i.e., if and only if $\mathbf{A}_2 = \mathbf{0}$. But $\mathbf{A}_2 \neq \mathbf{0}$ by hypothesis, so u_1 and u_2 cannot be independent.
(b) By Corollary 14.3.1.2a, u_1 and u_2 are independent if and only if $\mathbf{A}_2(\mathbf{I} + \mathbf{C}\mathbf{C}^T)\mathbf{A}_3 = \mathbf{0}$, i.e., if and only if $\mathbf{A}_2\mathbf{A}_3 + \mathbf{A}_2\mathbf{C}\mathbf{C}^T\mathbf{A}_3 = \mathbf{0}$. The second summand on the left-hand side of this equality is equal to $\mathbf{0}$ if we take $\mathbf{A}_2 = \mathbf{I} - \mathbf{P}_\mathbf{C}$, and then we can make the first summand equal to $\mathbf{0}$ by taking $\mathbf{A}_3 = \mathbf{P}_\mathbf{C}$.
(c) $\mathbf{A}_2\boldsymbol{\Sigma} = (\mathbf{I} - \mathbf{P}_\mathbf{C})(\mathbf{I} + \mathbf{C}\mathbf{C}^T) = \mathbf{I} - \mathbf{P}_\mathbf{C}$ which is idempotent, so by Corollary 14.2.6.1 $u_2 \sim \chi^2[n - \text{rank}(\mathbf{C}), NCP]$ where $NCP = \frac{1}{2}(\mu\mathbf{1})^T(\mathbf{I} - \mathbf{P}_\mathbf{C})(\mu\mathbf{1}) = (n/2)\mu^2$ because $\mathbf{C}^T\mathbf{1} = \mathbf{0}$. Furthermore, $\mathbf{A}_3\boldsymbol{\Sigma} = \mathbf{P}_\mathbf{C}(\mathbf{I} + \mathbf{C}\mathbf{C}^T) = \mathbf{P}_\mathbf{C} + \mathbf{C}\mathbf{C}^T$, which generally is not idempotent because $(\mathbf{P}_\mathbf{C} + \mathbf{C}\mathbf{C}^T)(\mathbf{P}_\mathbf{C} + \mathbf{C}\mathbf{C}^T) = \mathbf{P}_\mathbf{C} + \mathbf{C}\mathbf{C}^T + \mathbf{C}\mathbf{C}^T + \mathbf{C}\mathbf{C}^T\mathbf{C}\mathbf{C}^T$. Thus u_3 generally does not have a chi-square distribution.

▶ **Exercise 27** Suppose that the model $\{\mathbf{y}, \mathbf{X}\boldsymbol{\beta}\}$ is fit to $n > \text{rank}(\mathbf{X})$ observations by ordinary least squares. Let $\mathbf{c}^T\boldsymbol{\beta}$ be an estimable function under the model, and let $\hat{\boldsymbol{\beta}}$ be a solution to the normal equations. Furthermore, let $\hat{\sigma}^2$ denote the residual mean square from the fit, i.e., $\hat{\sigma}^2 = \mathbf{y}^T(\mathbf{I} - \mathbf{P}_\mathbf{X})\mathbf{y}/[n - \text{rank}(\mathbf{X})]$. Now, suppose that the distribution of $\mathbf{y}$ is $\mathrm{N}(\mathbf{X}\boldsymbol{\beta}, \boldsymbol{\Sigma})$ where $\boldsymbol{\beta}$ is an unknown vector of parameters and $\boldsymbol{\Sigma}$ is a specified positive definite matrix having one of the following four forms:

- Form 1: $\boldsymbol{\Sigma} = \sigma^2\mathbf{I}$.
- Form 2: $\boldsymbol{\Sigma} = \sigma^2(\mathbf{I} + \mathbf{P_X}\mathbf{A}\mathbf{P_X})$ for an arbitrary $n \times n$ matrix $\mathbf{A}$.
- Form 3: $\boldsymbol{\Sigma} = \sigma^2[\mathbf{I} + (\mathbf{I} - \mathbf{P_X})\mathbf{B}(\mathbf{I} - \mathbf{P_X})]$ for an arbitrary $n \times n$ matrix $\mathbf{B}$.
- Form 4: $\boldsymbol{\Sigma} = \sigma^2[\mathbf{I} + \mathbf{P_X}\mathbf{A}\mathbf{P_X} + (\mathbf{I} - \mathbf{P_X})\mathbf{B}(\mathbf{I} - \mathbf{P_X})]$ for arbitrary $n \times n$ matrices $\mathbf{A}$ and $\mathbf{B}$.

(a) For which of the four forms of $\boldsymbol{\Sigma}$, if any, will $\mathbf{c}^T\hat{\boldsymbol{\beta}}$ and the fitted residuals vector $\hat{\mathbf{e}}$ be independent? Justify your answer.
(b) For which of the four forms of $\boldsymbol{\Sigma}$, if any, will $[n - \mathrm{rank}(\mathbf{X})]\hat{\sigma}^2/\sigma^2$ have a noncentral chi-square distribution? Justify your answer, and for those forms for which the quantity does have a noncentral chi-square distribution, give the parameters of the distribution.
(c) Suppose that the ordinary least squares estimator of every estimable function $\mathbf{c}^T\boldsymbol{\beta}$ is a BLUE of that function. Will $\mathbf{c}^T\hat{\boldsymbol{\beta}}$ and $\hat{\mathbf{e}}$ be independent for every estimable function $\mathbf{c}^T\boldsymbol{\beta}$? Will $[n - \mathrm{rank}(\mathbf{X})]\hat{\sigma}^2/\sigma^2$ have a noncentral chi-square distribution? Justify your answers.

Solution

(a) Let $\mathbf{a}$ represent a vector such that $\mathbf{c}^T = \mathbf{a}^T\mathbf{X}$ (such a vector exists because $\mathbf{c}^T\boldsymbol{\beta}$ is estimable). Then $\mathbf{c}^T\hat{\boldsymbol{\beta}} = \mathbf{a}^T\mathbf{X}\hat{\boldsymbol{\beta}} = \mathbf{a}^T\mathbf{P_X}\mathbf{y}$, and $\hat{\mathbf{e}} = (\mathbf{I} - \mathbf{P_X})\mathbf{y}$. So $\mathbf{c}^T\hat{\boldsymbol{\beta}}$ and $\hat{\mathbf{e}}$ will be independent if and only if $\mathbf{a}^T\mathbf{P_X}\boldsymbol{\Sigma}(\mathbf{I} - \mathbf{P_X}) = \mathbf{0}^T$.
For Form 1,

$$\mathbf{a}^T\mathbf{P_X}(\sigma^2\mathbf{I})(\mathbf{I} - \mathbf{P_X}) = \sigma^2\mathbf{a}^T\mathbf{P_X}(\mathbf{I} - \mathbf{P_X}) = \mathbf{0}^T.$$

For Form 2,

$$\mathbf{a}^T\mathbf{P_X}[\sigma^2(\mathbf{I} + \mathbf{P_X}\mathbf{A}\mathbf{P_X})](\mathbf{I} - \mathbf{P_X}) = \sigma^2\mathbf{a}^T\mathbf{P_X}(\mathbf{I} - \mathbf{P_X}) = \mathbf{0}^T.$$

For Form 3,

$$\mathbf{a}^T\mathbf{P_X}\sigma^2[\mathbf{I} + (\mathbf{I} - \mathbf{P_X})\mathbf{B}(\mathbf{I} - \mathbf{P_X})](\mathbf{I} - \mathbf{P_X}) = \mathbf{0}^T.$$

For Form 4,

$$\mathbf{a}^T\mathbf{P_X}\sigma^2[\mathbf{I} + \mathbf{P_X}\mathbf{A}\mathbf{P_X} + (\mathbf{I} - \mathbf{P_X})\mathbf{B}(\mathbf{I} - \mathbf{P_X})](\mathbf{I} - \mathbf{P_X}) = \mathbf{0}^T.$$

Therefore, $\mathbf{c}^T\hat{\boldsymbol{\beta}}$ and $\hat{\mathbf{e}}$ will be independent for all four forms of $\boldsymbol{\Sigma}$.
(b) $[n - \mathrm{rank}(\mathbf{X})]\hat{\sigma}^2/\sigma^2$ will have a noncentral χ^2 distribution if and only if $(1/\sigma^2)(\mathbf{I} - \mathbf{P_X})\boldsymbol{\Sigma}$ is idempotent.
For Form 1, $(1/\sigma^2)(\mathbf{I} - \mathbf{P_X})\sigma^2\mathbf{I} = \mathbf{I} - \mathbf{P_X}$, which is idempotent.

For Form 2, $(1/\sigma^2)(\mathbf{I}-\mathbf{P_X})\sigma^2(\mathbf{I}+\mathbf{P_X}\mathbf{A}\mathbf{P_X}) = \mathbf{I}-\mathbf{P_X}$, which is idempotent.
For Form 3, $(1/\sigma^2)(\mathbf{I}-\mathbf{P_X})\sigma^2[\mathbf{I}+(\mathbf{I}-\mathbf{P_X})\mathbf{B}(\mathbf{I}-\mathbf{P_X})] = (\mathbf{I}-\mathbf{P_X}) + (\mathbf{I}-\mathbf{P_X})\mathbf{B}(\mathbf{I}-\mathbf{P_X})$, which is generally not idempotent.
For Form 4, $(1/\sigma^2)(\mathbf{I}-\mathbf{P_X})\sigma^2[\mathbf{I}+\mathbf{P_X}\mathbf{A}\mathbf{P_X}+(\mathbf{I}-\mathbf{P_X})\mathbf{B}(\mathbf{I}-\mathbf{P_X})] = (\mathbf{I}-\mathbf{P_X}) + (\mathbf{I}-\mathbf{P_X})\mathbf{B}(\mathbf{I}-\mathbf{P_X})$, which is generally not idempotent. Thus, $[n-\text{rank}(\mathbf{X})]\hat{\sigma}^2/\sigma^2$ will have a noncentral χ^2 distribution only for Forms 1 and 2. For both of those forms, $[n-\text{rank}(\mathbf{X})]\hat{\sigma}^2/\sigma^2$ will have a $\chi^2[\text{rank}(\mathbf{I}-\mathbf{P_X}), \boldsymbol{\beta}^T\mathbf{X}^T(\mathbf{I}-\mathbf{P_X})\mathbf{X}\boldsymbol{\beta}/2]$ distribution, i.e., $\chi^2[n-\text{rank}(\mathbf{X})]$.

(c) By Theorem 11.3.2, $\mathbf{W}$ has Form 4. Thus, $\mathbf{c}^T\hat{\boldsymbol{\beta}}$ and $\hat{\mathbf{e}}$ will be independent for every estimable function $\mathbf{c}^T\boldsymbol{\beta}$ by part (a), but $[n-\text{rank}(\mathbf{X})]\hat{\sigma}^2/\sigma^2$ will not necessarily have a noncentral χ^2 distribution.

▶ **Exercise 28** Suppose that $w \sim \text{F}(\nu_1, \nu_2, \lambda)$, where $\nu_2 > 2$. Find $\text{E}(w)$.

Solution By Definition 14.5.1, w has the same distribution as $\frac{u_1/\nu_1}{u_2/\nu_2}$ where u_1 and u_2 are distributed independently as $\chi^2(\nu_1, \lambda)$ and $\chi^2(\nu_2)$, respectively. Thus, by Theorem 14.2.7 and the discussion immediately after Corollary 14.2.3.1,

$$\text{E}(w) = \text{E}\left(\frac{u_1/\nu_1}{u_2/\nu_2}\right) = (\nu_2/\nu_1)\text{E}(u_1)\text{E}\left(\frac{1}{u_2}\right) = \frac{\nu_2[1+(2\lambda/\nu_1)]}{\nu_2-2}.$$

▶ **Exercise 29** Prove Theorem 14.5.1: If $t \sim \text{t}(\nu, \mu)$, then $t^2 \sim \text{F}(1, \nu, \mu^2/2)$.

Solution If $t \sim t(\nu, \mu)$, then by definition $t \sim \frac{z}{\sqrt{u/\nu}}$ where $z \sim \text{N}(\mu, 1)$, $u \sim \chi^2(\nu)$, and z and u are independent. By Theorem 14.2.3, $z^2 \sim \chi^2(1, \mu^2/2)$. Thus $t^2 \sim \frac{z^2}{u/\nu}$ where $z^2 \sim \chi^2(1, \mu/2)$, $u \sim \chi^2(\nu)$, and z^2 and u are independent, implying that $t^2 \sim \text{F}(1, \nu, \mu^2/2)$ by Definition 14.5.1.

▶ **Exercise 30** Prove Theorem 14.5.5: Consider the prediction of a vector of s predictable functions $\boldsymbol{\tau} = \mathbf{C}^T\boldsymbol{\beta} + \mathbf{u}$ under the normal prediction-extended positive definite Aitken model $\left\{\begin{pmatrix}\mathbf{y}\\ \mathbf{u}\end{pmatrix}, \begin{pmatrix}\mathbf{X}\boldsymbol{\beta}\\ \mathbf{0}\end{pmatrix}, \sigma^2\begin{pmatrix}\mathbf{W} & \mathbf{K}\\ \mathbf{K}^T & \mathbf{H}\end{pmatrix}\right\}$. Recall from Theorem 13.2.1 that $\tilde{\boldsymbol{\tau}} = \mathbf{C}^T\tilde{\boldsymbol{\beta}} + \mathbf{K}^T\mathbf{E}\mathbf{y}$ is the vector of BLUPs of the elements of $\boldsymbol{\tau}$, where $\tilde{\boldsymbol{\beta}}$ is any solution to the Aitken equations and $\mathbf{E} = \mathbf{W}^{-1} - \mathbf{W}^{-1}\mathbf{X}(\mathbf{X}^T\mathbf{W}^{-1}\mathbf{X})^-\mathbf{X}^T\mathbf{W}^{-1}$, and that $\text{var}(\tilde{\boldsymbol{\tau}} - \boldsymbol{\tau}) = \sigma^2\mathbf{Q}$, where an expression for $\mathbf{Q}$ was given in Theorem 13.2.2a. If $\mathbf{Q}$ is positive definite (as is the case under either of the conditions of Theorem 13.2.2b), then

$$\frac{(\tilde{\boldsymbol{\tau}}-\boldsymbol{\tau})^T\mathbf{Q}^{-1}(\tilde{\boldsymbol{\tau}}-\boldsymbol{\tau})}{s\tilde{\sigma}^2} \sim \text{F}(s, n-p^*).$$

If $s = 1$, then

$$\frac{\tilde{\tau} - \tau}{\tilde{\sigma}\sqrt{Q}} \sim \mathrm{t}(n - p^*),$$

where Q is the scalar version of $\mathbf{Q}$.

Solution First, by (14.6) $\tilde{\boldsymbol{\tau}} - \boldsymbol{\tau} \sim \mathrm{N}(\mathbf{0}, \sigma^2\mathbf{Q})$, so by Theorem 14.2.4

$$(\tilde{\boldsymbol{\tau}} - \boldsymbol{\tau})^T(\sigma^2\mathbf{Q})^{-1}(\tilde{\boldsymbol{\tau}} - \boldsymbol{\tau}) \sim \chi^2(s).$$

Second, as noted in Example 14.2-3,

$$(n - p^*)\tilde{\sigma}^2/\sigma^2 \sim \chi^2(n - p^*).$$

Third, as noted in Example 14.1-3, $\tilde{\boldsymbol{\tau}} - \boldsymbol{\tau}$ and $\tilde{\mathbf{e}}$ are independent, implying that $\tilde{\boldsymbol{\tau}} - \boldsymbol{\tau}$ and $\tilde{\mathbf{e}}^T\mathbf{W}^{-1}\tilde{\mathbf{e}}/(n - p^*) = \tilde{\sigma}^2$ are independent. Therefore,

$$\frac{(\tilde{\boldsymbol{\tau}} - \boldsymbol{\tau})^T(\sigma^2\mathbf{Q})^{-1}(\tilde{\boldsymbol{\tau}} - \boldsymbol{\tau})/s}{\frac{(n-p^*)\tilde{\sigma}^2/\sigma^2}{n-p^*}} = \frac{(\tilde{\boldsymbol{\tau}} - \boldsymbol{\tau})^T\mathbf{Q}^{-1}(\tilde{\boldsymbol{\tau}} - \boldsymbol{\tau})}{s\tilde{\sigma}^2} \sim \mathrm{F}(s, n - p^*).$$

When $s = 1$, this result simplifies to

$$\frac{\tilde{\tau} - \tau}{\tilde{\sigma}\sqrt{q}} \sim \mathrm{t}(n - p^*)$$

by Theorem 14.5.1.

▶ **Exercise 31** Let $\boldsymbol{\mu}_1$ and $\boldsymbol{\mu}_2$ be orthonormal nonrandom nonnull n-vectors. Suppose that $\mathbf{x} \sim \mathrm{N}_n(\boldsymbol{\mu}_1, \sigma^2\mathbf{I})$. Determine the distribution of $(\mathbf{x}^T\boldsymbol{\mu}_1\boldsymbol{\mu}_1^T\mathbf{x})/(\mathbf{x}^T\boldsymbol{\mu}_2\boldsymbol{\mu}_2^T\mathbf{x})$.

Solution

$$\left(\frac{1}{\sigma^2}\boldsymbol{\mu}_1\boldsymbol{\mu}_1^T\right)(\sigma^2\mathbf{I})\left(\frac{1}{\sigma^2}\boldsymbol{\mu}_1\boldsymbol{\mu}_1^T\right)(\sigma^2\mathbf{I}) = \boldsymbol{\mu}_1(\boldsymbol{\mu}_1^T\boldsymbol{\mu}_1)\boldsymbol{\mu}_1^T = \boldsymbol{\mu}_1\boldsymbol{\mu}_1^T = \left(\frac{1}{\sigma^2}\boldsymbol{\mu}_1\boldsymbol{\mu}_1^T\right)(\sigma^2\mathbf{I}),$$

i.e., $(\frac{1}{\sigma^2}\boldsymbol{\mu}_1\boldsymbol{\mu}_1^T)(\sigma^2\mathbf{I})$ is idempotent. Thus by Corollary 14.2.6.1, $\mathbf{x}^T(\frac{1}{\sigma^2}\boldsymbol{\mu}_1\boldsymbol{\mu}_1^T)\mathbf{x} \sim \chi^2(df, NCP)$ where $df = \mathrm{rank}(\boldsymbol{\mu}_1\boldsymbol{\mu}_1^T) = 1$ and $NCP = \frac{1}{2}\boldsymbol{\mu}_1^T(\frac{1}{\sigma^2}\boldsymbol{\mu}_1\boldsymbol{\mu}_1^T)\boldsymbol{\mu}_1 = \frac{1}{2\sigma^2}$. Furthermore,

$$\left(\frac{1}{\sigma^2}\boldsymbol{\mu}_2\boldsymbol{\mu}_2^T\right)(\sigma^2\mathbf{I})\left(\frac{1}{\sigma^2}\boldsymbol{\mu}_2\boldsymbol{\mu}_2^T\right)(\sigma^2\mathbf{I}) = \boldsymbol{\mu}_2(\boldsymbol{\mu}_2^T\boldsymbol{\mu}_2)\boldsymbol{\mu}_2^T = \boldsymbol{\mu}_2\boldsymbol{\mu}_2^T = \left(\frac{1}{\sigma^2}\boldsymbol{\mu}_2\boldsymbol{\mu}_2^T\right)(\sigma^2\mathbf{I}),$$

i.e., $(\frac{1}{\sigma^2}\boldsymbol{\mu}_2\boldsymbol{\mu}_2^T)(\sigma^2\mathbf{I})$ is idempotent. Thus by Corollary 14.2.6.1 again, $\mathbf{x}^T(\frac{1}{\sigma^2}\boldsymbol{\mu}_2\boldsymbol{\mu}_2^T)\mathbf{x} \sim \chi^2(df, NCP)$ where $df = \text{rank}(\boldsymbol{\mu}_2\boldsymbol{\mu}_2^T)) = 1$ and $NCP = \frac{1}{2}\boldsymbol{\mu}_1^T(\frac{1}{\sigma^2}\boldsymbol{\mu}_2\boldsymbol{\mu}_2^T)\boldsymbol{\mu}_1 = 0$ by the orthogonality of $\boldsymbol{\mu}_1$ and $\boldsymbol{\mu}_2$. Also, by Corollary 14.3.1.2a, $\mathbf{x}^T(\frac{1}{\sigma^2}\boldsymbol{\mu}_1\boldsymbol{\mu}_1^T)\mathbf{x}$ and $\mathbf{x}^T(\frac{1}{\sigma^2}\boldsymbol{\mu}_2\boldsymbol{\mu}_2^T)\mathbf{x}$ are independent because $\boldsymbol{\mu}_1\boldsymbol{\mu}_1^T(\sigma^2\mathbf{I})\boldsymbol{\mu}_2\boldsymbol{\mu}_2^T = \sigma^2\boldsymbol{\mu}_1(\boldsymbol{\mu}_1^T\boldsymbol{\mu}_2)\boldsymbol{\mu}_2^T = \mathbf{0}$. Therefore, by Definition 14.5.1, $(\mathbf{x}^T\boldsymbol{\mu}_1\boldsymbol{\mu}_1^T\mathbf{x})/(\mathbf{x}^T\boldsymbol{\mu}_2\boldsymbol{\mu}_2^T\mathbf{x}) \sim \text{F}(1, 1, \frac{1}{2\sigma^2})$.

▶ **Exercise 32** Suppose that $\mathbf{y}$ follows the normal Gauss–Markov model $\{\mathbf{y}, \mathbf{X}\boldsymbol{\beta}, \sigma^2\mathbf{I}\}$. Let $\mathbf{c}^T\boldsymbol{\beta}$ be an estimable function, and let $\mathbf{a}^T\mathbf{y}$ be a linear unbiased estimator (not necessarily the least squares estimator) of $\mathbf{c}^T\boldsymbol{\beta}$. Let $\hat{\sigma}^2 = \mathbf{y}^T(\mathbf{I} - \mathbf{P_X})\mathbf{y}/(n - p^*)$ be the residual mean square and let w represent a positive constant. Consider the quantity

$$t^* = \frac{\mathbf{a}^T\mathbf{y} - \mathbf{c}^T\boldsymbol{\beta}}{\hat{\sigma}w}.$$

(a) What general conditions on $\mathbf{a}$ and w are sufficient for t^* to be distributed as a central t random variable with $n - p^*$ degrees of freedom? Express the conditions in as simple a form as possible.
(b) If $\mathbf{a}^T\mathbf{y}$ was not unbiased but everything else was as specified above, and if the conditions found in part (a) held, what would be the distribution of t^*? Be as specific as possible.
(c) Repeat your answers to parts (a) and (b), but this time supposing that $\mathbf{y}$ follows a positive definite Aitken model (with variance–covariance matrix $\sigma^2\mathbf{W}$) and with $\tilde{\sigma}^2 = \mathbf{y}^T\mathbf{W}^{-1}(\mathbf{I} - \tilde{\mathbf{P}}_\mathbf{X})\mathbf{y}/(n - p^*)$ in place of $\hat{\sigma}^2$.

Solution

(a) By Theorems 14.1.6 and 6.1.1, $\mathbf{a}^T\mathbf{y} \sim \text{N}(\mathbf{a}^T\mathbf{X}\boldsymbol{\beta}, \mathbf{a}^T\sigma^2\mathbf{I}\mathbf{a}) = \text{N}(\mathbf{c}^T\boldsymbol{\beta}, \sigma^2\mathbf{a}^T\mathbf{a})$ because $\mathbf{a}^T\mathbf{y}$ is unbiased for $\mathbf{c}^T\boldsymbol{\beta}$. This implies that $\mathbf{a}^T\mathbf{y} - \mathbf{c}^T\boldsymbol{\beta} \sim \text{N}(0, \sigma^2\mathbf{a}^T\mathbf{a})$, so if $w = (\mathbf{a}^T\mathbf{a})^{\frac{1}{2}}$, then $(\mathbf{a}^T\mathbf{y} - \mathbf{c}^T\boldsymbol{\beta})/(\sigma w) \sim \text{N}(0, 1)$. Also, $(n - p^*)\hat{\sigma}^2/\sigma^2 \sim \chi^2(n - p^*)$ as usual. For the independence of $(\mathbf{a}^T\mathbf{y} - \mathbf{c}^T\boldsymbol{\beta})/(\sigma w)$ and $(n - p^*)\hat{\sigma}^2/\sigma^2$ to hold, a sufficient condition is (by Corollary 14.3.1.2b) $(\mathbf{I} - \mathbf{P_X})\mathbf{I}\mathbf{a} = \mathbf{0}$ or equivalently $\mathbf{a} \in \mathcal{C}(\mathbf{X})$. So sufficient conditions on $\mathbf{a}$ and w for t^* to be distributed as a central t random variable with $n - p^*$ degrees of freedom are (1) $\mathbf{a} \in \mathcal{C}(\mathbf{X})$ and (2) $w = (\mathbf{a}^T\mathbf{a})^{\frac{1}{2}}$.
(b) In this case $t^* \sim t\left(n - p^*, \frac{\mathbf{a}^T\mathbf{X}\boldsymbol{\beta} - \mathbf{c}^T\boldsymbol{\beta}}{\sigma(\mathbf{a}^T\mathbf{a})^{\frac{1}{2}}}\right)$.
(c) By arguments similar to those used in the solution to part (a), sufficient conditions on $\mathbf{a}$ and w for t^* to be distributed as a central t random variable with $n - p^*$ degrees of freedom are (1) $\mathbf{W}^{-1}(\mathbf{I} - \tilde{\mathbf{P}}_\mathbf{X})\mathbf{W}\mathbf{a} = \mathbf{0}$ or equivalently $(\mathbf{I} - \tilde{\mathbf{P}}_\mathbf{X}^T)\mathbf{a} = \mathbf{0}$ and (2) $w = (\mathbf{a}^T\mathbf{W}\mathbf{a})^{\frac{1}{2}}$. If these conditions hold but $\mathbf{a}^T\mathbf{y}$ is not unbiased, then $t^* \sim t\left(n - p^*, \frac{\mathbf{a}^T\mathbf{X}\boldsymbol{\beta} - \mathbf{c}^T\boldsymbol{\beta}}{\sigma(\mathbf{a}^T\mathbf{W}\mathbf{a})^{\frac{1}{2}}}\right)$.

▶ **Exercise 33** Consider the normal prediction-extended positive definite Aitken model

$$\left\{\begin{pmatrix}\mathbf{y}\\ \mathbf{u}\end{pmatrix}, \begin{pmatrix}\mathbf{X}\boldsymbol{\beta}\\ \mathbf{0}\end{pmatrix}, \sigma^2\begin{pmatrix}\mathbf{W} & \mathbf{K}\\ \mathbf{K}^T & \mathbf{H}\end{pmatrix}\right\}.$$

Let $\tilde{\boldsymbol{\beta}}$ be a solution to the Aitken equations; let $\boldsymbol{\tau} = \mathbf{C}^T\boldsymbol{\beta} + \mathbf{u}$ be a vector of s predictable functions; let $\tilde{\boldsymbol{\tau}}$ be the BLUP of $\boldsymbol{\tau}$; let $\tilde{\sigma}^2$ be the generalized residual mean square, i.e., $\tilde{\sigma}^2 = (\mathbf{y} - \mathbf{X}\tilde{\boldsymbol{\beta}})^T\mathbf{W}^{-1}(\mathbf{y} - \mathbf{X}\tilde{\boldsymbol{\beta}})/(n - p^*)$; and let $\mathbf{Q} = (1/\sigma^2)\text{var}(\tilde{\boldsymbol{\tau}} - \boldsymbol{\tau})$. Assume that $\mathbf{Q}$ and

$$\begin{pmatrix}\mathbf{W} & \mathbf{K}\\ \mathbf{K}^T & \mathbf{H}\end{pmatrix}$$

are positive definite. By Theorem 14.5.5,

$$F \equiv \frac{(\tilde{\boldsymbol{\tau}} - \boldsymbol{\tau})^T\mathbf{Q}^{-1}(\tilde{\boldsymbol{\tau}} - \boldsymbol{\tau})}{s\tilde{\sigma}^2} \sim \text{F}(s, n - p^*).$$

Suppose that $\mathbf{L}^T\mathbf{y}$ is some other linear unbiased predictor of $\boldsymbol{\tau}$; that is, $\mathbf{L}^T\mathbf{y}$ is a LUP but not the BLUP of $\boldsymbol{\tau}$. A quantity F^* analogous to F may be obtained by replacing $\tilde{\boldsymbol{\tau}}$ with $\mathbf{L}^T\mathbf{y}$ and replacing $\mathbf{Q}$ with $\mathbf{L}^T\mathbf{W}\mathbf{L} + \mathbf{H} - \mathbf{L}^T\mathbf{K} - \mathbf{K}^T\mathbf{L}$ in the definition of F, i.e.,

$$F^* \equiv \frac{(\mathbf{L}^T\mathbf{y} - \boldsymbol{\tau})^T(\mathbf{L}^T\mathbf{W}\mathbf{L} + \mathbf{H} - \mathbf{L}^T\mathbf{K} - \mathbf{K}^T\mathbf{L})^{-1}(\mathbf{L}^T\mathbf{y} - \boldsymbol{\tau})}{s\tilde{\sigma}^2}.$$

Does F^* have a central F-distribution in general? If so, verify it; if not, explain why not and give a necessary and sufficient condition on $\mathbf{L}^T$ for F^* to have a central F-distribution.

Solution Because $\mathbf{L}^T\mathbf{y}$ is a LUP, $\mathbf{L}^T\mathbf{X} = \mathbf{C}^T$, implying further that

$$\mathbf{L}^T\mathbf{y} - \boldsymbol{\tau} \sim \text{N}(\mathbf{0}, \sigma^2(\mathbf{L}^T\mathbf{W}\mathbf{L} + \mathbf{H} - \mathbf{L}^T\mathbf{K} - \mathbf{K}^T\mathbf{L})).$$

Also, of course, $(n - p^*)\tilde{\sigma}^2/\sigma^2 \sim \chi^2(n - p^*)$. Thus two of the three requirements are satisfied for F^* to have a central F-distribution. However, the third requirement, namely the independence of $\mathbf{L}^T\mathbf{y} - \boldsymbol{\tau}$ and $\tilde{\sigma}^2$, is not satisfied in general. By Corollary 14.3.1.2b, a necessary and sufficient condition for that independence is

$$\begin{pmatrix}\mathbf{L}^T & -\mathbf{I}\end{pmatrix}\begin{pmatrix}\mathbf{W} & \mathbf{K}\\ \mathbf{K}^T & \mathbf{H}\end{pmatrix}\begin{pmatrix}\mathbf{E} & \mathbf{0}\\ \mathbf{0} & \mathbf{0}\end{pmatrix} = \mathbf{0},$$

i.e., $(\mathbf{L}^T\mathbf{W} - \mathbf{K}^T)\mathbf{E} = \mathbf{0}$, i.e., $\mathbf{L}^T\mathbf{W}\mathbf{E} = \mathbf{K}^T\mathbf{E}$.

▶ **Exercise 34** Consider the normal Aitken model $\{\mathbf{y}, \mathbf{X}\boldsymbol{\beta}, \sigma^2\mathbf{W}\}$, for which point estimation of an estimable function $\mathbf{c}^T\boldsymbol{\beta}$ and the residual variance σ^2 were presented in Sect. 11.2. Suppose that $\text{rank}(\mathbf{W}, \mathbf{X}) > \text{rank}(\mathbf{X})$. Let $\begin{pmatrix} \tilde{\mathbf{t}} \\ \tilde{\boldsymbol{\lambda}} \end{pmatrix}$ be any solution to the BLUE equations

$$\begin{pmatrix} \mathbf{W} & \mathbf{X} \\ \mathbf{X}^T & \mathbf{0} \end{pmatrix} \begin{pmatrix} \mathbf{t} \\ \boldsymbol{\lambda} \end{pmatrix} = \begin{pmatrix} \mathbf{0} \\ \mathbf{c} \end{pmatrix}$$

for $\mathbf{c}^T\boldsymbol{\beta}$, and let $\tilde{\sigma}^2 = \mathbf{y}^T\mathbf{G}_{11}\mathbf{y}/[\text{rank}(\mathbf{W}, \mathbf{X}) - \text{rank}(\mathbf{X})]$ where $\mathbf{G}_{11}$ is the upper left $n \times n$ block of any symmetric generalized inverse $\begin{pmatrix} \mathbf{G}_{11} & \mathbf{G}_{12} \\ \mathbf{G}_{21} & \mathbf{G}_{22} \end{pmatrix}$ of the coefficient matrix of the BLUE equations. Show that

$$\frac{\tilde{\mathbf{t}}^T\mathbf{y}}{\tilde{\sigma}\sqrt{\tilde{\mathbf{t}}^T\mathbf{W}\tilde{\mathbf{t}}}}$$

has a noncentral t distribution with degrees of freedom equal to $\text{rank}(\mathbf{W}, \mathbf{X}) - \text{rank}(\mathbf{X})$ and noncentrality parameter $\frac{\mathbf{c}^T\boldsymbol{\beta}}{\sigma\sqrt{\tilde{\mathbf{t}}^T\mathbf{W}\tilde{\mathbf{t}}}}$.

Solution Because the BLUE equations arise by solving the problem of minimizing $\text{var}(\mathbf{t}^T\mathbf{y})$ among all unbiased estimators, $\tilde{\mathbf{t}}^T\mathbf{y}$ is unbiased and has variance $\sigma^2\tilde{\mathbf{t}}^T\mathbf{W}\tilde{\mathbf{t}}$. Thus $\tilde{\mathbf{t}}^T\mathbf{y} \sim \text{N}(\mathbf{c}^T\boldsymbol{\beta}, \sigma^2\tilde{\mathbf{t}}^T\mathbf{W}\tilde{\mathbf{t}})$. Partial standardization yields

$$\frac{\tilde{\mathbf{t}}^T\mathbf{y}}{\sigma\sqrt{\tilde{\mathbf{t}}^T\mathbf{W}\tilde{\mathbf{t}}}} \sim \text{N}\left(\frac{\mathbf{c}^T\boldsymbol{\beta}}{\sigma\sqrt{\tilde{\mathbf{t}}^T\mathbf{W}\tilde{\mathbf{t}}}}, 1\right).$$

Now,

$$\frac{[\text{rank}(\mathbf{W}, \mathbf{X}) - \text{rank}(\mathbf{X})]\tilde{\sigma}^2}{\sigma^2} = [(1/\sigma)\mathbf{y}]^T\mathbf{G}_{11}[(1/\sigma)\mathbf{y}].$$

Consider the conditions of Theorem 14.2.6 as they may apply to the quadratic form on the right-hand side of this last equation. We find that

$$\mathbf{W}\mathbf{G}_{11}\mathbf{W}\mathbf{G}_{11}\mathbf{W} = \mathbf{W}\mathbf{G}_{11}(\mathbf{W} - \mathbf{X}\mathbf{G}_{22}\mathbf{X}^T) = \mathbf{W}\mathbf{G}_{11}\mathbf{W}$$

by Theorem 3.3.8c, d (with $\mathbf{W}$ and $\mathbf{X}$ here playing the roles of $\mathbf{A}$ and $\mathbf{B}$ in the theorem);

$$\mathbf{W}\mathbf{G}_{11}\mathbf{W}\mathbf{G}_{11}[(1/\sigma)\mathbf{X}\boldsymbol{\beta}] = (\mathbf{W} - \mathbf{X}\mathbf{G}_{22}\mathbf{X}^T)\mathbf{G}_{11}\mathbf{X}\boldsymbol{\beta}(1/\sigma) = \mathbf{W}\mathbf{G}_{11}[(1/\sigma)\mathbf{X}\boldsymbol{\beta}],$$

also by Theorem 3.3.8c, d; and

$$[(1/\sigma)\mathbf{X}\boldsymbol{\beta}]^T\mathbf{G}_{11}\mathbf{W}\mathbf{G}_{11}[(1/\sigma)\mathbf{X}\boldsymbol{\beta}] = 0 = [(1/\sigma)\mathbf{X}\boldsymbol{\beta}]^T\mathbf{G}_{11}[(1/\sigma\mathbf{X}\boldsymbol{\beta}]$$

yet again by Theorem 3.3.8c. Thus all three conditions of Theorem 14.2.6 are satisfied, implying that

$$\frac{[\text{rank}(\mathbf{W},\mathbf{X}) - \text{rank}(\mathbf{X})]\tilde{\sigma}^2}{\sigma^2} \sim \chi^2(\text{tr}(\mathbf{G}_{11}\mathbf{W}), 0),$$

or equivalently (by Theorem 3.3.8c, e), $\chi^2[\text{rank}(\mathbf{W},\mathbf{X}) - \text{rank}(\mathbf{X})]$. Next consider the conditions of Theorem 14.3.1 as they apply here, with $\mathbf{G}_{11}, \mathbf{0}_{n\times n}, \mathbf{0}_n$, and $\tilde{\mathbf{t}}$ playing the roles of $\mathbf{A}_1, \mathbf{A}_2, \boldsymbol{\ell}_1$, and $\boldsymbol{\ell}_2$ in the theorem. We find that the first of the conditions in Theorem 14.3.1 is satisfied trivially for both $(i, j) = (1, 2)$ and $(i, j) = (2, 1)$; the second is satisfied when $(i, j) = (1, 2)$ because $\mathbf{W}\mathbf{G}_{11}\mathbf{W}\tilde{\mathbf{t}} = -\mathbf{W}\mathbf{G}_{11}\mathbf{X}\tilde{\boldsymbol{\lambda}} = -\mathbf{0}_{n\times p}\tilde{\boldsymbol{\lambda}} = \mathbf{0}$ by Theorem 3.3.8c, and is satisfied trivially when $(i, j) = (2, 1)$; and the third is satisfied for both $(i, j) = (1, 2)$ and $(i, j) = (2, 1)$ because $\tilde{\mathbf{t}}^T\mathbf{W}\mathbf{G}_{11}[(1/\sigma)\mathbf{X}\boldsymbol{\beta}] = 0$ by Theorem 3.3.8c. Thus, $\frac{\tilde{\mathbf{t}}^T\mathbf{y}}{\sigma\sqrt{\tilde{\mathbf{t}}^T\mathbf{W}\tilde{\mathbf{t}}}}$ and $[\text{rank}(\mathbf{W},\mathbf{X}) - \text{rank}(\mathbf{X})]\tilde{\sigma}^2/\sigma^2$ are independent. It follows immediately from Definition 14.5.2 that

$$\frac{\frac{\tilde{\mathbf{t}}^T\mathbf{y}}{\sigma\sqrt{\tilde{\mathbf{t}}^T\mathbf{W}\tilde{\mathbf{t}}}}}{\sqrt{\{[\text{rank}(\mathbf{W},\mathbf{X}) - \text{rank}(\mathbf{X})]\tilde{\sigma}^2/\sigma^2\}\Big/[\text{rank}(\mathbf{W},\mathbf{X}) - \text{rank}(\mathbf{X})]}} = \frac{\tilde{\mathbf{t}}^T\mathbf{y}}{\tilde{\sigma}\sqrt{\tilde{\mathbf{t}}\mathbf{W}\tilde{\mathbf{t}}}}$$

has the specified distribution.

▶ **Exercise 35** Let

$$\mathbf{M} = \begin{pmatrix} m_{11} & m_{12} \\ m_{21} & m_{22} \end{pmatrix}$$

represent a random matrix whose elements $m_{11}, m_{12}, m_{21}, m_{22}$ are independent N(0, 1) random variables.

(a) Show that $m_{11} - m_{22}$, $m_{12} + m_{21}$, and $m_{12} - m_{21}$ are mutually independent, and determine their distributions.
(b) Show that $\Pr[\{\text{tr}(\mathbf{M})\}^2 - 4|\mathbf{M}| > 0] = \Pr(W > \frac{1}{2})$, where $W \sim \text{F}(2, 1)$. [Hint: You may find it helpful to recall that $(u + v)^2 - (u - v)^2 = 4uv$ for all real numbers u and v.]

Solution

(a) Because $(m_{11}, m_{12}, m_{21}, m_{22})^T \sim \mathrm{N}(\mathbf{0}_4, \mathbf{I}_4)$ and

$$\begin{pmatrix} m_{11}-m_{22} \\ m_{12}+m_{21} \\ m_{12}-m_{21} \end{pmatrix} = \begin{pmatrix} 1 & 0 & 0 & -1 \\ 0 & 1 & 1 & 0 \\ 0 & 1 & -1 & 0 \end{pmatrix} \begin{pmatrix} m_{11} \\ m_{12} \\ m_{21} \\ m_{22} \end{pmatrix},$$

by Theorem 14.1.6

$$\begin{pmatrix} m_{11}-m_{22} \\ m_{12}+m_{21} \\ m_{12}-m_{21} \end{pmatrix} \sim \mathrm{N}(\mathbf{0}_3, \boldsymbol{\Sigma}),$$

where

$$\boldsymbol{\Sigma} = \begin{pmatrix} 1 & 0 & 0 & -1 \\ 0 & 1 & 1 & 0 \\ 0 & 1 & -1 & 0 \end{pmatrix} \mathbf{I} \begin{pmatrix} 1 & 0 & 0 \\ 0 & 1 & 1 \\ 0 & 1 & -1 \\ -1 & 0 & 0 \end{pmatrix} = 2\mathbf{I}_3.$$

Thus, by Theorem 14.1.8, $m_{11}-m_{22}$, $m_{12}+m_{21}$, and $m_{12}-m_{21}$ are independent, and by Theorem 14.1.7 each of them has a N(0, 2) distribution.

(b)

$$\begin{aligned} \{\mathrm{tr}(\mathbf{M})\}^2 - 4|\mathbf{M}| &= (m_{11}+m_{22})^2 - 4(m_{11}m_{22} - m_{12}m_{21}) \\ &= m_{11}^2 + m_{22}^2 - 2m_{11}m_{22} + 4m_{12}m_{21} \\ &= (m_{11}-m_{22})^2 + (m_{12}+m_{21})^2 - (m_{12}-m_{21})^2. \end{aligned}$$

By the distributional result in part (a), the three summands in this last expression are independent and each, divided by 2, has a $\chi^2(1)$ distribution. Thus,

$$\begin{aligned} \Pr[\{\mathrm{tr}(\mathbf{M})\}^2 - 4|\mathbf{M}| > 0] &= \Pr[(m_{11}-m_{22})^2 + (m_{12}+m_{21})^2 - (m_{12}-m_{21})^2 > 0] \\ &= \Pr\left[\frac{(m_{11}-m_{22})^2}{2} + \frac{(m_{12}+m_{21})^2}{2} > \frac{(m_{12}-m_{21})^2}{2}\right] \\ &= \Pr(U > V) \\ &= \Pr\left(\frac{U/2}{V} > \frac{1}{2}\right) \\ &= \Pr(W > \frac{1}{2}), \end{aligned}$$

where $U \sim \chi^2(2)$, $V \sim \chi^2(1)$, U and V are independent, and $W \sim \text{F}(2, 1)$ by Theorem 14.2.8 and Definition 14.5.1.

▶ **Exercise 36** Using the results of Example 14.4-2 show that the ratio of the Factor A mean square to the whole-plot error mean square, the ratio of the Factor B mean square to the split-plot error mean square, and the ratio of the AB interaction mean square to the split-plot error mean square have noncentral F distributions, and give the parameters of those distributions.

Solution The ratio of the Factor A mean square to the whole-plot error mean square is

$$\frac{rm\sum_{i=1}^{q}(\bar{y}_{i\cdot\cdot}-\bar{y}_{\cdot\cdot\cdot})^2/(q-1)}{m\sum_{i=1}^{q}\sum_{j=1}^{r}(\bar{y}_{ij\cdot}-\bar{y}_{i\cdot\cdot})^2/[q(r-1)]}$$

$$=\frac{\frac{rm}{\sigma^2+m\sigma_b^2}\sum_{i=1}^{q}(\bar{y}_{i\cdot\cdot}-\bar{y}_{\cdot\cdot\cdot})^2/(q-1)}{\frac{m}{\sigma^2+m\sigma_b^2}\sum_{i=1}^{q}\sum_{j=1}^{r}(\bar{y}_{ij\cdot}-\bar{y}_{i\cdot\cdot})^2/[q(r-1)]}$$

$$\sim \text{F}\left(q-1, q(r-1), \frac{rm}{2(q-1)(\sigma^2+m\sigma_b^2)}\sum_{i=1}^{q}[(\alpha_i-\bar{\alpha}_\cdot)+(\bar{\xi}_{i\cdot}-\bar{\xi}_{\cdot\cdot})]^2\right).$$

The ratio of the Factor B mean square to the split-plot error mean square is

$$\frac{qr\sum_{k=1}^{m}(\bar{y}_{\cdot\cdot k}-\bar{y}_{\cdot\cdot\cdot})^2/(m-1)}{\sum_{i=1}^{q}\sum_{j=1}^{r}\sum_{k=1}^{m}(y_{ijk}-\bar{y}_{ij\cdot}-\bar{y}_{i\cdot k}+\bar{y}_{i\cdot\cdot})^2/[q(r-1)(m-1)]}$$

$$=\frac{\frac{qr}{\sigma^2}\sum_{k=1}^{m}(\bar{y}_{\cdot\cdot k}-\bar{y}_{\cdot\cdot\cdot})^2/(m-1)}{\frac{1}{\sigma^2}\sum_{i=1}^{q}\sum_{j=1}^{r}\sum_{k=1}^{m}(y_{ijk}-\bar{y}_{ij\cdot}-\bar{y}_{i\cdot k}+\bar{y}_{i\cdot\cdot})^2/[q(r-1)(m-1)]}$$

$$\sim \text{F}\left(m-1, q(r-1)(m-1), \frac{qr}{2(m-1)\sigma^2}\sum_{k=1}^{m}[(\gamma_k-\bar{\gamma}_\cdot)+(\bar{\xi}_{\cdot k}-\bar{\xi}_{\cdot\cdot})]^2\right).$$

The ratio of the AB interaction mean square to the split-plot error mean square is

$$\frac{r\sum_{i=1}^{q}\sum_{k=1}^{m}(\bar{y}_{i\cdot k}-\bar{y}_{i\cdot\cdot}-\bar{y}_{\cdot\cdot k}+\bar{y}_{\cdot\cdot\cdot})^2/[(q-1)(m-1)]}{\sum_{i=1}^{q}\sum_{j=1}^{r}\sum_{k=1}^{m}(y_{ijk}-\bar{y}_{ij\cdot}-\bar{y}_{i\cdot k}+\bar{y}_{i\cdot\cdot})^2/[q(r-1)(m-1)]}$$

$$=\frac{\frac{r}{\sigma^2}\sum_{i=1}^{q}\sum_{k=1}^{m}(\bar{y}_{i\cdot k}-\bar{y}_{i\cdot\cdot}-\bar{y}_{\cdot\cdot k}+\bar{y}_{\cdot\cdot\cdot})^2/[(q-1)(m-1)]}{\frac{1}{\sigma^2}\sum_{i=1}^{q}\sum_{j=1}^{r}\sum_{k=1}^{m}(y_{ijk}-\bar{y}_{ij\cdot}-\bar{y}_{i\cdot k}+\bar{y}_{i\cdot\cdot})^2/[q(r-1)(m-1)]}$$

$$\sim \text{F}[(q-1)(m-1), q(r-1)(m-1)].$$

Reference

Casella, G. & Berger, R. L. (2002). *Statistical inference* (2nd ed.). Pacific Grove, CA: Duxbury.

Inference for Estimable and Predictable Functions

15

This chapter presents exercises on inference for estimable and predictable functions in linear models and provides solutions to those exercises.

▶ **Exercise 1** Consider the normal Gauss–Markov simple linear regression model for n observations, and suppose that in addition to the n responses that follow that model, there are m responses $y_1^*, y_2^*, \ldots, y_m^*$, all taken at an *unknown* value of x, say x^*, that also follow that model. Using the observation-pairs $(x_1, y_1), \ldots, (x_n, y_n)$ and the additional y-observations, an estimate of the unknown x^* can be obtained. One estimator of x^* that has been proposed is

$$\hat{x}^* = \frac{\bar{y}^* - \hat{\beta}_1}{\hat{\beta}_2},$$

where $\bar{y}^* = \frac{1}{m}\sum_{i=1}^m y_i^*$. This estimator is derived by equating $\bar{y}^*$ to $\hat{\beta}_1 + \hat{\beta}_2 x^*$ (because their expectations are both $\beta_1 + \beta_2 x^*$) and then solving for x^*. (Note that the estimator is not well defined when $\hat{\beta}_2 = 0$, but this is an event of probability zero.)

(a) Determine the distribution of $W \equiv \bar{y}^* - \hat{\beta}_1 - \hat{\beta}_2 x^*$.
(b) The random variable $W/(\hat{\sigma}c)$, where c is a nonrandom quantity, has a certain well-known distribution. Determine this distribution and the value of c.
(c) Let $0 < \xi < 1$. Based on your solution to part (b), derive an exact $100(1-\xi)\%$ confidence interval for x^* whose endpoints depend on the unknown x^* and then approximate the endpoints (by substituting $\hat{x}^*$ for x^*) to obtain an approximate $100(1-\xi)\%$ confidence interval for x^*. Is this confidence interval symmetric about $\hat{x}^*$?

D. L. Zimmerman, *Linear Model Theory*,
https://doi.org/10.1007/978-3-030-52074-8_15

Solution

(a) $\bar{y}^* \sim \text{N}(\beta_1+\beta_2 x^*, \sigma^2/m)$, and with the aid of Example 7.2-1 it is easily shown that $\hat{\beta}_1+\hat{\beta}_2 x^* \sim \text{N}(\beta_1+\beta_2 x^*, \sigma^2[\frac{1}{n}+\frac{(x^*-\bar{x})^2}{SXX}])$. Furthermore, $\bar{y}^*$ and $\hat{\beta}_1+\hat{\beta}_2 x^*$ are independent because $y_1^*,\ldots,y_m^*$ are independent of $y_1,\ldots,y_n$. Therefore, $W \sim \text{N}(0, \sigma^2[\frac{1}{m}+\frac{1}{n}+\frac{(x^*-\bar{x})^2}{SXX}])$.

(b) W and $\hat{\sigma}^2$ are independent because $(\hat{\beta}_1, \hat{\beta}_2, \bar{y}^*)^T$ and $\hat{\sigma}^2$ are independent. Therefore, letting $c = \sqrt{\frac{1}{m}+\frac{1}{n}+\frac{(x^*-\bar{x})^2}{SXX}}$, we have

$$\frac{W}{\hat{\sigma}c} = \frac{W/(\sigma c)}{\sqrt{(n-2)\hat{\sigma}^2/[\sigma^2(n-2)]}} \sim \text{t}(n-2).$$

(c) By part (b),

$$\begin{aligned}
1-\xi &= \Pr\left(-t_{\xi/2,n-2} \le \frac{\bar{y}^*-\hat{\beta}_1-\hat{\beta}_2 x^*}{\hat{\sigma}c} \le t_{\xi/2,n-2}\right) \\
&= \Pr\left(\hat{\beta}_1-\bar{y}^*-\hat{\sigma}ct_{\xi/2,n-2} \le -\hat{\beta}_2 x^* \le \hat{\beta}_1-\bar{y}^*+\hat{\sigma}ct_{\xi/2,n-2}\right) \\
&= \Pr\left(\frac{\bar{y}^*-\hat{\beta}_1-\hat{\sigma}ct_{\xi/2,n-2}}{\hat{\beta}_2} \le x^* \le \frac{\bar{y}^*-\hat{\beta}_1+\hat{\sigma}ct_{\xi/2,n-2}}{\hat{\beta}_2}\,\middle|\,\hat{\beta}_2>0\right)\cdot \Pr(\hat{\beta}_2>0) \\
&\quad +\Pr\left(\frac{\bar{y}^*-\hat{\beta}_1+\hat{\sigma}ct_{\xi/2,n-2}}{\hat{\beta}_2} \le x^* \le \frac{\bar{y}^*-\hat{\beta}_1-\hat{\sigma}ct_{\xi/2,n-2}}{\hat{\beta}_2}\,\middle|\,\hat{\beta}_2<0\right)\cdot \Pr(\hat{\beta}_2<0) \\
&= \Pr\left(\hat{x}^*-\frac{\hat{\sigma}ct_{\xi/2,n-2}}{\hat{\beta}_2} \le x^* \le \hat{x}^*+\frac{\hat{\sigma}ct_{\xi/2,n-2}}{\hat{\beta}_2}\,\middle|\,\hat{\beta}_2>0\right)\cdot \Pr(\hat{\beta}_2>0) \\
&\quad +\Pr\left(\hat{x}^*-\frac{\hat{\sigma}ct_{\xi/2,n-2}}{\hat{\beta}_2} \le x^* \le \hat{x}^*+\frac{\hat{\sigma}ct_{\xi/2,n-2}}{\hat{\beta}_2}\,\middle|\,\hat{\beta}_2<0\right)\cdot \Pr(\hat{\beta}_2<0).
\end{aligned}$$

Thus,

$$\hat{x}^* \pm \frac{\hat{\sigma}ct_{\xi/2,n-2}}{\hat{\beta}_2}$$

is an exact $100(1-\xi)\%$ confidence interval for x^*. Replacing c with $\hat{c} = \sqrt{\frac{1}{m}+\frac{1}{n}+\frac{(\hat{x}^*-\bar{x})^2}{SXX}}$ yields an approximate $100(1-\xi)\%$ confidence interval that is symmetric about $\hat{x}^*$.

▶ **Exercise 2** Consider the normal Gauss–Markov quadratic regression model

$$y_i = \beta_1+\beta_2 x_i+\beta_3 x_i^2+e_i \quad (i=1,\ldots,n),$$

where it is known that $\beta_3 \neq 0$. Assume that there are at least three distinct x_i's, so that $\mathbf{X}$ has full column rank. Let x_m be the value of x where the quadratic function $\mu(x) = \beta_1 + \beta_2 x + \beta_3 x^2$ is minimized or maximized (such an x-value exists because $\beta_3 \neq 0$).

(a) Verify that $x_m = -\beta_2/(2\beta_3)$.
(b) Define $\tau = \beta_2 + 2\beta_3 x_m$ $(= 0)$ and $\hat{\tau} = \hat{\beta}_2 + 2\hat{\beta}_3 x_m$, where $(\hat{\beta}_1, \hat{\beta}_2, \hat{\beta}_3)$ are the least squares estimators of $(\beta_1, \beta_2, \beta_3)$. Determine the distribution of $\hat{\tau}$ in terms of x_m, σ^2, and $\{c_{ij}\}$, where c_{ij} is the ijth element of $(\mathbf{X}^T\mathbf{X})^{-1}$.
(c) Based partly on your answer to part (b), determine a function of $\hat{\tau}$ and $\hat{\sigma}^2$ that has an F distribution, and give the parameters of this F distribution.
(d) Let $\xi \in (0, 1)$. Use the result from part (c) to find a quadratic function of x_m, say $q(x_m)$, such that $\Pr[q(x_m) \leq 0] = 1 - \xi$. When is this confidence set an interval?

Solution

(a) The stationary point of $\mu(x)$ occurs at the value of x, where $\frac{d}{dx}(\beta_1 + \beta_2 x + \beta_3 x^2) = 0$, i.e., where $\beta_2 + 2\beta_3 x = 0$, i.e., at $x_m = -\beta_2/(2\beta_3)$.
(b) $\hat{\beta}_2 + 2\hat{\beta}_3 x_m = \mathbf{c}^T\hat{\boldsymbol{\beta}}$, where $\mathbf{c}^T = (0, 1, 2x_m)$. Thus $\hat{\beta}_2 + 2\hat{\beta}_3 x_m$ has a normal distribution with mean $\beta_2 + 2\beta_3 x_m = 0$ and variance

$$\sigma^2 \begin{pmatrix} 0 & 1 & 2x_m \end{pmatrix} (\mathbf{X}^T\mathbf{X})^{-1} \begin{pmatrix} 0 \\ 1 \\ 2x_m \end{pmatrix} = \sigma^2(c_{22} + 4x_m c_{23} + 4x_m^2 c_{33}).$$

(c) $(n-3)\hat{\sigma}^2/\sigma^2 \sim \chi^2(n-3)$, and $\hat{\tau}$ and $\hat{\sigma}^2$ are independent, so

$$\frac{\hat{\tau}^2}{\hat{\sigma}^2(c_{22} + 4x_m c_{23} + 4x_m^2 c_{33})} \sim \mathrm{F}(1, n-3).$$

(d) By the result in part (c),

$$\begin{aligned} 1 - \xi &= \Pr\left(\frac{(\hat{\beta}_2 + 2\hat{\beta}_3 x_m)^2}{c_{22} + 4x_m c_{23} + 4x_m^2 c_{33}} \leq \hat{\sigma}^2 F_{\xi,1,n-3}\right) \\ &= \Pr[(4\hat{\beta}_3^2 - 4c_{33}\hat{\sigma}^2 F_{\xi,1,n-3})x_m^2 + (4\hat{\beta}_2\hat{\beta}_3 - 4c_{23}\hat{\sigma}^2 F_{\xi,1,n-3})x_m \\ &\quad + (\hat{\beta}_2^2 - c_{22}\hat{\sigma}^2 F_{\xi,1,n-3}) \leq 0]. \end{aligned}$$

Thus $q(x_m) = ax_m^2 + bx_m + c$, where $a = 4\hat{\beta}_3^2 - 4c_{33}\hat{\sigma}^2 F_{\xi,1,n-3}$, $b = 4\hat{\beta}_2\hat{\beta}_3 - 4c_{23}\hat{\sigma}^2 F_{\xi,1,n-3}$, and $c = \hat{\beta}_2^2 - c_{22}\hat{\sigma}^2 F_{\xi,1,n-3}$. The set $\{x_m : q(x_m) \leq 0\}$ is a $100(1-\xi)\%$ confidence set, but it is not necessarily an interval. It is an interval if $a > 0$ and $b^2 - 4ac > 0$.

▶ **Exercise 3** Consider the normal Gauss–Markov simple linear regression model

$$y_i = \beta_1 + \beta_2 x_i + e_i \quad (i = 1, \ldots, n).$$

In Example 12.1.1-1, it was shown that the mean squared error for the least squares estimator of the intercept from the fit of an (underspecified) intercept-only model is smaller than the mean squared error for the least squares estimator of the intercept from fitting the full simple linear regression model if and only if $\beta_2^2/[\sigma^2/SXX] \leq 1$. Consequently, it may be desirable to test the hypotheses

$$H_0 : \frac{\beta_2^2}{\sigma^2/SXX} \leq 1 \quad \text{versus} \quad H_a : \frac{\beta_2^2}{\sigma^2/SXX} > 1.$$

A natural test statistic for this hypothesis test is

$$\frac{\hat{\beta}_2^2}{\hat{\sigma}^2/SXX}.$$

Determine the distribution of this test statistic when

$$\frac{\beta_2^2}{\sigma^2/SXX} = 1.$$

Solution $\hat{\beta}_2 \sim \mathrm{N}(\beta_2, \sigma^2/SXX)$, implying that

$$\frac{\hat{\beta}_2^2}{\sigma^2/SXX} \sim \chi^2\left(1, \frac{\beta_2^2}{2\sigma^2/SXX}\right).$$

Also, we know that $(n-2)\hat{\sigma}^2/\sigma^2$ is distributed as $\chi^2(n-2)$, independently of $\hat{\beta}_2$. Thus,

$$\frac{\frac{\hat{\beta}_2^2}{\sigma^2/SXX}\Big/1}{\frac{(n-2)\hat{\sigma}^2}{\sigma^2}\Big/(n-2)} = \frac{\hat{\beta}_2^2}{\hat{\sigma}^2/SXX} \sim \mathrm{F}\left(1, n-2, \frac{\beta_2^2}{2\sigma^2/SXX}\right).$$

When $\frac{\beta_2^2}{\sigma^2/SXX} = 1$, this distribution specializes to $\mathrm{F}(1, n-2, \frac{1}{2})$.

▶ **Exercise 4** By completing the three parts of this exercise, verify the claim made in Example 15.2-3, i.e., that the power of the F-test for $H_0 : \beta_2 = 0$ versus $H_a : \beta_2 \neq 0$ in normal simple linear regression, when the explanatory variable is restricted to the interval $[a, b]$ and n is even, is maximized by setting half of the x-values equal to a and the other half equal to b.

(a) Show that $\sum_{i=1}^n (x_i - \bar{x})^2 \leq \sum_{i=1}^n \left(x_i - \frac{a+b}{2}\right)^2$. [Hint: Add and subtract $\frac{a+b}{2}$ within $\sum_{i=1}^n (x_i - \bar{x})^2$.]

(b) Show that $\sum_{i=1}^n \left(x_i - \frac{a+b}{2}\right)^2 \leq \frac{n(b-a)^2}{4}$.

(c) Show that $\sum_{i=1}^n \left(x_i - \frac{a+b}{2}\right)^2 = \frac{n(b-a)^2}{4}$ for the design that sets half of the x-values equal to a and the other half equal to b.

Solution

(a)

$$\begin{aligned}
\sum_{i=1}^n (x_i - \bar{x})^2 &= \sum_{i=1}^n \left[\left(x_i - \frac{a+b}{2}\right) - \left(\bar{x} - \frac{a+b}{2}\right)\right]^2 \\
&= \sum_{i=1}^n \left(x_i - \frac{a+b}{2}\right)^2 + n\left(\bar{x} - \frac{a+b}{2}\right)^2 \\
&\quad -2\left(\bar{x} - \frac{a+b}{2}\right)\sum_{i=1}^n \left(x_i - \frac{a+b}{2}\right) \\
&= \sum_{i=1}^n \left(x_i - \frac{a+b}{2}\right)^2 - n\left(\bar{x} - \frac{a+b}{2}\right)^2 \\
&\leq \sum_{i=1}^n \left(x_i - \frac{a+b}{2}\right)^2.
\end{aligned}$$

(b) Because $x_i - a \geq 0$ and $x_i - b \leq 0$,

$$(x_i - a) + (x_i - b) \leq (x_i - a) - (x_i - b) = b - a.$$

Therefore,

$$\sum_{i=1}^n \left(x_i - \frac{a+b}{2}\right)^2 = \sum_{i=1}^n (1/4)[(x_i - a) + (x_i - b)]^2 \leq \frac{n(b-a)^2}{4}.$$

(c) If half of the x-values are equal to a and the other half are equal to b, then $\bar{x} = \frac{a+b}{2}$, implying further that

$$\sum_{i=1}^n (x_i - \bar{x})^2 = \sum_{i=1}^n \left(x_i - \frac{a+b}{2}\right)^2 = \frac{n}{2}\left(a - \frac{a+b}{2}\right)^2 + \frac{n}{2}\left(b - \frac{a+b}{2}\right)^2 = \frac{n(b-a)^2}{4}.$$

Thus, by part (b) the design with half of the x-values equal to a and the other half equal to b maximizes $\sum_{i=1}^n (x_i - \bar{x})^2$.

Exercise 5 Consider the following modification of the normal Gauss–Markov model, which is called the normal Gauss–Markov variance shift outlier model. For this model, the model equation $\mathbf{y} = \mathbf{X}\boldsymbol{\beta} + \mathbf{e}$ can be written as

$$\begin{pmatrix} y_1 \\ \mathbf{y}_{-1} \end{pmatrix} = \begin{pmatrix} \mathbf{x}_1^T \\ \mathbf{X}_{-1} \end{pmatrix} \boldsymbol{\beta} + \begin{pmatrix} e_1 \\ \mathbf{e}_{-1} \end{pmatrix},$$

where

$$\begin{pmatrix} e_1 \\ \mathbf{e}_{-1} \end{pmatrix} \sim \mathrm{N}_n\left(\mathbf{0}, \mathrm{diag}\{\sigma^2 + \sigma_\delta^2, \sigma^2, \sigma^2, \ldots, \sigma^2\}\right).$$

Here, σ_δ^2 is an unknown nonnegative scalar parameter. Assume that $\mathrm{rank}(\mathbf{X}_{-1}) = \mathrm{rank}(\mathbf{X})\ (= p^*)$, i.e., the rank of $\mathbf{X}$ is not affected by deleting its first row, and define $\hat{e}_{1,-1}$, $\hat{\boldsymbol{\beta}}_{-1}$, $\hat{\sigma}_{-1}^2$, and $p_{11,-1}$ as in Methodological Interlude #9.

(a) Determine the distributions of $\hat{e}_{1,-1}$ and $(n - p^* - 1)\hat{\sigma}_{-1}^2/\sigma^2$.
(b) Determine the distribution of

$$\frac{\hat{e}_{1,-1}}{\hat{\sigma}_{-1}\sqrt{1 + p_{11,-1}}}.$$

(c) Explain how the hypotheses

$$H_0 : \sigma_\delta^2 = 0 \quad \text{vs} \quad H_a : \sigma_\delta^2 > 0$$

could be tested using the statistic in part (b), and explain how the value of $p_{11,-1}$ affects the power of the test.

Solution

(a) Because $\hat{\boldsymbol{\beta}}_{-1}$ is unbiased for $\boldsymbol{\beta}$ and does not depend on y_1, $\mathrm{E}(y_1 - \mathbf{x}_1^T\hat{\boldsymbol{\beta}}_{-1}) = \mathbf{x}_1^T\boldsymbol{\beta} - \mathbf{x}_1^T\boldsymbol{\beta} = 0$ and

$$\begin{aligned} \mathrm{var}(y_1 - \mathbf{x}_1^T\hat{\boldsymbol{\beta}}_{-1}) &= \mathrm{var}(y_1) + \mathbf{x}_1^T(\mathbf{X}_{-1}^T\mathbf{X}_{-1})^{-1}\mathbf{X}_{-1}^T(\sigma^2\mathbf{I})\mathbf{X}_{-1}(\mathbf{X}_{-1}^T\mathbf{X}_{-1})^{-1}\mathbf{x}_1 \\ &= \sigma^2 + \sigma_\delta^2 + \sigma^2\mathbf{x}_1^T(\mathbf{X}_{-1}^T\mathbf{X}_{-1})^{-1}\mathbf{x}_1 \\ &= \sigma^2(1 + p_{11,-1}) + \sigma_\delta^2. \end{aligned}$$

Thus $\hat{\mathbf{e}}_{1,-1} \sim \mathrm{N}(0, \sigma^2(1 + p_{11,-1}) + \sigma_\delta^2)$. Furthermore, $(n - p^* - 1)\hat{\sigma}_{-1}^2/\sigma^2 \sim \chi^2(n - p^* - 1)$ by applying (14.8) to the model without the first observation.

(b) Because

$$\begin{pmatrix} 0 & \mathbf{0}^T \\ \mathbf{0} & \mathbf{I}-\mathbf{P}_{\mathbf{X}_{-1}} \end{pmatrix} \begin{pmatrix} \sigma^2+\sigma_\delta^2 & \mathbf{0}^T \\ \mathbf{0} & \sigma^2\mathbf{I} \end{pmatrix} \begin{pmatrix} 1 \\ -\mathbf{X}_{-1}(\mathbf{X}_{-1}^T\mathbf{X}_{-1})^{-1}\mathbf{x}_1 \end{pmatrix} = \begin{pmatrix} 0 \\ \mathbf{0} \end{pmatrix},$$

$\hat{\mathbf{e}}_{1,-1}$ and $(n-p^*-1)\hat{\sigma}_{-1}^2/\sigma^2$ are independent by Corollary 14.3.1.2b. Thus,

$$\frac{\hat{\mathbf{e}}_{1,-1}}{\hat{\sigma}_{-1}\sqrt{1+p_{11,-1}}} = \frac{\hat{\mathbf{e}}_{1,-1}/\sqrt{\sigma^2(1+p_{11,-1})+\sigma_\delta^2}}{\sqrt{(n-p^*-1)\hat{\sigma}_{-1}^2/[\sigma^2(n-p^*-1)]}} \sim \mathrm{t}(n-p^*-1),$$

implying that

$$\frac{\hat{e}_{1,-1}}{\hat{\sigma}_{-1}\sqrt{1+p_{11,-1}}} \sim \sqrt{1+\frac{\sigma_\delta^2}{\sigma^2(1+p_{11,-1})}}\,\mathrm{t}(n-p^*-1).$$

(c) Under H_0, $\frac{\hat{e}_{1,-1}}{\hat{\sigma}_{-1}\sqrt{1+p_{11,-1}}} \sim \mathrm{t}(n-p^*-1)$, so a size-$\xi$ test of H_0 versus H_a is to reject H_0 if and only if

$$\frac{|\hat{e}_{1,-1}|}{\hat{\sigma}_{-1}\sqrt{1+p_{11,-1}}} > t_{\xi/2,n-p-1}.$$

The power of this test clearly increases as $\sqrt{1+\frac{\sigma_\delta^2}{\sigma^2(1+p_{11,-1})}}$ increases, so the power increases as $p_{11,-1}$ decreases.

▶ **Exercise 6** Consider the normal positive definite Aitken model

$$\mathbf{y} = \mathbf{1}_n\mu + \mathbf{e},$$

where $n \geq 2$, $\mathbf{e} \sim \mathrm{N}_n(\mathbf{0}, \sigma^2\mathbf{W})$, and $\mathbf{W} = (1-\rho)\mathbf{I}_n + \rho\mathbf{J}_n$. Here, σ^2 is unknown but $\rho \in \left(-\frac{1}{n-1}, 1\right)$ is a known constant. Let $\hat{\sigma}^2$ represent the sample variance of the responses, i.e., $\hat{\sigma}^2 = \sum_{i=1}^n (y_i - \bar{y})^2/(n-1)$.

(a) Determine the joint distribution of $\bar{y}$ and $\frac{(n-1)\hat{\sigma}^2}{\sigma^2(1-\rho)}$.
(b) Using results from part (a), derive a size-ξ test for testing $H_0 : \mu = 0$ versus $H_a : \mu \neq 0$.
(c) For the test you derived in part (b), describe how the magnitude of ρ affects the power.
(d) A statistician wants to test the null hypothesis $H_0 : \mu = 0$ versus the alternative hypothesis $H_a : \mu > 0$ at the ξ level of significance. However, rather than using the appropriate test obtained in part (b), (s)he wishes to ignore the correlations

among the observations and use a standard size-ξ t test. That is, (s)he will reject H_0 if and only if

$$\frac{\bar{y}}{\hat{\sigma}/\sqrt{n}} > t_{\xi,n-1}.$$

The size of this test is not necessarily equal to its nominal value, ξ. What is the actual size of the test, and how does it compare to ξ?

Solution

(a) First observe that $\bar{y} = (1/n)\mathbf{1}_n^T\mathbf{y}$ is normally distributed with mean $(1/n)\mathbf{1}_n^T(\mathbf{1}_n\mu) = \mu$ and variance

$$(1/n)\mathbf{1}^T(\sigma^2\mathbf{W})(1/n)\mathbf{1} = (\sigma^2/n^2)\mathbf{1}^T[(1-\rho)\mathbf{I}+\rho\mathbf{J}]\mathbf{1} = (\sigma^2/n)[1+(n-1)\rho].$$

Furthermore,

$$\frac{(n-1)\hat{\sigma}^2}{\sigma^2(1-\rho)} = (1/\sigma^2)\mathbf{y}^T\left[\frac{1}{1-\rho}(\mathbf{I}_n - \frac{1}{n}\mathbf{J}_n)\right]\mathbf{y}.$$

Since $[\frac{1}{1-\rho}(\mathbf{I}_n - \frac{1}{n}\mathbf{J}_n)]\mathbf{W} = [\frac{1}{1-\rho}(\mathbf{I}_n - \frac{1}{n}\mathbf{J}_n)][(1-\rho)\mathbf{I}_n + \rho\mathbf{J}_n] = \mathbf{I}_n - \frac{1}{n}\mathbf{J}_n$ is idempotent, it follows from Corollary 14.2.6.1 that $\frac{(n-1)\hat{\sigma}^2}{\sigma^2(1-\rho)} \sim \chi^2(df, NCP)$, where $df = \text{rank}[\frac{1}{1-\rho}(\mathbf{I}_n - \frac{1}{n}\mathbf{J}_n)] = n-1$ and $NCP = \frac{1}{2}(\mu/\sigma)^2\mathbf{1}_n^T[\frac{1}{1-\rho}(\mathbf{I}_n - \frac{1}{n}\mathbf{J}_n)]\mathbf{1}_n = 0$. Also, $\bar{y}$ and $\frac{(n-1)\hat{\sigma}^2}{\sigma^2(1-\rho)}$ are independent.

(b) By the distributional results obtained in the solution to part (a),

$$\frac{\bar{y}}{(\sigma/\sqrt{n})\sqrt{1+(n-1)\rho}} \sim \text{N}\left(\frac{\mu}{(\sigma/\sqrt{n})\sqrt{1+(n-1)\rho}}, 1\right)$$

and

$$\frac{\frac{\bar{y}}{(\sigma/\sqrt{n})\sqrt{1+(n-1)\rho}}}{\sqrt{\frac{(n-1)\hat{\sigma}^2}{\sigma^2(1-\rho)}\Big/(n-1)}} = \frac{\bar{y}}{\hat{\sigma}/\sqrt{n}} \cdot \frac{\sqrt{1-\rho}}{\sqrt{1+(n-1)\rho}} \sim \text{t}(n-1, \frac{\mu}{(\sigma/\sqrt{n})\sqrt{1+(n-1)\rho}}).$$

Thus, a size-ξ test of $H_0 : \mu = 0$ versus $H_a : \mu \neq 0$ is to reject H_0 if and only if $\frac{|\bar{y}|}{\hat{\sigma}/\sqrt{n}} \cdot \frac{\sqrt{1-\rho}}{\sqrt{1+(n-1)\rho}} > t_{\xi/2,n-1}$.

(c) As ρ increases from $\frac{-1}{n-1}$ to 1, the absolute value of the noncentrality parameter of the t distribution decreases, hence the power of the test in part (b) decreases.

(d) By part (b), the null distribution of $\frac{\bar{y}}{\hat{\sigma}/\sqrt{n}}$ is $\sqrt{\frac{1+(n-1)\rho}{1-\rho}}T$, where $T \sim t(n-1)$. It follows that the actual size of the standard size-ξ t test is

$$\Pr\left(\frac{\bar{y}}{\hat{\sigma}/\sqrt{n}} > t_{\xi,n-1}\right) = \Pr\left(\sqrt{\frac{1+(n-1)\rho}{1-\rho}}T > t_{\xi,n-1}\right)$$
$$= \Pr\left(T > \sqrt{\frac{1-\rho}{1+(n-1)\rho}}t_{\xi,n-1}\right).$$

Finally, it is easy to verify that $\sqrt{\frac{1-\rho}{1+(n-1)\rho}} > 1$ if $\rho \in (-\frac{1}{n-1}, 0)$; $\sqrt{\frac{1-\rho}{1+(n-1)\rho}} = 1$ if $\rho = 0$ and $\sqrt{\frac{1-\rho}{1+(n-1)\rho}} < 1$ if $\rho \in (0, 1)$. Hence the actual size satisfies

$$\Pr\left(\frac{\bar{y}}{\hat{\sigma}/\sqrt{n}} > t_{\xi,n-1}\right)\begin{cases} < \xi \text{ if } \rho \in (-\frac{1}{n-1}, 0) \\ = \xi \text{ if } \rho = 0 \\ > \xi \text{ if } \rho \in (0, 1). \end{cases}$$

▶ **Exercise 7** Consider a situation where observations are taken on some outcome in two groups (e.g., a control group and an intervention group). An investigator wishes to determine if exactly the same linear model applies to the two groups. Specifically, suppose that the model for n_1 observations in the first group is

$$\mathbf{y}_1 = \mathbf{X}_1\boldsymbol{\beta}_1 + \mathbf{e}_1$$

and the model for n_2 observations from the second group is

$$\mathbf{y}_2 = \mathbf{X}_2\boldsymbol{\beta}_2 + \mathbf{e}_2,$$

where $\mathbf{X}_1$ and $\mathbf{X}_2$ are model matrices whose p columns (each) correspond to the same p explanatory variables, and $\mathbf{e}_1$ and $\mathbf{e}_2$ are independent and normally distributed random vectors with mean $\mathbf{0}_{n_i}$ and variance–covariance matrix $\sigma^2\mathbf{I}_{n_i}$. Assume that $n_1 + n_2 > 2p$.

(a) Let $\mathbf{C}^T$ be an $s \times p$ matrix such that $\mathbf{C}^T\boldsymbol{\beta}_1$ is estimable under the first model and $\mathbf{C}^T\boldsymbol{\beta}_2$ is estimable under the second model. Show that the two sets of data can be combined into a single linear model of the form

$$\mathbf{y} = \mathbf{X}\boldsymbol{\beta} + \mathbf{e}$$

such that the null hypothesis H_0: $\mathbf{C}^T\boldsymbol{\beta}_1 = \mathbf{C}^T\boldsymbol{\beta}_2$ is testable under this model and can be expressed as H_0: $\mathbf{B}^T\boldsymbol{\beta} = \mathbf{0}$ for a suitably chosen matrix $\mathbf{B}$.

(b) By specializing Theorem 15.2.1, give the size-ξ F-test (in as simple a form as possible) for testing the null hypothesis of part (a) versus the alternative H_a: $\mathbf{C}^T\boldsymbol{\beta}_1 \neq \mathbf{C}^T\boldsymbol{\beta}_2$.

Solution

(a) $\begin{pmatrix} \mathbf{y}_1 \\ \mathbf{y}_2 \end{pmatrix} = \begin{pmatrix} \mathbf{X}_1 & \mathbf{0} \\ \mathbf{0} & \mathbf{X}_2 \end{pmatrix} \begin{pmatrix} \boldsymbol{\beta}_1 \\ \boldsymbol{\beta}_2 \end{pmatrix} + \begin{pmatrix} \mathbf{e}_1 \\ \mathbf{e}_2 \end{pmatrix}$. The null hypothesis $H_0 : \mathbf{C}^T\boldsymbol{\beta}_1 = \mathbf{C}^T\boldsymbol{\beta}_2$ may be written as $H_0 : \mathbf{B}^T\boldsymbol{\beta} = \mathbf{0}$, where $\mathbf{B}^T = (\mathbf{C}^T, -\mathbf{C}^T)$. Furthermore, by the given estimability conditions, $\mathcal{R}(\mathbf{C}^T) \subseteq \mathcal{R}(\mathbf{X}_1)$ and $\mathcal{R}(\mathbf{C}^T) \subseteq \mathcal{R}(\mathbf{X}_2)$, implying that $\mathcal{R}(\mathbf{B}^T) = \mathcal{R}(\mathbf{C}^T, -\mathbf{C}^T) \subseteq \mathcal{R}\begin{pmatrix} \mathbf{X}_1 & \mathbf{0} \\ \mathbf{0} & \mathbf{X}_2 \end{pmatrix} = \mathcal{R}(\mathbf{X})$. Thus the null hypothesis is testable under the single model.

(b) We may write an arbitrary solution to the normal equations as

$$\hat{\boldsymbol{\beta}} = \begin{pmatrix} \hat{\boldsymbol{\beta}}_2 \\ \hat{\boldsymbol{\beta}}_2 \end{pmatrix} = (\mathbf{X}^T\mathbf{X})^-\mathbf{X}^T\mathbf{y} = \begin{pmatrix} \mathbf{X}_1^T\mathbf{X}_1 & \mathbf{0} \\ \mathbf{0} & \mathbf{X}_2^T\mathbf{X}_2 \end{pmatrix}^- \begin{pmatrix} \mathbf{X}_1^T\mathbf{y}_1 \\ \mathbf{X}_2^T\mathbf{y}_2 \end{pmatrix} = \begin{pmatrix} (\mathbf{X}_1^T\mathbf{X}_1)^-\mathbf{X}_1^T\mathbf{y}_1 \\ (\mathbf{X}_2^T\mathbf{X}_2)^-\mathbf{X}_2^T\mathbf{y}_2 \end{pmatrix}.$$

Then $\mathbf{B}^T\hat{\boldsymbol{\beta}} = \mathbf{C}^T\hat{\boldsymbol{\beta}}_1 - \mathbf{C}^T\hat{\boldsymbol{\beta}}_2$, and

$$\begin{aligned} \text{var}(\mathbf{B}^T\hat{\boldsymbol{\beta}}) &= \sigma^2\mathbf{B}^T(\mathbf{X}^T\mathbf{X})^-\mathbf{B} = \sigma^2(\mathbf{C}^T, -\mathbf{C}^T)\begin{pmatrix} (\mathbf{X}_1^T\mathbf{X}_1)^- & \mathbf{0} \\ \mathbf{0} & (\mathbf{X}_2^T\mathbf{X}_2)^- \end{pmatrix}\begin{pmatrix} \mathbf{C} \\ -\mathbf{C} \end{pmatrix} \\ &= \sigma^2\mathbf{C}^T[(\mathbf{X}_1^T\mathbf{X}_1)^- + (\mathbf{X}_2^T\mathbf{X}_2)^-]\mathbf{C}. \end{aligned}$$

So, by Theorem 15.2.1 the size-ξ F-test of the specified hypotheses rejects H_0 if and only if

$$\frac{(\mathbf{C}^T\hat{\boldsymbol{\beta}}_1 - \mathbf{C}^T\hat{\boldsymbol{\beta}}_2)\{\mathbf{C}^T[(\mathbf{X}_1^T\mathbf{X}_1)^- + (\mathbf{X}_2^T\mathbf{X}_2)^-]\mathbf{C}\}^{-1}(\mathbf{C}^T\hat{\boldsymbol{\beta}}_1 - \mathbf{C}^T\hat{\boldsymbol{\beta}}_2)}{s\hat{\sigma}^2} > \mathrm{F}_{\xi,s,n_1+n_2-2p},$$

where $\hat{\sigma}^2 = (\mathbf{y} - \mathbf{X}_1\hat{\boldsymbol{\beta}}_1 - \mathbf{X}_2\hat{\boldsymbol{\beta}}_2)^T(\mathbf{y} - \mathbf{X}_1\hat{\boldsymbol{\beta}}_1 - \mathbf{X}_2\hat{\boldsymbol{\beta}}_2)/(n_1 + n_2 - 2p)$.

▶ **Exercise 8** Consider a situation in which data are available on a pair of variables (x, y) from individuals that belong to two groups. Let $\{(x_{1i}, y_{1i}) : i = 1, \ldots, n_1\}$ and $\{(x_{2i}, y_{2i}) : i = 1, \ldots, n_2\}$, respectively, denote the data from these two groups. Assume that the following model holds:

$$\begin{aligned} y_{1i} &= \beta_{11} + \beta_{21}(x_{1i} - \bar{x}_1) + e_{1i} \quad (i = 1, \ldots, n_1) \\ y_{2i} &= \beta_{12} + \beta_{22}(x_{2i} - \bar{x}_2) + e_{2i} \quad (i = 1, \ldots, n_2), \end{aligned}$$

where the e_{1i}'s and e_{2i}'s are independent $\mathrm{N}(0, \sigma^2)$ random variables, $n_1 \geq 2, n_2 \geq 2$, $x_{11} \neq x_{12}$, and $x_{21} \neq x_{22}$. Furthermore, define the following notation:

$$\bar{x}_1 = n_1^{-1}\sum_{i=1}^{n_1} x_{1i}, \quad \bar{x}_2 = n_2^{-1}\sum_{i=1}^{n_2} x_{2i}, \quad \bar{y}_1 = n_1^{-1}\sum_{i=1}^{n_1} y_{1i}, \quad \bar{y}_2 = n_2^{-1}\sum_{i=1}^{n_2} y_{2i},$$

$$SXX_1 = \sum_{i=1}^{n_1}(x_{1i} - \bar{x}_1)^2, \quad SXX_2 = \sum_{i=1}^{n_2}(x_{2i} - \bar{x}_2)^2,$$

$$SXY_1 = \sum_{i=1}^{n_1}(x_{1i} - \bar{x}_1)(y_{1i} - \bar{y}_1), \quad SXY_2 = \sum_{i=1}^{n_2}(x_{2i} - \bar{x}_2)(y_{2i} - \bar{y}_2).$$

Observe that this model can be written in matrix notation as the unordered 2-part model

$$\mathbf{y} = \begin{pmatrix} \mathbf{1}_{n_1} & \mathbf{0}_{n_1} \\ \mathbf{0}_{n_2} & \mathbf{1}_{n_2} \end{pmatrix} \begin{pmatrix} \beta_{11} \\ \beta_{12} \end{pmatrix} + \begin{pmatrix} \mathbf{x}_1 - \bar{x}_1\mathbf{1}_{n_1} & \mathbf{0} \\ \mathbf{0} & \mathbf{x}_2 - \bar{x}_2\mathbf{1}_{n_2} \end{pmatrix} \begin{pmatrix} \beta_{21} \\ \beta_{22} \end{pmatrix} + \mathbf{e}.$$

(a) Give nonmatrix expressions for the BLUEs of $\beta_{11}, \beta_{12}, \beta_{21}$, and β_{22}, and for the residual mean square, in terms of the quantities defined above. Also give an expression for the variance–covariance matrix of the BLUEs.
(b) Give the two-part ANOVA table for the ordered two-part model that coincides with the unordered two-part model written above. Give nonmatrix expressions for the degrees of freedom and sums of squares. Determine the distributions of the (suitably scaled) sums of squares in this table, giving nonmatrix expressions for any noncentrality parameters.
(c) Give a size-ξ test for equal intercepts, i.e., for $H_0 : \beta_{11} = \beta_{12}$ versus $H_a : \beta_{11} \neq \beta_{12}$.
(d) Give a size-ξ test for equal slopes (or "parallelism"), i.e., for $H_0 : \beta_{21} = \beta_{22}$ versus $H_a : \beta_{21} \neq \beta_{22}$.
(e) Give a size-ξ test for identical lines, i.e., for $H_0 : \beta_{11} = \beta_{12}, \beta_{21} = \beta_{22}$ versus H_a: either $\beta_{11} \neq \beta_{12}$ or $\beta_{21} \neq \beta_{22}$.
(f) Let c represent a specified real number and let $0 < \xi < 1$. Give a size-ξ test of the null hypothesis that the two groups' regression lines intersect at $x = c$ and that the slope for the first group is the negative of the slope for the second group versus the alternative hypothesis that at least one of these statements is false.
(g) Suppose that we desire to predict the value of a new observation $y_{1,n+1}$ from the first group corresponding to the value $x_{1,n+1}$ of x_1, and we also desire to predict the value of a new observation $y_{2,n+2}$ from the second group corresponding to the value $x_{2,n+2}$ of x_2. Give nonmatrix expressions for $100(1-\xi)\%$ prediction intervals for these two new observations, and give an expression for a $100(1-\xi)\%$ prediction ellipsoid for them. (In the latter expression, you may use vectors and matrices but if so, give nonmatrix expressions for each element of them.)

(h) Give confidence bands for the two lines for which the simultaneous coverage probability (for both lines and for all $x \in \mathbb{R}$) is at least $1-\xi$.

Solution

(a) $\mathbf{X} = \begin{pmatrix} \mathbf{1}_{n_1} & \mathbf{0} & \mathbf{x}_1 - \bar{x}_1\mathbf{1}_{n_1} & \mathbf{0} \\ \mathbf{0} & \mathbf{1}_{n_2} & \mathbf{0} & \mathbf{x}_2 - \bar{x}_2\mathbf{1}_{n_2} \end{pmatrix}$, where $\mathbf{x}_1 = (x_{11}, \ldots, x_{1n_1})^T$ and $\mathbf{x}_2 = (x_{21}, \ldots, x_{2n_2})^T$, $\mathbf{X}^T\mathbf{X} = \text{diag}(n_1, n_2, SXX_1, SXX_2)$, and $\mathbf{X}^T\mathbf{y} = (n_1\bar{y}_1, n_2\bar{y}_2, SXY_1, SXY_2)^T$. Therefore, the BLUEs of $\beta_{11}, \beta_{12}, \beta_{21}$, and β_{22} are given by $\bar{y}_1, \bar{y}_2, SXY_1/SXX_1$, and SXY_2/SXX_2, respectively. Furthermore,

$$\begin{aligned}\hat{\sigma}^2 &= \frac{(\mathbf{y}-\mathbf{X}\hat{\boldsymbol{\beta}})^T(\mathbf{y}-\mathbf{X}\hat{\boldsymbol{\beta}})}{n_1+n_2-\text{rank}(\mathbf{X})} \\ &= \frac{\sum_{i=1}^{n_1}(y_{1i}-\bar{y}_1)^2 + \sum_{i=1}^{n_2}(y_{2i}-\bar{y}_2)^2 - (SXY_1^2/SXX_1) - (SXY_2^2/SXX_2)}{n_1+n_2-4}\end{aligned}$$

and the variance–covariance matrix of the BLUEs is $\sigma^2(\mathbf{X}^T\mathbf{X})^{-1} = \sigma^2\text{diag}(1/n_1, 1/n_2, 1/SXX_1, 1/SXX_2)$.

(b)

Source	df	Sum of squares
Intercepts	2	$n_1\bar{y}_1^2 + n_2\bar{y}_2^2$
$\mathbf{x}_1, \mathbf{x}_2$|Intercepts	2	$(SXY_1^2/SXX_1) + (SXY_2^2/SXX_2)$
Residual	n_1+n_2-4	By subtraction
Total	n_1+n_2	$\sum_{i=1}^{n_1} y_{1i}^2 + \sum_{i=1}^{n_2} y_{2i}^2$

$$\frac{n_1\bar{y}_1^2 + n_2\bar{y}_2^2}{\sigma^2} \sim \chi^2\left(2, \frac{n_1\beta_{11}^2 + n_2\beta_{12}^2}{2\sigma^2}\right),$$

$$\frac{(SXY_1^2/SXX_1) + (SXY_2^2/SXX_2)}{\sigma^2} \sim \chi^2\left(2, \frac{\beta_{21}^2 SXX_1 + \beta_{22}^2 SXX_2}{2\sigma^2}\right),$$

$$\frac{(n_1+n_2-4)\hat{\sigma}^2}{\sigma^2} \sim \chi^2(n_1+n_2-4).$$

(c) Using the results from part (a), a size-ξ t test of the given hypotheses is to reject H_0 if and only if

$$\frac{|\bar{y}_1 - \bar{y}_2|}{\hat{\sigma}\sqrt{(1/n_1)+(1/n_2)}} \geq t_{\xi/2, n_1+n_2-4}.$$

(d) Using the results from part (a), a size-ξ t test of the given hypotheses is to reject H_0 if and only if

$$\frac{|\hat{\beta}_{21} - \hat{\beta}_{22}|}{\hat{\sigma}\sqrt{(1/SXX_1) + (1/SXX_2)}} \geq t_{\xi/2,n_1+n_2-4}.$$

(e) This null hypothesis may be written as $H_0 : \mathbf{C}^T\boldsymbol{\beta} = \mathbf{0}$, where $\mathbf{C}^T = \begin{pmatrix} 1 & -1 & 0 & 0 \\ 0 & 0 & 1 & -1 \end{pmatrix}$, and it is clearly testable. Furthermore,

$$\mathbf{C}^T\hat{\boldsymbol{\beta}} = \begin{pmatrix} 1 & -1 & 0 & 0 \\ 0 & 0 & 1 & -1 \end{pmatrix} \begin{pmatrix} \bar{y}_1 \\ \bar{y}_2 \\ SXY_1/SXX_1 \\ SXY_2/SXX_2 \end{pmatrix}$$

$$= \begin{pmatrix} \bar{y}_1 - \bar{y}_2 \\ (SXY_1/SXX_1) - (SXY_2/SXX_2) \end{pmatrix}$$

and

$$\mathbf{C}^T(\mathbf{X}^T\mathbf{X})^-\mathbf{C} = \begin{pmatrix} 1 & -1 & 0 & 0 \\ 0 & 0 & 1 & -1 \end{pmatrix} [\text{diag}(n_1, n_2, SXX_1, SXX_2)]^{-1} \begin{pmatrix} 1 & 0 \\ -1 & 0 \\ 0 & 1 \\ 0 & -1 \end{pmatrix}$$

$$= \begin{pmatrix} \frac{1}{n_1} + \frac{1}{n_2} & 0 \\ 0 & \frac{1}{SXX_1} + \frac{1}{SXX_2} \end{pmatrix}.$$

By Theorem 15.2.1, a size-ξ test of this null hypothesis versus the alternative $H_a : \mathbf{C}^T\boldsymbol{\beta} \neq \mathbf{0}$ is to reject the null hypothesis if and only if

$$\frac{(\mathbf{C}^T\hat{\boldsymbol{\beta}})^T[\mathbf{C}^T(\mathbf{X}^T\mathbf{X})^-\mathbf{C}]^{-1}\mathbf{C}^T\hat{\boldsymbol{\beta}}}{2\hat{\sigma}^2} \geq F_{\xi,2,n_1+n_2-4},$$

i.e., if and only if

$$\left[\frac{(\bar{y}_1 - \bar{y}_2)^2}{(1/n_1) + (1/n_2)} + \frac{[(SXY_1/SXX_1) - (SXY_2/SXX_2)]^2}{(1/SXX_1) + (1/SXX_2)}\right] \geq 2\hat{\sigma}^2 F_{\xi,2,n_1+n_2-4}.$$

(f) This null hypothesis may be written as $H_0 : \mathbf{C}^T\boldsymbol{\beta} = \mathbf{0}$, where $\mathbf{C}^T = \begin{pmatrix} 1 & -1 & c-\bar{x}_1 & \bar{x}_2 - c \\ 0 & 0 & 1 & 1 \end{pmatrix}$, and it is clearly testable. Furthermore,

$$\mathbf{C}^T\hat{\boldsymbol{\beta}} = \begin{pmatrix} 1 & -1 & c-\bar{x}_1 & \bar{x}_2 - c \\ 0 & 0 & 1 & 1 \end{pmatrix} \begin{pmatrix} \bar{y}_1 \\ \bar{y}_2 \\ SXY_1/SXX_1 \\ SXY_2/SXX_2 \end{pmatrix}$$

$$= \begin{pmatrix} \bar{y}_1 - \bar{y}_2 + (c-\bar{x}_1)(SXY_1/SXX_1) - (c-\bar{x}_2)(SXY_2/SXX_2) \\ (SXY_1/SXX_1) + (SXY_2/SXX_2) \end{pmatrix}$$

and

$$\mathbf{C}^T(\mathbf{X}^T\mathbf{X})^-\mathbf{C} = \begin{pmatrix} 1 & -1 & c-\bar{x}_1 & \bar{x}_2 - c \\ 0 & 0 & 1 & 1 \end{pmatrix} [\text{diag}(n_1, n_2, SXX_1, SXX_2)]^{-1} \begin{pmatrix} 1 & 0 \\ -1 & 0 \\ c-\bar{x}_1 & 1 \\ \bar{x}_2 - c & 1 \end{pmatrix}$$

$$= \begin{pmatrix} \frac{1}{n_1} + \frac{1}{n_2} + \frac{(c-\bar{x}_1)^2}{SXX_1} + \frac{(c-\bar{x}_2)^2}{SXX_2} & \frac{c-\bar{x}_1}{SXX_1} - \frac{c-\bar{x}_2}{SXX_2} \\ \frac{c-\bar{x}_1}{SXX_1} - \frac{c-\bar{x}_2}{SXX_2} & \frac{1}{SXX_1} + \frac{1}{SXX_2} \end{pmatrix}.$$

By Theorem 15.2.1, a size-ξ test of this null hypothesis versus the alternative $H_a : \mathbf{C}^T\boldsymbol{\beta} \neq \mathbf{0}$ is to reject the null hypothesis if and only if

$$\frac{(\mathbf{C}^T\hat{\boldsymbol{\beta}})^T[\mathbf{C}^T(\mathbf{X}^T\mathbf{X})^-\mathbf{C}]^{-1}\mathbf{C}^T\hat{\boldsymbol{\beta}}}{2\hat{\sigma}^2} \geq F_{\xi,2,n_1+n_2-4}.$$

(g) Define $\tau_1 = \mathbf{c}_1^T\boldsymbol{\beta} + e_{1,n+1}$ and $\tau_2 = \mathbf{c}_2^T\boldsymbol{\beta} + e_{2,n+2}$, where

$$\mathbf{c}_1^T = \begin{pmatrix} 1 & 0 & x_{1,n+1} - \bar{x}_1 & 0 \end{pmatrix}, \quad \mathbf{c}_2^T = \begin{pmatrix} 1 & 0 & x_{2,n+2} - \bar{x}_2 & 0 \end{pmatrix}.$$

The BLUPs of τ_1 and τ_2 are $\tilde{\tau}_1 = \hat{\beta}_{11} + \hat{\beta}_{21}(x_{1,n+1} - \bar{x}_1)$ and $\tilde{\tau}_2 = \hat{\beta}_{12} + \hat{\beta}_{22}(x_{2,n+2} - \bar{x}_2)$, respectively, and their prediction error variances are $\sigma^2\left[1 + \frac{1}{n_1} + \frac{(x_{1,n+1}-\bar{x}_1)^2}{SXX_1}\right] \equiv \sigma^2 Q_1$ and $\sigma^2\left[1 + \frac{1}{n_2} + \frac{(x_{2,n+2}-\bar{x}_2)^2}{SXX_2}\right] \equiv \sigma^2 Q_2$. Therefore, a $100(1-\xi)\%$ prediction interval for τ_1 is given by

$$\hat{\beta}_{11} + \hat{\beta}_{21}(x_{1,n+1} - \bar{x}_1) \pm t_{\xi/2,n_1+n_2+4}\hat{\sigma}\sqrt{Q_1}$$

and a $100(1-\xi)\%$ prediction interval for τ_2 is given by

$$\hat{\beta}_{12} + \hat{\beta}_{22}(x_{2,n+2} - \bar{x}_2) \pm t_{\xi/2,n_1+n_2+4}\hat{\sigma}\sqrt{Q_2}.$$

Because $\text{cov}[\hat{\beta}_{11} + \hat{\beta}_{21}(x_{1,n+1} - \bar{x}_1) - \tau_1, \hat{\beta}_{12} + \hat{\beta}_{22}(x_{2,n+2} - \bar{x}_2) - \tau_2] = 0$, a $100(1 - \xi)\%$ prediction ellipsoid for (τ_1, τ_2) is given by

$$\left\{(\tau_1, \tau_2) : \frac{(\tilde{\tau}_1 - \tau_1)^2}{Q_1} + \frac{(\tilde{\tau}_2 - \tau_2)^2}{Q_2} \leq 2\hat{\sigma}^2 F_{\xi,2,n_1+n_2-4}\right\}.$$

(h)

$$\left\{[\hat{\beta}_{11} + \hat{\beta}_{21}(x - \bar{x}_1)] \pm \sqrt{4F_{\xi,4,n_1+n_2-4}}\hat{\sigma}\sqrt{\frac{1}{n_1} + \frac{(x - \bar{x}_1)^2}{SXX_1}} \text{ for all } x \in \mathbb{R}\right\}$$

and

$$\left\{[\hat{\beta}_{12} + \hat{\beta}_{22}(x - \bar{x}_2)] \pm \sqrt{4F_{\xi,4,n_1+n_2-4}}\hat{\sigma}\sqrt{\frac{1}{n_2} + \frac{(x - \bar{x}_2)^2}{SXX_2}} \text{ for all } x \in \mathbb{R}\right\}.$$

► **Exercise 9** Consider a situation similar to that described in the previous exercise, except that the two lines are known to be parallel. Thus the model is

$$\begin{aligned} y_{1i} &= \beta_{11} + \beta_2(x_{1i} - \bar{x}_1) + e_{1i} \quad (i = 1, \ldots, n_1) \\ y_{2i} &= \beta_{12} + \beta_2(x_{2i} - \bar{x}_2) + e_{2i} \quad (i = 1, \ldots, n_2), \end{aligned}$$

where again the e_{1i}'s and e_{2i}'s are independent $\text{N}(0, \sigma^2)$ random variables, $n_1 \geq 2$, $n_2 \geq 2$, $x_{11} \neq x_{12}$, and $x_{21} \neq x_{22}$. Adopt the same notation as in the previous exercise, but also define the following:

$$SXX_{12} = SXX_1 + SXX_2, \quad SXY_{12} = SXY_1 + SXY_2.$$

Observe that this model can be written in matrix notation as the unordered 2-part model

$$\mathbf{y} = \begin{pmatrix} \mathbf{1}_{n_1} & \mathbf{0}_{n_1} \\ \mathbf{0}_{n_2} & \mathbf{1}_{n_2} \end{pmatrix} \begin{pmatrix} \beta_{11} \\ \beta_{12} \end{pmatrix} + \begin{pmatrix} \mathbf{x}_1 - \bar{x}_1 \mathbf{1}_{n_1} \\ \mathbf{x}_2 - \bar{x}_2 \mathbf{1}_{n_2} \end{pmatrix} \beta_2 + \mathbf{e}.$$

(a) Give nonmatrix expressions for the BLUEs of β_{11}, β_{12}, and β_2, and for the residual mean square, in terms of the quantities defined above. Also give an expression for the variance–covariance matrix of the BLUEs.
(b) Give the two-part ANOVA table for the ordered two-part model that coincides with the unordered two-part model written above. Give nonmatrix expressions for the degrees of freedom and sums of squares. Determine the distributions of the (suitably scaled) sums of squares in this table, giving nonmatrix expressions for any noncentrality parameters.

(c) Suppose that we wish to estimate the distance, δ, between the lines of the two groups, defined as the amount by which the line for Group 1 exceeds the line for Group 2. Verify that

$$\delta = \beta_{11} - \beta_{12} - \beta_2(\bar{x}_1 - \bar{x}_2)$$

and obtain a $100(1-\xi)\%$ confidence interval for δ.

(d) Obtain a size-ξ t test for $H_0 : \delta = 0$ versus $H_a : \delta \neq 0$. (This is a test for identical lines, given parallelism.)

(e) Suppose that enough resources (time, money, etc.) were available to take observations on a total of 20 observations (i.e., $n_1 + n_2 = 20$). From a design standpoint, how would you choose n_1, n_2, the x_{1i}'s and the x_{2i}'s in order to minimize the width of the confidence interval in part (c)?

(f) Give a size-ξ test for $H_0 : \beta_2 = 0$ versus $H_a : \beta_2 \neq 0$.

(g) Obtain the size-ξ F-test of $H_0 : \begin{pmatrix} \delta \\ \beta_2 \end{pmatrix} = \mathbf{0}$ versus H_a : not H_0.

(h) Let $\xi \in (0, 1)$. Give an expression for a $100(1-\xi)\%$ confidence ellipsoid for $\boldsymbol{\beta} \equiv (\beta_{11}, \beta_{12}, \beta_2)^T$.

(i) Let x_0 represent a specified x-value. Give confidence intervals for $\beta_{11} + \beta_2 x_0$ and $\beta_{12} + \beta_2 x_0$ whose simultaneous coverage probability is exactly $1 - \xi$.

(j) Give confidence bands for the two parallel lines whose simultaneous coverage probability (for both lines and for all $x \in \mathbb{R}$) is at least $1 - \xi$.

Solution

(a) $\mathbf{X}^T\mathbf{X} = \text{diag}(n_1, n_2, SXX_{12})$ and $\mathbf{X}^T\mathbf{y} = (n_1\bar{y}_1, n_2\bar{y}_2, SXY_{12})^T$. Therefore, the BLUEs of β_{11}, β_{12}, and β_2 are given by $\bar{y}_1, \bar{y}_2$, and SXY_{12}/SXX_{12}, respectively. Furthermore,

$$\hat{\sigma}^2 = \frac{\sum_{i=1}^{n_1}(y_{1i} - \bar{y}_1)^2 + \sum_{i=1}^{n_2}(y_{2i} - \bar{y}_2)^2 - (SXY_{12}^2/SXX_{12})}{n_1 + n_2 - 3}$$

and the variance–covariance matrix of the BLUEs is $\sigma^2\text{diag}(1/n_1, 1/n_2, 1/SXX_{12})$.

(b)

Source	df	Sum of squares
Intercepts	2	$n_1\bar{y}_1^2 + n_2\bar{y}_2^2$
$\mathbf{x}$|Intercepts	1	SXY_{12}^2/SXX_{12}
Residual	$n_1 + n_2 - 3$	By subtraction
Total	$n_1 + n_2$	$\sum_{i=1}^{n_1} y_{1i}^2 + \sum_{i=1}^{n_2} y_{2i}^2$

$$\frac{n_1\bar{y}_1^2 + n_2\bar{y}_2^2}{\sigma^2} \sim \chi^2\left(2, \frac{n_1\beta_{11}^2 + n_2\beta_{12}^2}{2\sigma^2}\right), \quad \frac{SXY_{12}^2}{\sigma^2 SXX_{12}} \sim \chi^2\left(1, \frac{\beta_2^2 SXX_{12}}{2\sigma^2}\right),$$

and

$$\frac{(n_1+n_2-3)\hat{\sigma}^2}{\sigma^2} \sim \chi^2(n_1+n_2-3).$$

(c) For any $x \in \mathbb{R}$, $\delta = \beta_{11}+\beta_{12}(x-\bar{x}_1)-[\beta_{12}+\beta_2(x-\bar{x}_2)] = \beta_{11}-\beta_{12}-\beta_2(\bar{x}_1-\bar{x}_2)$. The ordinary least squares estimator of δ is $\hat{\delta} = \hat{\beta}_{11} - \hat{\beta}_{12} - \hat{\beta}_2(\bar{x}_1 - \bar{x}_2)$, and

$$\hat{\delta} \pm t_{\xi/2,n_1+n_2-3}\hat{\sigma}\sqrt{(1/n_1)+(1/n_2)+(\bar{x}_1-\bar{x}_2)^2/SXX_{12}}$$

is a $100(1-\xi)\%$ confidence interval for δ.

(d) Reject H_0 if and only if

$$\frac{|\hat{\delta}|}{\hat{\sigma}\sqrt{(1/n_1)+(1/n_2)+(1/SXX_{12})(\bar{x}_1-\bar{x}_2)^2}} > t_{\xi/2,n_1+n_2-3}.$$

(e) To minimize the width of the confidence interval obtained in part (c) subject to $n_1+n_2=20$, we should minimize both of $(1/n_1)+(1/n_2)$ and $(\bar{x}_1-\bar{x}_2)^2$; the first of these is minimized at $n_1 = n_2 = 10$, and the latter is minimized when $\bar{x}_1 = \bar{x}_2$.

(f) Reject H_0 if and only if

$$\frac{|\hat{\beta}_2|}{\hat{\sigma}\sqrt{1/SXX_{12}}} > t_{\xi/2,n_1+n_2-3}.$$

(g) This null hypothesis may be written as $H_0 : \mathbf{C}^T\boldsymbol{\beta} = \mathbf{0}$, where $\mathbf{C}^T = \begin{pmatrix} 1 & -1 & -(\bar{x}_1-\bar{x}_2) \\ 0 & 0 & 1 \end{pmatrix}$, and it is clearly testable. Furthermore,

$$\mathbf{C}^T\hat{\boldsymbol{\beta}} = \begin{pmatrix} 1 & -1 & -(\bar{x}_1-\bar{x}_2) \\ 0 & 0 & 1 \end{pmatrix} \begin{pmatrix} \bar{y}_1 \\ \bar{y}_2 \\ SXY_{12}/SXX_{12} \end{pmatrix}$$

$$= \begin{pmatrix} \bar{y}_1 - \bar{y}_2 - (\bar{x}_1-\bar{x}_2)(SXY_{12}/SXX_{12}) \\ SXY_{12}/SXX_{12} \end{pmatrix}$$

and

$$\mathbf{C}^T(\mathbf{X}^T\mathbf{X})^-\mathbf{C} = \begin{pmatrix} 1 & -1 & -(\bar{x}_1-\bar{x}_2) \\ 0 & 0 & 1 \end{pmatrix} [\text{diag}(n_1, n_2, SXX_{12})]^{-1} \begin{pmatrix} 1 & 0 \\ -1 & 0 \\ -(\bar{x}_1-\bar{x}_2) & 1 \end{pmatrix}$$

$$= \begin{pmatrix} \frac{1}{n_1}+\frac{1}{n_2}+\frac{(\bar{x}_1-\bar{x}_2)^2}{SXX_{12}} & -\frac{\bar{x}_1-\bar{x}_2}{SXX_{12}} \\ -\frac{\bar{x}_1-\bar{x}_2}{SXX_{12}} & \frac{1}{SXX_{12}} \end{pmatrix}.$$

By Theorem 15.2.1, a size-ξ test of this null hypothesis versus the alternative $H_a : \mathbf{C}^T\boldsymbol{\beta} \neq \mathbf{0}$ is to reject the null hypothesis if and only if

$$\frac{(\mathbf{C}^T\hat{\boldsymbol{\beta}})^T[\mathbf{C}^T(\mathbf{X}^T\mathbf{X})^-\mathbf{C}]^{-1}\mathbf{C}^T\hat{\boldsymbol{\beta}}}{2\hat{\sigma}^2} \geq F_{\xi,2,n_1+n_2-3}.$$

(h) By (15.2) and the diagonality of $\mathbf{X}^T\mathbf{X}$, a $100(1-\xi)\%$ confidence ellipsoid for $\boldsymbol{\beta} = (\beta_{11}, \beta_{12}, \beta_2)^T$ is given by the set of such $\boldsymbol{\beta}$ for which

$$\{(\beta_{11}, \beta_{12}, \beta_2) : (\hat{\beta}_{11}-\beta_{11})^2 n_1 + (\hat{\beta}_{12}-\beta_{12})^2 n_2 + (\hat{\beta}_2-\beta_2)^2 SXX_{12} \leq 3\hat{\sigma}^2 F_{\xi,3,n_1+n_2-3}\}.$$

(i) The $100(1-\xi)\%$ multivariate t simultaneous confidence intervals for $\beta_{11}+\beta_2 x_0$ and $\beta_{12}+\beta_2 x_0$ are given by

$$[\hat{\beta}_{11} + \hat{\beta}_2(x_0 - \bar{x}_1)] \pm t_{\xi/2,2,n_1+n_2-3,\mathbf{R}}\hat{\sigma}\sqrt{(1/n_1) + x_0^2/SXX_{12}}$$

and

$$[\hat{\beta}_{12} + \hat{\beta}_2(x_0 - \bar{x}_2)] \pm t_{\xi/2,2,n_1+n_2-3,\mathbf{R}}\hat{\sigma}\sqrt{(1/n_2) + x_0^2/SXX_{12}},$$

where

$$\mathbf{R} = \begin{pmatrix} 1 & \frac{x_0^2/SXX_{12}}{\sqrt{(1/n_1)+(x_0^2/SXX_{12})}\sqrt{(1/n_2)+(x_0^2/SXX_{12})}} \\ & 1 \end{pmatrix}.$$

(j)

$$\left\{[\hat{\beta}_{11} + \hat{\beta}_2(x - \bar{x}_1)] \pm \sqrt{3F_{\xi,3,n_1+n_2-3}}\,\hat{\sigma}\sqrt{\frac{1}{n_1} + \frac{(x-\bar{x}_1)^2}{SXX_{12}}} \text{ for all } x \in \mathbb{R}\right\}$$

and

$$\left\{[\hat{\beta}_{21} + \hat{\beta}_2(x - \bar{x}_2)] \pm \sqrt{3F_{\xi,3,n_1+n_2-3}}\,\hat{\sigma}\sqrt{\frac{1}{n_2} + \frac{(x-\bar{x}_2)^2}{SXX_{12}}} \text{ for all } x \in \mathbb{R}\right\}.$$

▶ **Exercise 10** Consider the normal Gauss–Markov two-way main effects model with exactly one observation per cell, which has model equation

$$y_{ij} = \mu + \alpha_i + \gamma_j + e_{ij} \quad (i = 1, \ldots, q;\ j = 1, \ldots, m).$$

Define

$$f = \sum_{i=1}^{q}\sum_{j=1}^{m} b_{ij}(y_{ij} - \bar{y}_{i\cdot} - \bar{y}_{\cdot j} + \bar{y}_{\cdot\cdot}) = \mathbf{b}^T(\mathbf{I} - \mathbf{P_X})\mathbf{y},$$

where $\{b_{ij}\}$ is a set of real numbers, not all equal to 0, such that $\sum_{i=1}^{q} b_{ij} = 0$ for all j and $\sum_{j=1}^{m} b_{ij} = 0$ for all i.

(a) Show that

$$\frac{f^2}{\sigma^2\mathbf{b}^T\mathbf{b}} \quad \text{and} \quad (1/\sigma^2)\sum_{i=1}^{q}\sum_{j=1}^{m}(y_{ij} - \bar{y}_{i\cdot} - \bar{y}_{\cdot j} + \bar{y}_{\cdot\cdot})^2 - \frac{f^2}{\sigma^2\mathbf{b}^T\mathbf{b}}$$

are jointly distributed as independent $\chi^2(1)$ and $\chi^2(qm - q - m)$ random variables under this model.

(b) Let $\mathbf{y}_I$ represent the q-vector whose ith element is $\bar{y}_{i\cdot} - \bar{y}_{\cdot\cdot}$, let $\mathbf{y}_{II}$ represent the m-vector whose jth element is $\bar{y}_{\cdot j} - \bar{y}_{\cdot\cdot}$, and let $\mathbf{y}_{III}$ represent the qm-vector whose ijth element is $y_{ij} - \bar{y}_{i\cdot} - \bar{y}_{\cdot j} + \bar{y}_{\cdot\cdot}$. Show that $\mathbf{y}_I, \mathbf{y}_{II}$, and $\mathbf{y}_{III}$ are independent under this model.

(c) Now suppose that we condition on $\mathbf{y}_I$ and $\mathbf{y}_{II}$, and consider the expansion of the model to

$$y_{ij} = \mu + \alpha_i + \gamma_j + \lambda(\bar{y}_{i\cdot} - \bar{y}_{\cdot\cdot})(\bar{y}_{\cdot j} - \bar{y}_{\cdot\cdot}) + e_{ij} \quad (i = 1, \ldots, q;\; j = 1, \ldots, m).$$

Using part (a), show that conditional on $\mathbf{y}_I$ and $\mathbf{y}_{II}$,

$$\frac{\left(\sum_{i=1}^{q}\sum_{j=1}^{m}(\bar{y}_{i\cdot} - \bar{y}_{\cdot\cdot})(\bar{y}_{\cdot j} - \bar{y}_{\cdot\cdot})y_{ij}\right)^2}{\sigma^2\sum_{i=1}^{q}(\bar{y}_{i\cdot} - \bar{y}_{\cdot\cdot})^2\sum_{j=1}^{m}(\bar{y}_{\cdot j} - \bar{y}_{\cdot\cdot})^2}$$

and

$$(1/\sigma^2)\sum_{i=1}^{q}\sum_{j=1}^{m}(y_{ij} - \bar{y}_{i\cdot} - \bar{y}_{\cdot j} + \bar{y}_{\cdot\cdot})^2 - \frac{\left(\sum_{i=1}^{q}\sum_{j=1}^{m}(\bar{y}_{i\cdot} - \bar{y}_{\cdot\cdot})(\bar{y}_{\cdot j} - \bar{y}_{\cdot\cdot})y_{ij}\right)^2}{\sigma^2\sum_{i=1}^{q}(\bar{y}_{i\cdot} - \bar{y}_{\cdot\cdot})^2\sum_{j=1}^{m}(\bar{y}_{\cdot j} - \bar{y}_{\cdot\cdot})^2}$$

are jointly distributed as independent $\chi^2(1, \nu)$ and $\chi^2(qm - q - m)$ random variables, and determine ν. [Hint: First observe that $\sum_{i=1}^{q}\sum_{j=1}^{m}(\bar{y}_{i\cdot} - \bar{y}_{\cdot\cdot})(\bar{y}_{\cdot j} - \bar{y}_{\cdot\cdot})(y_{ij} - \bar{y}_{i\cdot} - \bar{y}_{\cdot j} + \bar{y}_{\cdot\cdot}) = \sum_{i=1}^{q}\sum_{j=1}^{m}(\bar{y}_{i\cdot} - \bar{y}_{\cdot\cdot})(\bar{y}_{\cdot j} - \bar{y}_{\cdot\cdot})y_{ij}$.]

(d) Consider testing $H_0 : \lambda = 0$ versus $H_a : \lambda \neq 0$ in the conditional model specified in part (c). Show that the noncentrality parameter of the chi-square distribution of the first quantity displayed above is equal to 0 under H_0, and then use parts (b) and (c) to determine the unconditional joint distribution of both quantities displayed above under H_0.

(e) Using part (d), obtain a size-ξ test of $H_0 : \lambda = 0$ versus $H_a : \lambda \neq 0$ in the unconditional statistical model (not a linear model according to our definition) given by

$$y_{ij} = \mu + \alpha_i + \gamma_j + \lambda(\bar{y}_{i\cdot} - \bar{y}_{\cdot\cdot})(\bar{y}_{\cdot j} - \bar{y}_{\cdot\cdot}) + e_{ij},$$

where the e_{ij}'s are independent $\mathrm{N}(0, \sigma^2)$ random variables.

Note: A nonzero λ in the model defined in part (e) implies that the effects of the two crossed factors are not additive. Consequently, the test ultimately obtained in part (e) tests for nonadditivity of a particular form. The test and its derivation (which is essentially equivalent to the steps in this exercise) are due to Tukey (1949), and the test is commonly referred to as "Tukey's one-degree-of-freedom test for nonadditivity."

Solution

(a) By the given properties of $\mathbf{b}$, we obtain

$$\begin{aligned}\mathbf{b}^T\mathbf{X} &= \mathbf{b}^T(\mathbf{1}_{qm}, \mathbf{I}_q \otimes \mathbf{1}_m, \mathbf{1}_q \otimes \mathbf{I}_m) \\ &= (b_{\cdot\cdot}, b_{1\cdot}, b_{2\cdot}, \ldots, b_{q\cdot}, b_{\cdot 1}, b_{\cdot 2}, \ldots, b_{\cdot m}) \\ &= \mathbf{0}^T_{1+q+m}.\end{aligned}$$

Thus, $\mathbf{P_X b} = \mathbf{0}$ and $f = \mathbf{b}^T\mathbf{y}$. Now, the two quantities whose joint distribution we seek may be written as $\mathbf{y}^T\mathbf{A}_1\mathbf{y}/\sigma^2$ and $\mathbf{y}^T\mathbf{A}_2\mathbf{y}/\sigma^2$, where

$$\mathbf{A}_1 = \frac{1}{\mathbf{b}^T\mathbf{b}}\mathbf{b}\mathbf{b}^T, \quad \mathbf{A}_2 = \mathbf{I} - \mathbf{P_X} - \frac{1}{\mathbf{b}^T\mathbf{b}}\mathbf{b}\mathbf{b}^T.$$

Observe that $\mathbf{A}_1$ and $\mathbf{A}_2$ sum to $\mathbf{I} - \mathbf{P_X}$, which is idempotent; that $\mathbf{A}_1$ is obviously idempotent; and that $\mathbf{A}_2$ is idempotent because

$$\left(\frac{1}{\mathbf{b}^T\mathbf{b}}\mathbf{b}\mathbf{b}^T\right)(\mathbf{I} - \mathbf{P_X}) = \frac{1}{\mathbf{b}^T\mathbf{b}}\mathbf{b}\mathbf{b}^T.$$

Therefore, by Theorem 14.4.1 (Cochran's theorem), the two quantities of interest are independent; the distribution of the first quantity is chi-square with one degree of freedom and noncentrality parameter

$$(1/2\sigma^2)\boldsymbol{\beta}^T\mathbf{X}^T\left(\frac{1}{\mathbf{b}^T\mathbf{b}}\mathbf{b}\mathbf{b}^T\right)\mathbf{X}\boldsymbol{\beta} = 0;$$

and the distribution of the second quantity is chi-square with $qm-q-m$ degrees of freedom and noncentrality parameter

$$(1/2\sigma^2)\boldsymbol{\beta}^T\mathbf{X}^T\left(\mathbf{I}-\mathbf{P_X}-\frac{1}{\mathbf{b}^T\mathbf{b}}\mathbf{b}\mathbf{b}^T\right)\mathbf{X}\boldsymbol{\beta}=0.$$

(b) The independence of $\mathbf{y}_I$ and $\mathbf{y}_{II}$ will be demonstrated; independence of the other two pairs of vectors is shown very similarly. Observe that

$$\mathbf{y}_I = [(\mathbf{I}_q \otimes (1/m)\mathbf{J}_m) - ((1/q)\mathbf{J}_q \otimes (1/m)\mathbf{J}_m)]\mathbf{y},$$
$$\mathbf{y}_{II} = [((1/q)\mathbf{J}_q \otimes \mathbf{I}_m) - ((1/q)\mathbf{J}_q \otimes (1/m)\mathbf{J}_m)]\mathbf{y}.$$

Also observe that

$$\begin{aligned}
&[(\mathbf{I}_q \otimes (1/m)\mathbf{J}_m) - ((1/q)\mathbf{J}_q \otimes (1/m)\mathbf{J}_m)][((1/q)\mathbf{J}_q \otimes \mathbf{I}_m) - ((1/q)\mathbf{J}_q \otimes (1/m)\mathbf{J}_m)]\\
&= ((1/q)\mathbf{J}_q \otimes (1/m)\mathbf{J}_m) - ((1/q)\mathbf{J}_q \otimes (1/m)\mathbf{J}_m) - ((1/q)\mathbf{J}_q \otimes (1/m)\mathbf{J}_m)\\
&\quad + ((1/q)\mathbf{J}_q \otimes (1/m)\mathbf{J}_m)\\
&= \mathbf{0}.
\end{aligned}$$

By Corollary 14.1.8.1, $\mathbf{y}_I$ and $\mathbf{y}_{II}$ are independent.

(c) The equality in the hint is satisfied because

$$\sum_{i=1}^{q}\sum_{j=1}^{m}(\bar{y}_{i\cdot}-\bar{y}_{\cdot\cdot})(\bar{y}_{\cdot j}-\bar{y}_{\cdot\cdot})\bar{y}_{i\cdot} = \sum_{i=1}^{q}(\bar{y}_{i\cdot}-\bar{y}_{\cdot\cdot})\bar{y}_{i\cdot}\sum_{j=1}^{m}(\bar{y}_{\cdot j}-\bar{y}_{\cdot\cdot}) = 0,$$

$$\sum_{i=1}^{q}\sum_{j=1}^{m}(\bar{y}_{i\cdot}-\bar{y}_{\cdot\cdot})(\bar{y}_{\cdot j}-\bar{y}_{\cdot\cdot})\bar{y}_{\cdot j} = \sum_{j=1}^{m}(\bar{y}_{\cdot j}-\bar{y}_{\cdot\cdot})\bar{y}_{\cdot j}\sum_{i=1}^{q}(\bar{y}_{i\cdot}-\bar{y}_{\cdot\cdot}) = 0,$$

$$\sum_{i=1}^{q}\sum_{j=1}^{m}(\bar{y}_{i\cdot}-\bar{y}_{\cdot\cdot})(\bar{y}_{\cdot j}-\bar{y}_{\cdot\cdot})\bar{y}_{\cdot\cdot} = \bar{y}_{\cdot\cdot}\sum_{i=1}^{q}(\bar{y}_{i\cdot}-\bar{y}_{\cdot\cdot})\sum_{j=1}^{m}(\bar{y}_{\cdot j}-\bar{y}_{\cdot\cdot}) = 0.$$

Similarly,

$$\sum_{i=1}^{q}(\bar{y}_{i\cdot}-\bar{y}_{\cdot\cdot})(\bar{y}_{\cdot j}-\bar{y}_{\cdot\cdot}) = (\bar{y}_{\cdot j}-\bar{y}_{\cdot\cdot})\sum_{i=1}^{q}(\bar{y}_{i\cdot}-\bar{y}_{\cdot\cdot}) = 0 \quad \text{for all } j$$

and

$$\sum_{j=1}^{m}(\bar{y}_{i\cdot}-\bar{y}_{\cdot\cdot})(\bar{y}_{\cdot j}-\bar{y}_{\cdot\cdot}) = (\bar{y}_{i\cdot}-\bar{y}_{\cdot\cdot})\sum_{j=1}^{m}(\bar{y}_{\cdot j}-\bar{y}_{\cdot\cdot}) = 0 \quad \text{for all } i,$$

so that $(\bar{y}_{i\cdot} - \bar{y}_{\cdot\cdot})(\bar{y}_{\cdot j} - \bar{y}_{\cdot\cdot})$ satisfies the requirements of the b_{ij}'s specified in part (a) (the condition that they are not all equal to 0 is satisfied with probability one). Moreover, by the independence established in part (b), the conditional distribution of $\mathbf{y}_{III}$, given $\mathbf{y}_I$ and $\mathbf{y}_{II}$, is identical to its unconditional distribution. Thus, by part (a), the conditional distributions, given $\mathbf{y}_I$ and $\mathbf{y}_{II}$, of the two quantities we seek are independent chi-square distributions with the same degrees of freedom as specified in part (a), but possibly different noncentrality parameters [because the mean structure of this model is different than that of the model in part (a)].

To obtain the new noncentrality parameters, let us write $\mathbf{b} = \mathbf{y}_I \otimes \mathbf{y}_{II}$. Now let $\mathbf{P_X}$ represent the orthogonal projection matrix onto the column space of this model, and let $\mathbf{P}_{\mathbf{X}_{ME}}$ represent the orthogonal projection matrix onto the column space of the main effects model considered in parts (a) and (b). For the current model, which is conditioned on $\mathbf{y}_I$ and $\mathbf{y}_{II}$,

$$\mathbf{X} = (\mathbf{1}_{qm},\ \mathbf{I}_q \otimes \mathbf{1}_m,\ \mathbf{1}_q \otimes \mathbf{I}_m, \mathbf{b}).$$

It was shown in the solution to part (a) that the last column of this matrix is orthogonal to the remaining columns, so

$$\mathbf{P_X} = \mathbf{P}_{\mathbf{X}_{ME}} + \frac{1}{\mathbf{b}^T\mathbf{b}}\mathbf{b}^T\mathbf{b}.$$

Thus

$$\mathbf{b}^T\mathbf{X} = (\mathbf{b}^T\mathbf{X}_{ME}, \mathbf{b}^T\mathbf{b}) = (\mathbf{0}^T_{qm-q-m}, \mathbf{b}^T\mathbf{b}),$$

yielding

$$\nu = (1/2\sigma^2)\boldsymbol{\beta}^T\mathbf{X}^T\left(\frac{1}{\mathbf{b}^T\mathbf{b}}\mathbf{b}\mathbf{b}^T\right)\mathbf{X}\boldsymbol{\beta} = (1/2\sigma^2)\lambda^2\mathbf{b}^T\mathbf{b}.$$

The noncentrality parameter of the other chi-square distribution is

$$\begin{aligned}(1/2\sigma^2)\boldsymbol{\beta}^T\mathbf{X}^T\left(\mathbf{P_X} - \mathbf{P}_{\mathbf{X}_{ME}} - \frac{1}{\mathbf{b}^T\mathbf{b}}\mathbf{b}\mathbf{b}^T\right)\mathbf{X}\boldsymbol{\beta} &= (1/2\sigma^2)\boldsymbol{\beta}^T\mathbf{X}^T\left(\frac{1}{\mathbf{b}^T\mathbf{b}}\mathbf{b}\mathbf{b}^T - \frac{1}{\mathbf{b}^T\mathbf{b}}\mathbf{b}\mathbf{b}^T\right)\mathbf{X}\boldsymbol{\beta} \\ &= 0.\end{aligned}$$

(d) Under H_0, $\nu = 0$ and the joint conditional distribution of the two quantities displayed in part (c) does not depend on the values conditioned upon. Thus, under H_0, the joint unconditional distribution of those quantities is the same as their joint conditional distribution.

(e) Let F denote the ratio of the first quantity to the second quantity, and observe that σ^2 cancels in this ratio so that F is a statistic. By part (d), under H_0 the unconditional distribution of F is $\mathrm{F}(1, qm - q - m)$. Thus, a size-$\xi$ test of H_0 versus H_a is to reject H_0 if and only if $F > F_{\xi,1,qm-q-m}$.

▶ **Exercise 11** Prove Facts 2, 3, and 4 listed in Sect. 15.3.

Solution Define $E_2 = \{\omega : A(\omega) \leq \tau(\omega) \leq B(\omega)\}$ and $F_2 = \{\omega : c^+A(\omega) + c^-B(\omega) \leq c\tau(\omega) \leq c^+B(\omega) + c^-A(\omega)$ for all $c \in \mathbb{R}\}$. If $\omega \in E_2$, then $A(\omega) \leq \tau(\omega) \leq B(\omega)$ and multiplication by any real number c yields $cA(\omega) \leq c\tau(\omega) \leq cB(\omega)$ if $c \geq 0$ and $cB(\omega) \leq c\tau(\omega) \leq cA(\omega)$ otherwise, or equivalently, $c^+A(\omega) + c^-B(\omega) \leq c\tau(\omega) \leq c^+B(\omega) + c^-A(\omega)$. Thus $\omega \in F_2$. Conversely, if $\omega \in F_2$, then consideration of the single interval corresponding to $c = 1$ reveals that $\omega \in E_2$. Thus $E_2 = F_2$, hence $\Pr(E_2) = \Pr(F_2)$. This proves Fact 2.

Define $E_3 = \{\omega : A_i(\omega) \leq \tau_i(\omega) \leq B_i(\omega)$ for $i = 1, \ldots, k\}$ and $F_3 = \{\omega : \sum_{i=1}^k A_i(\omega) \leq \sum_{i=1}^k \tau_i(\omega) \leq \sum_{i=1}^k B_i(\omega)\}$. If $\omega \in E_3$, then $A_i(\omega) \leq \tau_i(\omega) \leq B_i(\omega)$ for $i = 1, \ldots, k$, and by summing these inequalities we obtain the interval defined in F_3. Thus $\omega \in F_3$, implying that $E_3 \subseteq F_3$, implying further that $\Pr(E_3) \leq \Pr(F_3)$. This proves Fact 3.

Define $F_4 = \{\omega : \sum_{i=1}^k \left(c_i^+A_i(\omega) + c_i^-B_i(\omega)\right) \leq \sum_{i=1}^k c_i\tau_i(\omega) \leq \sum_{i=1}^k \left(c_i^+B_i(\omega) + c_i^-A_i(\omega)\right)$ for all $c_i \in \mathbb{R}\}$. If $\omega \in E_3$, then $A_i(\omega) \leq \tau_i(\omega) \leq B_i(\omega)$ for $i = 1, \ldots, k$, and multiplication of the ith of these intervals by the real number c_i and summing yields $\sum_{i=1}^k \left(c_i^+A_i(\omega) + c_i^-B_i(\omega)\right) \leq \sum_{i=1}^k c_i\tau_i(\omega) \leq \sum_{i=1}^k \left(c_i^+B_i(\omega) + c_i^-A_i(\omega)\right)$. Thus $\omega \in F_4$. Conversely, if $\omega \in F_4$, then consideration of the k intervals corresponding to $\{c_1, \ldots, c_k\} = \{1, 0, \ldots, 0\}, \{c_1, \ldots, c_k\} = \{0, 1, \ldots, 0\}, \ldots, \{c_1, \ldots, c_k\} = \{0, 0, \ldots, 1\}$ reveals that $\omega \in E_3$. Thus $E_3 = F_4$, hence $\Pr(E_3) = \Pr(F_4)$. This proves Fact 4.

▶ **Exercise 12** Prove (15.13), i.e., prove that for any $\xi \in (0, 1)$ and any $s^* > 0$, $\nu > 0$, and $l > 0$,

$$s^* F_{\xi,s^*,\nu} < (s^* + l)F_{\xi,s^*+l,\nu}.$$

[Hint: Consider three independent random variables $U \sim \chi^2(s^*)$, $V \sim \chi^2(l)$, and $W \sim \chi^2(\nu)$.]

Solution Let $U \sim \chi^2(s^*)$, $V \sim \chi^2(l)$, and $W \sim \chi^2(\nu)$ be independent. Then by Definition 14.5.1,

$$\frac{U/s^*}{W/\nu} \sim \mathrm{F}(s^*, \nu) \quad \text{and} \quad \frac{(U+V)/(s^*+l)}{W/\nu} \sim \mathrm{F}(s^* + l, \nu),$$

implying that

$$\frac{\nu U}{W} \sim s^*\mathrm{F}(s^*, \nu) \quad \text{and} \quad \frac{\nu(U+V)}{W} \sim (s^* + l)\mathrm{F}(s^* + l, \nu).$$

The latter distributional result implies that $\Pr\left(\frac{\nu(U+V)}{W} \le (s^*+l)F_{\xi,s^*+l,\nu}\right) = 1-\xi$. Now observe that

$$\frac{\nu U}{W} \le \frac{\nu U}{W} + \frac{\nu V}{W} = \frac{\nu(U+V)}{W},$$

where the inequality holds because $\nu V/W > 0$ with probability one. Therefore,

$$\Pr\left(\frac{\nu(U+V)}{W} \le s^* F_{\xi,s^*,\nu}\right) < \Pr\left(\frac{\nu U}{W} \le s^* F_{\xi,s^*,\nu}\right) = 1-\xi.$$

Therefore, it must be the case that $s^* F_{\xi,s^*,\nu} < (s^*+l)F_{\xi,s^*+l,\nu}$.

▶ **Exercise 13** Prove (15.14), i.e., prove that for any $\xi \in (0,1)$ and any $s > 0$ and $\nu > 0$,

$$t_{\xi/2,s,\nu,\mathbf{I}} \le \sqrt{sF_{\xi,s,\nu}}.$$

Solution Let $\mathbf{x} = (x_i) \sim \mathrm{N}_s(\mathbf{0}, \mathbf{I})$ and $w \sim \chi^2(\nu)$, and suppose that $\mathbf{x}$ and w are independent. Define $t_i = \frac{x_i}{\sqrt{w/\nu}}$ and $\mathbf{t} = (t_i)$. Then $\mathbf{t} \sim \mathrm{t}(s, \nu, \mathbf{I})$. Also, $\sum_{i=1}^s x_i^2 \sim \chi^2(s)$, and $\sum_{i=1}^s x_i^2$ and w are independent, implying that $\frac{\sum_{i=1}^s x_i^2/s}{w/\nu} \sim \mathrm{F}(s,\nu)$ or equivalently $\sum_{i=1}^s t_i^2 = \frac{\sum_{i=1}^s x_i^2}{w/\nu} \sim s\mathrm{F}(s,\nu)$. Thus,

$$\Pr\left(\sum_{i=1}^s t_i^2 \le sF_{\xi,s,\nu}\right) = 1-\xi.$$

But because $\max_i t_i^2 \le \sum_{i=1}^q t_i^2$,

$$\Pr\left(\sum_{i=1}^s t_i^2 \le t^2_{\xi/2,s,\nu,\mathbf{I}}\right) \le \Pr\left(\max_i t_i^2 \le t^2_{\xi/2,s,\nu,\mathbf{I}}\right) = 1-\xi.$$

Therefore, it must be the case that $t^2_{\xi/2,s,\nu,\mathbf{I}} \le sF_{\xi,s,\nu}$ or equivalently $t_{\xi/2,s,\nu,\mathbf{I}} \le \sqrt{sF_{\xi,s,\nu}}$.

▶ **Exercise 14** Prove Theorem 15.3.7: Under the normal prediction-extended positive definite Aitken model

$$\left\{\begin{pmatrix}\mathbf{y}\\ \mathbf{u}\end{pmatrix}, \begin{pmatrix}\mathbf{X}\boldsymbol{\beta}\\ \mathbf{0}\end{pmatrix}, \sigma^2\begin{pmatrix}\mathbf{W} & \mathbf{K}\\ \mathbf{K}^T & \mathbf{H}\end{pmatrix}\right\}$$

with $n > p^*$, if $\mathbf{C}^T\boldsymbol{\beta} + \mathbf{u}$ is an s-vector of predictable functions and $\mathbf{Q}$ is positive definite (as is the case under either condition of Theorem 13.2.2b), then the probability of simultaneous coverage of the infinite collection of intervals for $\{\mathbf{a}^T(\mathbf{C}^T\boldsymbol{\beta} + \mathbf{u}) : \mathbf{a} \in \mathbb{R}^s\}$ given by

$$\mathbf{a}^T(\mathbf{C}^T\tilde{\boldsymbol{\beta}} + \tilde{\mathbf{u}}) \pm \sqrt{sF_{\xi,s,n-p^*}\tilde{\sigma}^2(\mathbf{a}^T\mathbf{Q}\mathbf{a})} \quad \text{for all } \mathbf{a}$$

is $1 - \xi$.

Solution By Corollary 2.16.1.1,

$$\frac{[(\mathbf{C}^T\tilde{\boldsymbol{\beta}} + \tilde{\mathbf{u}}) - (\mathbf{C}^T\boldsymbol{\beta} + \mathbf{u})]^T\mathbf{Q}^{-1}[(\mathbf{C}^T\tilde{\boldsymbol{\beta}} + \tilde{\mathbf{u}}) - (\mathbf{C}^T\boldsymbol{\beta} + \mathbf{u})]}{s\tilde{\sigma}^2}$$
$$= \left(\frac{1}{s\tilde{\sigma}^2}\right)\max_{\mathbf{a}\neq\mathbf{0}}\left(\frac{\{\mathbf{a}^T[(\mathbf{C}^T\tilde{\boldsymbol{\beta}} + \tilde{\mathbf{u}}) - (\mathbf{C}^T\boldsymbol{\beta} + \mathbf{u})]\}^2}{\mathbf{a}^T\mathbf{Q}\mathbf{a}}\right).$$

Thus by Theorem 14.5.5,

$$1 - \xi = \Pr\left(\frac{[(\mathbf{C}^T\tilde{\boldsymbol{\beta}} + \tilde{\mathbf{u}}) - (\mathbf{C}^T\boldsymbol{\beta} + \mathbf{u})]^T\mathbf{Q}^{-1}[(\mathbf{C}^T\tilde{\boldsymbol{\beta}} + \tilde{\mathbf{u}}) - (\mathbf{C}^T\boldsymbol{\beta} + \mathbf{u})]}{s\tilde{\sigma}^2} \leq F_{\xi,s,n-p^*}\right)$$
$$= \Pr\left[\left(\frac{1}{s\tilde{\sigma}^2}\right)\max_{\mathbf{a}\neq\mathbf{0}}\left(\frac{\{\mathbf{a}^T[(\mathbf{C}^T\tilde{\boldsymbol{\beta}} + \tilde{\mathbf{u}}) - (\mathbf{C}^T\boldsymbol{\beta} + \mathbf{u})]\}^2}{\mathbf{a}^T\mathbf{Q}\mathbf{a}}\right) \leq F_{\xi,s,n-p^*} \text{ for all } \mathbf{a} \neq \mathbf{0}\right]$$
$$= \Pr\Big(\mathbf{a}^T(\mathbf{C}^T\tilde{\boldsymbol{\beta}} + \tilde{\mathbf{u}}) - \sqrt{sF_{\xi,s,n-p^*}\tilde{\sigma}^2(\mathbf{a}^T\mathbf{Q}\mathbf{a})} \leq \mathbf{a}^T(\mathbf{C}^T\tilde{\boldsymbol{\beta}} + \mathbf{u})$$
$$\leq \mathbf{a}^T(\mathbf{C}^T\tilde{\boldsymbol{\beta}} + \tilde{\mathbf{u}}) + \sqrt{sF_{\xi,s,n-p^*}\tilde{\sigma}^2(\mathbf{a}^T\mathbf{Q}\mathbf{a})} \text{ for all } \mathbf{a} \in \mathbb{R}^s\Big).$$

▶ **Exercise 15** Consider the mixed linear model for two observations specified in Exercise 13.18, and suppose that the joint distribution of b, d_1, and d_2 is multivariate normal. Let $0 < \xi < 1$.

(a) Obtain a $100(1 - \xi)\%$ confidence interval for β.
(b) Obtain a $100(1 - \xi)\%$ prediction interval for b.
(c) Obtain intervals for β and b whose simultaneous coverage probability is exactly $1 - \xi$.

Solution

(a) By (15.6) in Theorem 15.1.2, a $100(1 - \xi)\%$ confidence interval for β is given by

$$\tilde{\beta} \pm t_{\xi/2,1}\tilde{\sigma}\sqrt{Q_\beta},$$

where (by Theorem 13.4.4)

$$Q_\beta = (1\ 0)\begin{pmatrix} 5 & 5 \\ 5 & 12 \end{pmatrix}^{-1}\begin{pmatrix} 1 \\ 0 \end{pmatrix} = \frac{12}{35}.$$

(b) Again by (15.6) in Theorem 15.1.2, a $100(1-\xi)\%$ confidence interval for b is given by

$$\tilde{b} \pm t_{\xi/2,1}\tilde{\sigma}\sqrt{Q_b},$$

where (by Theorem 13.4.4)

$$Q_b = (0\ 1)\begin{pmatrix} 5 & 5 \\ 5 & 12 \end{pmatrix}^{-1}\begin{pmatrix} 0 \\ 1 \end{pmatrix} = \frac{1}{7}.$$

(c) By the discussion immediately following (15.15), $100(1-\xi)\%$ multivariate t simultaneous confidence/prediction intervals for β and b are given by

$$\tilde{\beta} \pm t_{\xi/2,2,1,\mathbf{R}}\tilde{\sigma}\sqrt{\frac{12}{35}} \quad \text{and} \quad \tilde{b} \pm t_{\xi/2,2,1,\mathbf{R}}\tilde{\sigma}\sqrt{\frac{1}{7}},$$

where, using Theorem 13.4.4 once more,

$$\mathbf{R} = \begin{pmatrix} 1 & \text{corr}(\tilde{\beta}, \tilde{b}-b) \\ \text{corr}(\tilde{\beta}, \tilde{b}-b) & 1 \end{pmatrix}$$

and $\text{corr}(\tilde{\beta}, \tilde{b}-b) = \frac{-5/35}{\sqrt{(12/35)(1/7)}} = -\sqrt{\frac{5}{12}}$.

▶ **Exercise 16** Generalize all four types of simultaneous confidence intervals presented in Sect. 15.3 (Bonferroni, Scheffé, multivariate t, Tukey) to be suitable for use under a normal positive definite Aitken model.

Solution Let $\mathbf{M} = (m_{ij}) = \mathbf{C}^T(\mathbf{X}^T\mathbf{W}^{-1}\mathbf{X})^-\mathbf{C}$. By Theorem 15.1.1, the $100(1-\xi)\%$ Bonferroni intervals and the $100(1-\xi)\%$ Scheffé intervals are

$$\mathbf{c}_i^T\tilde{\boldsymbol{\beta}} \pm t_{\xi/(2s),n-p^*}\tilde{\sigma}\sqrt{m_{ii}} \quad (i = 1, \ldots, s)$$

and

$$\mathbf{c}^T\tilde{\boldsymbol{\beta}} \pm \sqrt{s^* F_{\xi,s^*,n-p^*}}\tilde{\sigma}\sqrt{\mathbf{c}^T(\mathbf{X}^T\mathbf{W}^{-1}\mathbf{X})^-\mathbf{c}} \quad \text{for all } \mathbf{c}^T \in \mathbf{R}(\mathbf{C}^T).$$

Because

$$\frac{(\mathbf{c}_i^T\tilde{\boldsymbol{\beta}} - \mathbf{c}_i^T\boldsymbol{\beta})/(\sigma\sqrt{m_{ii}})}{\sqrt{\frac{(n-p^*)\tilde{\sigma}^2/\sigma^2}{n-p^*}}} = \frac{\mathbf{c}_i^T\tilde{\boldsymbol{\beta}} - \mathbf{c}_i^T\boldsymbol{\beta}}{\tilde{\sigma}\sqrt{m_{ii}}} \sim \mathrm{t}(s, n-p^*, \mathbf{R}),$$

where

$$\mathbf{R} = \mathrm{var}\left(\frac{\mathbf{c}_1^T\tilde{\boldsymbol{\beta}}}{\sigma\sqrt{m_{11}}}, \ldots, \frac{\mathbf{c}_s^T\tilde{\boldsymbol{\beta}}}{\sigma\sqrt{m_{ss}}}\right)^T,$$

the $100(1-\xi)\%$ multivariate t intervals are

$$\mathbf{c}_i^T\tilde{\boldsymbol{\beta}} \pm t_{\xi/2,s,n-p^*,\mathbf{R}}\tilde{\sigma}\sqrt{m_{ii}} \quad (i = 1, \ldots, s).$$

Letting $\tilde{\psi}_1, \ldots, \tilde{\psi}_k$ denote the generalized least squares estimators of the estimable functions $\psi_1, \ldots, \psi_k$, and assuming that $\mathrm{var}(\tilde{\psi}_1, \ldots, \tilde{\psi}_k)^T = c^2\sigma^2\mathbf{I}$ for some $c^2 > 0$, we find that the $100(1-\xi)\%$ Tukey intervals for all pairwise differences among the ψ_i's are

$$(\tilde{\psi}_i - \tilde{\psi}_j) \pm c\tilde{\sigma}q^*_{\xi,k,n-p^*}.$$

▶ **Exercise 17** Obtain specialized expressions for $\mathbf{R}$ for the multivariate t-based simultaneous confidence and simultaneous prediction intervals presented in Example 15.3-1.

Solution For the multivariate t simultaneous confidence intervals for β_1 and β_2,

$$\mathbf{R} = \begin{pmatrix} 1 & -\frac{\bar{x}}{\sqrt{\bar{x}^2+(SXX/n)}} \\ & 1 \end{pmatrix}.$$

Next consider the multivariate t-based simultaneous confidence intervals for $(\beta_1 + \beta_2 x_{n+1}, \ldots, \beta_1 + \beta_2 x_{n+s})$. In this case,

$$\mathbf{R} = \mathrm{var}\left(\frac{\hat{\beta}_1 + \hat{\beta}_2 x_{n+1}}{\sigma\sqrt{(1/n) + (x_{n+1} - \bar{x})^2/SXX}}, \ldots, \frac{\hat{\beta}_1 + \hat{\beta}_2 x_{n+s}}{\sigma\sqrt{(1/n) + (x_{n+s} - \bar{x})^2/SXX}}\right)^T.$$

The (i, j)-th element of this $\mathbf{R}$ is equal to 1 if $i = j$ and is equal to

$$\mathrm{cov}\left(\frac{\hat{\beta}_1 + \hat{\beta}_2 x_{n+i}}{\sigma\sqrt{(1/n) + (x_{n+i} - \bar{x})^2/SXX}}, \frac{\hat{\beta}_1 + \hat{\beta}_2 x_{n+j}}{\sigma\sqrt{(1/n) + (x_{n+j} - \bar{x})^2/SXX}}\right)$$
$$= \frac{(1/n) + (x_{n+i} - \bar{x})(x_{n+j} - \bar{x})/SXX}{\sqrt{(1/n) + (x_{n+i} - \bar{x})^2/SXX}\sqrt{(1/n) + (x_{n+j} - \bar{x})^2/SXX}}$$

if $i \neq j$.

Finally, consider the multivariate t-based simultaneous prediction intervals for $(y_{n+1}, \ldots, y_{n+s})$. In this case,

$$\mathbf{R} = \mathrm{var}\left(\frac{\hat{\beta}_1 + \hat{\beta}_2 x_{n+1} - e_{n+1}}{\sigma\sqrt{1 + (1/n) + (x_{n+1} - \bar{x})^2/SXX}}, \ldots, \frac{\hat{\beta}_1 + \hat{\beta}_2 x_{n+s} - e_{n+s}}{\sigma\sqrt{1 + (1/n) + (x_{n+s} - \bar{x})^2/SXX}}\right)^T.$$

The (i, j)-th element of this $\mathbf{R}$ is equal to 1 if $i = j$ and is equal to

$$\mathrm{cov}\left(\frac{\hat{\beta}_1 + \hat{\beta}_2 x_{n+i} - e_{n+i}}{\sigma\sqrt{1 + (1/n) + (x_{n+i} - \bar{x})^2/SXX}}, \frac{\hat{\beta}_1 + \hat{\beta}_2 x_{n+j} - e_{n+j}}{\sigma\sqrt{1 + (1/n) + (x_{n+j} - \bar{x})^2/SXX}}\right)$$
$$= \frac{(1/n) + (x_{n+i} - \bar{x})(x_{n+j} - \bar{x})/SXX}{\sqrt{1 + (1/n) + (x_{n+i} - \bar{x})^2/SXX}\sqrt{1 + (1/n) + (x_{n+j} - \bar{x})^2/SXX}}$$

if $i \neq j$.

▶ **Exercise 18** Extend the expressions for simultaneous confidence intervals presented in Example 15.3-2 to the case of unbalanced data, if applicable.

Solution The $100(1 - \xi)\%$ Bonferroni simultaneous confidence intervals are

$$(\bar{y}_{i\cdot} - \bar{y}_{i'\cdot}) \pm t_{\xi/[q(q-1)],n-q}\hat{\sigma}\sqrt{\frac{1}{n_i} + \frac{1}{n_{i'}}};$$

the $100(1 - \xi)\%$ multivariate t simultaneous confidence intervals are

$$(\bar{y}_{i\cdot} - \bar{y}_{i'\cdot}) \pm t_{\xi/2,q(q-1)/2,n-q,\mathbf{R}}\hat{\sigma}\sqrt{\frac{1}{n_i} + \frac{1}{n_{i'}}},$$

where

$$\mathbf{R} = \text{var}\left(\frac{\bar{y}_{1\cdot} - \bar{y}_{2\cdot}}{\sigma\sqrt{(1/n_1) + (1/n_2)}}, \frac{\bar{y}_{1\cdot} - \bar{y}_{3\cdot}}{\sigma\sqrt{(1/n_1) + (1/n_3)}}, \ldots, \frac{\bar{y}_{(q-1)\cdot} - \bar{y}_{q\cdot}}{\sigma\sqrt{(1/n_{q-1}) + (1/n_q)}}\right);$$

and the $100(1-\xi)\%$ Scheffé simultaneous confidence intervals are

$$(\bar{y}_{i\cdot} - \bar{y}_{i'\cdot}) \pm \sqrt{(q-1)F_{\xi,q-1,n-q}}\hat{\sigma}\sqrt{\frac{1}{n_i} + \frac{1}{n_{i'}}}$$

for $i' > i = 1, \ldots, q$. Tukey's method is not applicable.

▶ **Exercise 19** Show that the simultaneous coverage probability of the classical confidence band given by (15.17) is $1-\xi$ despite the restriction that the first element of $\mathbf{c}$ is equal to 1.

Solution Consider the normal full-rank Gauss–Markov regression model in its centered form, which can be written in the two equivalent forms

$$y_i = (1, \mathbf{x}_{i,-1}^T)\begin{pmatrix} \beta_1 \\ \boldsymbol{\beta}_{-1} \end{pmatrix} + e_i \quad (i = 1, \ldots, n)$$

and

$$\mathbf{y} = \mathbf{X}\boldsymbol{\beta} + \mathbf{e} = \begin{pmatrix} \mathbf{1}_n, & \mathbf{X}_{-1} \end{pmatrix}\begin{pmatrix} \beta_1 \\ \boldsymbol{\beta}_{-1} \end{pmatrix} + \mathbf{e},$$

where $\mathbf{x}_{i,-1}^T = (x_{i2} - \bar{x}_2, x_{i3} - \bar{x}_3, \ldots, x_{ip} - \bar{x}_p)$ and

$$\mathbf{X}_{-1} = \begin{pmatrix} \mathbf{x}_{1,-1}^T \\ \vdots \\ \mathbf{x}_{n,-1}^T \end{pmatrix}.$$

Observe that $\mathbf{X}^T\mathbf{X} = \begin{pmatrix} n & \mathbf{0}_{n-1}^T \\ \mathbf{0}_{n-1} & \mathbf{X}_{-1}^T\mathbf{X}_{-1} \end{pmatrix}$. By Corollary 2.16.1.1, the maximum of

$$\frac{\left[\mathbf{c}^T\begin{pmatrix} \hat{\beta}_1 - \beta_1 \\ \hat{\boldsymbol{\beta}}_{-1} - \boldsymbol{\beta}_{-1} \end{pmatrix}\right]^2}{\mathbf{c}^T(\mathbf{X}^T\mathbf{X})^{-1}\mathbf{c}}$$

over all $\mathbf{c} = (c_i) \neq \mathbf{0}$ is attained at only those $\mathbf{c}$ that are proportional to $\mathbf{X}^T\mathbf{X}\begin{pmatrix} \hat{\beta}_1 - \beta_1 \\ \hat{\boldsymbol{\beta}}_{-1} - \boldsymbol{\beta}_{-1} \end{pmatrix}$, i.e., only those $\mathbf{c}$ for which

$$\mathbf{c} = c\begin{pmatrix} n(\hat{\beta}_1 - \beta_1) \\ \mathbf{X}_{-1}^T\mathbf{X}_{-1}(\hat{\boldsymbol{\beta}}_{-1} - \boldsymbol{\beta}_{-1}) \end{pmatrix}$$

for some $c \neq 0$. This establishes that, at the maximum, $c_1 \neq 0$ with probability one. Therefore, with probability one,

$$\begin{aligned}
\max_{\mathbf{c}\neq\mathbf{0}}\left(\frac{\left[\mathbf{c}^T\begin{pmatrix} \hat{\beta}_1 - \beta_1 \\ \hat{\boldsymbol{\beta}}_{-1} - \boldsymbol{\beta}_{-1} \end{pmatrix}\right]^2}{\mathbf{c}^T(\mathbf{X}^T\mathbf{X})^{-1}\mathbf{c}}\right) &= \max_{c_1\neq 0,\mathbf{c}_2\in\mathbb{R}^{p-1}}\left(\frac{\left[(c_1, \mathbf{c}_2^T)\begin{pmatrix} \hat{\beta}_1 - \beta_1 \\ \hat{\boldsymbol{\beta}}_{-1} - \boldsymbol{\beta}_{-1} \end{pmatrix}\right]^2}{(c_1, \mathbf{c}_2^T)(\mathbf{X}^T\mathbf{X})^{-1}\begin{pmatrix} c_1 \\ \mathbf{c}_2 \end{pmatrix}}\right) \\
&= \max_{c_1\neq 0,\mathbf{c}_2\in\mathbb{R}^{p-1}}\left(\frac{\left[[1, (1/c_1)\mathbf{c}_2^T]\begin{pmatrix} \hat{\beta}_1 - \beta_1 \\ \hat{\boldsymbol{\beta}}_{-1} - \boldsymbol{\beta}_{-1} \end{pmatrix}\right]^2}{[1, (1/c_1)\mathbf{c}_2^T](\mathbf{X}^T\mathbf{X})^{-1}\begin{pmatrix} 1 \\ (1/c_1)\mathbf{c}_2 \end{pmatrix}}\right) \\
&= \max_{\mathbf{x}_{-1}\in\mathbb{R}^{p-1}}\left(\frac{\left[(1, \mathbf{x}_{-1}^T)\begin{pmatrix} \hat{\beta}_1 - \beta_1 \\ \hat{\boldsymbol{\beta}}_{-1} - \boldsymbol{\beta}_{-1} \end{pmatrix}\right]^2}{(1, \mathbf{x}_{-1}^T)(\mathbf{X}^T\mathbf{X})^{-1}\begin{pmatrix} 1 \\ \mathbf{x}_{-1} \end{pmatrix}}\right).
\end{aligned}$$

Thus

$$\begin{aligned}
1-\xi &= \Pr\Big(\mathbf{c}^T\hat{\boldsymbol{\beta}} - \sqrt{pF_{\xi,p,n-p}\hat{\sigma}^2\mathbf{c}^T(\mathbf{X}^T\mathbf{X})^{-1}\mathbf{c}} \leq \mathbf{c}^T\boldsymbol{\beta} \leq \mathbf{c}^T\hat{\boldsymbol{\beta}} \\
&\quad + \sqrt{pF_{\xi,p,n-p}\hat{\sigma}^2\mathbf{c}^T(\mathbf{X}^T\mathbf{X})^{-1}\mathbf{c}} \text{ for all } \mathbf{c}\in\mathbb{R}^p\Big) \\
&= \Pr\Bigg((\hat{\beta}_1 + \mathbf{x}_{-1}^T\hat{\boldsymbol{\beta}}_{-1}) - \sqrt{pF_{\xi,p,n-p}\hat{\sigma}^2(1, \mathbf{x}_{-1}^T)(\mathbf{X}^T\mathbf{X})^{-1}\begin{pmatrix} 1 \\ \mathbf{x}_{-1} \end{pmatrix}} \leq \beta_1 + \mathbf{x}_{-1}^T\boldsymbol{\beta}_{-1} \\
&\quad \leq (\hat{\beta}_1 + \mathbf{x}_{-1}^T\hat{\boldsymbol{\beta}}_{-1}) + \sqrt{pF_{\xi,p,n-p}\hat{\sigma}^2(1, \mathbf{x}_{-1}^T)(\mathbf{X}^T\mathbf{X})^{-1}\begin{pmatrix} 1 \\ \mathbf{x}_{-1} \end{pmatrix}} \text{ for all } \mathbf{x}_{-1}\in\mathbb{R}^{p-1}\Bigg).
\end{aligned}$$

▶ **Exercise 20** Consider the normal random-slope, fixed-intercept simple linear regression model

$$y_i = \beta + bz_i + d_i \quad (i = 1, \ldots, n),$$

described previously in more detail in Exercise 13.13, where var($\mathbf{y}$), among other quantities, was determined. The model equation may be written in matrix form as

$$\mathbf{y} = \beta\mathbf{1} + b\mathbf{z} + \mathbf{d}.$$

(a) Consider the sums of squares that arise in the following ordered two-part mixed-model ANOVA:

Source	df	Sum of squares
$\mathbf{z}$	1	$\mathbf{y}^T\mathbf{P}_\mathbf{z}\mathbf{y}$
$\mathbf{1}\vert\mathbf{z}$	1	$\mathbf{y}^T(\mathbf{P}_{1,\mathbf{z}} - \mathbf{P}_\mathbf{z})\mathbf{y}$
Residual	$n-2$	$\mathbf{y}^T(\mathbf{I} - \mathbf{P}_{1,\mathbf{z}})\mathbf{y}$

Here $\mathbf{P}_\mathbf{z} = \mathbf{z}\mathbf{z}^T/(\sum_{i=1}^n z_i^2)$. Determine which two of these three sums of squares, when divided by σ^2, have noncentral chi-square distributions under the mixed simple linear regression model defined above. For those two, give nonmatrix expressions for the parameters of the distributions.

(b) It is possible to use the sums of squares from the ANOVA above to test H_0 : $\beta = 0$ versus H_a : $\beta \neq 0$ by an F-test. Derive the appropriate F-statistic and give its distribution under each of H_a and H_0.

(c) Let $\boldsymbol{\tau} = (\beta, b)^T$ and let $\tilde{\boldsymbol{\tau}} = (\tilde{\beta}, \tilde{b})^T$ be the BLUP of $\boldsymbol{\tau}$. Let $\sigma^2\mathbf{Q} = \text{var}(\tilde{\boldsymbol{\tau}} - \boldsymbol{\tau}) = \sigma^2 \begin{pmatrix} q_{11} & q_{12} \\ q_{12} & q_{22} \end{pmatrix}$ and assume that this matrix is positive definite. Starting with the probability statement

$$1 - \xi = \Pr\left(\frac{(\tilde{\boldsymbol{\tau}} - \boldsymbol{\tau})^T\mathbf{Q}^{-1}(\tilde{\boldsymbol{\tau}} - \boldsymbol{\tau})}{s\tilde{\sigma}^2} \leq F_{\xi,s,n-p^*}\right),$$

for a particular choice of s and p^*, Scheffé's method can be used to derive a $100(1 - \xi)\%$ simultaneous prediction band for $\{\beta + bz : z \in \mathbb{R}\}$, where $\xi \in (0, 1)$. This band is of the form $(\tilde{\beta} + \tilde{b}z) \pm g(z)$, for some function $g(z)$. Determine $g(z)$; you may leave your answer in terms of q_{11}, q_{12}, and q_{22} (and other quantities) but determine numerical values for s and p^*.

Solution

(a) According to the solution to Exercise 13.13, $\text{var}(\mathbf{y}) = \sigma^2[\mathbf{I} + (\sigma_b^2/\sigma^2)\mathbf{z}\mathbf{z}^T]$. Now,

$$(1/\sigma^2)\mathbf{P}_\mathbf{z}\text{var}(\mathbf{y}) = \left(\sum_{i=1}^n z_i^2\right)^{-1} \mathbf{z}\mathbf{z}^T[\mathbf{I} + (\sigma_b^2/\sigma^2)\mathbf{z}\mathbf{z}^T] = \left[\left(\sum_{i=1}^n z_i^2\right)^{-1} + (\sigma_b^2/\sigma^2)\right]\mathbf{z}\mathbf{z}^T,$$

which is not idempotent. But

$$(1/\sigma^2)(\mathbf{P}_{1,\mathbf{z}} - \mathbf{P}_\mathbf{z})\text{var}(\mathbf{y}) = (\mathbf{P}_{1,\mathbf{z}} - \mathbf{P}_\mathbf{z})[\mathbf{I} + (\sigma_b^2/\sigma^2)\mathbf{z}\mathbf{z}^T] = \mathbf{P}_{1,\mathbf{z}} - \mathbf{P}_\mathbf{z},$$

which is idempotent, and

$$(1/\sigma^2)(\mathbf{I} - \mathbf{P}_{1,\mathbf{z}})\text{var}(\mathbf{y}) = (\mathbf{I} - \mathbf{P}_{1,\mathbf{z}})[\mathbf{I} + (\sigma_b^2/\sigma^2)\mathbf{z}\mathbf{z}^T] = \mathbf{I} - \mathbf{P}_{1,\mathbf{z}},$$

which is also idempotent. Therefore,

$$(1/\sigma^2)\mathbf{y}^T(\mathbf{P}_{1,\mathbf{z}} - \mathbf{P}_\mathbf{z})\mathbf{y} \sim \chi^2(1, NCP_1),$$

where

$$NCP_1 = (\beta\mathbf{1})^T(\mathbf{P}_{1,\mathbf{z}} - \mathbf{P}_\mathbf{z})(\beta\mathbf{1})/(2\sigma^2) = \frac{\beta^2 nSZZ}{2\sigma^2\sum_{i=1}^n z_i^2},$$

and

$$(1/\sigma^2)\mathbf{y}^T(\mathbf{I} - \mathbf{P}_{1,\mathbf{z}})\mathbf{y} \sim \chi^2(n-2, NCP_2),$$

where

$$NCP_2 = (\beta\mathbf{1})^T(\mathbf{I} - \mathbf{P}_{1,\mathbf{z}})(\beta\mathbf{1})/(2\sigma^2) = 0.$$

(b) The two quadratic forms in part (a) that have chi-square distributions are independent because

$$\begin{aligned}(1/\sigma^2)(\mathbf{P}_{1,\mathbf{z}} - \mathbf{P}_\mathbf{z})\text{var}(\mathbf{y})(1/\sigma^2)(\mathbf{I} - \mathbf{P}_{1,\mathbf{z}}) &= (1/\sigma^2)(\mathbf{P}_{1,\mathbf{z}} - \mathbf{P}_\mathbf{z})(\mathbf{I} - \mathbf{P}_{1,\mathbf{z}}) \\ &= (1/\sigma^2)(\mathbf{P}_{1,\mathbf{z}} - \mathbf{P}_\mathbf{z} - \mathbf{P}_{1,\mathbf{z}} + \mathbf{P}_\mathbf{z}) \\ &= \mathbf{0}.\end{aligned}$$

Therefore,

$$F \equiv \frac{\mathbf{y}^T(\mathbf{P}_{1,\mathbf{z}} - \mathbf{P}_{\mathbf{z}})\mathbf{y}}{\mathbf{y}^T(\mathbf{I} - \mathbf{P}_{1,\mathbf{z}})\mathbf{y}/(n-2)} = \frac{\mathbf{y}^T(\mathbf{P}_{1,\mathbf{z}} - \mathbf{P}_{\mathbf{z}})\mathbf{y}/\sigma^2}{\mathbf{y}^T(\mathbf{I} - \mathbf{P}_{1,\mathbf{z}})\mathbf{y}/(n-2)\sigma^2} \sim \mathrm{F}\left(1, n-2, \frac{\beta^2 nSZZ}{2\sigma^2 \sum_{i=1}^n z_i^2}\right).$$

Under H_0, $F \sim \mathrm{F}(1, n-2)$, and under H_a, $F \sim \mathrm{F}\left(1, n-2, \frac{\beta^2 nSZZ}{2\sigma^2 \sum_{i=1}^n z_i^2}\right)$.

(c) Here $s = 2$ and $p^* = 1$. Thus, by Corollary 2.16.1.1,

$$\begin{aligned} 1 - \xi &= \Pr\left(\frac{(\tilde{\boldsymbol{\tau}} - \boldsymbol{\tau})^T \mathbf{Q}^{-1}(\tilde{\boldsymbol{\tau}} - \boldsymbol{\tau})}{2\tilde{\sigma}^2} \le F_{\xi,2,n-1}\right) \\ &= \Pr\left(\left(\frac{1}{2\tilde{\sigma}^2}\right)\frac{[\mathbf{a}^T(\tilde{\boldsymbol{\tau}} - \boldsymbol{\tau})]^2}{\mathbf{a}^T\mathbf{Q}\mathbf{a}} \le F_{\xi,2,n-1} \text{ for all } \mathbf{a} \ne \mathbf{0}\right) \\ &= \Pr\left(\mathbf{a}^T\tilde{\boldsymbol{\tau}} - \sqrt{2F_{\xi,2,n-1}\tilde{\sigma}^2(\mathbf{a}^T\mathbf{Q}\mathbf{a})} \le \mathbf{a}^T\boldsymbol{\tau} \le \mathbf{a}^T\tilde{\boldsymbol{\tau}} + \sqrt{2F_{\xi,2,n-1}\tilde{\sigma}^2(\mathbf{a}^T\mathbf{Q}\mathbf{a})} \text{ for all } \mathbf{a}\right). \end{aligned}$$

Thus

$$\Pr[(\tilde{\beta} + \tilde{b}z) - g(z) \le \beta + bz \le (\tilde{\beta} + \tilde{b}z) + g(z) \text{ for all } z] = 1 - \xi,$$

where $g(z) = \sqrt{2F_{\xi,2,n-1}\tilde{\sigma}^2(q_{11} + 2q_{12}z + q_{22}z^2)}$. The same argument as that used in the solution of Exercise 15.19 establishes that the simultaneous coverage probability is exactly (rather than at least) $1 - \xi$.

▶ **Exercise 21** Consider the normal Gauss–Markov simple linear regression model and suppose that $n = 22$ and $x_1 = x_2 = 5$, $x_3 = x_4 = 6, \ldots, x_{21} = x_{22} = 15$.

(a) Determine the expected squared lengths of the Bonferroni and multivariate t 95% simultaneous confidence intervals for β_1 and β_2. (Expressions for these intervals may be found in Example 15.3-1.)
(b) Let $x_{n+1} = 5$, $x_{n+2} = 10$, and $x_{n+3} = 15$, and suppose that the original responses and the responses corresponding to these three "new" values of x follow the normal Gauss–Markov prediction-extended simple linear regression model. Give an expression for the multivariate t 95% simultaneous prediction intervals for $\{\beta_1 + \beta_2 x_{n+i} + e_{n+i} : i = 1, 2, 3\}$. Determine how much wider (proportionally) these intervals are than the multivariate t 95% simultaneous confidence intervals for $\{\beta_1 + \beta_2 x_{n+i} : i = 1, 2, 3\}$.
(c) Obtain the 95% trapezoidal (actually it is a parallelogram) confidence band described in Example 15.3.3 for the line over the bounded interval $[a = 5, b = 15]$, and compare the expected squared length of any interval in that band to the expected squared length of the intervals in the 95% classical confidence band at $x = 5$, $x = 10$, and $x = 15$.

Solution

(a) In this scenario, $\bar{x} = 10$, $SXX = 220$, and (according to Example 7.2-1) $\text{corr}(\hat{\beta}_1, \hat{\beta}_2) = -\bar{x}/\sqrt{\bar{x}^2 + SXX} = -10/\sqrt{110}$. The expected squared lengths of the Bonferroni and multivariate t 95% simultaneous confidence intervals for β_1 and β_2 are listed in the following table:

Intervals	β_1	β_2
Bonferroni	$4t^2_{0.0125,20}\sigma^2\left(\frac{1}{n} + \frac{\bar{x}^2}{SXX}\right) = 11.743\sigma^2$	$4t^2_{0.0125,20}\sigma^2/SXX = 0.10675\sigma^2$
Multivariate t	$4t^2_{0.025,2,20,\mathbf{R}}\sigma^2\left(\frac{1}{n} + \frac{\bar{x}^2}{SXX}\right) = 9.749\sigma^2$	$4t^2_{0.025,2,20,\mathbf{R}}\sigma^2/SXX = 0.08863\sigma^2$

To obtain the results for the multivariate t intervals, we used the fact that, according to Example 15.3-1, $\mathbf{R} = \begin{pmatrix} 1 & \text{corr}(\hat{\beta}_1, \hat{\beta}_2) \\ \text{corr}(\hat{\beta}_1, \hat{\beta}_2) & 1 \end{pmatrix} = \begin{pmatrix} 1 & \frac{-10}{\sqrt{110}} \\ \frac{-10}{\sqrt{110}} & 1 \end{pmatrix}$. Then, using the `qmvt` function in the `mvtnorm` package of `R` with this $\mathbf{R}$-matrix, we obtain $t_{0.025,2,20,\mathbf{R}} = 2.207874$.

(b) The multivariate t 95% simultaneous prediction intervals for $y_{n+1}, y_{n+2}, y_{n+3}$ are given by

$$(\hat{\beta}_1 + \hat{\beta}_2 x_{n+i}) \pm t_{0.025,3,n-2,\mathbf{R}}\hat{\sigma}\sqrt{1 + \frac{1}{n} + \frac{(x_{n+i} - \bar{x})^2}{SXX}} \qquad (i = 1, 2, 3),$$

where

$$\mathbf{R} = \text{var}\left(\frac{\hat{\beta}_1 + \hat{\beta}_2 x_{n+1} - y_{n+1}}{\sigma\sqrt{q_{11}}}, \frac{\hat{\beta}_1 + \hat{\beta}_2 x_{n+2} - y_{n+2}}{\sigma\sqrt{q_{22}}}, \frac{\hat{\beta}_1 + \hat{\beta}_2 x_{n+3} - y_{n+3}}{\sigma\sqrt{q_{33}}}\right)^T = (r_{ij}),$$

$r_{ij} = \left(\frac{1}{\sqrt{q_{ii}}\sqrt{q_{jj}}}\right)\left(\frac{1}{n} + \frac{(x_{n+i}-\bar{x})(x_{n+j}-\bar{x})}{SXX}\right)$ for $i \neq j = 1, 2, 3$, and $q_{ii} = 1 + \frac{1}{n} + \frac{(x_{n+i}-\bar{x})^2}{SXX}$ for $i = 1, 2, 3$. Furthermore, the multivariate t 95% simultaneous confidence intervals for $\{\beta_1 + \beta_2 x_{n+i} : i = 1, 2, 3\}$ are given by

$$(\hat{\beta}_1 + \hat{\beta}_2 x_{n+i}) \pm t_{0.025,3,n-2,\mathbf{R}^*}\hat{\sigma}\sqrt{\frac{1}{n} + \frac{(x_{n+i} - \bar{x})^2}{SXX}} \qquad (i = 1, 2, 3),$$

where

$$\mathbf{R}^* = \text{var}\left(\frac{\hat{\beta}_1 + \hat{\beta}_2 x_{n+1}}{\sigma\sqrt{m_{11}}}, \frac{\hat{\beta}_1 + \hat{\beta}_2 x_{n+2}}{\sigma\sqrt{m_{22}}}, \frac{\hat{\beta}_1 + \hat{\beta}_2 x_{n+3}}{\sigma\sqrt{m_{33}}}\right)^T$$
$$= (r^*_{ij}),$$

$r^*_{ij} = \left(\frac{1}{\sqrt{m_{ii}}\sqrt{m_{jj}}}\right)\left(\frac{1}{n} + \frac{(x_{n+i}-\bar{x})(x_{n+j}-\bar{x})}{SXX}\right)$ for $i \neq j = 1, 2, 3$, and $m_{ii} = \frac{1}{n} + \frac{(x_{n+i}-\bar{x})^2}{SXX}$ for $i = 1, 2, 3$. Finally, the ratio of the width of the multivariate t 95% simultaneous prediction intervals for $y_{n+1}, y_{n+2}, y_{n+3}$ to the width of the confidence intervals for $\beta_1 + \beta_2 x_{n+i}$ $(i = 1, 2, 3)$ is

$$\frac{2t_{0.025,3,n-2,\mathbf{R}}\hat{\sigma}\sqrt{1+\frac{1}{n}+\frac{(x_{n+i}-\bar{x})^2}{SXX}}}{2t_{0.025,3,n-2,\mathbf{R}^*}\hat{\sigma}\sqrt{\frac{1}{n}+\frac{(x_{n+i}-\bar{x})^2}{SXX}}} = \frac{t_{0.025,3,20,\mathbf{R}}\sqrt{1+\frac{1}{22}+\frac{(x_{n+i}-10)^2}{220}}}{t_{0.025,3,20,\mathbf{R}^*}\sqrt{\frac{1}{22}+\frac{(x_{n+i}-10)^2}{220}}}$$

$$= \begin{cases} 2.770 \text{ if } i = 1 \text{ or } i = 3, \\ 4.920 \text{ if } i = 2. \end{cases}$$

(c) Because $\bar{x} = 10$ lies at the midpoint of $[a, b] = [5, 15]$, the trapezoidal confidence region is actually a parallelogram. Therefore, the expected squared length of the interval in the 95% trapezoidal confidence band corresponding to any $x \in [5, 15]$ is

$$\left(2t_{0.025,2,20,\mathbf{R}}\sqrt{\frac{1}{22}+\frac{25}{220}}\right)^2\sigma^2 = 4t^2_{0.025,2,20,\mathbf{R}}\left(\frac{1}{22}+\frac{25}{220}\right)\sigma^2,$$

where $\mathbf{R}$ is a 2×2 correlation matrix with off-diagonal element equal to

$$\frac{m_{ab}}{\sqrt{m_{aa}m_{bb}}} = \frac{\frac{1}{22}+(5-10)(15-10)/220}{\sqrt{\left(\frac{1}{22}+\frac{(5-10)^2}{220}\right)\left(\frac{1}{22}+\frac{(15-10)^2}{220}\right)}} = -\frac{3}{7}.$$

We find that $t_{0.025,2,20,\mathbf{R}} = 2.38808$, so the expected squared length is $4(2.38808)^2[(1/22) + (25/220)]\sigma^2 = 3.6291\sigma^2$. For the classical 95% confidence band, the expected squared length of the interval for the line's ordinate at x is

$$\left[2\sqrt{2F_{0.05,2,20}\sigma^2\left(\frac{1}{22}+\frac{(x-10)^2}{SXX}\right)}\right]^2 = 8F_{0.05,2,20}\sigma^2\left(\frac{1}{22}+\frac{(x-10)^2}{SXX}\right).$$

When $x = 5, 10, 15$, this is equal to $4.445\sigma^2, 1.270\sigma^2$, and $4.445\sigma^2$, respectively.

Exercise 22 Consider the two-way main effects model with equal cell frequencies, i.e.,

$$y_{ijk} = \mu + \alpha_i + \gamma_j + e_{ijk} \quad (i = 1, \ldots, q;\ j = 1, \ldots, m;\ k = 1, \ldots, r).$$

Here μ, the α_i's, and the γ_j's are unknown parameters, while the e_{ijk}'s are independent normally distributed random variables having zero means and common variances σ^2. Let $0 < \xi < 1$.

(a) Obtain an expression for a $100(1-\xi)\%$ "one-at-a-time" (symmetric) confidence interval for $\alpha_i - \alpha_{i'}$ ($i' > i = 1, \ldots, q$). Obtain the analogous expression for a $100(1-\xi)\%$ one-at-a-time confidence interval for $\gamma_j - \gamma_{j'}$ ($j' > j = 1, \ldots, m$).
(b) Obtain an expression for a $100(1 - \xi)\%$ confidence ellipsoid for the subset of Factor A differences $\{\alpha_1 - \alpha_i : i = 2, \ldots, q\}$; likewise, obtain an expression for a $100(1 - \xi)\%$ confidence ellipsoid for the subset of Factor B differences $\{\gamma_1 - \gamma_j : j = 2, \ldots, m\}$. Your expressions can involve vectors and matrices provided that you give nonmatrix expressions for each of the elements of those vectors and matrices.
(c) Use each of the Bonferroni, Scheffé, and multivariate-t methods to obtain confidence intervals for all $[q(q - 1)/2] + m(m - 1)/2]$ differences $\alpha_i - \alpha_{i'}$ ($i' > i = 1, \ldots, q$) and $\gamma_j - \gamma_{j'}$ ($j' > j = 1, \ldots, m$) such that the probability of simultaneous coverage is at least $1 - \xi$.
(d) Discuss why Tukey's method cannot be used directly to obtain simultaneous confidence intervals for $\alpha_i - \alpha_{i'}$ ($i' > i = 1, \ldots, q$) and $\gamma_j - \gamma_{j'}$ ($j' > j = 1, \ldots, m$) such that the probability of simultaneous coverage is at least $1 - \xi$. Show, however, that Tukey's method could be used directly to obtain confidence intervals for $\alpha_i - \alpha_{i'}$ ($i' > i = 1, \ldots, q$) such that the probability of simultaneous coverage is at least $1 - \xi_1$, and again to obtain confidence intervals for $\gamma_j - \gamma_{j'}$ ($j' > j = 1, \ldots, m$) such that the probability of simultaneous coverage is at least $1 - \xi_2$; then combine these via the Bonferroni inequality to get intervals for the union of all of these differences whose simultaneous coverage probability is at least $1 - \xi_1 - \xi_2$.
(e) For each of the four sets of intervals for the α-differences and γ-differences obtained in parts (a) and (c), determine $\mathrm{E}(L^2)/\sigma^2$, where L represents the length of each interval in the set.
(f) Compute $[\mathrm{E}(L^2)/\sigma^2]^{\frac{1}{2}}$ for each of the four sets of confidence intervals when $q = 4, m = 5, r = 1$, and $\xi = .05$, and using $\mathbf{I}$ in place of $\mathbf{R}$ for the multivariate t intervals. Which of the three sets of simultaneous confidence intervals would you recommend in this case, assuming that your interest is in only the Factor A differences and Factor B differences?

Solution

(a) According to the solution to Exercise 7.11, for $i \leq s, i' > i = 1, \ldots, q$, and $s' > s = 1, \ldots, q$,

$$\text{cov}(\bar{y}_{i\cdot\cdot} - \bar{y}_{i'\cdot\cdot}, \bar{y}_{s\cdot\cdot} - \bar{y}_{s'\cdot\cdot}) = \begin{cases} \frac{2\sigma^2}{mr} & \text{if } i = s \text{ and } i' = s', \\ \frac{\sigma^2}{mr} & \text{if } i = s \text{ and } i' \neq s', \quad \text{or } i \neq s \text{ and } i' = s', \\ -\frac{\sigma^2}{mr} & \text{if } i' = s, \\ 0, & \text{otherwise;} \end{cases}$$

for $j \leq t, j' > j = 1, \ldots, m$, and $t' > t = 1, \ldots, m$,

$$\text{cov}(\bar{y}_{\cdot j\cdot} - \bar{y}_{\cdot j'\cdot}, \bar{y}_{\cdot t\cdot} - \bar{y}_{\cdot t'\cdot}) = \begin{cases} \frac{2\sigma^2}{qr} & \text{if } j = t \text{ and } j' = t', \\ \frac{\sigma^2}{qr} & \text{if } j = t \text{ and } j' \neq t', \quad \text{or } j \neq t \text{ and } j' = t', \\ -\frac{\sigma^2}{qr} & \text{if } j' = t, \\ 0, & \text{otherwise;} \end{cases}$$

and for $i' > i = 1, \ldots, q$ and $j' > j = 1, \ldots, m$,

$$\text{cov}(\bar{y}_{i\cdot\cdot} - \bar{y}_{i'\cdot\cdot}, \bar{y}_{\cdot j\cdot} - \bar{y}_{\cdot j'\cdot}) = 0.$$

Therefore, using (15.1),

$$(\bar{y}_{i\cdot\cdot} - \bar{y}_{i'\cdot\cdot}) \pm t_{\xi/2,qmr-q-m+1}\hat{\sigma}\sqrt{\frac{2}{mr}}$$

is a $100(1-\xi)\%$ confidence interval for $\alpha_i - \alpha_{i'}$ $(i' > i = 1, \ldots, q)$. Similarly,

$$(\bar{y}_{\cdot j\cdot} - \bar{y}_{\cdot j'\cdot}) \pm t_{\xi/2,qmr-q-m+1}\hat{\sigma}\sqrt{\frac{2}{qr}}$$

is a $100(1-\xi)\%$ confidence interval for $\gamma_j - \gamma_{j'}$ $(j' > j = 1, \ldots, m)$.

(b) Define the $(q-1) \times (1+q+m)$ matrix $\mathbf{C}_1$ as follows:

$$\mathbf{C}_1^T = \begin{pmatrix} 0 & 1 & -1 & 0 & \cdots & 0 & 0 & \cdots & 0 \\ 0 & 1 & 0 & -1 & \cdots & 0 & 0 & \cdots & 0 \\ \vdots & & & & & & & & \\ 0 & 1 & 0 & 0 & \cdots & -1 & 0 & \cdots & 0 \end{pmatrix}.$$

Then $\mathbf{C}_1^T\boldsymbol{\beta} = (\alpha_1 - \alpha_2, \alpha_1 - \alpha_3, \ldots, \alpha_1 - \alpha_q)^T$ and $\mathbf{C}_1^T\hat{\boldsymbol{\beta}} = (\bar{y}_{1\cdot\cdot} - \bar{y}_{2\cdot\cdot}, \bar{y}_{1\cdot\cdot} - \bar{y}_{3\cdot\cdot}, \ldots, \bar{y}_{1\cdot\cdot} - \bar{y}_{q\cdot\cdot})^T$. Therefore, by Theorem 15.1.1, a $100(1-\xi)\%$ confidence ellipsoid for $\{\alpha_1 - \alpha_j : j = 2, \ldots, q\}$ is given by

$$\{\boldsymbol{\beta} : (\mathbf{C}_1^T\hat{\boldsymbol{\beta}} - \mathbf{C}_1^T\boldsymbol{\beta})^T[\mathbf{C}_1^T(\mathbf{X}^T\mathbf{X})^-\mathbf{C}_1]^{-1}(\mathbf{C}_1^T\hat{\boldsymbol{\beta}} - \mathbf{C}_1^T\boldsymbol{\beta}) \le (q-1)\hat{\sigma}^2 F_{\xi,q-1,qmr-q-m+1}\}.$$

Observe that the (i, i')th element of $\mathbf{C}_1^T(\mathbf{X}^T\mathbf{X})^-\mathbf{C}_1$ is

$$\left(\frac{1}{\sigma^2}\right)\text{cov}(\bar{y}_{1\cdot\cdot} - \bar{y}_{(i+1)\cdot\cdot}, \bar{y}_{1\cdot\cdot} - \bar{y}_{(i'+1)\cdot\cdot}) = \begin{cases} 2/mr & \text{if } i = i', \\ 1/mr & \text{if } i \neq i', \end{cases}$$

for $i, i' = 1, \ldots, q-1$.

Similarly, define the $(m-1) \times (1+q+m)$ matrix $\mathbf{C}_2$ as follows:

$$\mathbf{C}_2^T = \begin{pmatrix} 0\,0 \cdots 0\;1 & -1 & 0 & \cdots & 0 \\ 0\,0 \cdots 0\;1 & 0 & -1 & \cdots & 0 \\ \vdots & & & & \\ 0\,0 \cdots 0\;1 & 0 & 0 & \cdots & -1 \end{pmatrix}.$$

Then $\mathbf{C}_2^T\boldsymbol{\beta} = (\gamma_1 - \gamma_2, \gamma_1 - \gamma_3, \ldots, \gamma_1 - \gamma_m)^T$ and $\mathbf{C}_2^T\hat{\boldsymbol{\beta}} = (\bar{y}_{\cdot 1\cdot} - \bar{y}_{\cdot 2\cdot}, \bar{y}_{\cdot 1\cdot} - \bar{y}_{\cdot 3\cdot}, \ldots, \bar{y}_{\cdot 1\cdot} - \bar{y}_{\cdot m\cdot})^T$. Therefore, by Theorem 15.1.1, a $100(1-\xi)\%$ confidence ellipsoid for $\{\gamma_1 - \gamma_j : j = 2, \ldots, m\}$ is given by

$$\{\boldsymbol{\beta} : (\mathbf{C}_2^T\hat{\boldsymbol{\beta}} - \mathbf{C}_2^T\boldsymbol{\beta})^T[\mathbf{C}_2^T(\mathbf{X}^T\mathbf{X})^-\mathbf{C}_2]^{-1}(\mathbf{C}_2^T\hat{\boldsymbol{\beta}} - \mathbf{C}_2^T\boldsymbol{\beta}) \le (m-1)\hat{\sigma}^2 F_{\xi,m-1,qmr-q-m+1}\}.$$

Observe that the (j, j')th element of $\mathbf{C}_2^T(\mathbf{X}^T\mathbf{X})^-\mathbf{C}_2$ is

$$\left(\frac{1}{\sigma^2}\right)\text{cov}(\bar{y}_{\cdot 1\cdot} - \bar{y}_{\cdot(j+1)\cdot}, \bar{y}_{\cdot 1\cdot} - \bar{y}_{\cdot(j'+1)\cdot}) = \begin{cases} 2/qr & \text{if } j = j', \\ 1/qr & \text{if } j \neq j', \end{cases}$$

for $j, j' = 1, \ldots, m-1$.

(c) The Bonferroni simultaneous confidence intervals for $\alpha_i - \alpha_{i'}$ $(i' > i = 1, \ldots, q)$ and $\gamma_j - \gamma_{j'}$ $(j' > j = 1, \ldots, m)$ are

$$(\bar{y}_{i\cdot\cdot} - \bar{y}_{i'\cdot\cdot}) \pm t_{\xi/[q(q-1)+m(m-1)],qmr-q-m+1}\hat{\sigma}\sqrt{\frac{2}{mr}} \quad \text{and}$$

$$(\bar{y}_{\cdot j\cdot} - \bar{y}_{\cdot j'\cdot}) \pm t_{\xi/[q(q-1)+m(m-1)],qmr-q-m+1}\hat{\sigma}\sqrt{\frac{2}{qr}}.$$

For the Scheffé method, take

$$\mathbf{C} = \begin{pmatrix} 0\ 1\ -1\ 0\ 0 \cdots 0\ 0\ 1\ -1\ 0\ 0 \cdots 0\ 0 \\ 0\ 0\ 1\ -1\ 0 \cdots 0\ 0\ 0\ 1\ -1\ 0 \cdots 0\ 0 \\ \vdots \\ 0\ 0\ 0\ 0\ 0 \cdots 1\ -1\ 0\ 0\ 0\ 0 \cdots 1\ -1 \end{pmatrix}.$$

Observe that $\text{rank}(\mathbf{C}) = q + m - 2$. Thus, the Scheffé simultaneous confidence intervals for $\alpha_i - \alpha_{i'}$ ($i' > i = 1, \ldots, q$) and $\gamma_j - \gamma_{j'}$ ($j' > j = 1, \ldots, m$) are

$$(\bar{y}_{i\cdot\cdot} - \bar{y}_{i'\cdot\cdot}) \pm \hat{\sigma}\sqrt{\frac{2}{mr}}\sqrt{(q + m - 2)F_{\xi,q+m-2,qmr-q-m+1}}$$

and

$$(\bar{y}_{\cdot j\cdot} - \bar{y}_{\cdot j'\cdot}) \pm \hat{\sigma}\sqrt{\frac{2}{qr}}\sqrt{(q + m - 2)F_{\xi,q+m-2,qmr-q-m+1}}.$$

The multivariate t simultaneous confidence intervals for $\alpha_i - \alpha_{i'}$ ($i' > i = 1, \ldots, q$) and $\gamma_j - \gamma_{j'}$ ($j' > j = 1, \ldots, m$) (jointly) are

$$(\bar{y}_{i\cdot\cdot} - \bar{y}_{i'\cdot\cdot}) \pm t_{\xi/2,[q(q-1)+m(m-1)]/2,qmr-q-m+1,\mathbf{R}}\hat{\sigma}\sqrt{\frac{2}{mr}}$$

and

$$(\bar{y}_{\cdot j\cdot} - \bar{y}_{\cdot j'\cdot}) \pm t_{\xi/2,[q(q-1)+m(m-1)]/2,qmr-q-m+1,\mathbf{R}}\hat{\sigma}\sqrt{\frac{2}{qr}},$$

respectively, where

$$\mathbf{R} = \text{var}\left(\frac{\bar{y}_{1\cdot\cdot} - \bar{y}_{2\cdot\cdot}}{\sigma\sqrt{2/mr}}, \ldots, \frac{\bar{y}_{(q-1)\cdot\cdot} - \bar{y}_{q\cdot\cdot}}{\sigma\sqrt{2/mr}}, \frac{\bar{y}_{\cdot 1\cdot} - \bar{y}_{\cdot 2\cdot}}{\sigma\sqrt{2/qr}}, \ldots, \frac{\bar{y}_{\cdot(m-1)\cdot} - \bar{y}_{\cdot m\cdot}}{\sigma\sqrt{2/qr}}\right)^T.$$

(d) Because the differences $\alpha_i - \alpha_{i'}$ ($i' > i = 1, \ldots, q$) and $\gamma_j - \gamma_{j'}$ ($j' > j = 1, \ldots, m$) cannot be represented as all possible pairwise differences of a set of linearly independent estimable functions, Tukey's method cannot be applied directly. However, observe that $\psi_i \equiv \mu + \alpha_i + \frac{1}{m}\sum_{j=1}^m \gamma_j$ ($i = 1, \ldots, q$) are estimable and $\alpha_i - \alpha_{i'}$ ($i' > i = 1, \ldots, q$) are all possible differences among $\psi_1, \ldots, \psi_q$. The least squares estimator of ψ_i is

$$\hat{\psi}_i = \bar{y}_{\cdots} + (\bar{y}_{i\cdot\cdot} - \bar{y}_{\cdots}) + \frac{1}{m}\sum_{j=1}^m (\bar{y}_{\cdot j\cdot} - \bar{y}_{\cdots}) = \bar{y}_{i\cdot\cdot} \quad (i = 1, \ldots, q)$$

and

$$\text{var}[(\hat{\psi}_1, \ldots, \hat{\psi}_q)^T] = \left(\frac{\sigma^2}{mr}\right)\mathbf{I}.$$

Therefore, the $100(1-\xi_1)\%$ simultaneous Tukey intervals for $\alpha_i - \alpha_{i'}$ $(i' > i = 1, \ldots, q)$ are given by

$$(\bar{y}_{i\cdot\cdot} - \bar{y}_{i'\cdot\cdot}) \pm q^*_{\xi_1,q,qmr-q-m+1}\hat{\sigma}\sqrt{1/mr}.$$

By a very similar argument, the $100(1-\xi_2)\%$ simultaneous Tukey intervals for $\gamma_j - \gamma_{j'}$ $(j' > j = 1, \ldots, m)$ are given by

$$(\bar{y}_{\cdot j\cdot} - \bar{y}_{\cdot j'\cdot}) \pm q^*_{\xi_2,m,qmr-q-m+1}\hat{\sigma}\sqrt{1/qr}.$$

Finally, let

$$E_1 = \{(\alpha_i - \alpha_{i'}) \in (\bar{y}_{i\cdot\cdot} - \bar{y}_{i'\cdot\cdot}) \pm q^*_{\xi_1,q,qmr-q-m+1}\hat{\sigma}\sqrt{1/mr} \quad (i' > i = 1, \ldots, q)\},$$
$$E_2 = \{(\gamma_j - \gamma_{j'}) \in (\bar{y}_{\cdot j\cdot} - \bar{y}_{\cdot j'\cdot}) \pm q^*_{\xi_2,m,qmr-q-m+1}\hat{\sigma}\sqrt{1/qr} \quad (j' > j = 1, \ldots, m)\}.$$

By the Bonferroni inequality, $\Pr(E_1 \cap E_2) \geq 1 - \xi_1 - \xi_2$.

(e)

Intervals	$E(L^2)/\sigma^2$ for α-difference	$E(L^2)/\sigma^2$ for γ-difference
One-at-a-time	$(8/mr)t^2_{\xi/2,qmr-q-m+1}$	$(8/qr)t^2_{\xi/2,qmr-q-m+1}$
Bonferroni	$(8/mr)t^2_{\xi/[q(q-1)+m(m-1)],qmr-q-m+1}$	$(8/qr)t^2_{\xi/[q(q-1)+m(m-1)],qmr-q-m+1}$
Scheffé	$(8/mr)(q+m-2)F_{\xi,q+m-2,qmr-q-m+1}$	$(8/qr)(q+m-2)F_{\xi,q+m-2,qmr-q-m+1}$
Multivariate t	$(8/mr)t^2_{\xi/2,[q(q-1)+m(m-1)]/2,qmr-q-m+1,\mathbf{R}}$	$(8/qr)t^2_{\xi/2,[q(q-1)+m(m-1)]/2,qmr-q-m+1,\mathbf{R}}$

(f)

Intervals	$[E(L^2)/\sigma^2]^{\frac{1}{2}}$ for α-difference	$[E(L^2)/\sigma^2]^{\frac{1}{2}}$ for γ-difference
One-at-a-time	2.756	3.081
Bonferroni	4.660	5.210
Scheffé	5.712	6.386
Multivariate t (using $\mathbf{R} = \mathbf{I}$)	4.517	5.050

The multivariate t intervals are recommended, despite using **I** in place of **R**, because they have the shortest expected squared length.

▶ **Exercise 23** Consider data in a 3 × 3 layout that follow a normal Gauss–Markov two-way model with interaction, and suppose that each cell contains exactly r

observations. Thus, the model is

$$y_{ijk} = \mu + \alpha_i + \gamma_j + \xi_{ij} + e_{ijk} \quad (i = 1, \ldots, 3;\ j = 1, \ldots, 3;\ k = 1, \ldots, r),$$

where the e_{ijk}'s are independent $\mathrm{N}(0, \sigma^2)$ random variables.

(a) List the "essentially different" interaction contrasts $\xi_{ij} - \xi_{ij'} - \xi_{i'j} + \xi_{i'j'}$ within this layout and their ordinary least squares estimators. (A list of interaction contrasts is "essentially different" if no contrast in the list is a scalar multiple of another.)
(b) Give expressions for one-at-a-time $100(1 - \xi)\%$ confidence intervals for the essentially different interaction contrasts.
(c) Use each of the Bonferroni, Scheffé, and multivariate t methods to obtain confidence intervals for the essentially different interaction contrasts that have probability of simultaneous coverage at least $1 - \xi$. For the multivariate t method, give the numerical entries of $\mathbf{R}$.
(d) Explain why Tukey's method cannot be used to obtain simultaneous confidence intervals for the essentially different interaction contrasts that have probability of simultaneous coverage at least $1 - \xi$.
(e) For each of the four sets of intervals for the essentially different error contrasts obtained in parts (b) and (c), determine $\mathrm{E}(L^2)/\sigma^2$, where L represents the length of each interval in the set.
(f) Compute $[\mathrm{E}(L^2)/\sigma^2]^{\frac{1}{2}}$ for each of the four sets of confidence intervals when $\xi = 0.05$ and $r = 3$. Which of the three sets of simultaneous confidence intervals would you recommend in this case?

Solution

(a) There are nine essentially different interaction contrasts, as follows: $\xi_{11} - \xi_{12} - \xi_{21} + \xi_{22}$, $\xi_{11} - \xi_{13} - \xi_{21} + \xi_{23}$, $\xi_{11} - \xi_{12} - \xi_{31} + \xi_{32}$, $\xi_{11} - \xi_{13} - \xi_{31} + \xi_{33}$, $\xi_{12} - \xi_{13} - \xi_{22} + \xi_{23}$, $\xi_{12} - \xi_{13} - \xi_{32} + \xi_{33}$, $\xi_{21} - \xi_{22} - \xi_{31} + \xi_{32}$, $\xi_{21} - \xi_{23} - \xi_{31} + \xi_{33}$, $\xi_{22} - \xi_{23} - \xi_{32} + \xi_{33}$. The ordinary least squares estimator of $\xi_{ij} - \xi_{ij'} - \xi_{i'j} + \xi_{i'j'}$ is $\hat{\tau}_{iji'j'} \equiv \bar{y}_{ij} - \bar{y}_{ij'} - \bar{y}_{i'j} + \bar{y}_{i'j'}$.
(b) One-at-a-time $100(1 - \xi)\%$ confidence intervals for the essentially different interaction contrasts are given by

$$\hat{\tau}_{iji'j'} \pm t_{\xi/2,9(r-1)}\hat{\sigma}\sqrt{4/r}.$$

(c) The Bonferroni intervals are given by

$$\hat{\tau}_{iji'j'} \pm t_{\xi/18,9(r-1)}\hat{\sigma}\sqrt{4/r}.$$

To determine the Scheffé intervals, we must find a basis for the essentially different interaction contrasts. The first four interaction contrasts in the list given in the solution to part (a) are linearly independent because ξ_{22}, ξ_{23}, ξ_{32}, and ξ_{33} appear only once among those four contrasts. Furthermore, each of the other five contrasts can be written as a linear combination of the first four, as follows:

$$\xi_{12} - \xi_{13} - \xi_{22} + \xi_{23} = (\xi_{11} - \xi_{13} - \xi_{21} + \xi_{23}) - (\xi_{11} - \xi_{12} - \xi_{21} + \xi_{22})$$

$$\xi_{12} - \xi_{13} - \xi_{32} + \xi_{33} = (\xi_{11} - \xi_{13} - \xi_{31} + \xi_{33}) - (\xi_{11} - \xi_{12} - \xi_{31} + \xi_{32})$$

$$\xi_{21} - \xi_{22} - \xi_{31} + \xi_{32} = (\xi_{11} - \xi_{12} - \xi_{31} + \xi_{32}) - (\xi_{11} - \xi_{12} - \xi_{21} + \xi_{22})$$

$$\xi_{21} - \xi_{23} - \xi_{31} + \xi_{33} = (\xi_{11} - \xi_{13} - \xi_{31} + \xi_{33}) - (\xi_{11} - \xi_{13} - \xi_{21} + \xi_{23})$$

$$\xi_{22} - \xi_{23} - \xi_{32} + \xi_{33} = (\xi_{11} - \xi_{13} - \xi_{31} + \xi_{33}) - (\xi_{11} - \xi_{12} - \xi_{31} + \xi_{32}) \\ -(\xi_{11} - \xi_{13} - \xi_{21} + \xi_{23}) + (\xi_{11} - \xi_{12} - \xi_{21} + \xi_{22}).$$

Thus the number of vectors in a basis for all nine essentially different interaction contrasts is four, and the Scheffé intervals are given by

$$\hat{\tau}_{iji'j'} \pm \sqrt{4F_{\xi,4,9(r-1)}}\hat{\sigma}\sqrt{4/r}.$$

The multivariate t intervals are given by

$$\hat{\tau}_{iji'j'} \pm t_{\xi/2,9,9(r-1),\mathbf{R}}\hat{\sigma}\sqrt{4/r},$$

where

$$\mathbf{R} = \begin{pmatrix} 1.00 & 0.50 & 0.50 & 0.25 & 0.50 & 0.25 & 0.50 & 0.25 & 0.25 \\ 0.50 & 1.00 & 0.25 & 0.50 & 0.50 & 0.25 & 0.25 & 0.50 & 0.25 \\ 0.50 & 0.25 & 1.00 & 0.50 & 0.25 & 0.50 & 0.50 & 0.25 & 0.25 \\ 0.25 & 0.50 & 0.50 & 1.00 & 0.25 & 0.50 & 0.25 & 0.50 & 0.25 \\ 0.50 & 0.50 & 0.25 & 0.25 & 1.00 & 0.50 & 0.25 & 0.25 & 0.50 \\ 0.25 & 0.25 & 0.50 & 0.50 & 0.50 & 1.00 & 0.50 & 0.25 & 0.50 \\ 0.50 & 0.25 & 0.50 & 0.25 & 0.25 & 0.50 & 1.00 & 0.50 & 0.50 \\ 0.25 & 0.50 & 0.25 & 0.50 & 0.25 & 0.25 & 0.50 & 1.00 & 0.50 \\ 0.25 & 0.25 & 0.25 & 0.25 & 0.50 & 0.50 & 0.50 & 0.50 & 1.00 \end{pmatrix}.$$

(d) Tukey's method is not applicable because it is not possible to write the essentially different interaction contrasts as all pairwise differences of estimable functions whose least squares estimators have a variance–covariance matrix equal to a scalar multiple of an identity matrix.

(e)

Intervals	Expected squared length/σ^2
One-at-a-time	$(16/r)t^2_{\xi/2,9(r-1)}$
Bonferroni	$(16/r)t^2_{\xi/18,9(r-1)}$
Scheffé	$(64/r)F_{\xi,4,9(r-1)}$
Multivariate t	$(16/r)t^2_{\xi/2,9,9(r-1),\mathbf{R}}$

(f)

Intervals	$[E(L^2)/\sigma^2]^{\frac{1}{2}}$
One-at-a-time	4.85
Bonferroni	7.27
Scheffé	7.90
Multivariate t	6.96

The multivariate t intervals are recommended because they have the shortest expected squared length among the three sets of simultaneous confidence intervals.

▶ **Exercise 24** Consider the two-factor nested model

$$y_{ijk} = \mu + \alpha_i + \gamma_{ij} + e_{ijk} \quad (i = 1, \ldots, q; \quad j = 1, \ldots, m_i; \quad k = 1, \ldots, n_{ij}),$$

where the e_{ijk}'s are independent N$(0, \sigma^2)$ random variables. Suppose that $100(1-\xi)\%$ simultaneous confidence intervals for all of the differences $\gamma_{ij} - \gamma_{ij'}(i = 1, \ldots, q;\ j' > j = 1, \ldots, m_i)$ are desired.

(a) Give $100(1-\xi)\%$ one-at-a-time confidence intervals for $\gamma_{ij} - \gamma_{ij'}$ $(j' > j = 1, \ldots, m_i;\ i = 1, \ldots, q)$.
(b) For each of the Bonferroni, Scheffé, and multivariate t approaches to this problem, the desired confidence intervals can be expressed as

$$(\hat{\gamma}_{ij} - \hat{\gamma}_{ij'}) \pm a\hat{\sigma}\sqrt{v_{ijj'}},$$

where $\hat{\gamma}_{ij} - \hat{\gamma}_{ij'}$ is the least squares estimator of $\gamma_{ij} - \gamma_{ij'}$, $\hat{\sigma}^2$ is the residual mean square, $v_{ijj'} = \text{var}(\hat{\gamma}_{ij} - \hat{\gamma}_{ij'})/\sigma^2$, and a is a percentage point from an appropriate distribution. For each of the three approaches, give a (indexed by its tail probability, degree(s) of freedom, and any other parameters used to index the appropriate distribution). Note: You need not give expressions for the elements of the correlation matrix $\mathbf{R}$ for the multivariate t method.
(c) Consider the case $q = 2$ and suppose that n_{ij} is constant (equal to r) across j. Although Tukey's method is not directly applicable to this problem, it can be

used to obtain $100(1-\xi_1)\%$ simultaneous confidence intervals for $\gamma_{1j}-\gamma_{1j'}$ ($j' > j = 1,\ldots,m_1$) and then used a second time to obtain $100(1-\xi_2)\%$ simultaneous confidence intervals for $\gamma_{2j}-\gamma_{2j'}$ ($j' > j = 1,\ldots,m_2$). If ξ_1 and ξ_2 are chosen appropriately, these intervals can then be combined, using Bonferroni's inequality, to obtain a set of intervals for all of the differences $\gamma_{ij}-\gamma_{ij'}$ ($j' > j$) whose simultaneous coverage probability is at least $1-\xi_1-\xi_2$. Give appropriate cutoff point(s) for this final set of intervals. (Note: The cutoff point for some of the intervals in this final set may not be the same as the cutoff point for other intervals in the set. Be sure to indicate the intervals to which each cutoff point corresponds.)

(d) For each of the four sets of intervals for the γ-differences obtained in parts (a) and (b), determine an expression for $\mathrm{E}(L^2/\sigma^2)$, where L represents the length of each interval in the set. Do likewise for the method described in part (c), with $\xi_1 = \xi_2 = \xi/2$. For the multivariate t approach, replace $\mathbf{R}$ with $\mathbf{I}$ to obtain conservative intervals.

(e) Evaluate $[\mathrm{E}(L^2)/\sigma^2]^{\frac{1}{2}}$ for each of the four sets of confidence intervals when $q = 2$, $m_1 = m_2 = 5$, $n_{ij} = 2$ for all i and j, and $\xi = .05$. Based on these results, which set of simultaneous confidence intervals would you recommend in this case?

Solution

(a) According to the solution to Exercise 7.13, for $i \le s, j \le t, j' > j = 1,\ldots,m_i, t' > t = 1,\ldots,m_s$,

$$\mathrm{cov}(\bar{y}_{ij\cdot}-\bar{y}_{ij'\cdot}, \bar{y}_{st\cdot}-\bar{y}_{st'\cdot}) = \begin{cases} \sigma^2(\frac{1}{n_{ij}}+\frac{1}{n_{ij'}}) & \text{if } i=s,\ j=t,\ j'=t', \\ \frac{\sigma^2}{n_{ij}} & \text{if } i=s,\ j=t,\ j'\neq t', \\ \frac{\sigma^2}{n_{ij'}} & \text{if } i=s,\ j\neq t,\ j'=t', \\ -\frac{\sigma^2}{n_{ij}} & \text{if } i=s \text{ and } j=t', \\ -\frac{\sigma^2}{n_{ij'}} & \text{if } i=s \text{ and } j'=t, \\ 0 & \text{otherwise.} \end{cases}$$

Therefore, using (15.1),

$$(\bar{y}_{ij\cdot}-\bar{y}_{ij'\cdot}) \pm t_{\xi/2,n-\sum_{i=1}^q m_i}\hat{\sigma}\sqrt{\frac{1}{n_{ij}}+\frac{1}{n_{ij'}}}$$

is a $100(1-\xi)\%$ confidence interval for $\gamma_{ij}-\gamma_{ij'}$ ($i = 1,\ldots,q;\ j' > j = 1,\ldots,m_i$).

(b) For the Bonferroni approach,

$$a = t_{\xi/\sum_{i=1}^q m_i(m_i-1),n-\sum_{i=1}^q m_i}.$$

For the Scheffé approach, observe that for fixed i the $m_i(m_i - 1)/2$ differences $\{\gamma_{ij} - \gamma_{ij'} : j' > j = 1, \ldots, m_i\}$ are linearly dependent, but any basis for them consists of $m_i - 1$ linearly independent differences, and for different i these bases are linearly independent. It follows that for the Scheffé approach,

$$a = \sqrt{\sum_{i=1}^{q}(m_i - 1)F_{\xi, \sum_{i=1}^{q}(m_i-1), n-\sum_{i=1}^{q} m_i}}.$$

For the multivariate t approach,

$$a = t_{\xi/2, \sum_{i=1}^{q} m_i(m_i-1)/2, n-\sum_{i=1}^{q} m_i, \mathbf{R}},$$

where

$$\mathbf{R} = \text{var}\left(\frac{\bar{y}_{11\cdot} - \bar{y}_{12\cdot}}{\sigma\sqrt{(2/n_{11}) + (2/n_{12})}}, \ldots, \frac{\bar{y}_{11\cdot} - \bar{y}_{1m_1\cdot}}{\sigma\sqrt{(2/n_{11}) + (2/n_{1m_1})}}, \right.$$

$$\left. \frac{\bar{y}_{21\cdot} - \bar{y}_{22\cdot}}{\sigma\sqrt{(2/n_{21}) + (2/n_{22})}}, \ldots, \frac{\bar{y}_{q,(m_q-1)\cdot} - \bar{y}_{qm_q\cdot}}{\sigma\sqrt{(2/n_{q,m_q-1}) + (2/n_{qm_q})}}\right)^T.$$

(c) $\{(\gamma_{1j} - \gamma_{1j'}) : j' > j = 1, \ldots, m_1\}$ can be represented as all pairwise differences of the linearly independent estimable functions $\psi_{11}, \psi_{12}, \ldots, \psi_{1m_1}$, where $\psi_{1j} = \mu + \alpha_1 + \gamma_{1j}$ $(j = 1, \ldots, m_1)$. The least squares estimator of ψ_{1j} is $\bar{y}_{1j\cdot}$ and

$$\text{var}[(\bar{y}_{11}, \ldots, \bar{y}_{1m_1})^T] = \left(\frac{\sigma^2}{r}\right)\mathbf{I}.$$

Therefore, the $100(1 - \xi_1)\%$ simultaneous Tukey intervals for $\gamma_{1j} - \gamma_{1j'}$ $(j' > j = 1, \ldots, m_1)$ are given by

$$(\bar{y}_{1j\cdot} - \bar{y}_{1j'\cdot}) \pm q^*_{\xi_1, m_1-1, n-\sum_{i=1}^{q} m_i}\hat{\sigma}\sqrt{1/r} \quad (j' > j = 1, \ldots, m_1).$$

By a very similar argument, the $100(1 - \xi_2)\%$ simultaneous Tukey intervals for $\gamma_{2j} - \gamma_{2j'}$ $(j' > j = 1, \ldots, m_2)$ are given by

$$(\bar{y}_{2j\cdot} - \bar{y}_{2j'\cdot}) \pm q^*_{\xi_2, m_2-1, n-\sum_{i=1}^{q} m_i}\hat{\sigma}\sqrt{1/r} \quad (j' > j = 1, \ldots, m_2).$$

Finally, let

$$E_1 = \{(\gamma_{1j} - \gamma_{1j'}) \in (\bar{y}_{1j\cdot} - \bar{y}_{1j'\cdot}) \pm q^*_{\xi_1, m_1-1, n-\sum_{i=1}^{q} m_i}\hat{\sigma}\sqrt{1/r} \quad (j' > j = 1, \ldots, m_1)\},$$

$$E_2 = \{(\gamma_{2j} - \gamma_{2j'}) \in (\bar{y}_{2j\cdot} - \bar{y}_{2j'\cdot}) \pm q^*_{\xi_2, m_2-1, n-\sum_{i=1}^{q} m_i}\hat{\sigma}\sqrt{1/r} \quad (j' > j = 1, \ldots, m_2)\}.$$

By the Bonferroni inequality, $\Pr(E_1 \cap E_2) \geq 1 - \xi_1 - \xi_2$.

(d)

Intervals	$E(L^2)/\sigma^2$
One-at-a-time	$4t^2_{\xi/2,n-\sum_{i=1}^q m_i}\left(\frac{1}{n_{ij}}+\frac{1}{n_{ij'}}\right)$
Bonferroni	$4t^2_{\xi/\sum_{i=1}^q m_i(m_i-1),n-\sum_{i=1}^q m_i}\left(\frac{1}{n_{ij}}+\frac{1}{n_{ij'}}\right)$
Scheffé	$4\left(\sum_{i=1}^q(m_i-1)\right)F_{\xi,\sum_{i=1}^q(m_i-1),n-\sum_{i=1}^q m_i}\left(\frac{1}{n_{ij}}+\frac{1}{n_{ij'}}\right)$
Multivariate t (with **I** in place of **R**)	$4t^2_{\xi/2,\sum_{i=1}^q m_i(m_i-1)/2,n-\sum_{i=1}^q m_i,\mathbf{I}}\left(\frac{1}{n_{ij}}+\frac{1}{n_{ij'}}\right)$
Tukey (with $q=2$ and $n_{ij}\equiv r$)	$4q^{*2}_{\xi/2,m_i-1,n-(m_1+m_2)}\left(\frac{1}{r}\right)$

(e)

Intervals	$[E(L^2)/\sigma^2]^{\frac{1}{2}}$
One-at-a-time	4.456
Bonferroni	8.009
Scheffé	9.914
Multivariate t (with **I** in place of **R**)	7.645
Tukey	6.990

The Tukey intervals are recommended because they have the shortest expected squared length.

▶ **Exercise 25** Under the normal Gauss–Markov two-way partially crossed model introduced in Example 5.1.4-1 with one observation per cell:

(a) Give $100(1-\xi)\%$ one-at-a-time confidence intervals for $\alpha_i-\alpha_j$ and $\mu+\alpha_i-\alpha_j$.
(b) Give $100(1-\xi)\%$ Bonferroni, Scheffé, and multivariate t simultaneous confidence intervals for $\{\alpha_i-\alpha_j : j>i=1,\ldots,q\}$.

Solution

(a) Using results from the solution to Exercise 7.14, we find that a $100(1-\xi)\%$ confidence interval for $\alpha_i-\alpha_j$ is given by

$$\left(\frac{y_{i\cdot}-y_{\cdot i}-y_{j\cdot}+y_{\cdot j}}{2q}\right)\pm t_{\xi/2,q(q-2)}\hat{\sigma}\sqrt{\frac{1}{q}},$$

and that a $100(1-\xi)\%$ confidence interval for $\mu+\alpha_i-\alpha_j$ is given by

$$\left(\bar{y}_{\cdot\cdot}+\frac{y_{i\cdot}-y_{\cdot i}-y_{j\cdot}+y_{\cdot j}}{2q}\right)\pm t_{\xi/2,q(q-2)}\hat{\sigma}\sqrt{\frac{1}{q-1}}$$

because $\frac{1}{q(q-1)}+\frac{1}{q}=\frac{1}{q-1}$.

(b) The $100(1-\xi)\%$ Bonferroni intervals for all differences $\alpha_i-\alpha_j$ ($j>i=1,\ldots,q$) may be obtained from the one-at-a-time intervals for the same quantities by replacing $t_{\xi/2,q(q-2)}$ with $t_{\xi/[(q-1)q],q(q-2)}$; the Scheffé intervals may be obtained by replacing the same quantity with $\sqrt{(q-1)F_{\xi,q-1,q(q-2)}}$; and the multivariate t intervals may be obtained by replacing the same quantities with $t_{\xi/2,(q-1)q/2,q(q-2),\mathbf{R}}$, where, using the solution to Exercise 7.14b, the (ij,st)th element of $\mathbf{R}$ ($i\le s$, $j>i=1,\ldots,q$, $t>s=1,\ldots,q$) is given by

$$\text{corr}\left(\frac{y_{i\cdot}-y_{\cdot i}-y_{j\cdot}+y_{\cdot j}}{2q},\ \frac{y_{s\cdot}-y_{\cdot s}-y_{t\cdot}+y_{\cdot t}}{2q}\right)$$

$$=\begin{cases}1 & \text{if } i=s \text{ and } j=t,\\ 0.5 & \text{if } i=s \text{ and } j\neq t, \text{ or } i\neq s \text{ and } j=t,\\ -0.5 & \text{if } j=s,\\ 0 & \text{otherwise.}\end{cases}$$

► **Exercise 26** Under the normal Gauss–Markov Latin square model with q treatments introduced in Exercise 6.21:

(a) Give $100(1-\xi)\%$ one-at-a-time confidence intervals for $\tau_k-\tau_{k'}$ ($k'>k=1,\ldots,q$).
(b) Give $100(1-\xi)\%$ Bonferroni, Scheffé, and multivariate t simultaneous confidence intervals for $\{\tau_k-\tau_{k'}: k'>k=1,\ldots,q\}$.

Solution

(a) Using results from the solution to Exercise 7.15, we find that a $100(1-\xi)\%$ confidence interval for $\tau_k-\tau_{k'}$ is given by

$$\bar{y}_{\cdot\cdot k}-\bar{y}_{\cdot\cdot k'}\pm t_{\xi/2,(q-1)(q-2)}\hat{\sigma}\sqrt{\frac{2}{q}}.$$

(b) The $100(1-\xi)\%$ Bonferroni intervals for all differences $\tau_k-\tau_{k'}$ may be obtained from the one-at-a-time intervals for the same quantities by replacing $t_{\xi/2,(q-1)(q-2)}$ with $t_{\xi/[(q-1)q],(q-1)(q-2)}$; the Scheffé intervals may be obtained by replacing the same quantity with $\sqrt{(q-1)F_{\xi,q-1,(q-1)(q-2)}}$; and the multivariate t intervals may be obtained by replacing the same quantities with

$t_{\xi/2,(q-1)q/2,(q-1)(q-2),\mathbf{R}}$, where the (kk', ll')th element of $\mathbf{R}$ (for $k \leq l, k' > k = 1, \ldots, q, l' > l = 1, \ldots, q$) is given by

$$\text{corr}\,(\bar{y}_{\cdot\cdot k} - \bar{y}_{\cdot\cdot k'}, \bar{y}_{\cdot\cdot l} - \bar{y}_{\cdot\cdot l'}) = \begin{cases} 1 & \text{if } k = l \text{ and } k' = l', \\ 0.5 & \text{if } k = l \text{ and } k' \neq l', \text{ or } k \neq l \text{ and } k' = l', \\ -0.5 & \text{if } k' = l, \\ 0 & \text{otherwise.} \end{cases}$$

Exercise 27 Consider the normal split-plot model introduced in Example 13.4.5-1, but with $\psi \equiv \sigma_b^2/\sigma^2 > 0$ known, and recall the expressions obtained for variances of various mean differences obtained in Exercise 13.25. Let $\tilde{\sigma}^2$ denote the generalized residual mean square.

(a) Obtain $100(1-\xi)\%$ one-at-a-time confidence intervals for the functions in the sets $\{(\alpha_i - \alpha_{i'}) + (\bar{\xi}_{i\cdot} - \bar{\xi}_{i'\cdot}) : i' > i = 1, \ldots, q\}$ and $\{(\gamma_k - \gamma_{k'}) + (\bar{\xi}_{\cdot k} - \bar{\xi}_{\cdot k'}) : k' > k = 1, \ldots, m\}$. [Note that $\bar{y}_{i\cdot\cdot} - \bar{y}_{i'\cdot\cdot}$ is the least squares estimator of $(\alpha_i - \alpha_{i'}) + (\bar{\xi}_{i\cdot} - \bar{\xi}_{i'\cdot})$, and $\bar{y}_{\cdot\cdot k} - \bar{y}_{\cdot\cdot k'}$ is the least squares estimator of $(\gamma_k - \gamma_{k'}) + (\bar{\xi}_{\cdot k} - \bar{\xi}_{\cdot k'})$.]

(b) Using the Bonferroni method, obtain a set of confidence intervals whose simultaneous coverage probability for all of the estimable functions listed in part (a) is at least $1 - \xi$.

(c) Using the Scheffé method, obtain a set of confidence intervals whose simultaneous coverage probability for the functions $\{(\alpha_i - \alpha_{i'}) + (\bar{\xi}_{i\cdot} - \bar{\xi}_{i'\cdot}) : i' > i = 1, \ldots, q\}$ and all linear combinations of those functions is $1 - \xi$.

Solution

(a) Using various results in Example 13.4.5-1 and Exercise 13.25, we find that $100(1-\xi)\%$ one-at-a-time confidence intervals for the functions in the sets $\{(\alpha_i - \alpha_{i'}) + (\bar{\xi}_{i\cdot} - \bar{\xi}_{i'\cdot}) : i' > i = 1, \ldots, q\}$ are given by

$$(\bar{y}_{i\cdot\cdot} - \bar{y}_{i'\cdot\cdot}) \pm t_{\xi/2,q(r-1)(m-1)}\tilde{\sigma}\sqrt{\frac{2}{rm}(m\psi + 1)}.$$

Similarly, $100(1-\xi)\%$ one-at-a-time confidence intervals for the functions in the sets $\{(\gamma_k - \gamma_{k'}) + (\bar{\xi}_{\cdot k} - \bar{\xi}_{\cdot k'}) : k' > k = 1, \ldots, m\}$ are given by

$$(\bar{y}_{\cdot\cdot k} - \bar{y}_{\cdot\cdot k'}) \pm t_{\xi/2,q(r-1)(m-1)}\tilde{\sigma}\sqrt{\frac{2}{qr}}.$$

(b) The estimable functions listed in part (a) number $q(q-1) + m(m-1)$ in total. In order to obtain simultaneous confidence intervals for all of them using the Bonferroni method, it suffices to replace the t-multiplier $t_{\xi/2,q(r-1)(m-1)}$ used in part (a) by $t_{\xi/(2s),q(r-1)(m-1)}$, where $s = q(q-1) + m(m-1)$.

(c) Note that a basis for the set of estimable functions $\{(\alpha_i - \alpha_{i'}) + (\bar{\xi}_{i\cdot} - \bar{\xi}_{i'\cdot}) : i' > i = 1, \ldots, q\}$ is given by the $q - 1$ linearly independent estimable functions $\{(\alpha_i - \alpha_1) + (\bar{\xi}_{i\cdot} - \bar{\xi}_{1\cdot}) : i = 2, \ldots, q\}$. Hence, the $100(1 - \xi)\%$ Scheffé intervals are

$$\mathbf{c}^T \hat{\boldsymbol{\beta}} \pm \sqrt{(q-1)F_{\xi, q-1, q(r-1)(m-1)} \tilde{\sigma}^2 \{\mathbf{c}^T (\mathbf{X}^T\mathbf{X})^- \mathbf{c}\}},$$

where $\mathbf{c}^T \boldsymbol{\beta}$ is any linear combination of functions in the set $\{(\alpha_i - \alpha_{i'}) + (\bar{\xi}_{i\cdot} - \bar{\xi}_{i'\cdot}) : i' > i = 1, \ldots, q\}$.

▶ **Exercise 28** Consider the normal Gauss–Markov no-intercept simple linear regression model

$$y_i = \beta x_i + e_i \qquad (i = 1, \ldots, n),$$

where $n \geq 2$.

(a) Find the Scheffé-based $100(1 - \xi)\%$ confidence band for the regression line, i.e., for $\{\beta x : x \in \mathbb{R}\}$.
(b) Describe the behavior of this band as a function of x (for example, where is it narrowest and how narrow is it at its narrowest point?). Compare this to the behavior of the Scheffé-based $100(1 - \xi)\%$ confidence band for the regression line in normal simple linear regression.

Solution

(a) $\{\hat{\beta}x \pm t_{\xi/2, n-1}\hat{\sigma}|x|/\sqrt{\sum_{i=1}^n x_i^2}$ for all $x \in \mathbb{R}\}$.
(b) This band is narrowest at $x = 0$, where its width is equal to 0. Its width is a monotone increasing function of $|x|$. These behaviors contrast with those of the band for the regression line in simple linear regression with an intercept, which is narrowest at $x = \bar{x}$, where its width is $\sqrt{2F_{\xi,2,n-2}\hat{\sigma}^2/n}$, and whose width is a monotone increasing function of $|x - \bar{x}|$.

▶ **Exercise 29** Consider the normal prediction-extended Gauss–Markov full-rank multiple regression model

$$y_i = \mathbf{x}_i^T \boldsymbol{\beta} + e_i \quad (i = 1, \ldots, n + s),$$

where $\mathbf{x}_1, \ldots, \mathbf{x}_n$ are known p-vectors of explanatory variables (possibly including an intercept) but $\mathbf{x}_{n+1}, \ldots, \mathbf{x}_{n+s}$ $(s \geq 1)$ are p-vectors of the same explanatory variables that cannot be ascertained prior to actually observing $y_{n+1}, \ldots, y_{n+s}$ and must therefore be treated as unknown. This exercise considers the problem of obtaining prediction intervals for $y_{n+1}, \ldots, y_{n+s}$ that have simultaneous coverage probability at least $1 - \xi$, where $0 < \xi < 1$, in this scenario.

(a) Apply the Scheffé method to show that one solution to the problem consists of intervals of form

$$\left\{ \mathbf{x}_{n+i}^T \hat{\boldsymbol{\beta}} \pm \hat{\sigma}\sqrt{(p+s)F_{\xi,p+s,n-p}\mathbf{x}_{n+i}^T(\mathbf{X}^T\mathbf{X})^{-1}\mathbf{x}_{n+i} + 1} \right.$$
$$\left. \text{for all } \mathbf{x}_{n+i} \in \mathbb{R}^p \text{ and all } i = 1, \ldots, s \right\}.$$

(b) Another possibility is to decompose y_{n+i} into its two components $\mathbf{x}_{n+i}^T\boldsymbol{\beta}$ and e_{n+i}, obtain confidence or prediction intervals separately for each of these components by the Scheffé method, and then combine them via the Bonferroni inequality to achieve the desired simultaneous coverage probability. Verify that the resulting intervals are of form

$$\left\{ \mathbf{x}_{n+i}^T \hat{\boldsymbol{\beta}} \pm \hat{\sigma}[a(\mathbf{x}_{n+i}) + b] \quad \text{for all } \mathbf{x}_{n+i} \in \mathbb{R}^p \text{ and all } i = 1, \ldots, s \right\},$$

where $a(\mathbf{x}_{n+i}) = \sqrt{pF_{\xi^*,p,n-p}\mathbf{x}_{n+i}^T(\mathbf{X}^T\mathbf{X})^{-1}\mathbf{x}_{n+i}}$ and $b = \sqrt{sF_{\xi-\xi^*,s,n-p}}$, and $0 < \xi^* < \xi$.

(c) Still another possibility is based on the fact that $(1/\hat{\sigma})\mathbf{e}_+$, where $\mathbf{e}_+ = (e_{n+1}, \ldots, e_{n+s})^T$, has a certain multivariate t distribution under the assumed model. Prove this fact and use it to obtain a set of intervals for $y_{n+1}(\mathbf{x}_{n+1}), \cdots, y_{n+s}(\mathbf{x}_{n+s})$ of the form

$$\left\{ \mathbf{x}_{n+i}^T \hat{\boldsymbol{\beta}} \pm \hat{\sigma}[a(\mathbf{x}_{n+i}) + c] \quad \text{for all } \mathbf{x}_{n+i} \in \mathbb{R}^p \text{ and all } i = 1, \ldots, s \right\}$$

whose simultaneous coverage probability is at least $1 - \xi$, where c is a quantity you should determine.

(d) Are the intervals obtained in part (b) uniformly narrower than the intervals obtained in part (c) or vice versa? Explain.

(e) The assumptions that $e_1, \ldots, e_{n+s}$ are mutually independent can be relaxed somewhat without affecting the validity of intervals obtained in part (c). In precisely what manner can this independence assumption be relaxed?

Note: The approaches described in parts (a) and (b) were proposed by Carlstein (1986), and the approach described in part (c) was proposed by Zimmerman (1987).

Solution

(a) By Corollary 2.16.1.1,

$$\left(\frac{1}{(p+s)\hat{\sigma}^2}\right)\max_{\mathbf{a}\neq\mathbf{0}}\left(\frac{\left[\mathbf{a}^T\begin{pmatrix}\hat{\boldsymbol{\beta}}-\boldsymbol{\beta}\\ \mathbf{e}_+-\mathbf{0}_s\end{pmatrix}\right]^2}{\mathbf{a}^T\begin{pmatrix}(\mathbf{X}^T\mathbf{X})^{-1} & \mathbf{0}\\ \mathbf{0} & \mathbf{I}_s\end{pmatrix}\mathbf{a}}\right)=\frac{\begin{pmatrix}\hat{\boldsymbol{\beta}}-\boldsymbol{\beta}\\ \mathbf{e}_+\end{pmatrix}^T\begin{pmatrix}\mathbf{X}^T\mathbf{X} & \mathbf{0}\\ \mathbf{0} & \mathbf{I}_s\end{pmatrix}\begin{pmatrix}\hat{\boldsymbol{\beta}}-\boldsymbol{\beta}\\ \mathbf{e}_+\end{pmatrix}}{(p+s)\hat{\sigma}^2}$$

$$\sim \mathrm{F}(p+s, n-p)$$

where the distributional result holds because $\begin{pmatrix}\hat{\boldsymbol{\beta}}-\boldsymbol{\beta}\\ \mathbf{e}_+\end{pmatrix}^T\begin{pmatrix}\mathbf{X}^T\mathbf{X} & \mathbf{0}\\ \mathbf{0} & \mathbf{I}_s\end{pmatrix}\begin{pmatrix}\hat{\boldsymbol{\beta}}-\boldsymbol{\beta}\\ \mathbf{e}_+\end{pmatrix}$ $\sim \chi^2(p+s)$ and $\begin{pmatrix}\hat{\boldsymbol{\beta}}\\ \mathbf{e}_+\end{pmatrix}$ is independent of $\hat{\sigma}^2$. Consideration of the subset of vectors $\mathbf{a} \in \mathbb{R}^{p+s}$ such that $\mathbf{a}^T = \{(\mathbf{x}_{n+i}^T, \mathbf{u}_i^T) : \mathbf{x}_{n+i} \in \mathbb{R}^p, i = 1, \ldots, s\}$ (where $\mathbf{u}_i$ is the ith unit s-vector) yields the specified prediction intervals, which have simultaneous coverage probability at least $1-\xi$.

(b) By Corollary 2.16.1.1,

$$\left(\frac{1}{p\hat{\sigma}^2}\right)\max_{\mathbf{a}\neq\mathbf{0}}\left(\frac{[\mathbf{a}^T(\hat{\boldsymbol{\beta}}-\boldsymbol{\beta})]^2}{\mathbf{a}^T(\mathbf{X}^T\mathbf{X})^{-1}\mathbf{a}}\right)=\frac{(\hat{\boldsymbol{\beta}}-\boldsymbol{\beta})^T\mathbf{X}^T\mathbf{X}(\hat{\boldsymbol{\beta}}-\boldsymbol{\beta})}{p\hat{\sigma}^2}\sim \mathrm{F}(p, n-p)$$

and

$$\left(\frac{1}{s\hat{\sigma}^2}\right)\max_{\mathbf{b}\neq\mathbf{0}}\left(\frac{[\mathbf{b}^T(\mathbf{e}_+-\mathbf{0}_s)]^2}{\mathbf{b}^T\mathbf{I}^{-1}\mathbf{b}}\right)=\frac{\mathbf{e}_+^T\mathbf{e}_+}{s\hat{\sigma}^2}\sim \mathrm{F}(s, n-p).$$

Rewriting s choices of $\mathbf{a} \in \mathbb{R}^p$ as $\mathbf{x}_{n+1}, \ldots, \mathbf{x}_{n+s}$ and consideration of the subset of vectors $\mathbf{b} \in \mathbb{R}^s$ given by $\mathbf{b} = \{\mathbf{u}_i : i = 1, \ldots, s\}$ yield the specified separate intervals for $\{\mathbf{x}_{n+i}^T\boldsymbol{\beta} : \mathbf{x}_{n+i} \in \mathbb{R}^p, i = 1, \ldots, s\}$ and the elements of $\mathbf{e}_+$, which have simultaneous coverage probability at least $1-\xi^*$ and at least $1-\xi+\xi^*$, respectively. Applying Bonferroni's inequality yields the final set of intervals, which have simultaneous coverage probability at least $1-\xi$.

(c) $(1/\sigma)\mathbf{e}_+ \sim \mathrm{N}(\mathbf{0}, \mathbf{I})$ and, of course, $(n-p)\hat{\sigma}^2/\sigma^2$ is distributed independently as $\chi^2(n-p)$. Therefore,

$$(1/\hat{\sigma})\mathbf{e}_+ = \frac{(1/\sigma)\mathbf{e}_+}{\sqrt{\frac{(n-p)\hat{\sigma}^2}{\sigma^2(n-p)}}} \sim \mathrm{t}(s, n-p, \mathbf{I}).$$

Consequently, for $\xi^* \in (0, \xi)$ we have

$$1 - \xi + \xi^* = \Pr(|e_{n+i}| \leq \hat{\sigma} t_{(\xi-\xi^*)/2,s,n-p,\mathbf{I}} \text{ for all } i = 1, \ldots, s).$$

By the same argument used in the solution to part (b), we obtain

$$\Pr(y_{n+i} \in \mathbf{x}_{n+i}^T \hat{\boldsymbol{\beta}} \pm \hat{\sigma}[a(\mathbf{x}_{n+i}) + c] \text{ for all } \mathbf{x}_{n+i} \in \mathbb{R}^p \text{ and all } i = 1, \ldots, s) \geq 1 - \xi,$$

where $c = t_{(\xi-\xi^*)/2,s,n-p,\mathbf{I}}$.

(d) By (15.14), $t_{(\xi-\xi^*)/2,s,n-p,\mathbf{I}} \leq \sqrt{sF_{\xi-\xi^*,s,n-p}}$, so the intervals obtained in part (c) are uniformly narrower than the intervals obtained in part (b).

(e) The condition that $e_{n+1}, e_{n+2}, \ldots, e_{n+s}$ are independent can be relaxed to allow arbitrary dependence among these variables.

▶ **Exercise 30** Consider the one-factor random effects model with balanced data

$$y_{ij} = \mu + b_i + d_{ij} \quad (i = 1, \ldots, q; \quad j = 1, \ldots, r),$$

described previously in more detail in Example 13.4.2-2. Recall from that example that the BLUP of $b_i - b_{i'}$ $(i \neq i')$ is

$$\widetilde{b_i - b_{i'}} = \frac{r\psi}{r\psi + 1}(\bar{y}_{i\cdot} - \bar{y}_{i'\cdot}).$$

Let $\tilde{\sigma}^2$ be the generalized residual mean square and suppose that the joint distribution of the b_i's and d_{ij}'s is multivariate normal.

(a) Give $100(1-\xi)\%$ one-at-a-time prediction intervals for $b_i - b_{i'}$ $(i' > i = 1, \ldots, q)$.

(b) Obtain multivariate t-based prediction intervals for all $q(q-1)/2$ pairwise differences $b_i - b_{i'}$ $(i' > i = 1, \ldots, q)$ whose simultaneous coverage probability is $1 - \xi$. Determine the elements of the appropriate matrix $\mathbf{R}$.

(c) Explain why Tukey's method is not directly applicable to the problem of obtaining simultaneous prediction intervals for all $q(q-1)/2$ pairwise differences $b_i - b_{i'}$ $(i' > i = 1, \ldots, q)$.

Solution

(a) According to the solution to Exercise 13.16a,

$$\text{var}[\widetilde{b_i - b_{i'}} - (b_i - b_{i'})] = \frac{2\psi\sigma^2}{r\psi + 1}.$$

Thus, by Theorem 15.1.2,

$$\widetilde{b_i - b_{i'}} \pm t_{\xi/2,qr-1}\tilde{\sigma}\sqrt{\frac{2\psi}{r\psi + 1}}$$

is a $100(1-\xi)\%$ prediction interval for $b_i - b_{i'}$.

(b) Multivariate t-based prediction intervals for all $q(q-1)/2$ pairwise differences $b_i - b_{i'}$ ($i' > i = 1, \ldots, q$) whose simultaneous coverage probability is $1 - \xi$ are given by

$$\widetilde{b_i - b_{i'}} \pm t_{\xi/2,q(q-1)/2,qr-1,\mathbf{R}}\tilde{\sigma}\sqrt{\frac{2\psi}{r\psi + 1}},$$

where

$$\mathbf{R} = \left(r_{(i,i'),(j,j')} : i' > i = 1, \ldots, q;\ j' > j = 1, \ldots, q\right)$$
$$= \text{var}\left(\frac{\widetilde{b_1 - b_2} - (b_1 - b_2)}{\sigma\sqrt{2\psi/(r\psi+1)}}, \frac{\widetilde{b_1 - b_3} - (b_1 - b_3)}{\sigma\sqrt{2\psi/(r\psi+1)}}, \ldots, \frac{\widetilde{b_{q-1} - b_q} - (b_{q-1} - b_q)}{\sigma\sqrt{2\psi/(r\psi+1)}},\right).$$

Using the variance–covariance matrix of the prediction errors obtained in Exercise 13.16a, we find that for $i \leq j, i' > i = 1, \ldots, q, j' > j = 1, \ldots, q$,

$$r_{(i,i'),(j,j')} = \begin{cases} 1 \text{ if } i = j \text{ and } i' = j', \\ 1/2 \text{ if } i = j \text{ and } i' \neq j', \text{ or } i \neq j \text{ and } i' = j', \\ -1/2 \text{ if } i' = j, \\ 0 \text{ otherwise.} \end{cases}$$

(c) Tukey's method is not directly applicable because the variance–covariance matrix of the $(\tilde{b}_i - b_i)$'s, from which the $\widetilde{b_i - b_{i'}} - (b_i - b_{i'})$'s $(i < j)$ are obtained as all possible pairwise differences, is not a scalar multiple of the identity matrix. To see this, first observe from Example 13.4.2-2 that

$$\begin{aligned} \tilde{b}_i - b_i &= \left(\frac{r\psi}{r\psi + 1}\right)(\bar{y}_{i\cdot} - \bar{y}_{\cdot\cdot}) - b_i \\ &= \left(\frac{r\psi}{r\psi + 1}\right)[(b_i + \bar{d}_{i\cdot}) - (\bar{b}_\cdot + \bar{d}_{\cdot\cdot})] - b_i \\ &= \left(\frac{r\psi}{r\psi + 1}\right)[(b_i - \bar{b}_\cdot) + (\bar{d}_{i\cdot} - \bar{d}_{\cdot\cdot})] - b_i. \end{aligned}$$

Hence for $i' > i = 1, \ldots, q$,

$$\begin{aligned}
\text{cov}(\tilde{b}_i - b_i, \tilde{b}_{i'} - b_{i'}) &= \text{cov}\Bigg\{\left(\frac{r\psi}{r\psi+1}\right)[(b_i - \bar{b}_\cdot) + (\bar{d}_{i\cdot} - \bar{d}_{\cdot\cdot})] - b_i, \\
&\left(\frac{r\psi}{r\psi+1}\right)[(b_{i'} - \bar{b}_\cdot) + (\bar{d}_{i'\cdot} - \bar{d}_{\cdot\cdot})] - b_{i'}\Bigg\} \\
&= \left(\frac{r\psi}{r\psi+1}\right)^2 \text{cov}[(b_i - \bar{b}_\cdot), (b_{i'} - \bar{b}_\cdot)] \\
&+ \left(\frac{r\psi}{r\psi+1}\right)^2 \text{cov}[(d_{i\cdot} - \bar{d}_{\cdot\cdot}), (d_{i'\cdot} - \bar{d}_{\cdot\cdot})] \\
&+ \left(\frac{r\psi}{r\psi+1}\right) \text{cov}[(b_i - \bar{b}_\cdot), -b_{i'}] + \left(\frac{r\psi}{r\psi+1}\right) \text{cov}[-b_i, (b_{i'} - \bar{b}_\cdot)] \\
&= \left(\frac{r\psi}{r\psi+1}\right)^2 \left(\frac{-\sigma_b^2}{q}\right) + \left(\frac{r\psi}{r\psi+1}\right)^2 \left(\frac{-\sigma^2}{qr}\right) + \left(\frac{r\psi}{r\psi+1}\right)\left(\frac{2\sigma_b^2}{q}\right) \\
&= \frac{r\psi(r\psi+2)\sigma_b^2 - r\psi^2\sigma^2}{q(r\psi+1)^2} \\
&\neq 0.
\end{aligned}$$

▶ **Exercise 31** Let $a_1, \ldots, a_k$ represent real numbers. It can be shown that $|a_i - a_{i'}| \leq 1$ for all i and i' if and only if $|\sum_{i=1}^k c_i a_i| \leq \frac{1}{2}\sum_{i=1}^k |c_i|$ for all c_i such that $\sum_{i=1}^k c_i = 0$.

(a) Using this result, extend Tukey's method for obtaining $100(1-\xi)\%$ simultaneous confidence intervals for all possible differences among k linearly independent estimable functions $\psi_1, \ldots, \psi_k$, to all functions of the form $\sum_{i=1}^k d_i \psi_i$ with $\sum_{i=1}^k d_i = 0$ (i.e., to all contrasts).
(b) Specialize the intervals obtained in part (a) to the case of a balanced one-factor model.

Solution

(a) The Tukey intervals satisfy

$$1 - \xi = \Pr[|(\hat{\psi}_i - \hat{\psi}_{i'}) - (\psi_i - \psi_{i'})| \leq c\hat{\sigma} q^*_{\xi,k,n-p^*} \text{ for all } i' > i = 1, \ldots, k].$$

Using the result given in this exercise, we may manipulate the event in the probability statement above as follows:

$$
\begin{aligned}
1-\xi &= \Pr[|(\hat{\psi}_i - \hat{\psi}_{i'}) - (\psi_i - \psi_{i'})| \leq c\hat{\sigma} q^*_{\xi,k,n-p^*} \text{ for all } i' > i = 1,\ldots,k] \\
&= \Pr\left[\left|\frac{(\hat{\psi}_i - \psi_i) - (\hat{\psi}_{i'} - \psi_{i'})}{c\hat{\sigma} q^*_{\xi,k,n-p^*}}\right| \leq 1 \text{ for all } i' > i = 1,\ldots,k\right] \\
&= \Pr\left[\left|\sum_{i=1}^k d_i(\hat{\psi}_i - \hat{\psi}_{i'})\right| \leq \frac{1}{2} c\hat{\sigma} q^*_{\xi,k,n-p^*} \sum_{i=1}^k |d_i| \text{ for all } d_i \text{ such that } \sum_{i=1}^k d_i = 0\right] \\
&= \Pr\Big[\sum_{i=1}^k d_i\hat{\psi}_i - \frac{1}{2} c\hat{\sigma} q^*_{\xi,k,n-p^*} \sum_{i=1}^k |d_i| \leq \sum_{i=1}^k d_i\psi_i \leq \sum_{i=1}^k d_i\hat{\psi}_i + \frac{1}{2} c\hat{\sigma} q^*_{\xi,k,n-p^*} \sum_{i=1}^k |d_i| \\
&\qquad \text{for all } d_i \text{ such that } \sum_{i=1}^k d_i = 0\Big].
\end{aligned}
$$

Thus, $100(1-\xi)\%$ simultaneous confidence intervals for all contrasts among the ψ_i's are given by

$$\left(\sum_{i=1}^k d_i\hat{\psi}_i\right) \pm \frac{1}{2} c\hat{\sigma} q^*_{\xi,k,n-p^*} \sum_{i=1}^k |d_i|.$$

(b) In the case of a balanced one-factor model, we have, from Example 15.3-2, $\psi_i = \mu + \alpha_i$, $\hat{\psi}_i = \bar{y}_{i\cdot}$, $c = \sqrt{\frac{1}{r}}$, $k = q$, and $n - p^* = q(r-1)$. Thus, intervals of the type obtained in part (a) specialize to

$$\left(\sum_{i=1}^q d_i\bar{y}_{i\cdot}\right) \pm \frac{1}{2}\hat{\sigma}\sqrt{\frac{1}{r}} q^*_{\xi,q,q(r-1)} \sum_{i=1}^q |d_i|.$$

These are $100(1-\xi)\%$ simultaneous confidence intervals for all contrasts among the α_i's.

▶ **Exercise 32** Consider the normal Gauss–Markov model $\{\mathbf{y}, \mathbf{X}\boldsymbol{\beta}, \sigma^2\mathbf{I}\}$, and suppose that $\mathbf{X}$ is $n \times 3$ (where $n > 3$) and $\text{rank}(\mathbf{X}) = 3$. Let $(\mathbf{X}^T\mathbf{X})^{-1} = (c_{ij})$, $\boldsymbol{\beta} = (\beta_j)$, $\hat{\boldsymbol{\beta}} = (\hat{\boldsymbol{\beta}}_j) = (\mathbf{X}^T\mathbf{X})^{-1}\mathbf{X}^T\mathbf{y}$, and $\hat{\sigma}^2 = (\mathbf{y} - \mathbf{X}\hat{\boldsymbol{\beta}})^T(\mathbf{y} - \mathbf{X}\hat{\boldsymbol{\beta}})/(n-3)$.

(a) Each of the Bonferroni, Scheffé, and multivariate t methods can be used to obtain a set of intervals for $\beta_1, \beta_2, \beta_3, \beta_1+\beta_2, \beta_1+\beta_3, \beta_2+\beta_3$, and $\beta_1+\beta_2+\beta_3$ whose simultaneous coverage probability is equal to, or larger than, $1-\xi$ (where $0 < \xi < 1$). Give the set of intervals corresponding to each method.
(b) Obtain another set of intervals for the same functions listed in part (a) whose simultaneous coverage probability is equal to $1-\xi$, but which are constructed by adding the endpoints of $100(1-\xi)\%$ multivariate t intervals for β_1, β_2, and β_3 only.

(c) Obtain a $100(1-\xi)\%$ simultaneous confidence band for $\{\beta_1 x_1 + \beta_2 x_2\}$, i.e., an infinite collection of intervals $\{(L(x_1, x_2), U(x_1, x_2)) \colon (x_1, x_2) \in \mathbb{R}^2\}$ such that

$$\Pr[\beta_1 x_1 + \beta_2 x_2 \in (L(x_1, x_2), U(x_1, x_2)) \text{ for all } (x_1, x_2) \in \mathbb{R}^2] = 1 - \xi.$$

Solution

(a) The Bonferroni intervals are given by

$$\hat{\beta}_j \pm t_{\xi/14,n-3}\hat{\sigma}\sqrt{c_{jj}} \quad (j = 1, 2, 3),$$
$$(\hat{\beta}_j + \hat{\beta}_k) \pm t_{\xi/14,n-3}\hat{\sigma}\sqrt{c_{jj} + c_{kk} + 2c_{jk}} \quad (k > j = 1, 2, 3), \text{ and}$$
$$(\hat{\beta}_1 + \hat{\beta}_2 + \hat{\beta}_3) \pm t_{\xi/14,n-3}\hat{\sigma}\sqrt{c_{11} + c_{22} + c_{33} + 2c_{12} + 2c_{13} + 2c_{23}}.$$

The Scheffé and multivariate t intervals are identical except that $t_{\xi/14,n-3}$ is replaced by $\sqrt{3F_{\xi,3,n-3}}$ or $t_{\xi/2,7,n-3,\mathbf{R}}$, respectively, where

$$\mathbf{R} = \text{var}\Bigg(\frac{\hat{\beta}_1}{\sigma\sqrt{c_{11}}}, \frac{\hat{\beta}_2}{\sigma\sqrt{c_{22}}}, \frac{\hat{\beta}_3}{\sigma\sqrt{c_{33}}}, \frac{\hat{\beta}_1 + \hat{\beta}_2}{\sigma\sqrt{c_{11} + c_{22} + 2c_{12}}}, \frac{\hat{\beta}_1 + \hat{\beta}_3}{\sigma\sqrt{c_{11} + c_{33} + 2c_{13}}},$$
$$\frac{\hat{\beta}_2 + \hat{\beta}_3}{\sigma\sqrt{c_{22} + c_{33} + 2c_{23}}}, \frac{\hat{\beta}_1 + \hat{\beta}_2 + \hat{\beta}_3}{\sigma\sqrt{c_{11} + c_{22} + c_{33} + 2c_{12} + 2c_{13} + 2c_{23}}}\Bigg)^T.$$

(b) $100(1-\xi)\%$ multivariate t intervals for β_1, β_2, and β_3 only are given by

$$\hat{\beta}_j \pm t_{\xi/2,3,n-3,\mathbf{R}^*}\hat{\sigma}\sqrt{c_{jj}} \quad (j = 1, 2, 3),$$

where $\mathbf{R}^*$ is the upper left 3×3 block of $\mathbf{R}$. Then, using Fact 4,

$$1 - \xi = \Pr\Big[\hat{\beta}_j - t_{\xi/2,3,n-3,\mathbf{R}^*}\hat{\sigma}\sqrt{c_{jj}} \le \beta_j \le \hat{\beta}_j + t_{\xi/2,3,n-3,\mathbf{R}^*}\hat{\sigma}\sqrt{c_{jj}} \; (j = 1, 2, 3);$$
$$(\hat{\beta}_j + \hat{\beta}_k) - t_{\xi/2,3,n-3,\mathbf{R}^*}\hat{\sigma}(\sqrt{c_{jj}} + \sqrt{c_{kk}}) \le \beta_j + \beta_k \le (\hat{\beta}_j + \hat{\beta}_k)$$
$$+ t_{\xi/2,3,n-3,\mathbf{R}^*}\hat{\sigma}(\sqrt{c_{jj}} + \sqrt{c_{kk}}) \; (k > j = 1, 2, 3);$$
$$(\hat{\beta}_1 + \hat{\beta}_2 + \hat{\beta}_3) - t_{\xi/2,3,n-3,\mathbf{R}^*}\hat{\sigma}(\sqrt{c_{11}} + \sqrt{c_{22}} + \sqrt{c_{33}})$$
$$\le \beta_1 + \beta_2 + \beta_3 \le (\hat{\beta}_1 + \hat{\beta}_2 + \hat{\beta}_3)$$
$$+ t_{\xi/2,3,n-3,\mathbf{R}^*}\hat{\sigma}(\sqrt{c_{11}} + \sqrt{c_{22}} + \sqrt{c_{33}})\Big].$$

(c) The desired confidence band is

$$\left\{(\hat{\beta}_1 x_1 + \hat{\beta}_2 x_2) \pm \sqrt{2F_{\xi,2,n-3}}\hat{\sigma}\sqrt{x_1^2 c_{11} + x_2^2 c_{22} + 2x_1 x_2 c_{12}} \text{ for all } (x_1, x_2) \in \mathbb{R}^2\right\}.$$

▶ **Exercise 33** Consider the normal Gauss–Markov model $\{\mathbf{y}, \mathbf{X}\boldsymbol{\beta}, \sigma^2\mathbf{I}\}$, let $\tau_i = \mathbf{c}_i^T\boldsymbol{\beta}$ $(i = 1, \ldots, q)$, and let $\tau_{q+1} = \sum_{i=1}^q \tau_i = (\sum_{i=1}^q \mathbf{c}_i)^T\boldsymbol{\beta}$. Give the interval for τ_{q+1} belonging to each of the following sets of simultaneous confidence intervals:

(a) $100(1-\xi)\%$ Scheffé intervals that include intervals for $\tau_1, \ldots, \tau_q$.
(b) The infinite collection of all intervals generated by taking linear combinations of the $100(1-\xi)\%$ multivariate t intervals for $\tau_1, \ldots, \tau_q$.
(c) The finite set of multivariate t intervals for $\tau_1, \ldots, \tau_q, \tau_{q+1}$ that have exact simultaneous coverage probability equal to $1-\xi$.

Solution

(a) $(\sum_{i=1}^q \mathbf{c}_i)^T\hat{\boldsymbol{\beta}} \pm \hat{\sigma}\sqrt{q^* F_{\xi,q^*,n-p^*}(\sum_{i=1}^q \mathbf{c}_i)^T(\mathbf{X}^T\mathbf{X})^-(\sum_{i=1}^q \mathbf{c}_i)}$, where

$$q^* = \text{rank}\begin{pmatrix} \mathbf{c}_1^T \\ \vdots \\ \mathbf{c}_q^T \end{pmatrix}.$$

(b) $(\sum_{i=1}^q \mathbf{c}_i)^T\hat{\boldsymbol{\beta}} \pm t_{\xi/2,q,n-p^*,\mathbf{R}}\hat{\sigma}\sum_{i=1}^q \sqrt{\mathbf{c}_i^T(\mathbf{X}^T\mathbf{X})^-\mathbf{c}_i}$, where

$$\mathbf{R} = \text{var}\left(\frac{\mathbf{c}_1^T\hat{\boldsymbol{\beta}}}{\sigma\sqrt{\mathbf{c}_1^T(\mathbf{X}^T\mathbf{X})^-\mathbf{c}_1}}, \ldots, \frac{\mathbf{c}_q^T\hat{\boldsymbol{\beta}}}{\sigma\sqrt{\mathbf{c}_q^T(\mathbf{X}^T\mathbf{X})^-\mathbf{c}_q}}\right)^T.$$

(c) $(\sum_{i=1}^q \mathbf{c}_i)^T\hat{\boldsymbol{\beta}} \pm t_{\xi,q+1,n-p^*,\mathbf{R}^*}\hat{\sigma}\sqrt{(\sum_{i=1}^q \mathbf{c}_i)^T(\mathbf{X}^T\mathbf{X})^-(\sum_{i=1}^q \mathbf{c}_i)}$, where

$$\mathbf{R}^* = \text{var}\left(\frac{\mathbf{c}_1^T\hat{\boldsymbol{\beta}}}{\sigma\sqrt{\mathbf{c}_1^T(\mathbf{X}^T\mathbf{X})^-\mathbf{c}_1}}, \ldots, \frac{\mathbf{c}_{q+1}^T\hat{\boldsymbol{\beta}}}{\sigma\sqrt{\mathbf{c}_{q+1}^T(\mathbf{X}^T\mathbf{X})^-\mathbf{c}_{q+1}}}\right)^T.$$

▶ **Exercise 34** Consider the normal Gauss–Markov no-intercept simple linear regression model

$$y_i = \beta x_i + e_i \qquad (i = 1, \ldots, n),$$

where $n \geq 2$. Let y_{n+1} and y_{n+2} represent responses not yet observed at x-values x_{n+1} and x_{n+2}, respectively, and suppose that

$$y_i = \beta x_i + e_i \qquad (i = n+1, n+2),$$

where

$$\begin{pmatrix} e_{n+1} \\ e_{n+2} \end{pmatrix} \sim \mathrm{N}_2 \left(\mathbf{0}_2, \sigma^2 \begin{pmatrix} x_{n+1}^2 & 0 \\ 0 & x_{n+2}^2 \end{pmatrix} \right),$$

and $(e_{n+1}, e_{n+2})^T$ and $(e_1, e_2, \ldots, e_n)^T$ are independent.

(a) Obtain the vector of BLUPs of $(y_{n+1}, y_{n+2})^T$, giving each element of the vector by an expression free of matrices.
(b) Obtain the variance–covariance matrix of prediction errors corresponding to the vector of BLUPs you obtained in part (a). Again, give expressions for the elements of this matrix that do not involve matrices.
(c) Using the results of parts (a) and (b), obtain multivariate-t prediction intervals for $(y_{n+1}, y_{n+2})^T$ having simultaneous coverage probability $1 - \xi$ (where $0 < \xi < 1$).
(d) Obtain a $100(1 - \xi)\%$ prediction interval for $(y_{n+1} + y_{n+2})/2$.

Solution

(a) By Theorem 13.2.1, the BLUP of $\begin{pmatrix} y_{n+1} \\ y_{n+2} \end{pmatrix}$ is

$$\begin{pmatrix} x_{n+1} \\ x_{n+2} \end{pmatrix} \hat{\beta} = \begin{pmatrix} x_{n+1} \sum_{i=1}^n x_i y_i / \sum_{i=1}^n x_i^2 \\ x_{n+2} \sum_{i=1}^n x_i y_i / \sum_{i=1}^n x_i^2 \end{pmatrix}.$$

(b) By Theorem 13.2.2, the variance–covariance matrix of the BLUP's prediction errors is

$$\sigma^2 \left[\begin{pmatrix} x_{n+1} \\ x_{n+2} \end{pmatrix} (\sum_{i=1}^n x_i^2)^{-1} \begin{pmatrix} x_{n+1} \\ x_{n+2} \end{pmatrix}^T + \begin{pmatrix} x_{n+1}^2 & 0 \\ 0 & x_{n+2}^2 \end{pmatrix} \right]$$

$$= \sigma^2 \begin{pmatrix} x_{n+1}^2[1 + (1/\sum_{i=1}^n x_i^2)] & x_{n+1}x_{n+2}/\sum_{i=1}^n x_i^2 \\ x_{n+1}x_{n+2}/\sum_{i=1}^n x_i^2 & x_{n+2}^2[1 + (1/\sum_{i=1}^n x_i^2)] \end{pmatrix}.$$

(c) $100(1 - \xi)\%$ multivariate t prediction intervals for y_{n+1} and y_{n+2} are given by

$$x_{n+1}\hat{\beta} \pm t_{\xi/2,2,n-1,\mathbf{R}} \hat{\sigma} \sqrt{x_{n+1}^2[1 + (1/\sum_{i=1}^n x_i^2)]}$$

and

$$x_{n+2}\hat{\beta} \pm t_{\xi/2,2,n-1,\mathbf{R}}\hat{\sigma}\sqrt{x_{n+2}^2[1+(1/\sum_{i=1}^{n}x_i^2)]},$$

where $\hat{\sigma} = \sqrt{\frac{1}{n-1}\sum_{i=1}^{n}(y_i - \hat{\beta}x_i)^2}$ and the off-diagonal element of $\mathbf{R}$ is

$$\frac{x_{n+1}x_{n+2}/\sum_{i=1}^{n}x_i^2}{\sqrt{[1+(x_{n+1}^2/\sum_{i=1}^{n}x_i^2)][1+(x_{n+2}^2/\sum_{i=1}^{n}x_i^2)]}}.$$

(d) By Corollary 13.2.1.1, the BLUP of $(y_{n+1}+y_{n+2})/2$ is $(x_{n+1}+x_{n+1})\hat{\beta}/2$, and using the solution to part (b) we find that its prediction error variance is

$$\sigma^2\{(x_{n+1}^2+x_{n+2}^2)[1+(1/\sum_{i=1}^{n}x_i^2)]+(2x_{n+1}x_{n+2}/\sum_{i=1}^{n}x_i^2)\}/4 \equiv \sigma^2 Q.$$

Therefore, the desired prediction interval is given by

$$(x_{n+1}+x_{n+2})\hat{\beta}/2 \pm t_{\xi/2,n-1}\hat{\sigma}\sqrt{Q}.$$

▶ **Exercise 35** Suppose that observations $\{(x_i, y_i) : i = 1, \ldots, n\}$ follow the normal Gauss–Markov simple linear regression model

$$y_i = \beta_1 + \beta_2 x_i + e_i,$$

where $n \geq 3$. Let $\hat{\beta}_1$ and $\hat{\beta}_2$ be the ordinary least squares estimators of β_1 and β_2, respectively; let $\hat{\sigma}^2$ be the usual residual mean square; let $\bar{x} = (\sum_{i=1}^{n} x_i)/n$; and let $SXX = \sum_{i=1}^{n}(x_i - \bar{x})^2$. Recall that in this setting,

$$(\hat{\beta}_1 + \hat{\beta}_2 x_{n+1}) \pm t_{\xi/2,n-2}\hat{\sigma}\sqrt{1+\frac{1}{n}+\frac{(x_{n+1}-\bar{x})^2}{SXX}}$$

is a $100(1-\xi)\%$ prediction interval for an unobserved y-value to be taken at a specified x-value x_{n+1}. Suppose that it is desired to predict the values of three unobserved y-values, say y_{n+1}, y_{n+2}, and y_{n+3}, which are all to be taken at the same known x-value, say x^*. Assume that the unobserved values of y follow the same model as the observed data; that is,

$$y_i = \beta_1 + \beta_2 x^* + e_i \quad (i = n+1,\ n+2,\ n+3),$$

where $e_{n+1}, e_{n+2}, e_{n+3}$ are independent $\mathrm{N}(0, \sigma^2)$ random variables and are independent of $e_1, \ldots, e_n$.

(a) Give expressions for Bonferroni prediction intervals for $y_{n+1}, y_{n+2}, y_{n+3}$ whose simultaneous coverage probability is at least $1 - \xi$.
(b) Give expressions for Scheffé prediction intervals for $y_{n+1}, y_{n+2}, y_{n+3}$ whose simultaneous coverage probability is at least $1 - \xi$. (Note: The three requested intervals are part of an infinite collection of intervals, but give expressions for just those three.)
(c) Give expressions for multivariate t prediction intervals for $y_{n+1}, y_{n+2}, y_{n+3}$ whose simultaneous coverage probability is exactly $1 - \xi$. Note: The off-diagonal elements of the correlation matrix **R** referenced by the multivariate t quantiles in your prediction intervals are all equal to each other; give an expression for this common correlation coefficient.
(d) Give expressions for Bonferroni prediction intervals for all pairwise differences among $y_{n+1}, y_{n+2}, y_{n+3}$ whose simultaneous coverage probability is at least $1 - \xi$.
(e) Give expressions for multivariate t prediction intervals for all pairwise differences among $y_{n+1}, y_{n+2}, y_{n+3}$ whose simultaneous coverage probability is exactly $1 - \xi$. Give expressions for the off-diagonal elements of the correlation matrix **R** referenced by the multivariate t quantiles in your prediction intervals.
(f) Explain why Tukey's method is applicable to the problem of obtaining simultaneous prediction intervals for all pairwise differences among $y_{n+1}, y_{n+2}, y_{n+3}$ and obtain such intervals.

Solution

(a) Each of the three intervals is given by

$$(\hat{\beta}_1 + \hat{\beta}_2 x^*) \pm t_{\xi/2,n-2}\hat{\sigma}\sqrt{1 + \frac{1}{n} + \frac{(x^* - \bar{x})^2}{SXX}}.$$

(b) Using Theorem 15.3.7, we find that each of the three intervals is given by

$$(\hat{\beta}_1 + \hat{\beta}_2 x^*) \pm \hat{\sigma}\sqrt{3F_{\xi,3,n-2}\left(1 + \frac{1}{n} + \frac{(x^* - \bar{x})^2}{SXX}\right)}.$$

(c) Each of the three intervals is given by

$$(\hat{\beta}_1 + \hat{\beta}_2 x^*) \pm t_{\xi/2,3,n-2,\mathbf{R}}\hat{\sigma}\sqrt{1 + \frac{1}{n} + \frac{(x^* - \bar{x})^2}{SXX}},$$

where $\mathbf{R} = \text{corr}\begin{pmatrix} \hat{y}_{n+1} - y_{n+1} \\ \hat{y}_{n+2} - y_{n+2} \\ \hat{y}_{n+3} - y_{n+3} \end{pmatrix}$, which has common off-diagonal element (correlation coefficient) equal to

$$\frac{\frac{1}{n} + \frac{(x^*-\bar{x})^2}{SXX}}{1 + \frac{1}{n} + \frac{(x^*-\bar{x})^2}{SXX}}.$$

(d) Because $\hat{y}_{n+1} = \hat{y}_{n+2} = \hat{y}_{n+3} = \hat{\beta}_1 + \hat{\beta}_2 x^*$, the BLUPs of $y_{n+1} - y_{n+2}$, $y_{n+1} - y_{n+3}$, and $y_{n+2} - y_{n+3}$ are all equal to 0 and their prediction error variances are all equal to $\text{var}\{[(\hat{\beta}_1 + \hat{\beta}_2 x^*) - y_{n+1}] - [(\hat{\beta}_1 + \hat{\beta}_2 x^*) - y_{n+2}]\} = 2\sigma^2$. Therefore, each of the desired intervals is $0 \pm t_{\xi/6,n-2}\hat{\sigma}\sqrt{2}$.

(e) Each of the desired intervals is $0 \pm t_{\xi/2,3,n-2,\mathbf{R}^*}\hat{\sigma}\sqrt{2}$, where

$$\mathbf{R}^* = \text{corr}\begin{pmatrix} \hat{y}_{n+1} - \hat{y}_{n+2} - (y_{n+1} - y_{n+2}) \\ \hat{y}_{n+1} - \hat{y}_{n+3} - (y_{n+1} - y_{n+3}) \\ \hat{y}_{n+2} - \hat{y}_{n+3} - (y_{n+2} - y_{n+3}) \end{pmatrix}$$

$$= \text{corr}\begin{pmatrix} y_{n+2} - y_{n+1} \\ y_{n+3} - y_{n+1} \\ y_{n+3} - y_{n+2} \end{pmatrix} = \begin{pmatrix} 1 & 0.5 & -0.5 \\ 0.5 & 1 & 0.5 \\ -0.5 & 0.5 & 1 \end{pmatrix}.$$

(f) Tukey's method is applicable because $\hat{y}_{n+i} - \hat{y}_{n+j} - (y_{n+i} - y_{n+j}) = y_{n+j} - y_{n+i}$ for $j \neq i = 1, 2, 3$, which comprise all pairwise differences among the predictable functions $y_{n+1}, y_{n+2}, y_{n+3}$, and $\text{var}\begin{pmatrix} y_{n+1} \\ y_{n+2} \\ y_{n+3} \end{pmatrix} = c^2\sigma^2\mathbf{I}$, where $c = 1$. Therefore, by Theorem 15.3.6 the Tukey simultaneous confidence intervals for all pairwise differences among $y_{n+1}, y_{n+2}, y_{n+3}$ are given by

$$(\hat{y}_{n+j} - \hat{y}_{n+i}) \pm \hat{\sigma} q^*_{\xi,3,n-2} \quad (i \neq j = 1, 2, 3).$$

► **Exercise 36** Consider the normal Gauss–Markov one-factor model with balanced data:

$$y_{ij} = \mu + \alpha_i + e_{ij} \qquad (i = 1, \ldots, q; \quad j = 1, \ldots, r).$$

Recall from Example 15.3-2 that a $100(1 - \xi)\%$ confidence interval for a single level difference $\alpha_i - \alpha_{i'}$ $(i' > i = 1, \ldots, q)$ is given by

$$(\bar{y}_{i\cdot} - \bar{y}_{i'\cdot}) \pm t_{\xi/2,q(r-1)}\hat{\sigma}\sqrt{2/r},$$

where $\hat{\sigma}^2 = \sum_{i=1}^{q} \sum_{j=1}^{r} (y_{ij} - \bar{y}_{i\cdot})^2 / [q(r-1)]$. Also recall that in the same example, several solutions were given to the problem of obtaining confidence intervals for the level differences $\{\alpha_i - \alpha_{i'} : i' > i = 1, \ldots, q\}$ whose simultaneous coverage probability is at least $1 - \xi$.

Now, however, consider a slightly different problem under the same model. Suppose that the factors represent treatments, one of which is a "control" treatment or placebo, and that the investigator has considerably less interest in estimating differences involving the control treatment than in estimating differences not involving the control treatment. That is, letting α_1 correspond to the control treatment, the investigator has less interest in $\alpha_1 - \alpha_2, \alpha_1 - \alpha_3, \ldots, \alpha_1 - \alpha_q$ than in $\{\alpha_i - \alpha_{i'} : i' > i = 2, \ldots, q\}$. Thus, rather than using any of the simultaneous confidence intervals obtained in Example 15.3-2, the investigator decides to use "new" confidence intervals for the treatment differences $\{\alpha_i - \alpha_{i'} : i' > i = 1, \ldots, q\}$ that satisfy the following requirements:

(I) The new intervals for $\{\alpha_i - \alpha_{i'} : i' > i = 1, \ldots, q\}$, like the intervals in Example 15.3-2, have simultaneous coverage probability at least $1 - \xi$.

(II) Each of the new intervals for $\{\alpha_i - \alpha_{i'} : i' > i = 2, \ldots, q\}$ obtained by a given method is narrower than the interval in Example 15.3-2 for the same treatment difference, obtained using the same method.

(III) Each of the new intervals for $\alpha_1 - \alpha_2, \alpha_1 - \alpha_3, \ldots, \alpha_1 - \alpha_q$ is no more than twice as wide as each of the new intervals for $\{\alpha_i - \alpha_{i'} : i' > i = 2, \ldots, q\}$.

(IV) All of the new intervals for $\{\alpha_i - \alpha_{i'} : i' > i = 2, \ldots, q\}$ are of equal width and as narrow as possible, subject to the first three rules.

(a) If it is possible to use the Scheffé method to obtain new intervals that satisfy the prescribed rules, give such a solution. Otherwise, explain why it is not possible.

(b) If it is possible to use the multivariate t method to obtain new intervals that satisfy the prescribed rules, give such a solution. Otherwise, explain why it is not possible.

Note: You may not use the Bonferroni method in formulating solutions, but you may use the notion of linear combinations of intervals.

Solution

(a) It is not possible. The original Scheffé intervals

$$(\bar{y}_{i\cdot} - \bar{y}_{i'\cdot}) \pm \sqrt{(q-1)F_{\xi,q-1,q(r-1)}}\,\hat{\sigma}\sqrt{2/r} \quad (i' > i = 1, \ldots, q)$$

were obtained by taking

$$\mathbf{C}^T = \begin{pmatrix} 0 & 1 & -1 & 0 & 0 & \cdots & 0 \\ 0 & 1 & 0 & -1 & 0 & \cdots & 0 \\ 0 & 1 & 0 & 0 & -1 & \cdots & 0 \\ \vdots & & & & & & \\ 0 & 1 & 0 & 0 & 0 & \cdots & -1 \end{pmatrix}.$$

To satisfy Requirement II we must eliminate at least one row from $\mathbf{C}^T$, but then the $\mathbf{c}^T$ of $\mathbf{c}^T\boldsymbol{\beta}$ corresponding to some of the level differences will not be an element of $\mathcal{R}(\mathbf{C}^T)$.

(b) Start with the multivariate t intervals for $\alpha_i - \alpha_{i'}$ ($i' > i = 2, \ldots, q$) and $\alpha_1 - \alpha_2$ given by

$$\left\{ \begin{array}{l} (\bar{y}_{i\cdot} - \bar{y}_{i'\cdot}) \pm t_{\xi/2,(q-1)(q-2)/2+1,q(r-1),\mathbf{R}}\hat{\sigma}\sqrt{2/r} \text{ if } i' > i = 2, \ldots, q, \\ (\bar{y}_{1\cdot} - \bar{y}_{2\cdot}) \pm t_{\xi/2,(q-1)(q-2)/2+1,q(r-1),\mathbf{R}}\hat{\sigma}\sqrt{2/r} \end{array} \right\}$$

where

$$\mathbf{R} = \text{var}\left(\frac{\bar{y}_{2\cdot} - \bar{y}_{3\cdot}}{\sigma\sqrt{2/r}}, \frac{\bar{y}_{2\cdot} - \bar{y}_{4\cdot}}{\sigma\sqrt{2/r}}, \ldots, \frac{\bar{y}_{(q-1)\cdot} - \bar{y}_{q\cdot}}{\sigma\sqrt{2/r}}, \frac{\bar{y}_{1\cdot} - \bar{y}_{2\cdot}}{\sigma\sqrt{2/r}} \right)^T.$$

These intervals satisfy Requirements I and II. Now observe that $\alpha_1 - \alpha_{i'} = (\alpha_1 - \alpha_2) + (\alpha_2 - \alpha_{i'})$ for $i' = 3, \ldots, q$. Therefore, we can get intervals for $\{\alpha_1 - \alpha_{i'} : i' = 3, \ldots, q\}$ from the intervals just given above via the extension of the multivariate t method to linear combinations of estimable functions. The resulting intervals are

$$(\bar{y}_{1\cdot} - \bar{y}_{i'\cdot}) \pm 2t_{\xi/2,(q-1)(q-2)/2+1,q(r-1),\mathbf{R}}\hat{\sigma}\sqrt{2/r} \quad (i' = 3, \ldots, q).$$

Together, this last set of intervals and the multivariate t intervals listed above satisfy all the requirements.

▶ **Exercise 37** Consider a normal prediction-extended linear model

$$\begin{pmatrix} \mathbf{y} \\ y_{n+1} \\ y_{n+2} \end{pmatrix} = \begin{pmatrix} \mathbf{X} \\ \mathbf{x}_{n+1}^T \\ \mathbf{x}_{n+2}^T \end{pmatrix} \boldsymbol{\beta} + \begin{pmatrix} \mathbf{e} \\ e_{n+1} \\ e_{n+2} \end{pmatrix},$$

where $\begin{pmatrix} \mathbf{e} \\ e_{n+1} \\ e_{n+2} \end{pmatrix}$ satisfies Gauss–Markov assumptions except that $\text{var}(e_{n+1}) = 2\sigma^2$ and $\text{var}(e_{n+2}) = 3\sigma^2$.

(a) Give a $100(1-\xi)\%$ prediction interval for y_{n+1}.
(b) Give a $100(1-\xi)\%$ prediction interval for y_{n+2}.
(c) Give the $100(1-\xi)\%$ Bonferroni simultaneous prediction intervals for y_{n+1} and y_{n+2}.
(d) The intervals you obtained in part (c) do not have the same width. Indicate how the Bonferroni method could be used to obtain prediction intervals for y_{n+1} and y_{n+2} that have the same width yet have simultaneous coverage probability at least $1-\xi$.
(e) Obtain the $100(1-\xi)\%$ multivariate t simultaneous prediction intervals for y_{n+1} and y_{n+2}.

Solution

(a) $\mathbf{x}_{n+1}^T\hat{\boldsymbol\beta} \pm t_{\xi/2,n-p^*}\hat\sigma\sqrt{\mathbf{x}_{n+1}^T(\mathbf{X}^T\mathbf{X})^-\mathbf{x}_{n+1}+2}$.
(b) $\mathbf{x}_{n+2}^T\hat{\boldsymbol\beta} \pm t_{\xi/2,n-p^*}\hat\sigma\sqrt{\mathbf{x}_{n+2}^T(\mathbf{X}^T\mathbf{X})^-\mathbf{x}_{n+2}+3}$.
(c) $\{\mathbf{x}_{n+1}^T\hat{\boldsymbol\beta} \pm t_{\xi/4,n-p^*}\hat\sigma\sqrt{\mathbf{x}_{n+1}^T(\mathbf{X}^T\mathbf{X})^-\mathbf{x}_{n+1}+2},\ \mathbf{x}_{n+2}^T\hat{\boldsymbol\beta} \pm t_{\xi/4,n-p^*}\hat\sigma\sqrt{\mathbf{x}_{n+2}^T(\mathbf{X}^T\mathbf{X})^-\mathbf{x}_{n+2}+3}$.
(d) To obtain Bonferroni-based intervals that have the same width yet have simultaneous coverage probability at least $1-\xi$, we must find ξ_1 and ξ_2 that satisfy $\xi_1+\xi_2=\xi$ and

$$\sqrt{\frac{\mathbf{x}_{n+1}^T(\mathbf{X}^T\mathbf{X})^-\mathbf{x}_{n+1}+2}{\mathbf{x}_{n+2}^T(\mathbf{X}^T\mathbf{X})^-\mathbf{x}_{n+2}+3}} = \frac{t_{\xi_2/2,n-p^*}}{t_{\xi_1/2,n-p^*}}.$$

A solution can be found "by trial and error."
(e) The two intervals are

$$\mathbf{x}_{n+1}^T\hat{\boldsymbol\beta} \pm t_{\xi/2,2,n-p^*,\mathbf{R}}\hat\sigma\sqrt{\mathbf{x}_{n+1}^T(\mathbf{X}^T\mathbf{X})^-\mathbf{x}_{n+1}+2} \quad \text{and}$$

$$\mathbf{x}_{n+2}^T\hat{\boldsymbol\beta} \pm t_{\xi/2,2,n-p^*,\mathbf{R}}\hat\sigma\sqrt{\mathbf{x}_{n+2}^T(\mathbf{X}^T\mathbf{X})^-\mathbf{x}_{n+2}+3},$$

where

$$\mathbf{R} = \begin{pmatrix} 1 & g \\ g & 1 \end{pmatrix}$$

and

$$\begin{aligned} g &= \mathrm{corr}\left(\mathbf{x}_{n+1}^T\hat{\boldsymbol\beta} - y_{n+1},\ \mathbf{x}_{n+2}^T\hat{\boldsymbol\beta} - y_{n+2}\right) \\ &= \frac{\mathbf{x}_{n+1}^T(\mathbf{X}^T\mathbf{X})^-\mathbf{x}_{n+2}}{\sqrt{\mathbf{x}_{n+1}^T(\mathbf{X}^T\mathbf{X})^-\mathbf{x}_{n+1}+2}\sqrt{\mathbf{x}_{n+2}^T(\mathbf{X}^T\mathbf{X})^-\mathbf{x}_{n+2}+3}}. \end{aligned}$$

Exercise 38 Consider the normal Gauss–Markov one-factor analysis-of-covariance model for balanced data,

$$y_{ij} = \mu + \alpha_i + \gamma x_{ij} + e_{ij} \quad (i = 1, \ldots, q, j = 1, \ldots, r),$$

where $r \geq 2$ and $x_{ij} \neq x_{ij'}$ for $j \neq j'$ and $i = 1, \ldots, q$. Let $\mathbf{y} = (y_{11}, y_{12}, \ldots, y_{qr})^T$, let $\mathbf{X}$ be the corresponding model matrix, and let $\boldsymbol{\beta} = (\mu, \alpha_1, \ldots, \alpha_q, \gamma)^T$. Let $\hat{\boldsymbol{\beta}}$ be a solution to the normal equations and let $\hat{\sigma}^2$ be the residual mean square. Furthermore, let $\mathbf{c}_0$, $\mathbf{c}_i$, and $\mathbf{c}_{ix}$ be such that

$$\mathbf{c}_0^T \boldsymbol{\beta} = \gamma, \quad \mathbf{c}_i^T \boldsymbol{\beta} = \mu + \alpha_i, \quad \text{and} \quad \mathbf{c}_{ix}^T \boldsymbol{\beta} = \mu + \alpha_i + \gamma x \quad \text{where } -\infty < x < \infty.$$

Also let $\mathbf{c}_{ii'} = \mathbf{c}_i - \mathbf{c}_{i'}$ for $i' > i = 1, \ldots, q$. Let $0 < \xi < 1$. For each of the following four parts, give a confidence interval or a set of confidence intervals that satisfy the stated criteria. Express these confidence intervals in terms of the following quantities: $\mathbf{X}$, $\hat{\boldsymbol{\beta}}$, $\hat{\sigma}$, σ, $\mathbf{c}_0$, $\mathbf{c}_i$, $\mathbf{c}_{ii'}$, $\mathbf{c}_{ix}$, ξ, q, r, and appropriate cutoff points from an appropriate distribution.

(a) A $100(1-\xi)\%$ confidence interval for γ.
(b) A set of confidence intervals for $\{\mu + \alpha_i \colon i = 1, \ldots, q\}$ whose simultaneous coverage probability is exactly $1 - \xi$.
(c) A set of confidence intervals for $\{\mu + \alpha_i + \gamma x \colon i = 1, \ldots, q$ and all $x \in (-\infty, \infty)\}$ whose simultaneous coverage probability is at least $1 - \xi$.
(d) A set of confidence intervals for $\{\alpha_i - \alpha_{i'} \colon i' > i = 1, \ldots, q\}$ whose simultaneous coverage probability is exactly $1 - \xi$.

Solution

(a) $\mathbf{c}_0^T \hat{\boldsymbol{\beta}} \pm t_{\xi/2, q(r-1)-1} \hat{\sigma} \sqrt{\mathbf{c}_0^T (\mathbf{X}^T\mathbf{X})^- \mathbf{c}_0}$.
(b) $\{\mathbf{c}_i^T \hat{\boldsymbol{\beta}} \pm t_{\xi/2, q, q(r-1)-1, \mathbf{R}} \hat{\sigma} \sqrt{\mathbf{c}_i^T (\mathbf{X}^T\mathbf{X})^- \mathbf{c}_i} : i = 1, \ldots, q\}$, where

$$\mathbf{R} = \text{var}\left(\frac{\mathbf{c}_1^T \hat{\boldsymbol{\beta}}}{\sigma \sqrt{\mathbf{c}_1^T (\mathbf{X}^T\mathbf{X})^- \mathbf{c}_1}}, \ldots, \frac{\mathbf{c}_q^T \hat{\boldsymbol{\beta}}}{\sigma \sqrt{\mathbf{c}_q^T (\mathbf{X}^T\mathbf{X})^- \mathbf{c}_q}} \right)^T.$$

(c) $\text{rank}(\mathbf{X}) = q + 1$, so the desired intervals are

$$\left\{ \mathbf{c}_{ix}^T \hat{\boldsymbol{\beta}} \pm \sqrt{(q+1) F_{\xi, q+1, q(r-1)-1}} \hat{\sigma} \sqrt{\mathbf{c}_{ix}^T (\mathbf{X}^T\mathbf{X})^- \mathbf{c}_{ix}} \text{ for all } i = 1, \ldots, q \text{ and all } x \in \mathbb{R} \right\}.$$

(d) $\left\{\mathbf{c}_{ii'}^T\hat{\boldsymbol{\beta}} \pm t_{\xi/2,q(q-1)/2,q(r-1)-1,\mathbf{R}^*}\hat{\sigma}\sqrt{\mathbf{c}_{ii'}^T(\mathbf{X}^T\mathbf{X})^-\mathbf{c}_{ii'}} \text{ for all } i' > i = 1,\ldots,q\right\}$,
where

$$\mathbf{R}^* = \text{var}\left(\frac{\mathbf{c}_{12}^T\hat{\boldsymbol{\beta}}}{\sigma\sqrt{\mathbf{c}_{12}^T(\mathbf{X}^T\mathbf{X})^-\mathbf{c}_{12}}}, \frac{\mathbf{c}_{13}^T\hat{\boldsymbol{\beta}}}{\sigma\sqrt{\mathbf{c}_{13}^T(\mathbf{X}^T\mathbf{X})^-\mathbf{c}_{13}}}, \ldots, \frac{\mathbf{c}_{q,q-1}^T\hat{\boldsymbol{\beta}}}{\sigma\sqrt{\mathbf{c}_{q,q-1}^T(\mathbf{X}^T\mathbf{X})^-\mathbf{c}_{q,q-1}}}\right)^T.$$

▶ **Exercise 39** Consider a situation in which a response variable, y, and a single explanatory variable, x, are measured on n subjects. Suppose that the response for each subject having a nonnegative value of x is related to x through a simple linear regression model without an intercept, and the response for each subject having a negative value of x is also related to x through a simple linear regression without an intercept; however, the slopes of the two regression models are possibly different. That is, assume that the observations follow the model

$$y_i = \begin{cases} \beta_1 x_i + e_i & \text{if } x_i < 0, \\ \beta_2 x_i + e_i & \text{if } x_i \geq 0, \end{cases}$$

for $i = 1, \ldots, n$. Suppose further that the e_i's are independent $\text{N}(0, \sigma^2)$ variables; that $x_i \neq 0$ for all i; and that there is at least one x_i less than 0 and at least one x_i greater than 0. Let n_1 denote the number of subjects whose x-value is less than 0, and let $n_2 = n - n_1$. Finally, let $0 < \xi < 1$.

(a) Give expressions for the elements of the model matrix $\mathbf{X}$. (Note: To make things easier, assume that the elements of the response vector $\mathbf{y}$ are arranged in such a way that the first n_1 elements of $\mathbf{y}$ correspond to those subjects whose x-value is less than 0.) Also obtain nonmatrix expressions for the least squares estimators of β_1 and β_2, and obtain the variance–covariance matrix of those two estimators. Finally, obtain a nonmatrix expression for $\hat{\sigma}^2$.
(b) Give confidence intervals for β_1 and β_2 whose simultaneous coverage probability is exactly $1 - \xi$.
(c) Using the Scheffé method, obtain a $100(1 - \xi)\%$ simultaneous confidence band for $\text{E}(y)$; that is, obtain expressions (in as simple a form as possible) for functions $a_1(x)$, $b_1(x)$, $a_2(x)$, and $b_2(x)$ such that

$$\Pr[a_1(x) \leq \beta_1 x \leq b_1(x) \text{ for all } x < 0 \text{ and } a_2(x) \leq \beta_2 x \leq b_2(x) \text{ for all } x \geq 0] \geq 1 - \xi.$$

(d) Consider the confidence band obtained in part (c). Under what circumstances will the band's width at x equal the width at $-x$ (for all x)?

Solution

(a) $\mathbf{X} = \begin{pmatrix} \mathbf{x}_1 & \mathbf{0} \\ \mathbf{0} & \mathbf{x}_2 \end{pmatrix}$ where $\mathbf{x}_1 = (x_1, x_2, \ldots, x_{n_1})^T$ and $\mathbf{x}_2 = (x_{n_1+1}, x_{n_1+2}, \ldots, x_{n_1+n_2})^T$,

$$\hat{\boldsymbol{\beta}} = \begin{pmatrix} \hat{\beta}_1 \\ \hat{\beta}_2 \end{pmatrix} = \begin{pmatrix} \sum_{i=1}^{n_1} x_i^2 & 0 \\ 0 & \sum_{i=n_1+1}^{n_1+n_2} x_i^2 \end{pmatrix}^{-1} \begin{pmatrix} \sum_{i=1}^{n_1} x_i y_i \\ \sum_{i=n_1+1}^{n_1+n_2} x_i y_i \end{pmatrix}$$

$$= \begin{pmatrix} \sum_{i=1}^{n_1} x_i y_i / \sum_{i=1}^{n_1} x_i^2 \\ \sum_{i=n_1+1}^{n_1+n_2} x_i y_i / \sum_{i=n_1+1}^{n_1+n_2} x_i^2 \end{pmatrix},$$

and

$$\text{var}(\hat{\boldsymbol{\beta}}) = \sigma^2 \begin{pmatrix} 1/\sum_{i=1}^{n_1} x_i^2 & 0 \\ 0 & 1/\sum_{i=n_1+1}^{n_1+n_2} x_i^2 \end{pmatrix}.$$

Finally, $\hat{\sigma}^2 = \left[\sum_{i=1}^{n_1}(y_i - \hat{\beta}_1 x_i)^2 + \sum_{i=n_1+1}^{n_1+n_2}(y_i - \hat{\beta}_2 x_i)^2\right]/(n_1 + n_2)$.

(b) $\hat{\beta}_1 \pm t_{\xi/2,2,n-2,\mathbf{R}}\hat{\sigma}/\sqrt{\sum_{i=1}^{n_1} x_i^2}$ and $\hat{\beta}_2 \pm t_{\xi/2,2,n-2,\mathbf{R}}\hat{\sigma}/\sqrt{\sum_{i=n_1+1}^{n_1+n_2} x_i^2}$, where

$$\mathbf{R} = \text{var}\left(\frac{\hat{\beta}_1}{\sigma/\sqrt{\sum_{i=1}^{n_1} x_i^2}}, \frac{\hat{\beta}_2}{\sigma/\sqrt{\sum_{i=n_1+1}^{n_1+n_2} x_i^2}}\right)^T = \mathbf{I}.$$

(c) Take $\mathbf{C}^T = \begin{pmatrix} -x & 0 \\ 0 & x \end{pmatrix}$, where $x > 0$, and observe that $\text{rank}(\mathbf{C}^T) = 2$. Thus, the Scheffé intervals are

$$\{\mathbf{c}^T\hat{\boldsymbol{\beta}} \pm \sqrt{2F_{\xi,2,n_1+n_2-2}}\hat{\sigma}\sqrt{\mathbf{c}^T(\mathbf{X}^T\mathbf{X})^-\mathbf{c}} \text{ for all } \mathbf{c}^T \in \mathcal{R}(\mathbf{C}^T)\},$$

i.e.,

$$\left\{\hat{\beta}_1 x \pm \sqrt{2F_{\xi,2,n_1+n_2-2}}\hat{\sigma}\sqrt{\frac{x^2}{\sum_{i=1}^{n_1} x_i^2}} \text{ for all } x < 0,\right.$$

$$\left.\hat{\beta}_2 x \pm \sqrt{2F_{\xi,2,n_1+n_2-2}}\hat{\sigma}\sqrt{\frac{x^2}{\sum_{i=n_1+1}^{n_1+n_2} x_i^2}} \text{ for all } x \geq 0\right\}.$$

(d) From examination of the intervals derived in part (c), the width at x is equal to the width at $-x$ if and only if $\sum_{i=1}^{n_1} x_i^2 = \sum_{i=n_1+1}^{n_1+n_2} x_i^2$.

▶ **Exercise 40** Consider a full-rank normal Gauss–Markov analysis-of-covariance model in a setting where there are one or more factors of classification and one or more regression variables. Let q represent the number of combinations of the factor levels. Suppose further that at each such combination the model is a regression model with the same number, p_c, of regression variables (including the intercept in this number); see part (c) for two examples. For $i = 1, \ldots, q$ let $\mathbf{x}_i$ represent an arbitrary p_c-vector of the regression variables (including the intercept) for the ith combination of classificatory explanatory variables, and let $\boldsymbol{\beta}_i$ and $\hat{\boldsymbol{\beta}}_i$ represent the corresponding vectors of regression coefficients and their least squares estimators from the fit of the complete model, respectively. Note that an arbitrary p-vector $\mathbf{x}$ of the explanatory variables consists of $\mathbf{x}_i$ for some i, padded with zeroes in appropriate places.

(a) For $i = 1, \ldots, q$ let $\mathbf{G}_{ii}$ represent the submatrix of $(\mathbf{X}^T\mathbf{X})^{-1}$ obtained by deleting the rows and columns that correspond to the elements of $\mathbf{x}$ excluded to form $\mathbf{x}_i$. Show that

$$\Pr\left[\mathbf{x}_i^T\boldsymbol{\beta}_i \in \mathbf{x}_i^T\hat{\boldsymbol{\beta}}_i \pm \hat{\sigma}\sqrt{p_c F_{\xi/q,p_c,n-p}\mathbf{x}_i^T\mathbf{G}_{ii}\mathbf{x}_i} \text{ for all } \mathbf{x}_i \in \mathbb{R}^{p_c} \text{ and all } i = 1, \ldots, q\right] \geq 1 - \xi.$$

(Hint: First apply Scheffé's method to obtain a confidence band for $\mathbf{x}_i^T\boldsymbol{\beta}_i$ for fixed $i \in \{1, \ldots, q\}$.)

(b) Obtain an expression for the ratio of the width of the interval for $\mathbf{x}_i^T\boldsymbol{\beta}_i$ in the collection of $100(1-\xi)\%$ simultaneous confidence intervals determined in part (a), to the width of the interval for the same $\mathbf{x}_i^T\boldsymbol{\beta}_i$ in the collection of standard $100(1-\xi)\%$ Scheffé confidence intervals

$$\mathbf{x}^T\hat{\boldsymbol{\beta}} \pm \hat{\sigma}\sqrt{pF_{\xi,p,n-p}\mathbf{x}^T(\mathbf{X}^T\mathbf{X})^{-1}\mathbf{x}}, \quad \text{for all } \mathbf{x} \in \mathbb{R}^p.$$

(c) For each of the following special cases of models, evaluate the ratio obtained in part (b) when $\xi = .05$:

(i) The three-group simple linear regression model

$$y_{ij} = \beta_{i1} + \beta_{i2}x_{ij} + e_{ij} \quad (i = 1, 2, 3;\ j = 1, \ldots, 10).$$

(ii) The three-group common-slope simple linear regression model

$$y_{ij} = \beta_{i1} + \beta_{2}x_{ij} + e_{ij} \quad (i = 1, 2, 3;\ j = 1, \ldots, 10).$$

Note: This exercise was inspired by Lane and Dumouchel (1994).

Solution

(a) For the ith combination of the classificatory explanatory variables, $\mathbf{x}^T\boldsymbol{\beta} = \mathbf{x}_i^T\boldsymbol{\beta}_i$ and $\mathbf{x}^T(\mathbf{X}^T\mathbf{X})^{-1}\mathbf{x} = \mathbf{x}_i^T\mathbf{G}_{ii}\mathbf{x}_i$. Thus, the $100(1-\xi/q)\%$ Scheffé-based confidence band for $\mathbf{x}_i^T\boldsymbol{\beta}_i$ is given by

$$\mathbf{x}_i^T\hat{\boldsymbol{\beta}}_i \pm \hat{\sigma}\sqrt{p_c F_{\xi/q,p_c,n-p}\mathbf{x}_i^T\mathbf{G}_{ii}\mathbf{x}_i} \text{ for all } \mathbf{x}_i \in \mathbb{R}^{p_c}.$$

Applying the Bonferroni inequality to these intervals yields the specified intervals for $\mathbf{x}_i^T\boldsymbol{\beta}_i$ for all $\mathbf{x}_i \in \mathbb{R}^{p_c}$ and all $i = 1, \ldots, q$ whose simultaneous coverage probability is at least $1-\xi$.

(b) $\sqrt{\frac{p_c F_{\xi/q,p_c,n-p}}{p F_{\xi,p,n-p}}}$.

(c) (i) $\sqrt{\frac{2F_{.05/3,2,24}}{6F_{.05,6,24}}} = 0.8053$.

(ii) $\sqrt{\frac{2F_{.05/3,2,26}}{4F_{.05,4,26}}} = 0.9367$.

References

Carlstein, E. (1986). Simultaneous confidence regions for predictions. *The American Statistician, 40*, 277–279.

Lane, T. P. & Dumouchel, W. H. (1994). Simultaneous confidence intervals in multiple regression. *The American Statistician, 48*, 315–321.

Tukey, J. W. (1949). One degree of freedom for non-additivity. *Biometrics, 5*, 232–242.

Zimmerman, D. L. (1987). Simultaneous confidence regions for predictions based on the multivariate t distribution. *The American Statistician, 41*, 247.

Inference for Variance–Covariance Parameters

16

This chapter presents exercises on inference for the variance–covariance parameters of a linear model and provides solutions to those exercises.

▶ **Exercise 1** Consider a two-way layout with two rows and two columns. Suppose that there are two observations, labelled as y_{111} and y_{112}, in the upper left cell; one observation, labelled as y_{121}, in the upper right cell; and one observation, labelled as y_{211}, in the lower left cell, as in the fourth layout displayed in Example 7.1-3 (the lower right cell is empty). Suppose that the observations follow the normal two-way mixed main effects model

$$y_{ijk} = \mu + \alpha_i + b_j + d_{ijk} \qquad (i, j, k) \in \{(1, 1, 1),\ (1, 1, 2),\ (1, 2, 1),\ (2, 1, 1)\},$$

where

$$\mathrm{E}\begin{pmatrix} b_1 \\ b_2 \\ d_{111} \\ d_{112} \\ d_{121} \\ d_{211} \end{pmatrix} = \mathbf{0} \quad \text{and} \quad \mathrm{var}\begin{pmatrix} b_1 \\ b_2 \\ d_{111} \\ d_{112} \\ d_{121} \\ d_{211} \end{pmatrix} = \begin{pmatrix} \sigma_b^2 \mathbf{I}_2 & \mathbf{0} \\ \mathbf{0} & \sigma^2 \mathbf{I}_4 \end{pmatrix}.$$

Here, the variance components $\sigma_b^2 \geq 0$ and $\sigma^2 > 0$ are unknown.

(a) Obtain quadratic unbiased estimators of the variance components.
(b) Let $\xi \in (0, 1)$. Obtain a $100(1-\xi)\%$ confidence interval for $\psi \equiv \sigma_b^2/\sigma^2$, which depends on the data only through $y_{111} - y_{112}$ and $y_{121} - (y_{111} + y_{112})/2$. (Hint: First obtain the joint distribution of these two linear functions of the data, and then obtain the joint distribution of their squares, suitably scaled.)

D. L. Zimmerman, *Linear Model Theory*,
https://doi.org/10.1007/978-3-030-52074-8_16

Solution

(a) $\mathrm{E}[(y_{111}-y_{112})^2] = \mathrm{var}(y_{111}-y_{112}) + [\mathrm{E}(y_{111}-y_{112})]^2 = 2\sigma^2$, so $(y_{111}-y_{112})^2/2$ is a quadratic unbiased estimator of σ^2. Also, $\mathrm{E}[(y_{111}-y_{121})^2] = \mathrm{var}(y_{111}-y_{121})+[\mathrm{E}(y_{111}-y_{121})]^2 = \mathrm{var}(b_1+d_{111}-b_2-d_{121}) = 2\sigma_b^2+2\sigma^2$. Thus, $\frac{1}{2}(y_{111}-y_{121})^2 - \frac{1}{2}(y_{111}-y_{112})^2$ is a quadratic unbiased estimator of σ_b^2.

(b) The joint distribution of $\begin{pmatrix} y_{111}-y_{112} \\ y_{121}-(y_{111}+y_{112})/2 \end{pmatrix}$ is bivariate normal with mean vector $\mathbf{0}_2$ and variance–covariance matrix $\begin{pmatrix} 2\sigma^2 & 0 \\ 0 & 2\sigma_b^2+\frac{3}{2}\sigma^2 \end{pmatrix}$ because $\mathrm{var}(y_{111}-y_{112}) = 2\sigma^2$ according to part (a), $\mathrm{var}[y_{121}-(y_{111}+y_{112})/2] = \mathrm{var}[b_2+d_{121}-b_1-(d_{111}+d_{112})/2] = 2\sigma_b^2+\frac{3}{2}\sigma^2$, and $\mathrm{cov}[y_{111}-y_{112}, y_{121}-(y_{111}+y_{112})/2] = \mathrm{cov}(y_{111}, y_{121}) - \mathrm{cov}(y_{112}, y_{121}) - \mathrm{cov}[y_{111}, (y_{111}+y_{112})/2] + \mathrm{cov}[y_{112}, (y_{111}+y_{112})/2] = 0 - 0 - \mathrm{cov}[b_1+d_{111}, b_1+(d_{111}+d_{112})/2]+\mathrm{cov}[b_1+d_{112}, b_1+(d_{111}+d_{112})/2] = -(\sigma_b^2+\frac{1}{2}\sigma^2)+\sigma_b^2+\frac{1}{2}\sigma^2 = 0$. Next define $S_1^2 = \frac{1}{2}(y_{111}-y_{112})^2$ and $S_2^2 = \frac{1}{2}[y_{121}-(y_{111}+y_{112})/2]^2$. Then,

$$\frac{S_1^2}{\sigma^2} \sim \chi^2(1) \quad \text{and} \quad \frac{S_2^2}{\sigma_b^2+\frac{3}{4}\sigma^2} \sim \chi^2(1),$$

and these two random variables are independent because $y_{111}-y_{112}$ and $y_{121}-(y_{111}+y_{112})/2$ are independent by Theorem 14.1.8.

Now define

$$U = \frac{S_2^2}{S_1^2} \quad \text{and} \quad U^* = \frac{S_2^2/(\sigma_b^2+\frac{3}{4}\sigma^2)}{S_1^2/\sigma^2} = \left(\frac{\sigma^2}{\sigma_b^2+\frac{3}{4}\sigma^2}\right)U.$$

$U^* \sim \mathrm{F}(1, 1)$ by Definition 14.5.1, so

$$\begin{aligned} 1-\xi &= \Pr(F_{1-(\xi/2),1,1} \le U^* \le F_{\xi/2,1,1}) \\ &= \Pr\left(F_{1-(\xi/2),1,1} \le \frac{\sigma^2}{\sigma_b^2+\frac{3}{4}\sigma^2}U \le F_{\xi/2,1,1}\right) \\ &= \Pr\left(F_{1-(\xi/2),1,1} \le \frac{1}{\psi+\frac{3}{4}}U \le F_{\xi/2,1,1}\right) \\ &= \Pr\left(\frac{U}{F_{\xi/2,1,1}} - \frac{3}{4} \le \psi \le \frac{U}{F_{1-(\xi/2),1,1}} - \frac{3}{4}\right). \end{aligned}$$

Thus,

$$\left[\frac{U}{F_{\xi/2,1,1}} - \frac{3}{4}, \frac{U}{F_{1-(\xi/2),1,1}} - \frac{3}{4}\right]$$

is a $100(1-\xi)\%$ confidence interval for ψ. If any endpoints of the interval are negative, they can be set equal to zero.

Exercise 2 Consider the normal two-way main effects components-of-variance model with only one observation per cell,

$$y_{ij} = \mu + a_i + b_j + d_{ij} \quad (i = 1, \ldots, q;\ j = 1, \ldots, m)$$

where $a_i \sim \mathrm{N}(0, \sigma_a^2)$, $b_j \sim \mathrm{N}(0, \sigma_b^2)$, and $d_{ij} \sim \mathrm{N}(0, \sigma^2)$ for all i and j, and where $\{a_i\}$, $\{b_j\}$, and $\{d_{ij}\}$ are mutually independent. The corrected two-part mixed-model ANOVA table (with Factor A fitted first) is given below, with an additional column giving the distributions of suitably scaled mean squares under the model.

Source	df	Sum of squares	Mean square
Factor A	$q-1$	$m\sum_{i=1}^{q}(\bar{y}_{i\cdot} - \bar{y}_{\cdot\cdot})^2$	S_1^2
Factor B	$m-1$	$q\sum_{j=1}^{m}(\bar{y}_{\cdot j} - \bar{y}_{\cdot\cdot})^2$	S_2^2
Residual	$(q-1)(m-1)$	$\sum_{i=1}^{q}\sum_{j=1}^{m}(y_{ij} - \bar{y}_{i\cdot} - \bar{y}_{\cdot j} + \bar{y}_{\cdot\cdot})^2$	S_3^2
Total	$qm-1$	$\sum_{i=1}^{q}\sum_{j=1}^{m}(y_{ij} - \bar{y}_{\cdot\cdot})^2$	

Here, $f_1 S_1^2 \sim \chi^2(q-1)$, $f_2 S_2^2 \sim \chi^2(m-1)$, and $f_3 S_3^2 \sim \chi^2[(q-1)(m-1)]$, where

$$f_1 = \frac{q-1}{\sigma^2 + m\sigma_a^2}, \quad f_2 = \frac{m-1}{\sigma^2 + q\sigma_b^2}, \quad f_3 = \frac{(q-1)(m-1)}{\sigma^2}.$$

Furthermore, S_1^2, S_2^2, and S_3^2 are mutually independent. Let $\xi \in (0, 1)$.

(a) Verify the distributions given in the last column of the table, and verify that S_1^2, S_2^2, and S_3^2 are mutually independent.
(b) Obtain minimum variance quadratic unbiased estimators of the variance components.
(c) Obtain a $100(1-\xi)\%$ confidence interval for $\sigma^2 + q\sigma_b^2$.
(d) Obtain a $100(1-\xi)\%$ confidence interval for σ_b^2/σ^2.
(e) Obtain a confidence interval for $(\sigma_a^2 - \sigma_b^2)/\sigma^2$ which has coverage probability at least $1-\xi$. [Hint: first combine the interval from part (d) with a similarly constructed interval for σ_a^2/σ^2 via the Bonferroni inequality.]

(f) Use Satterthwaite's method to obtain an approximate $100(1-\xi)\%$ confidence interval for σ_a^2.

(g) Obtain size-ξ tests of $H_0{:}\,\sigma_a^2 = 0$ versus $H_a{:}\,\sigma_a^2 > 0$ and $H_0{:}\,\sigma_b^2 = 0$ versus $H_a{:}\,\sigma_b^2 > 0$.

Solution

(a) Here $f_1 S_1^2 = \mathbf{y}^T\mathbf{A}_1\mathbf{y}$, $f_2 S_2^2 = \mathbf{y}^T\mathbf{A}_2\mathbf{y}$, and $f_3 S_3^2 = \mathbf{y}^T\mathbf{A}_3\mathbf{y}$, where

$$\mathbf{A}_1 = \left(\frac{1}{\sigma^2 + m\sigma_a^2}\right)[(\mathbf{I}_q - \frac{1}{q}\mathbf{J}_q) \otimes \frac{1}{m}\mathbf{J}_m],$$

$$\mathbf{A}_2 = \left(\frac{1}{\sigma^2 + q\sigma_b^2}\right)[\frac{1}{q}\mathbf{J}_q \otimes (\mathbf{I}_m - \frac{1}{m}\mathbf{J}_m)],$$

$$\mathbf{A}_3 = \left(\frac{1}{\sigma^2}\right)[(\mathbf{I}_q - \frac{1}{q}\mathbf{J}_q) \otimes (\mathbf{I}_m - \frac{1}{m}\mathbf{J}_m)].$$

Here the expression for $\mathbf{A}_1$ was obtained by exploiting the similarity of this model up to and including the first random effect to the one-factor random model of Example 16.1.4-1; the expression for $\mathbf{A}_2$ was obtained by symmetry; and the expression for $\mathbf{A}_3$ was obtained by subtraction of the first two sums of squares from the corrected total sum of squares followed by the repeated use of Theorem 2.17.2, i.e.,

$$\mathbf{y}^T\{(\mathbf{I}_q \otimes \mathbf{I}_m) - (\frac{1}{q}\mathbf{J}_q \otimes \frac{1}{m}\mathbf{J}_m) - [(\mathbf{I}_q - \frac{1}{q}\mathbf{J}_q) \otimes \frac{1}{m}\mathbf{J}_m] - [\frac{1}{q}\mathbf{J}_q \otimes (\mathbf{I}_m - \frac{1}{m}\mathbf{J}_m)]\}\mathbf{y}$$

$$= \mathbf{y}^T\{(\mathbf{I}_q \otimes \mathbf{I}_m) - (\mathbf{I}_q \otimes \frac{1}{m}\mathbf{J}_m) - [\frac{1}{q}\mathbf{J}_q \otimes (\mathbf{I}_m - \frac{1}{m}\mathbf{J}_m)]\}\mathbf{y}$$

$$= \mathbf{y}^T\{[\mathbf{I}_q \otimes (\mathbf{I}_m - \frac{1}{m}\mathbf{J}_m) - [\frac{1}{q}\mathbf{J}_q \otimes (\mathbf{I}_m - \frac{1}{m}\mathbf{J}_m)]\}\mathbf{y}$$

$$= \mathbf{y}^T[(\mathbf{I}_q - \frac{1}{q}\mathbf{J}_q) \otimes (\mathbf{I}_m - \frac{1}{m}\mathbf{J}_m)]\mathbf{y}.$$

Furthermore,

$$\boldsymbol{\Sigma} \equiv \mathrm{var}(\mathbf{y}) = \sigma_a^2(\mathbf{I}_q \otimes \mathbf{J}_m) + \sigma_b^2(\mathbf{J}_q \otimes \mathbf{I}_m) + \sigma^2(\mathbf{I}_q \otimes \mathbf{I}_m).$$

Now,

$$\mathbf{A}_1\boldsymbol{\Sigma} = \left(\frac{1}{\sigma^2 + m\sigma_a^2}\right)\{\sigma_a^2[(\mathbf{I}_q - \frac{1}{q}\mathbf{J}_q) \otimes \mathbf{J}_m] + \sigma_b^2[\mathbf{0}_{q\times q} \otimes \frac{1}{m}\mathbf{J}_m]$$
$$+\sigma^2[(\mathbf{I}_q - \frac{1}{q}\mathbf{J}_q) \otimes \frac{1}{m}\mathbf{J}_m\}$$

$$= (\mathbf{I}_q - \frac{1}{q}\mathbf{J}_q) \otimes \frac{1}{m}\mathbf{J}_m,$$

$$\mathbf{A}_2\boldsymbol{\Sigma} = \left(\frac{1}{\sigma^2 + q\sigma_b^2}\right) \{\sigma_a^2[\frac{1}{q}\mathbf{J}_q \otimes \mathbf{0}_{m\times m}] + \sigma_b^2[\mathbf{J}_q \otimes (\mathbf{I}_m - \frac{1}{m}\mathbf{J}_m)]$$

$$+\sigma^2[\frac{1}{q}\mathbf{J}_q \otimes (\mathbf{I}_m - \frac{1}{m}\mathbf{J}_m)]\}$$

$$= \frac{1}{q}\mathbf{J}_q \otimes (\mathbf{I}_m - \frac{1}{m}\mathbf{J}_m),$$

$$\mathbf{A}_3\boldsymbol{\Sigma} = \left(\frac{1}{\sigma^2}\right) \{\sigma_a^2[(\mathbf{I}_q - \frac{1}{q}\mathbf{J}_q) \otimes \mathbf{0}_{m\times m}] + \sigma_b^2[\mathbf{0}_{q\times q} \otimes (\mathbf{I}_m - \frac{1}{m}\mathbf{J}_m)]$$

$$+\sigma^2[(\mathbf{I}_q - \frac{1}{q}\mathbf{J}_q) \otimes (\mathbf{I}_m - \frac{1}{m}\mathbf{J}_m)]\}$$

$$= (\mathbf{I}_q - \frac{1}{q}\mathbf{J}_q) \otimes (\mathbf{I}_m - \frac{1}{m}\mathbf{J}_m).$$

Observe that $\mathbf{A}_1\boldsymbol{\Sigma}$, $\mathbf{A}_2\boldsymbol{\Sigma}$, and $\mathbf{A}_3\boldsymbol{\Sigma}$ are idempotent (because each term in each Kronecker product is idempotent). Also observe that $\text{rank}(\mathbf{A}_1) = q - 1$, $\text{rank}(\mathbf{A}_2) = m - 1$, and $\text{rank}(\mathbf{A}_3) = (q - 1)(m - 1)$ by Theorem 2.17.8, and that $\mathbf{A}_i\mathbf{1}\mu = \mu\mathbf{A}_i(\mathbf{1}_q \otimes \mathbf{1}_m) = \mathbf{0}$ for $i = 1, 2, 3$. Thus, by Corollary 14.2.6.1, $f_1S_1^2 \sim \chi^2(q - 1)$, $f_2S_2^2 \sim \chi^2(m - 1)$, and $f_3S_3^2 \sim \chi^2[(q - 1)(m - 1)]$. Moreover,

$$\mathbf{A}_1\boldsymbol{\Sigma}\mathbf{A}_2 = \left(\frac{1}{\sigma^2 + q\sigma_b^2}\right) (\mathbf{0}_{q\times q} \otimes \mathbf{0}_{m\times m}) = \mathbf{0},$$

$$\mathbf{A}_1\boldsymbol{\Sigma}\mathbf{A}_3 = \left(\frac{1}{\sigma^2}\right) [(\mathbf{I}_q - \frac{1}{q}\mathbf{J}_q) \otimes \mathbf{0}_{m\times m}] = \mathbf{0},$$

$$\mathbf{A}_2\boldsymbol{\Sigma}\mathbf{A}_3 = \left(\frac{1}{\sigma^2}\right) [\mathbf{0}_{q\times q} \otimes (\mathbf{I}_m - \frac{1}{m}\mathbf{J}_m)] = \mathbf{0},$$

so by Corollary 14.3.1.1 S_1^2, S_2^2, and S_3^2 are independent.

(b) Using the distributional results from part (a), $\text{E}\left(\frac{(q-1)S_1^2}{\sigma^2+m\sigma_a^2}\right) = q - 1$ so $\text{E}(S_1^2) = \sigma^2+m\sigma_a^2$; $\text{E}\left(\frac{(m-1)S_2^2}{\sigma^2+q\sigma_b^2}\right) = m-1$ so $\text{E}(S_2^2) = \sigma^2+q\sigma_b^2$; and $\text{E}\left(\frac{(q-1)(m-1)S_3^2}{\sigma^2}\right) = (q - 1)(m - 1)$, so $\text{E}(S_3^2) = \sigma^2$. Thus, minimum variance quadratic unbiased estimators are

$$\hat{\sigma}^2 = S_3^2, \quad \hat{\sigma}_a^2 = (S_1^2 - S_3^2)/m, \quad \hat{\sigma}_b^2 = (S_2^2 - S_3^2)/q.$$

(c)

$$1-\xi = \Pr\left(\chi^2_{1-(\xi/2),m-1} \le \frac{(m-1)S_2^2}{\sigma^2+q\sigma_b^2} \le \chi^2_{\xi/2,m-1}\right)$$
$$= \Pr\left(\frac{(m-1)S_2^2}{\chi^2_{\xi/2,m-1}} \le \sigma^2+q\sigma_b^2 \le \frac{(m-1)S_2^2}{\chi^2_{1-(\xi/2),m-1}}\right).$$

Therefore, $\left[\frac{(m-1)S_2^2}{\chi^2_{\xi/2,m-1}}, \frac{(m-1)S_2^2}{\chi^2_{1-(\xi/2),m-1}}\right]$ is a $100(1-\xi)\%$ confidence interval for $\sigma^2+q\sigma_b^2$.

(d) Define

$$U = \frac{S_2^2}{S_3^2}, \quad U^* = \frac{S_2^2/(\sigma^2+q\sigma_b^2)}{S_3^2/\sigma^2}, \quad \text{and} \quad \psi = \frac{\sigma_b^2}{\sigma^2},$$

and observe that $U^* \sim \mathrm{F}[m-1, (q-1)(m-1)]$. Therefore,

$$1-\xi = \Pr\left(F_{1-(\xi/2),m-1,(q-1)(m-1)} \le \frac{\sigma^2}{\sigma^2+q\sigma_b^2}U \le F_{\xi/2,m-1,(q-1)(m-1)}\right)$$
$$= \Pr\left(F_{1-(\xi/2),m-1,(q-1)(m-1)} \le \frac{1}{1+q\psi}U \le F_{\xi/2,m-1,(q-1)(m-1)}\right)$$
$$= \Pr\left(\frac{U}{F_{\xi/2,m-1,(q-1)(m-1)}} \le 1+q\psi \le \frac{U}{F_{1-(\xi/2),m-1,(q-1)(m-1)}}\right)$$
$$= \Pr\left(\frac{U/F_{\xi/2,m-1,(q-1)(m-1)}-1}{q} \le \psi \le \frac{U/F_{1-(\xi/2),m-1,(q-1)(m-1)}-1}{q}\right).$$

Therefore,

$$\left[\frac{U/F_{\xi/2,m-1,(q-1)(m-1)}-1}{q}, \frac{U/F_{1-(\xi/2),m-1,(q-1)(m-1)}-1}{q}\right]$$

is a $100(1-\xi)\%$ confidence interval for σ_b^2/σ^2. If any endpoints of the interval are negative, they can be set equal to zero.

(e) Denote the interval obtained by replacing $\xi/2$ with $\xi/4$ in part (d) by $[L_1, R_1]$. A similar interval could be constructed for σ_a^2/σ^2; denote that interval by $[L_2, R_2]$. Let A denote the event $\left\{L_1 \le \frac{\sigma_b^2}{\sigma^2} \le R_1 \text{ and } L_2 \le \frac{\sigma_a^2}{\sigma^2} \le R_2\right\}$, which is equivalent to the event $\left\{-R_1 \le -\frac{\sigma_b^2}{\sigma^2} \le -L_1 \text{ and } L_2 \le \frac{\sigma_a^2}{\sigma^2} \le R_2\right\}$.

Then, by the Bonferroni inequality, $\Pr(A) \geq 1 - \xi$. Now by Fact 15.3,

$$\Pr(A) \leq \Pr\left(L_2 - R_1 \leq \frac{\sigma_a^2 - \sigma_b^2}{\sigma^2} \leq R_2 - L_1\right).$$

Thus, $[L_2 - R_1, R_2 - L_1]$ is an interval for $(\sigma_a^2 - \sigma_b^2)/\sigma^2$, which has coverage probability at least $1 - \xi$.

(f) In the notation used to describe Satterthwaite's approximation, let $U_1 = \frac{n_1 S_1^2}{\alpha_1^2}$, where $n_1 = q - 1$ and $\alpha_1^2 = \sigma^2 + m\sigma_a^2$. Similarly, let $U_3 = \frac{n_3 S_3^2}{\alpha_3^2}$, where $n_3 = (q-1)(m-1)$ and $\alpha_3^2 = \sigma^2$. We want a confidence interval for $\sigma_a^2 = \frac{1}{m}(\sigma^2 + m\sigma_a^2) - \frac{\sigma^2}{m}$, so we let $c_1 = \frac{1}{m}$ and $c_3 = -\frac{1}{m}$. Then, specializing (16.4) and (16.5), an approximate $100(1-\xi)\%$ confidence interval for σ_a^2 is given by

$$\left[\frac{\hat{t}(S_1^2 - S_3^2)}{m\chi^2_{\xi/2,\hat{t}}}, \frac{\hat{t}(S_1^2 - S_3^2)}{m\chi^2_{1-(\xi/2),\hat{t}}}\right]$$

where

$$\hat{t} = \frac{(\frac{1}{m}S_1^2 - \frac{1}{m}S_3^2)^2}{\frac{(S_1^2/m)^2}{q-1} + \frac{(-S_3^2/m)^2}{(q-1)(m-1)}} = \frac{(S_1^2 - S_3^2)^2}{\frac{S_1^4}{q-1} + \frac{S_3^4}{(q-1)(m-1)}}.$$

(g) $\frac{S_1^2/(\sigma^2+m\sigma_a^2)}{S_3^2/\sigma^2} \sim \mathrm{F}[q-1, (q-1)(m-1)]$ so $\frac{S_1^2}{S_3^2} \sim \frac{\sigma^2+m\sigma_a^2}{\sigma^2}\mathrm{F}[q-1, (q-1)(m-1)]$. Thus, a size-$\xi$ test of $H_0 : \sigma_a^2 = 0$ versus $H_a : \sigma_a^2 > 0$ is to reject H_0 if and only if $\frac{S_1^2}{S_3^2} > F_{\xi,q-1,(q-1)(m-1)}$. Similarly, $\frac{S_2^2/(\sigma^2+q\sigma_b^2)}{S_3^2/\sigma^2} \sim \mathrm{F}[(m-1), (q-1)(m-1)]$ so $\frac{S_2^2}{S_3^2} \sim \frac{\sigma^2+q\sigma_b^2}{\sigma^2}\mathrm{F}[(m-1), (q-1)(m-1)]$. Thus, a size-$\xi$ test of $H_0 : \sigma_b^2 = 0$ versus $H_a : \sigma_b^2 > 0$ is to reject H_0 if and only if $\frac{S_2^2}{S_3^2} > F_{\xi,m-1,(q-1)(m-1)}$.

▶ **Exercise 3** Consider the normal two-way components-of-variance model with interaction and balanced data:

$$y_{ijk} = \mu + a_i + b_j + c_{ij} + d_{ijk} \quad (i = 1, \ldots, q;\ j = 1, \ldots, m;\ k = 1, \cdots, r)$$

where $a_i \sim \mathrm{N}(0, \sigma_a^2)$, $b_j \sim \mathrm{N}(0, \sigma_b^2)$, $c_{ij} \sim \mathrm{N}(0, \sigma_c^2)$, and $d_{ijk} \sim \mathrm{N}(0, \sigma^2)$ for all i, j, and k, and where $\{a_i\}$, $\{b_j\}$, $\{c_{ij}\}$, and $\{d_{ijk}\}$ are mutually independent. The corrected three-part mixed-model ANOVA table (corresponding to fitting the interaction terms after the main effects) is given below, with an additional column giving the distributions of suitably scaled mean squares under the model.

Source	df	Sum of squares	Mean square
Factor A	$q-1$	$mr\sum_{i=1}^{q}(\bar{y}_{i\cdot\cdot}-\bar{y}_{\cdots})^2$	S_1^2
Factor B	$m-1$	$qr\sum_{j=1}^{m}(\bar{y}_{\cdot j\cdot}-\bar{y}_{\cdots})^2$	S_2^2
Interaction	$(q-1)(m-1)$	$r\sum_{i=1}^{q}\sum_{j=1}^{m}(\bar{y}_{ij\cdot}-\bar{y}_{i\cdot\cdot}-\bar{y}_{\cdot j\cdot}+\bar{y}_{\cdots})^2$	S_3^2
Residual	$qm(r-1)$	$\sum_{i=1}^{q}\sum_{j=1}^{m}\sum_{k=1}^{r}(y_{ijk}-\bar{y}_{ij\cdot})^2$	S_4^2
Total	$qmr-1$	$\sum_{i=1}^{q}\sum_{j=1}^{m}\sum_{k=1}^{r}(y_{ijk}-\bar{y}_{\cdots})^2$	

Here $f_1S_1^2 \sim \chi^2(q-1)$, $f_2S_2^2 \sim \chi^2(m-1)$, $f_3S_3^2 \sim \chi^2[(q-1)(m-1)]$, $f_4S_4^2 \sim \chi^2[qm(r-1)]$, where

$$f_1 = \frac{q-1}{\sigma^2+r\sigma_c^2+mr\sigma_a^2}, \quad f_2 = \frac{m-1}{\sigma^2+r\sigma_c^2+qr\sigma_b^2},$$

$$f_3 = \frac{(q-1)(m-1)}{\sigma^2+r\sigma_c^2}, \quad f_4 = \frac{qm(r-1)}{\sigma^2}.$$

Furthermore, S_1^2, S_2^2, S_3^2, and S_4^2 are mutually independent. Let $\xi \in (0,1)$.

(a) Verify the distributions given in the last column of the table, and verify that S_1^2, S_2^2, S_3^2, and S_4^2 are mutually independent.
(b) Obtain minimum variance quadratic unbiased estimators of the variance components.
(c) Obtain a $100(1-\xi)\%$ confidence interval for $\frac{\sigma_b^2}{\sigma^2+r\sigma_c^2}$. (Hint: consider the distribution of S_2^2/S_3^2, suitably scaled.)
(d) Obtain a $100(1-\xi)\%$ confidence interval for $\frac{\sigma_a^2}{\sigma^2+r\sigma_c^2}$. [Hint: exploit the similarity with part (c).]
(e) Using the results of part (c) and (d), obtain a confidence interval for $\frac{\sigma_a^2+\sigma_b^2}{\sigma^2+r\sigma_c^2}$, which has coverage probability at least $1-\xi$.
(f) Obtain size-ξ tests of $H_0{:}\,\sigma_a^2 = 0$ versus $H_a{:}\,\sigma_a^2 > 0$, $H_0{:}\,\sigma_b^2 = 0$ versus $H_a{:}\,\sigma_b^2 > 0$, and $H_0{:}\,\sigma_c^2 = 0$ versus $H_a{:}\,\sigma_c^2 > 0$.

Solution

(a) Here $f_1S_1^2 = \mathbf{y}^T\mathbf{A}_1\mathbf{y}$, $f_2S_2^2 = \mathbf{y}^T\mathbf{A}_2\mathbf{y}$, $f_3S_3^2 = \mathbf{y}^T\mathbf{A}_3\mathbf{y}$, and $f_4S_4^2 = \mathbf{y}^T\mathbf{A}_4\mathbf{y}$, where

$$\mathbf{A}_1 = \left(\frac{1}{\sigma^2+r\sigma_c^2+mr\sigma_a^2}\right)[(\mathbf{I}_q-\frac{1}{q}\mathbf{J}_q)\otimes\frac{1}{m}\mathbf{J}_m\otimes\frac{1}{r}\mathbf{J}_r],$$

$$\mathbf{A}_2 = \left(\frac{1}{\sigma^2+r\sigma_c^2+qr\sigma_b^2}\right)[\frac{1}{q}\mathbf{J}_q\otimes(\mathbf{I}_m-\frac{1}{m}\mathbf{J}_m)\otimes\frac{1}{r}\mathbf{J}_r],$$

$$\mathbf{A}_3 = \left(\frac{1}{\sigma^2 + r\sigma_c^2}\right)[(\mathbf{I}_q - \frac{1}{q}\mathbf{J}_q) \otimes (\mathbf{I}_m - \frac{1}{m}\mathbf{J}_m) \otimes \frac{1}{r}\mathbf{J}_r],$$

$$\mathbf{A}_4 = \left(\frac{1}{\sigma^2}\right)[\mathbf{I}_q \otimes \mathbf{I}_m \otimes (\mathbf{I}_r - \frac{1}{r}\mathbf{J}_r)]$$

Here the expressions for $\mathbf{A}_1$, $\mathbf{A}_2$, and $\mathbf{A}_3$ were obtained by extending expressions for matrices labelled with the same symbols in the solution to Exercise 16.2 to account for replication; obtaining the expression for $\mathbf{A}_4$ is trivial. Furthermore,

$$\boldsymbol{\Sigma} \equiv \text{var}(\mathbf{y}) = \sigma_a^2(\mathbf{I}_q \otimes \mathbf{J}_m \otimes \mathbf{J}_r) + \sigma_b^2(\mathbf{J}_q \otimes \mathbf{I}_m \otimes \mathbf{J}_r) + \sigma_c^2(\mathbf{I}_q \otimes \mathbf{I}_m \otimes \mathbf{J}_r) + \sigma^2(\mathbf{I}_q \otimes \mathbf{I}_m \otimes \mathbf{I}_r).$$

Now,

$$\begin{aligned}
\mathbf{A}_1\boldsymbol{\Sigma} &= \left(\frac{1}{\sigma^2 + r\sigma_c^2 + mr\sigma_a^2}\right)\{\sigma_a^2[(\mathbf{I}_q - \frac{1}{q}\mathbf{J}_q) \otimes \mathbf{J}_m \otimes \mathbf{J}_r] + \sigma_b^2[\mathbf{0}_{q\times q} \otimes \frac{1}{m}\mathbf{J}_m \otimes \mathbf{J}_r] \\
&\quad + \sigma_c^2[(\mathbf{I}_q - \frac{1}{q}\mathbf{J}_q) \otimes \frac{1}{m}\mathbf{J}_m \otimes \mathbf{J}_r] + \sigma^2[\mathbf{I}_q - \frac{1}{q}\mathbf{J}_q) \otimes \frac{1}{m}\mathbf{J}_m \otimes \frac{1}{r}\mathbf{J}_r]\} \\
&= (\mathbf{I}_q - \frac{1}{q}\mathbf{J}_q) \otimes \frac{1}{m}\mathbf{J}_m \otimes \frac{1}{r}\mathbf{J}_r, \\
\mathbf{A}_2\boldsymbol{\Sigma} &= \left(\frac{1}{\sigma^2 + r\sigma_c^2 + qr\sigma_b^2}\right)\{\sigma_a^2[\frac{1}{q}\mathbf{J}_q \otimes \mathbf{0}_{m\times m} \otimes \mathbf{J}_r] + \sigma_b^2[\mathbf{J}_q \otimes (\mathbf{I}_m - \frac{1}{m}\mathbf{J}_m) \otimes \mathbf{J}_r] \\
&\quad + \sigma_c^2[\frac{1}{q}\mathbf{J}_q \otimes (\mathbf{I}_m - \frac{1}{m}\mathbf{J}_m) \otimes \mathbf{J}_r] + \sigma^2[\frac{1}{q}\mathbf{J}_q \otimes (\mathbf{I}_m - \frac{1}{m}\mathbf{J}_m) \otimes \mathbf{J}_r]\} \\
&= \frac{1}{q}\mathbf{J}_q \otimes (\mathbf{I}_m - \frac{1}{m}\mathbf{J}_m) \otimes \frac{1}{r}\mathbf{J}_r, \\
\mathbf{A}_3\boldsymbol{\Sigma} &= \left(\frac{1}{\sigma^2 + r\sigma_c^2}\right)\{\sigma_a^2[(\mathbf{I}_q - \frac{1}{q}\mathbf{J}_q) \otimes \mathbf{0}_{m\times m} \otimes \mathbf{J}_r] + \sigma_b^2[\mathbf{0}_{q\times q} \otimes (\mathbf{I}_m - \frac{1}{m}\mathbf{J}_m) \otimes \mathbf{J}_r] \\
&\quad + \sigma_c^2[(\mathbf{I}_q - \frac{1}{q}\mathbf{J}_q) \otimes (\mathbf{I}_m - \frac{1}{m}\mathbf{J}_m) \otimes \mathbf{J}_r] \\
&\quad + \sigma^2[(\mathbf{I}_q - \frac{1}{q}\mathbf{J}_q) \otimes (\mathbf{I}_m - \frac{1}{m}\mathbf{J}_m) \otimes \frac{1}{r}\mathbf{J}_r]\} \\
&= (\mathbf{I}_q - \frac{1}{q}\mathbf{J}_q) \otimes (\mathbf{I}_m - \frac{1}{m}\mathbf{J}_m) \otimes \frac{1}{r}\mathbf{J}_r, \\
\mathbf{A}_4\boldsymbol{\Sigma} &= \left(\frac{1}{\sigma^2}\right)\{\sigma_a^2[\mathbf{I}_q \otimes \mathbf{J}_m \otimes \mathbf{0}_{r\times r}] + \sigma_b^2[\mathbf{J}_q \otimes \mathbf{I}_m \otimes \mathbf{0}_{r\times r}] + \sigma_c^2[\mathbf{I}_q \otimes \mathbf{I}_m \otimes \mathbf{0}_{r\times r}] \\
&\quad + \sigma^2[\mathbf{I}_q \otimes \mathbf{I}_m \otimes (\mathbf{I}_r - \frac{1}{r}\mathbf{J}_r)]\} \\
&= \mathbf{I}_q \otimes \mathbf{I}_m \otimes (\mathbf{I}_r - \frac{1}{r}\mathbf{J}_r).
\end{aligned}$$

Observe that $\mathbf{A}_1\boldsymbol{\Sigma}$, $\mathbf{A}_2\boldsymbol{\Sigma}$, $\mathbf{A}_3\boldsymbol{\Sigma}$, and $\mathbf{A}_4\boldsymbol{\Sigma}$ are idempotent (because each term in each Kronecker product is idempotent). Also observe that $\text{rank}(\mathbf{A}_1) = q - 1$, $\text{rank}(\mathbf{A}_2) = m - 1$, $\text{rank}(\mathbf{A}_3) = (q - 1)(m - 1)$, and $\text{rank}(\mathbf{A}_4) = qm(r - 1)$ by Theorem 2.17.8, and that $\mathbf{A}_i\mathbf{1}\mu = \mu\mathbf{A}_i(\mathbf{1}_q \otimes \mathbf{1}_m \otimes \mathbf{1}_r) = \mathbf{0}$ for $i = 1, 2, 3, 4$. Thus, by Corollary 14.2.6.1, $f_1S_1^2 \sim \chi^2(q - 1)$, $f_2S_2^2 \sim \chi^2(m - 1)$, $f_3S_3^2 \sim \chi^2[(q - 1)(m - 1)]$, and $f_4S_4^2 \sim \chi^2[qm(r - 1)]$. Moreover,

$$\mathbf{A}_1\boldsymbol{\Sigma}\mathbf{A}_2 = \left(\frac{1}{\sigma^2 + r\sigma_c^2 + qr\sigma_b^2}\right)(\mathbf{0}_{q\times q} \otimes \mathbf{0}_{m\times m} \otimes \frac{1}{r}\mathbf{J}_r) = \mathbf{0},$$

$$\mathbf{A}_1\boldsymbol{\Sigma}\mathbf{A}_3 = \left(\frac{1}{\sigma^2 + r\sigma_c^2}\right)[(\mathbf{I}_q - \frac{1}{q}\mathbf{J}_q) \otimes \mathbf{0}_{m\times m} \otimes \frac{1}{r}\mathbf{J}_r] = \mathbf{0},$$

$$\mathbf{A}_1\boldsymbol{\Sigma}\mathbf{A}_4 = \left(\frac{1}{\sigma^2}\right)[(\mathbf{I}_q - \frac{1}{q}\mathbf{J}_q) \otimes \frac{1}{m}\mathbf{J}_m \otimes \mathbf{0}_{r\times r}] = \mathbf{0},$$

$$\mathbf{A}_2\boldsymbol{\Sigma}\mathbf{A}_3 = \left(\frac{1}{\sigma^2 + r\sigma_c^2}\right)[\mathbf{0}_{q\times q} \otimes (\mathbf{I}_m - \frac{1}{m}\mathbf{J}_m) \otimes \frac{1}{r}\mathbf{J}_r] = \mathbf{0},$$

$$\mathbf{A}_2\boldsymbol{\Sigma}\mathbf{A}_4 = \left(\frac{1}{\sigma^2}\right)[\frac{1}{q}\mathbf{J}_q \otimes (\mathbf{I}_m - \frac{1}{m}\mathbf{J}_m) \otimes \mathbf{0}_{r\times r}] = \mathbf{0},$$

$$\mathbf{A}_3\boldsymbol{\Sigma}\mathbf{A}_4 = \left(\frac{1}{\sigma^2}\right)[(\mathbf{I}_q - \frac{1}{q}\mathbf{J}_q) \otimes (\mathbf{I}_m - \frac{1}{m}\mathbf{J}_m) \otimes \mathbf{0}_{r\times r}] = \mathbf{0},$$

so by Corollary 14.3.1.1 S_1^2, S_2^2, S_3^2, and S_4^2 are independent.

(b) Using the distributional results from part (a), $\text{E}\left(\frac{(q-1)S_1^2}{\sigma^2+r\sigma_c^2+mr\sigma_a^2}\right) = q - 1$ so $\text{E}(S_1^2) = \sigma^2 + r\sigma_c^2 + mr\sigma_a^2$; $\text{E}\left(\frac{(m-1)S_2^2}{\sigma^2+r\sigma_c^2+qr\sigma_b^2}\right) = m - 1$ so $\text{E}(S_2^2) = \sigma^2 + r\sigma_c^2 + qr\sigma_b^2$; $\text{E}\left(\frac{(q-1)(m-1)S_3^2}{\sigma^2+r\sigma_c^2}\right) = (q - 1)(m - 1)$ so $\text{E}(S_3^2) = \sigma^2 + r\sigma_c^2$; and $\text{E}\left(\frac{qm(r-1)S_4^2}{\sigma^2}\right) = qm(r - 1)$ so $\text{E}(S_4^2) = \sigma^2$. Thus, minimum variance quadratic unbiased estimators are

$$\hat{\sigma}^2 = S_4^2, \quad \hat{\sigma}_c^2 = \frac{S_3^2 - S_4^2}{r}, \quad \hat{\sigma}_b^2 = \frac{S_2^2 - S_3^2}{qr}, \quad \hat{\sigma}_a^2 = \frac{S_1^2 - S_3^2}{mr}.$$

(c) Define

$$U = \frac{S_2^2}{S_3^2}, \quad U^* = \frac{S_2^2/(\sigma^2 + r\sigma_c^2 + qr\sigma_b^2)}{S_3^2/(\sigma^2 + r\sigma_c^2)}, \quad \text{and} \quad \psi_1 = \frac{\sigma_b^2}{\sigma^2 + r\sigma_c^2}.$$

Observe that $U^* \sim \mathrm{F}[m-1, (q-1)(m-1)]$. Therefore,

$$1-\xi = \Pr\left(F_{1-(\xi/2),m-1,(q-1)(m-1)} \leq \frac{\sigma^2+r\sigma_c^2}{\sigma^2+r\sigma_c^2+qr\sigma_b^2}U \leq F_{\xi/2,m-1,(q-1)(m-1)}\right)$$

$$= \Pr\left(F_{1-(\xi/2),m-1,(q-1)(m-1)} \leq \frac{1}{1+qr\psi_1}U \leq F_{\xi/2,m-1,(q-1)(m-1)}\right)$$

$$= \Pr\left(\frac{(U/F_{\xi/2,m-1,(q-1)(m-1)})-1}{qr} \leq \psi_1 \leq \frac{(U/F_{1-(\xi/2),m-1,(q-1)(m-1)})-1}{qr}\right).$$

So

$$\left[\frac{(U/F_{\xi/2,m-1,(q-1)(m-1)})-1}{qr}, \frac{(U/F_{1-(\xi/2),m-1,(q-1)(m-1)})-1}{qr}\right]$$

is a $100(1-\xi)\%$ confidence interval for $\sigma_b^2/(\sigma^2+r\sigma_c^2)$. If any endpoints of the interval are negative, they can be set equal to zero.

(d) Define

$$V = S_1^2/S_3^2, \quad V^* = \frac{S_1^2/(\sigma^2+r\sigma_c^2+mr\sigma_a^2)}{S_3^2/(\sigma^2+r\sigma_c^2)}, \quad \text{and} \quad \psi_2 = \frac{\sigma_a^2}{\sigma^2+r\sigma_c^2}.$$

Observe that $V^* \sim \mathrm{F}[q-1, (q-1)(m-1)]$. By the symmetry,

$$\left[\frac{(V/F_{\xi/2,q-1,(q-1)(m-1)}))-1}{mr}, \frac{(V/F_{1-(\xi/2),q-1,(q-1)(m-1)})-1}{mr}\right]$$

is a $100(1-\xi)\%$ confidence interval for $\sigma_a^2/(\sigma^2+r\sigma_c^2)$. If any of the endpoints of the interval are negative, they can be set equal to zero.

(e) Define

$$L_1 = \frac{(U/F_{\xi/4,q-1,(q-1)(m-1)})-1}{mr}, \qquad R_1 = \frac{(U/F_{1-\xi/4,q-1,(q-1)(m-1)})-1}{mr}$$

$$L_2 = \frac{(V/F_{\xi/4,m-1,(q-1)(m-1)})-1}{qr}, \qquad R_2 = \frac{(U/F_{1-\xi/4,m-1,(q-1)(m-1)})-1}{qr}.$$

Then $\Pr(L_1 \leq \frac{\sigma_b^2}{\sigma^2+r\sigma_c^2} \leq R_1) = 1-(\xi/2)$ and $\Pr(L_2 \leq \frac{\sigma_a^2}{\sigma^2+r\sigma_c^2} \leq R_2) = 1-(\xi/2)$; furthermore, by the Bonferroni inequality, $\Pr(L_1 \leq \frac{\sigma_b^2}{\sigma^2+r\sigma_c^2} \leq R_1$ and $L_2 \leq \frac{\sigma_a^2}{\sigma^2+r\sigma_c^2} \leq R_2) \geq 1-\xi$. Now, if $L_1 \leq \frac{\sigma_b^2}{\sigma^2+r\sigma_c^2} \leq R_1$ and $L_2 \leq$

$\frac{\sigma_a^2}{\sigma^2+r\sigma_c^2} \leq R_2$, then $L_1 + L_2 \leq \frac{\sigma_a^2+\sigma_b^2}{\sigma^2+r\sigma_c^2} \leq R_1 + R_2$. By Fact 15.3 it follows that

$$\Pr\left(L_1 + L_2 \leq \frac{\sigma_a^2 + \sigma_b^2}{\sigma^2 + r\sigma_c^2} \leq R_1 + R_2\right) \geq 1 - \xi.$$

(f) $\frac{S_1^2/(\sigma^2+r\sigma_c^2+mr\sigma_a^2)}{S_3^2/(\sigma^2+r\sigma_c^2)} \sim \mathrm{F}[q-1,(q-1)(m-1)]$ so $\frac{S_1^2}{S_3^2} \sim \frac{\sigma^2+r\sigma_c^2+mr\sigma_a^2}{\sigma^2+r\sigma_c^2}\mathrm{F}[q-1,(q-1)(m-1)]$. Thus, a size-$\xi$ test of $H_0: \sigma_a^2 = 0$ versus $H_a: \sigma_a^2 > 0$ is to reject H_0 if and only if $\frac{S_1^2}{S_3^2} > F_{\xi,q-1,(q-1)(m-1)}$. Similarly, $\frac{S_2^2/(\sigma^2+r\sigma_c^2+qr\sigma_b^2)}{S_3^2/(\sigma^2+r\sigma_c^2)} \sim \mathrm{F}[(m-1),(q-1)(m-1)]$ so $\frac{S_2^2}{S_3^2} \sim \frac{\sigma^2+r\sigma_c^2+qr\sigma_b^2}{\sigma^2+r\sigma_c^2}\mathrm{F}[(m-1),(q-1)(m-1)]$. Thus, a size-$\xi$ test of $H_0: \sigma_b^2 = 0$ versus $H_a: \sigma_b^2 > 0$ is to reject H_0 if and only if $\frac{S_2^2}{S_3^2} > F_{\xi,m-1,(q-1)(m-1)}$. Finally, $\frac{S_3^2/(\sigma^2+r\sigma_c^2)}{S_4^2/\sigma^2} \sim \mathrm{F}[(q-1)(m-1),qm(r-1)]$ so $\frac{S_3^2}{S_4^2} \sim \frac{\sigma^2+r\sigma_c^2}{\sigma^2}\mathrm{F}[(q-1)(m-1),qm(r-1)]$. Thus, a size-$\xi$ test of $H_0: \sigma_c^2 = 0$ versus $H_a: \sigma_c^2 > 0$ is to reject H_0 if and only if $\frac{S_3^2}{S_4^2} > F_{\xi,(q-1)(m-1),qm(r-1)}$.

▶ **Exercise 4** Consider the normal two-factor nested components-of-variance model with balanced data:

$$y_{ijk} = \mu + a_i + b_{ij} + d_{ijk} \quad (i = 1,\ldots,q;\ j = 1,\ldots,m;\ k = 1,\ldots,r)$$

where $a_i \sim \mathrm{N}(0,\sigma_a^2)$, $b_{ij} \sim \mathrm{N}(0,\sigma_b^2)$, and $d_{ijk} \sim \mathrm{N}(0,\sigma^2)$ for all i, j, and k, and where $\{a_i\}$, $\{b_j\}$, and $\{d_{ijk}\}$ are mutually independent. The corrected two-part mixed-model ANOVA table (with Factor A fitted first) is given below, with an additional column giving the distributions of suitably scaled mean squares under the model.

Source	df	Sum of squares	Mean square
Factor A	$q-1$	$mr\sum_{i=1}^{q}(\bar{y}_{i\cdot\cdot} - \bar{y}_{\cdot\cdot\cdot})^2$	S_1^2
Factor B within A	$q(m-1)$	$r\sum_{i=1}^{q}\sum_{j=1}^{m}(\bar{y}_{ij\cdot} - \bar{y}_{i\cdot\cdot})^2$	S_2^2
Residual	$qm(r-1)$	$\sum_{i=1}^{q}\sum_{j=1}^{m}\sum_{k=1}^{r}(y_{ijk} - \bar{y}_{ij\cdot})^2$	S_3^2
Total	$qmr-1$	$\sum_{i=1}^{q}\sum_{j=1}^{m}\sum_{k=1}^{r}(y_{ijk} - \bar{y}_{\cdot\cdot\cdot})^2$	

Here $f_1 S_1^2 \sim \chi^2(q-1)$, $f_2 S_2^2 \sim \chi^2[q(m-1)]$, and $f_3 S_3^2 \sim \chi^2[qm(r-1)]$, where

$$f_1 = \frac{q-1}{\sigma^2 + r\sigma_b^2 + mr\sigma_a^2}, \quad f_2 = \frac{q(m-1)}{\sigma^2 + r\sigma_b^2}, \quad f_3 = \frac{qm(r-1)}{\sigma^2}.$$

Furthermore, S_1^2, S_2^2, and S_3^2 are mutually independent. Let $\xi \in (0, 1)$.

(a) Verify the distributions given in the last column of the table, and verify that S_1^2, S_2^2, and S_3^2 are mutually independent.
(b) Obtain minimum variance quadratic unbiased estimators of the variance components.
(c) Obtain a $100(1-\xi)\%$ confidence interval for σ^2.
(d) Obtain a $100(1-\xi)\%$ confidence interval for $\frac{\sigma_b^2+m\sigma_a^2}{\sigma^2}$.
(e) Use Satterthwaite's method to obtain an approximate $100(1-\xi)\%$ confidence interval for σ_a^2.
(f) Obtain size-ξ tests of $H_0\colon \sigma_a^2 = 0$ versus $H_a\colon \sigma_a^2 > 0$ and $H_0\colon \sigma_b^2 = 0$ versus $H_a\colon \sigma_b^2 > 0$.

Solution

(a) Here $f_1 S_1^2 = \mathbf{y}^T \mathbf{A}_1 \mathbf{y}$, $f_2 S_2^2 = \mathbf{y}^T \mathbf{A}_2 \mathbf{y}$, and $f_3 S_3^2 = \mathbf{y}^T \mathbf{A}_3 \mathbf{y}$, where

$$\mathbf{A}_1 = \left(\frac{1}{\sigma^2 + r\sigma_b^2 + mr\sigma_a^2}\right)[(\mathbf{I}_q - \frac{1}{q}\mathbf{J}_q) \otimes \frac{1}{m}\mathbf{J}_m \otimes \frac{1}{r}\mathbf{J}_r],$$

$$\mathbf{A}_2 = \left(\frac{1}{\sigma^2 + r\sigma_b^2}\right)[\mathbf{I}_q \otimes (\mathbf{I}_m - \frac{1}{m}\mathbf{J}_m) \otimes \frac{1}{r}\mathbf{J}_r],$$

$$\mathbf{A}_3 = \left(\frac{1}{\sigma^2}\right)[\mathbf{I}_q \otimes \mathbf{I}_m \otimes (\mathbf{I}_r - \frac{1}{r}\mathbf{J}_r)],$$

and

$$\boldsymbol{\Sigma} \equiv \text{var}(\mathbf{y}) = \sigma_a^2(\mathbf{I}_q \otimes \mathbf{J}_m \otimes \mathbf{J}_r) + \sigma_b^2(\mathbf{I}_q \otimes \mathbf{I}_m \otimes \mathbf{J}_r) + \sigma^2(\mathbf{I}_q \otimes \mathbf{I}_m \otimes \mathbf{I}_r).$$

Now,

$$\begin{aligned}
\mathbf{A}_1\boldsymbol{\Sigma} &= \left(\frac{1}{\sigma^2 + r\sigma_b^2 + mr\sigma_a^2}\right)\{\sigma_a^2[(\mathbf{I}_q - \frac{1}{q}\mathbf{J}_q) \otimes \mathbf{J}_m \otimes \mathbf{J}_r] \\
&\quad + \sigma_b^2[(\mathbf{I}_q - \frac{1}{q}\mathbf{J}_q) \otimes \frac{1}{m}\mathbf{J}_m \otimes \mathbf{J}_r] \\
&\quad + \sigma^2[(\mathbf{I}_q - \frac{1}{q}\mathbf{J}_q) \otimes \frac{1}{m}\mathbf{J}_m \otimes \frac{1}{r}\mathbf{J}_r] \\
&= (\mathbf{I}_q - \frac{1}{q}\mathbf{J}_q) \otimes \frac{1}{m}\mathbf{J}_m \otimes \frac{1}{r}\mathbf{J}_r,
\end{aligned}$$

$$\mathbf{A}_2\boldsymbol{\Sigma} = \left(\frac{1}{\sigma^2 + r\sigma_b^2}\right)\{\sigma_a^2[\mathbf{I}_q \otimes \mathbf{0}_{m\times m} \otimes \mathbf{J}_r] + \sigma_b^2[\mathbf{I}_q \otimes (\mathbf{I}_m - \frac{1}{m}\mathbf{J}_m) \otimes \mathbf{J}_r]$$
$$+\sigma^2[\mathbf{I}_q \otimes (\mathbf{I}_m - \frac{1}{m}\mathbf{J}_m) \otimes \frac{1}{r}\mathbf{J}_r]\}$$
$$= \mathbf{I}_q \otimes (\mathbf{I}_m - \frac{1}{m}\mathbf{J}_m) \otimes \frac{1}{r}\mathbf{J}_r,$$
$$\mathbf{A}_3\boldsymbol{\Sigma} = \left(\frac{1}{\sigma^2}\right)\{\sigma_a^2[\mathbf{I}_q \otimes \mathbf{J}_m \otimes \mathbf{0}_{r\times r}] + \sigma_b^2[\mathbf{I}_q \otimes \mathbf{I}_m \otimes \mathbf{0}_{r\times r}]$$
$$+\sigma^2[\mathbf{I}_q \otimes \mathbf{I}_m \otimes (\mathbf{I}_r - \frac{1}{r}\mathbf{J}_r)]\}$$
$$= \mathbf{I}_q \otimes \mathbf{I}_m \otimes (\mathbf{I}_r - \frac{1}{r}\mathbf{J}_r).$$

Observe that $\mathbf{A}_1\boldsymbol{\Sigma}$, $\mathbf{A}_2\boldsymbol{\Sigma}$, and $\mathbf{A}_3\boldsymbol{\Sigma}$ are idempotent (because each term in each Kronecker product is idempotent). Also observe that $\text{rank}(\mathbf{A}_1) = q-1$, $\text{rank}(\mathbf{A}_2) = q(m-1)$, and $\text{rank}(\mathbf{A}_3) = qm(r-1)$ by Theorem 2.17.8, and that $\mathbf{A}_i\mathbf{1}\mu = \mu\mathbf{A}_i(\mathbf{1}_q \otimes \mathbf{1}_m \otimes \mathbf{1}_r) = \mathbf{0}$ for $i = 1, 2, 3$. Thus, by Corollary 14.2.6.1, $f_1S_1^2 \sim \chi^2(q-1)$, $f_2S_2^2 \sim \chi^2[q(m-1)]$, and $f_3S_3^2 \sim \chi^2[qm(r-1)]$. Moreover,

$$\mathbf{A}_1\boldsymbol{\Sigma}\mathbf{A}_2 = \left(\frac{1}{\sigma^2 + r\sigma_b^2}\right)[(\mathbf{I}_q - \frac{1}{q}\mathbf{J}_q) \otimes \mathbf{0}_{m\times m} \otimes \frac{1}{r}\mathbf{J}_r] = \mathbf{0},$$
$$\mathbf{A}_1\boldsymbol{\Sigma}\mathbf{A}_3 = \left(\frac{1}{\sigma^2}\right)[(\mathbf{I}_q - \frac{1}{q}\mathbf{J}_q) \otimes \frac{1}{m}\mathbf{J}_m \otimes \mathbf{0}_{r\times r}] = \mathbf{0},$$
$$\mathbf{A}_2\boldsymbol{\Sigma}\mathbf{A}_3 = \left(\frac{1}{\sigma^2}\right)[\mathbf{I}_q \otimes (\mathbf{I}_m - \frac{1}{m}\mathbf{J}_m) \otimes \mathbf{0}_{r\times r}] = \mathbf{0},$$

so by Corollary 14.3.1.1 S_1^2, S_2^2, and S_3^2 are independent.

(b) $\text{E}\left(\frac{qm(r-1)}{\sigma^2}S_3^2\right) = qm(r-1)$, so $\hat{\sigma}^2 = S_3^2$; $\text{E}\left(\frac{q(m-1)}{\sigma^2+r\sigma_b^2}S_2^2\right) = q(m-1)$, so $\hat{\sigma}^2 + r\hat{\sigma}_b^2 = S_2^2$, i.e., $\hat{\sigma}_b^2 = (S_2^2 - S_3^2)/r$; $\text{E}\left(\frac{q-1}{\sigma^2+r\sigma_b^2+mr\sigma_a^2}S_1^2\right) = q-1$, so $\hat{\sigma}^2 + r\hat{\sigma}_b^2 + mr\hat{\sigma}_a^2 = S_1^2$, i.e., $\hat{\sigma}_a^2 = (S_1^2 - S_2^2)/mr$.

(c)

$$1 - \xi = \Pr\left(\chi^2_{1-(\xi/2),qm(r-1)} \le \frac{qm(r-1)S_3^2}{\sigma^2} \le \chi^2_{\xi/2,qm(r-1)}\right)$$
$$= \Pr\left(\frac{qm(r-1)S_3^2}{\chi^2_{\xi/2,qm(r-1)}} \le \sigma^2 \le \frac{qm(r-1)S_3^2}{\chi^2_{1-(\xi/2),qm(r-1)}}\right),$$

so $\left[\frac{qm(r-1)S_3^2}{\chi^2_{\xi/2,qm(r-1)}}, \frac{qm(r-1)S_3^2}{\chi^2_{1-(\xi/2),qm(r-1)}}\right]$ is a $100(1-\xi)\%$ confidence interval for σ^2.

(d) Define

$$U = \frac{S_1^2}{S_3^2}, \quad U^* = \frac{S_1^2/(\sigma^2 + r\sigma_b^2 + mr\sigma_a^2)}{S_3^2/\sigma^2}, \quad \text{and} \quad \psi = \frac{\sigma_b^2 + m\sigma_a^2}{\sigma^2}.$$

Observe that $U^* \sim \mathrm{F}[q-1, qm(r-1)]$. Therefore,

$$\begin{aligned} 1-\xi &= \Pr\left(F_{1-(\xi/2),q-1,qm(r-1)} \leq \frac{\sigma^2}{\sigma^2 + r\sigma_b^2 + mr\sigma_a^2} U \leq F_{\xi/2,q-1,qm(r-1)}\right) \\ &= \Pr\left(\frac{1}{F_{\xi/2,q-1,qm(r-1)}} \leq \frac{1}{1+r\psi} U \leq \frac{1}{F_{1-(\xi/2),q-1,qm(r-1)}}\right) \\ &= \Pr\left(\frac{(U/F_{\xi/2,q-1,qm(r-1)})-1}{r} \leq \psi \leq \frac{(U/F_{1-(\xi/2),q-1,qm(r-1)})-1}{r}\right). \end{aligned}$$

Therefore,

$$\left[\frac{(U/F_{\xi/2,q-1,qm(r-1)})-1}{r}, \frac{(U/F_{1-(\xi/2),q-1,qm(r-1)})-1}{r}\right]$$

is a $100(1-\xi)\%$ confidence interval for $\frac{\sigma_b^2+m\sigma_a^2}{\sigma^2}$. If any endpoints of the interval are negative, they can be set equal to zero.

(e) $\sigma_a^2 = [(\sigma^2 + r\sigma_b^2 + mr\sigma_a^2) - (\sigma^2 + r\sigma_b^2)]/mr \equiv (\alpha_1^2/mr) - (\alpha_2^2/mr)$. Thus, by Satterthwaite's method an approximate $100(1-\xi)\%$ confidence interval for σ_a^2 is given by

$$\left[\frac{\hat{t}\left(\frac{S_1^2}{mr} - \frac{S_2^2}{mr}\right)}{\chi^2_{\xi/2,\hat{t}}}, \frac{\hat{t}\left(\frac{S_1^2}{mr} - \frac{S_2^2}{mr}\right)}{\chi^2_{1-(\xi/2),\hat{t}}}\right]$$

where

$$\hat{t} = \frac{(S_1^2 - S_2^2)^2}{\frac{S_1^4}{q-1} + \frac{S_2^4}{q(m-1)}}.$$

(f) $\frac{S_1^2/(\sigma^2+r\sigma_b^2+mr\sigma_a^2)}{S_2^2/(\sigma^2+r\sigma_b^2)} \sim \mathrm{F}[q-1, q(m-1)]$ so $\frac{S_1^2}{S_2^2} \sim \frac{\sigma^2+r\sigma_b^2+mr\sigma_a^2}{\sigma^2+r\sigma_b^2}\mathrm{F}[q-1, q(m-1)]$. Thus, a size-$\xi$ test of $H_0 : \sigma_a^2 = 0$ versus $H_a : \sigma_a^2 > 0$ is to reject H_0 if and only if $\frac{S_1^2}{S_2^2} > F_{\xi,q-1,q(m-1)}$. Similarly, $\frac{S_2^2/(\sigma^2+r\sigma_b^2)}{S_3^2/\sigma^2} \sim \mathrm{F}[q(m-1), qm(r-1)]$

so $\frac{S_2^2}{S_3^2} \sim \frac{\sigma^2+r\sigma_b^2}{\sigma^2} F[q(m-1), qm(r-1)]$. Thus, a size-$\xi$ test of $H_0 : \sigma_b^2 = 0$ versus $H_a : \sigma_b^2 > 0$ is to reject H_0 if and only if $\frac{S_2^2}{S_3^2} > F_{\xi,q(m-1),qm(r-1)}$.

▶ **Exercise 5** In a certain components-of-variance model, there are four unknown variance components: σ^2, σ_1^2, σ_2^2, and σ_3^2. Suppose that four quadratic forms S_i^2 ($i = 1, 2, 3, 4$) whose expectations are linearly independent linear combinations of the variance components are obtained from an ANOVA table. Furthermore, suppose that S_1^2, S_2^2, S_3^2, and S_4^2 are mutually independent and that the following distributional results hold:

$$\frac{f_1 S_1^2}{\sigma^2 + \sigma_1^2 + \sigma_2^2} \sim \chi^2(f_1)$$

$$\frac{f_2 S_2^2}{\sigma^2 + \sigma_1^2 + \sigma_3^2} \sim \chi^2(f_2)$$

$$\frac{f_3 S_3^2}{\sigma^2 + \sigma_2^2 + \sigma_3^2} \sim \chi^2(f_3)$$

$$\frac{f_4 S_4^2}{\sigma^2} \sim \chi^2(f_4)$$

Let $\xi \in (0, 1)$.

(a) Obtain quadratic unbiased estimators of the four variance components.
(b) Obtain a $100(1-\xi)\%$ confidence interval for $\sigma^2 + \sigma_1^2 + \sigma_2^2$.
(c) Obtain a confidence interval for $\sigma_2^2 - \sigma_3^2$, which has coverage probability at least $1-\xi$. (Hint: Using the Bonferroni inequality, combine the interval from part [b] with a similarly constructed interval for another quantity.)
(d) Use Satterthwaite's method to obtain an approximate $100(1-\xi)\%$ confidence interval for σ_1^2.

Solution

(a)

$$E(S_1^2) = \sigma^2 + \sigma_1^2 + \sigma_2^2, \quad E(S_2^2) = \sigma^2 + \sigma_1^2 + \sigma_3^2,$$
$$E(S_3^2) = \sigma^2 + \sigma_2^2 + \sigma_3^2, \quad E(S_4^2) = \sigma^2.$$

So quadratic unbiased estimators of the variance components can be obtained by solving the equations:

$$S_1^2 = \sigma^2 + \sigma_1^2 + \sigma_2^2, \quad S_2^2 = \sigma^2 + \sigma_1^2 + \sigma_3^2,$$
$$S_3^2 = \sigma^2 + \sigma_2^2 + \sigma_3^2, \quad S_4^2 = \sigma^2.$$

A solution will satisfy

$$\hat{\sigma}^2 = S_4^2, \quad \hat{\sigma}_1^2 + \hat{\sigma}_2^2 = S_1^2 - S_4^2, \quad \hat{\sigma}_2^2 + \hat{\sigma}_3^2 = S_3^2 - S_4^2,$$
$$\hat{\sigma}_2^2 - \hat{\sigma}_3^2 = S_1^2 - S_2^2, \quad \hat{\sigma}_1^2 - \hat{\sigma}_2^2 = S_2^2 - S_3^2.$$

The first of these equations yields the estimator of σ^2; the second and fifth equations yield $\hat{\sigma}_1^2 = (S_1^2 + S_2^2 - S_3^2 - S_4^2)/2$; the third and fourth equations yield $\hat{\sigma}_2^2 = (S_1^2 - S_2^2 + S_3^2 - S_4^2)/2$; and back-solving yields $\hat{\sigma}_3^2 = (-S_1^2 + S_2^2 + S_3^2 - S_4^2)/2$.

(b)

$$1 - \xi = \Pr\left(\chi^2_{1-(\xi/2), f_1} \le \frac{f_1 S_1^2}{\sigma^2 + \sigma_1^2 + \sigma_2^2} \le \chi^2_{\xi/2, f_1}\right)$$
$$= \Pr\left(L_\xi \le \sigma^2 + \sigma_1^2 + \sigma_2^2 \le R_\xi\right)$$

where $L_\xi = \frac{f_1 S_1^2}{\chi^2_{\xi/2, f_1}}$ and $R_\xi = \frac{f_1 S_1^2}{\chi^2_{1-(\xi/2), f_1}}$. Thus $[L_\xi, R_\xi]$ is a $100(1-\xi)\%$ confidence interval for $\sigma^2 + \sigma_1^2 + \sigma_2^2$.

(c) Using the same method as in part (b), we obtain

$$1 - \xi = \Pr\left(L_\xi^* \le \sigma^2 + \sigma_1^2 + \sigma_3^2 \le R_\xi^*\right)$$

where $L_\xi^* = \frac{f_2 S_2^2}{\chi^2_{\xi/2, f_2}}$ and $R_\xi^* = \frac{f_2 S_2^2}{\chi^2_{1-(\xi/2), f_2}}$. Thus, by Bonferroni's inequality,

$$1 - \xi \le \Pr(L_{\xi/2} \le \sigma^2 + \sigma_1^2 + \sigma_2^2 \le R_{\xi/2} \text{ and } L_{\xi/2}^* \le \sigma^2 + \sigma_1^2 + \sigma_3^2 \le R_{\xi/2}^*)$$
$$= \Pr(L_{\xi/2} \le \sigma^2 + \sigma_1^2 + \sigma_2^2 \le R_{\xi/2} \text{ and } - R_{\xi/2}^* \le -\sigma^2 - \sigma_1^2 - \sigma_3^2 \le -L_{\xi/2}^*),$$

implying further (by Fact 15.3) that $\Pr(L_{\xi/2} - R_{\xi/2}^* \le \sigma_2^2 - \sigma_3^2 \le R_{\xi/2} - L_{\xi/2}^*) \ge 1 - \xi$.

(d) $\sigma_1^2 = (1/2)(\sigma^2 + \sigma_1^2 + \sigma_2^2) + (1/2)(\sigma^2 + \sigma_1^2 + \sigma_3^2) - (1/2)(\sigma^2 + \sigma_2^2 + \sigma_3^2) - (1/2)\sigma^2$. Thus, the Satterthwaite-method approximate $100(1-\xi)\%$ confidence interval for σ_1^2 is given by

$$\left[\frac{(1/2)\hat{t}(S_1^2 + S_2^2 - S_3^2 - S_4^2)}{\chi^2_{\xi/2,\hat{t}}}, \frac{(1/2)\hat{t}(S_1^2 + S_2^2 - S_3^2 - S_4^2)}{\chi^2_{1-(\xi/2),\hat{t}}}\right]$$

where

$$\hat{t} = \frac{(S_1^2 + S_2^2 - S_3^2 - S_4^2)^2}{(S_1^4/f_1) + (S_2^4/f_2) + (S_3^4/f_3) + (S_4^4/f_4)}.$$

If $S_1^2 + S_2^2 - S_3^2 - S_4^2 < 0$, then set both endpoints equal to 0.

▶ **Exercise 6** Consider once again the normal one-factor components-of-variance model with balanced data that was considered in Examples 16.1.3-2 and 16.1.4-1. Recall that $\hat{\sigma}_b^2 = (S_2^2 - S_3^2)/r$ is the minimum variance quadratic unbiased estimator of σ_b^2 under this model. Suppose that $\sigma_b^2/\sigma^2 = 1$.

(a) Express $\Pr(\hat{\sigma}_b^2 < 0)$ as $\Pr(F < c)$, where F is a random variable having an $\mathrm{F}(\nu_1, \nu_2)$ distribution and c is a constant. Determine ν_1, ν_2, and c in as simple a form as possible.
(b) Using the `pf` function in R, obtain a numerical value for $\Pr(\hat{\sigma}_b^2 < 0)$ when $q = 5$ and $r = 4$.

Solution

(a) By Definition 14.5.1,

$$\begin{aligned}\Pr(\hat{\sigma}_b^2 < 0) &= \Pr[(S_2^2 - S_3^2)/r < 0) = \Pr(S_2^2 < S_3^2) = \Pr(S_2^2/S_3^2 < 1) \\ &= \Pr\left(\frac{S_2^2/(r\sigma_b^2 + \sigma^2)}{S_3^2/\sigma^2} < \frac{\sigma^2}{r\sigma_b^2 + \sigma^2}\right) = \Pr(F < c)\end{aligned}$$

where $F \sim \mathrm{F}[q-1, q(r-1)]$ and $c = \frac{\sigma^2}{r\sigma_b^2+\sigma^2} = \frac{1}{r(\sigma_b^2/\sigma^2)+1} = \frac{1}{r+1}$.

(b) Using the `pf` function in R, we find, by specializing the result of part (a) to the case $q = 5$ and $r = 4$, that

$$\Pr(\hat{\sigma}_b^2 < 0) = \Pr[\mathrm{F}(4, 15) < 1/5] \doteq 0.066.$$

▶ **Exercise 7** Consider once again the normal one-factor components-of-variance model with balanced data that was considered in Examples 16.1.3-2 and 16.1.4-1. Simulate 10,000 response vectors $\mathbf{y}$ that follow the special case of such a model in which $q = 4$, $r = 5$, and $\sigma_b^2 = \sigma^2 = 1$, and obtain the following:

(a) the proportion of simulated vectors for which the quadratic unbiased estimator of σ_b^2 given in Example 16.1.3-2 is negative;
(b) the average width and empirical coverage probability of the approximate 95% confidence interval for σ_b^2 derived in Example 16.1.4-1 using Satterthwaite's approach;
(c) the average width and empirical coverage probability of the approximate 95% confidence intervals for σ_b^2 derived in Example 16.1.4-1 using Williams' approach.

Based on your results for parts (b) and (c), which of the two approaches for obtaining an approximate 95% confidence interval for σ_b^2 performs best? (Note: In parts [b] and [c], if any endpoints of the intervals are negative, you should modify your code to set them equal to 0.)

Solution

(a) Relevant R code and the results obtained using that code are as follows:

```
library(magic)
set.seed(10)
q <- 4; r <- 5; sig_b2 <- 1; sig_2 <- 1
result <- replicate(1e4, {y <- rnorm(q*r,0,sqrt(sig_2)) +
rep(rnorm(q,0,sqrt(sig_b2)), each = r);
S2 <- t(y) %*% (do.call(adiag, replicate(q,matrix(1/r,r,r),simplify = FALSE))
 - matrix(1/(q*r), q*r, q*r)) %*% y/(q-1);
S3 <- t(y) %*% (diag(q*r) - do.call(adiag,
replicate(q,matrix(1/r,r,r),simplify = FALSE))) %*% y/(q*(r-1));
sigmabhat <- (S2 - S3)/r;
},simplify = TRUE)
> mean(result < 0)
[1] 0.0851
```

In this set of simulations, the quadratic unbiased estimate of σ_b^2 obtained in Example 16.1.3-2 was negative in 8.51% of the simulations.

(b) Relevant R code and the results obtained using that code are as follows:

```
xi <- 0.05
result <- replicate(1e4, {
y <- rnorm(q*r,0,sqrt(sig_2)) + rep(rnorm(q,0,sqrt(sig_b2)), each = r);
S2 <- t(y) %*% (do.call(adiag, replicate(q,matrix(1/r,r,r),simplify = FALSE))
 - matrix(1/(q*r), q*r, q*r)) %*% y/(q-1);
S3 <- t(y) %*% (diag(q*r) -
do.call(adiag, replicate(q,matrix(1/r,r,r),simplify = FALSE)))
 %*% y/(q*(r-1));
dhat <- (S2 - S3)^2/(S2^2/(q-1) + S3^2/(q*(r-1)));
c(max(dhat*(S2 - S3)/(r*qchisq(1-xi/2,df = dhat)),0),
 max(dhat*(S2 - S3)/(r*qchisq(xi/2,df = dhat)),0));
},simplify = TRUE)

> mean((result[2,] >=1) & (result[1,] <= 1))
[1] 0.8815
> mean(result[2,] -result[1,], na.rm = TRUE)
[1] Inf
> mean((result[2,] -result[1,])[is.finite(result[2,] -result[1,])])
[1] 1.063704e+301
```

From this set of simulations, we can make the following observations:

1. The empirical coverage probability of the Satterthwaite approximate confidence interval was less than the nominal level by a substantial amount (88.15% compared to 95%).
2. The average width of the Satterthwaite approximate confidence interval was very large (even after removing Inf values). There is a small probability that simulated data will lead to a very small estimated degrees of freedom $\widehat{t}$, for which the corresponding denominators $\chi^2_{0.025,\hat{t}}$ and $\chi^2_{0.975,\hat{t}}$ are very small, producing very large endpoints. In this sense, the Satterthwaite's approximate confidence interval is not very stable.

(c) Relevant R code and the results obtained using that code are as follows:

```
xi <- 0.025
result <- replicate(1e4, {
y <- rnorm(q*r,0,sqrt(sig_2)) + rep(rnorm(q,0,sqrt(sig_b2)), each = r);
S2 <- t(y) %*% (do.call(adiag, replicate(q,matrix(1/r,r,r),simplify = FALSE))
 - matrix(1/(q*r), q*r, q*r)) %*% y/(q-1);
S3 <- t(y) %*% (diag(q*r)
- do.call(adiag, replicate(q,matrix(1/r,r,r),simplify = FALSE)))
 %*% y/(q*(r-1));
L <- max((q-1)*(S2-S3*qf(xi/2, q-1,q*(r-1),lower.tail = FALSE))
/(r*qchisq(xi/2, q-1,lower.tail = FALSE)),0);
R <- max((q-1)*(S2-S3*qf(xi/2, q-1,q*(r-1),lower.tail = TRUE))
/(r*qchisq(xi/2, q-1,lower.tail = TRUE)),0);
c(L, R);
},simplify = TRUE)

> mean((result[2,] >=1) & (result[1,] <= 1))
[1] 0.9789
> mean(result[2,] -result[1,])
[1] 26.20318
```

From the R output for these simulations, we can observe that the empirical coverage probability of this confidence interval was higher than the nominal level (97.89% compared to 95%); furthermore, the average length is reasonable. Together with the results from part (b), it appears that Williams' approach to constructing an approximate $100(1-\xi)\%$ confidence interval for σ_b^2 performs better than Satterthwaite's approach.

▶ **Exercise 8** Consider the normal general mixed linear model $\{\mathbf{y}, \mathbf{X}\boldsymbol{\beta}, \mathbf{V}(\boldsymbol{\theta})\}$ with model matrix

$$\mathbf{X} = \begin{pmatrix} 1 & 0 \\ 1 & 0 \\ 1 & 0 \\ 0 & 1 \\ 0 & 1 \end{pmatrix}.$$

Specify, in terms of the elements $y_1, \ldots, y_5$ of $\mathbf{y}$, the elements of a vector $\mathbf{w}$ of error contrasts (of appropriate dimension) whose likelihood, when maximized over the parameter space for $\boldsymbol{\theta}$, yields a REML estimate of $\boldsymbol{\theta}$.

Solution $n = 5$ and $p^* = 2$ so we require $5 - 2 = 3$ linearly independent error contrasts. One of many possibilities is $\mathbf{w} = (y_1 - y_2, y_1 - y_3, y_4 - y_5)$.

▶ **Exercise 9** Consider the normal Gauss–Markov model with $n > p^*$, and define $\hat{\sigma}^2$ and $\bar{\sigma}^2$ as in (16.13) and (16.12), respectively.

(a) Show that $\hat{\sigma}^2$ is the REML estimator of σ^2.
(b) Show that the mean square error of $\hat{\sigma}^2$ is less than or equal to that of $\bar{\sigma}^2$ if $p^* > 5$ and $n > p^*(p^* - 2)/(p^* - 4)$, and that if $p^* \geq 13$, then $n > p^* + 2$ suffices.

Solution

(a) Substituting $\sigma^2\mathbf{I}$ for $\mathbf{V}$ in Theorem 16.2.3 yields the REML log-likelihood function

$$\begin{aligned}
L_1(\sigma^2; \mathbf{y}) &= -\frac{1}{2}\log|\sigma^2\mathbf{I}| - \frac{1}{2}\log|\breve{\mathbf{X}}^T(\sigma^2\mathbf{I})^{-1}\breve{\mathbf{X}}| \\
&\quad -\frac{1}{2}\mathbf{y}^T\{(\sigma^2\mathbf{I})^{-1} - (\sigma^2\mathbf{I})^{-1}\mathbf{X}[\mathbf{X}^T(\sigma^2\mathbf{I})^{-1}\mathbf{X}]^-\mathbf{X}^T(\sigma^2\mathbf{I})^{-1}\}\mathbf{y} \\
&= -\frac{n}{2}\log\sigma^2 + \frac{p^*}{2}\log\sigma^2 - \frac{1}{2}\log|\breve{\mathbf{X}}^T\breve{\mathbf{X}}| - \frac{1}{2\sigma^2}\mathbf{y}^T(\mathbf{I} - \mathbf{P_X})\mathbf{y} \\
&= -\frac{n - p^*}{2}\log\sigma^2 - \frac{1}{2\sigma^2}\mathbf{y}^T(\mathbf{I} - \mathbf{P_X})\mathbf{y} - \frac{1}{2}\log|\breve{\mathbf{X}}^T\breve{\mathbf{X}}|.
\end{aligned}$$

Thus, the REML equation is

$$\frac{dL_1}{d\sigma^2} = -\frac{1}{2}\left(\frac{n - p^*}{\sigma^2} - \frac{1}{(\sigma^2)^2}\mathbf{y}^T(\mathbf{I} - \mathbf{P_X})\mathbf{y}\right) = 0.$$

The unique solution to this equation is given by

$$\hat{\sigma}^2 = \frac{\mathbf{y}^T(\mathbf{I} - \mathbf{P_X})\mathbf{y}}{n - p^*}.$$

Observe further that

$$\frac{dL_1}{d\sigma^2} = -\frac{n - p^*}{2(\sigma^2)^2}\left(\sigma^2 - \frac{\mathbf{y}^T(\mathbf{I} - \mathbf{P_X})\mathbf{y}}{n - p^*}\right) = \begin{cases} > 0 \text{ if } \sigma^2 < \frac{\mathbf{y}^T(\mathbf{I}-\mathbf{P_X})\mathbf{y}}{n-p^*} \\ = 0 \text{ if } \sigma^2 = \frac{\mathbf{y}^T(\mathbf{I}-\mathbf{P_X})\mathbf{y}}{n-p^*} \\ < 0 \text{ if } \sigma^2 > \frac{\mathbf{y}^T(\mathbf{I}-\mathbf{P_X})\mathbf{y}}{n-p^*}. \end{cases}$$

Therefore, $\hat{\sigma}^2$ is the unique global maximizer of $L_1(\sigma^2; \mathbf{y})$, so it is the REML estimator of σ^2.

(b) $E(\hat{\sigma}^2) = \sigma^2$, $E(\bar{\sigma}^2) = \sigma^2(1 - \frac{p^*}{n})$ as shown immediately after (16.13), $\text{var}(\hat{\sigma}^2) = \frac{2\sigma^4}{n-p^*}$ according to Theorem 8.2.3, and

$$\text{var}(\bar{\sigma}^2) = \text{var}\left(\frac{n-p^*}{n}\hat{\sigma}^2\right) = \frac{2(n-p^*)\sigma^4}{n^2}.$$

Therefore,

$$\begin{aligned} MSE(\bar{\sigma}^2) - MSE(\hat{\sigma}^2) &= \text{var}(\bar{\sigma}^2) + [E(\bar{\sigma}^2) - \sigma^2]^2 - \text{var}(\hat{\sigma}^2) - [E(\hat{\sigma}^2) - \sigma^2]^2 \\ &= \frac{2(n-p^*)\sigma^4}{n^2} + \left(-\frac{p^*\sigma^2}{n}\right)^2 - \frac{2\sigma^4}{n-p^*} \\ &= \frac{\sigma^4}{n^2(n-p^*)}[2(n-p^*)^2 + (p^*)^2(n-p^*) - 2n^2] \\ &= \frac{\sigma^4 p^*}{n^2(n-p^*)}[n(p^*-4) - p^*(p^*-2)]. \end{aligned}$$

If $p^* > 5$ and $n > p^*(p^*-2)/(p^*-4)$, then $n > p^*$ and

$$MSE(\bar{\sigma}^2) - MSE(\hat{\sigma}^2) = \frac{\sigma^4 p^*(p^*-4)}{n^2(n-p^*)}[n - p^*(p^*-2)/(p^*-4)] \geq 0,$$

hence $MSE(\hat{\sigma}^2) \leq MSE(\bar{\sigma}^2)$. If $p^* \geq 13$ and $n > p^* + 2$, then

$$n(p^*-4) \geq (p^*+3)(p^*-4) = (p^*)^2 - p^* - 12 > (p^*)^2 - 2p^* = p^*(p^*-2),$$

implying that $n > \frac{p^*(p^*-2)}{p^*-4}$, i.e., that the first condition is satisfied. Thus $MSE(\bar{\sigma}^2) - MSE(\hat{\sigma}^2) \geq 0$.

▶ **Exercise 10** Show that, under the general mixed linear model, solving the REML equations

$$\frac{\partial L_1}{\partial \theta_i} = 0 \quad (i = 1, \cdots, m)$$

is equivalent to equating m quadratic forms in $\mathbf{y}$ to their expectations. Also show that the profile likelihood equations, defined as

$$\frac{\partial L_0}{\partial \theta_i} = 0 \quad (i = 1, \cdots, m),$$

do not have this property.

Solution Using (16.26), the equation $\frac{\partial L_1}{\partial \theta_i} = 0$ is

$$-\frac{1}{2}\text{tr}\left(\mathbf{E}\frac{\partial \mathbf{V}}{\partial \theta_i}\right) + \frac{1}{2}\mathbf{y}^T\mathbf{E}\left(\frac{\partial \mathbf{V}}{\partial \theta_i}\right)\mathbf{E}\mathbf{y} = 0,$$

which is equivalent to

$$\mathbf{y}^T\mathbf{E}\left(\frac{\partial \mathbf{V}}{\partial \theta_i}\right)\mathbf{E}\mathbf{y} = \text{tr}\left(\mathbf{E}\frac{\partial \mathbf{V}}{\partial \theta_i}\right).$$

Now by Theorems 4.2.4, 2.10.3, and 11.1.6d, we obtain

$$\begin{aligned}
\text{E}\left[\mathbf{y}^T\mathbf{E}\left(\frac{\partial \mathbf{V}}{\partial \theta_i}\right)\mathbf{E}\mathbf{y}\right] &= \boldsymbol{\beta}^T\mathbf{X}^T\mathbf{E}\left(\frac{\partial \mathbf{V}}{\partial \theta_i}\right)\mathbf{E}\mathbf{X}\boldsymbol{\beta} + \text{tr}\left[\mathbf{E}\left(\frac{\partial \mathbf{V}}{\partial \theta_i}\right)\mathbf{E}\mathbf{V}\right] \\
&= 0 + \text{tr}\left(\frac{\partial \mathbf{V}}{\partial \theta_i}\mathbf{E}\mathbf{V}\mathbf{E}\right) \\
&= \text{tr}\left(\mathbf{E}\frac{\partial \mathbf{V}}{\partial \theta_i}\right).
\end{aligned}$$

This establishes the equivalence between the REML equations and equating m quadratic forms in $\mathbf{y}$ to their expectations. In contrast, by (16.21) the ith of the profile likelihood equations is

$$-\frac{1}{2}\text{tr}\left(\mathbf{V}^{-1}\frac{\partial \mathbf{V}}{\partial \theta_i}\right) + \frac{1}{2}\mathbf{y}^T\mathbf{E}\left(\frac{\partial \mathbf{V}}{\partial \theta_i}\right)\mathbf{E}\mathbf{y} = 0,$$

or equivalently

$$\text{tr}\left(\mathbf{V}^{-1}\frac{\partial \mathbf{V}}{\partial \theta_i}\right) = \mathbf{y}^T\mathbf{E}\left(\frac{\partial \mathbf{V}}{\partial \theta_i}\right)\mathbf{E}\mathbf{y}.$$

The left-hand side of this equation is not the expectation of the right-hand side.

▶ **Exercise 11** Verify the expressions for $\frac{\partial L_0}{\partial \theta_i}$, $\frac{\partial^2 L_0}{\partial \theta_i \partial \theta_j}$, and $\text{E}\left(\frac{\partial^2 L_0}{\partial \theta_i \partial \theta_j}\right)$ given by (16.21)–(16.25).

Solution First, using Theorems 2.21.1 and 11.1.6c we obtain $\frac{\partial \mathbf{E}}{\partial \theta_i}$ (for $i = 1, \ldots, m$):

$$\frac{\partial \mathbf{E}}{\partial \theta_i} = \frac{\partial \mathbf{V}^{-1}}{\partial \theta_i} - \frac{\partial[\mathbf{V}^{-1}\breve{\mathbf{X}}(\breve{\mathbf{X}}^T\mathbf{V}^{-1}\breve{\mathbf{X}})^{-1}\breve{\mathbf{X}}^T\mathbf{V}^{-1}]}{\partial \theta_i}$$

$$= \frac{\partial \mathbf{V}^{-1}}{\partial \theta_i} - \left(\frac{\partial \mathbf{V}^{-1}}{\partial \theta_i}\right)\breve{\mathbf{X}}(\breve{\mathbf{X}}^T\mathbf{V}^{-1}\breve{\mathbf{X}})^{-1}\breve{\mathbf{X}}^T\mathbf{V}^{-1} - \mathbf{V}^{-1}\breve{\mathbf{X}}\left[\frac{\partial(\breve{\mathbf{X}}^T\mathbf{V}^{-1}\breve{\mathbf{X}})^{-1}}{\partial \theta_i}\right]\breve{\mathbf{X}}^T\mathbf{V}^{-1}$$

$$-\mathbf{V}^{-1}\breve{\mathbf{X}}(\breve{\mathbf{X}}^T\mathbf{V}^{-1}\breve{\mathbf{X}})^{-1}\breve{\mathbf{X}}^T\left(\frac{\partial \mathbf{V}^{-1}}{\partial \theta_i}\right)$$

$$= -\mathbf{V}^{-1}\left(\frac{\partial \mathbf{V}}{\partial \theta_i}\right)\mathbf{V}^{-1} + \mathbf{V}^{-1}\left(\frac{\partial \mathbf{V}}{\partial \theta_i}\right)\mathbf{V}^{-1}\breve{\mathbf{X}}(\breve{\mathbf{X}}^T\mathbf{V}^{-1}\breve{\mathbf{X}})^{-1}\breve{\mathbf{X}}^T\mathbf{V}^{-1}$$

$$+\mathbf{V}^{-1}\breve{\mathbf{X}}(\breve{\mathbf{X}}^T\mathbf{V}^{-1}\breve{\mathbf{X}})^{-1}\breve{\mathbf{X}}^T\left(\frac{\partial \mathbf{V}^{-1}}{\partial \theta_i}\right)\breve{\mathbf{X}}(\breve{\mathbf{X}}^T\mathbf{V}^{-1}\breve{\mathbf{X}})^{-1}\breve{\mathbf{X}}^T\mathbf{V}^{-1}$$

$$+\mathbf{V}^{-1}\breve{\mathbf{X}}(\breve{\mathbf{X}}^T\mathbf{V}^{-1}\breve{\mathbf{X}})^{-1}\breve{\mathbf{X}}^T\mathbf{V}^{-1}\left(\frac{\partial \mathbf{V}}{\partial \theta_i}\right)\mathbf{V}^{-1}$$

$$= -\mathbf{V}^{-1}\left(\frac{\partial \mathbf{V}}{\partial \theta_i}\right)\mathbf{E} + \mathbf{V}^{-1}\breve{\mathbf{X}}(\breve{\mathbf{X}}^T\mathbf{V}^{-1}\breve{\mathbf{X}})^{-1}\breve{\mathbf{X}}^T\mathbf{V}^{-1}\left(\frac{\partial \mathbf{V}}{\partial \theta_i}\right)\mathbf{E}$$

$$= -\mathbf{E}\left(\frac{\partial \mathbf{V}}{\partial \theta_i}\right)\mathbf{E}.$$

Thus, using Theorem 2.21.1 again,

$$\frac{\partial L_0}{\partial \theta_i} = -\frac{1}{2}\mathrm{tr}\left(\mathbf{V}^{-1}\frac{\partial \mathbf{V}}{\partial \theta_i}\right) + \frac{1}{2}\mathbf{y}^T\mathbf{E}\left(\frac{\partial \mathbf{V}}{\partial \theta_i}\right)\mathbf{E}\mathbf{y}.$$

Using Theorem 2.21.1 yet again, along with repeated use of the chain rule, we obtain

$$\begin{aligned}
\frac{\partial^2 L_0}{\partial \theta_i \partial \theta_j} &= \frac{\partial}{\partial \theta_j}\frac{\partial L_0}{\partial \theta_i} \\
&= \frac{\partial}{\partial \theta_j}\left[-\frac{1}{2}\mathrm{tr}\left(\mathbf{V}^{-1}\frac{\partial \mathbf{V}}{\partial \theta_i}\right) + \frac{1}{2}\mathbf{y}^T\mathbf{E}\left(\frac{\partial \mathbf{V}}{\partial \theta_i}\right)\mathbf{E}\mathbf{y}\right] \\
&= -\frac{1}{2}\mathrm{tr}\left[-\mathbf{V}^{-1}\left(\frac{\partial \mathbf{V}}{\partial \theta_j}\right)\mathbf{V}^{-1}\left(\frac{\partial \mathbf{V}}{\partial \theta_i}\right) + \mathbf{V}^{-1}\left(\frac{\partial^2 \mathbf{V}}{\partial \theta_i \partial \theta_j}\right)\right] \\
&\quad -\frac{1}{2}\mathbf{y}^T\mathbf{E}\left(\frac{\partial \mathbf{V}}{\partial \theta_j}\right)\mathbf{E}\left(\frac{\partial \mathbf{V}}{\partial \theta_i}\right)\mathbf{E}\mathbf{y} + \frac{1}{2}\mathbf{y}^T\mathbf{E}\left(\frac{\partial^2 \mathbf{V}}{\partial \theta_i \partial \theta_j}\right)\mathbf{E}\mathbf{y} \\
&\quad -\frac{1}{2}\mathbf{y}^T\mathbf{E}\left(\frac{\partial \mathbf{V}}{\partial \theta_i}\right)\mathbf{E}\left(\frac{\partial \mathbf{V}}{\partial \theta_j}\right)\mathbf{E}\mathbf{y} \\
&= \frac{1}{2}\mathrm{tr}\left[\mathbf{V}^{-1}\left(\frac{\partial \mathbf{V}}{\partial \theta_i}\right)\mathbf{V}^{-1}\left(\frac{\partial \mathbf{V}}{\partial \theta_j}\right)\right] - \frac{1}{2}\mathrm{tr}\left[\mathbf{V}^{-1}\left(\frac{\partial^2 \mathbf{V}}{\partial \theta_i \partial \theta_j}\right)\right] \\
&\quad +\frac{1}{2}\mathbf{y}^T\mathbf{E}\left(\frac{\partial^2 \mathbf{V}}{\partial \theta_i \partial \theta_j}\right)\mathbf{E}\mathbf{y} - \mathbf{y}^T\mathbf{E}\left(\frac{\partial \mathbf{V}}{\partial \theta_i}\right)\mathbf{E}\left(\frac{\partial \mathbf{V}}{\partial \theta_j}\right)\mathbf{E}\mathbf{y}.
\end{aligned}$$

Finally, by Theorems 4.2.4, 11.1.6d, and 2.10.3 we obtain

$$\mathrm{E}\left(\frac{\partial^2 L_0}{\partial\theta_i\partial\theta_j}\right) = \frac{1}{2}\mathrm{tr}\left[\mathbf{V}^{-1}\left(\frac{\partial\mathbf{V}}{\partial\theta_i}\right)\mathbf{V}^{-1}\left(\frac{\partial\mathbf{V}}{\partial\theta_j}\right)\right] - \frac{1}{2}\mathrm{tr}\left[\mathbf{V}^{-1}\left(\frac{\partial^2\mathbf{V}}{\partial\theta_i\partial\theta_j}\right)\right]$$
$$+\frac{1}{2}\mathrm{tr}\left[\mathbf{E}\left(\frac{\partial^2\mathbf{V}}{\partial\theta_i\partial\theta_j}\right)\right] - \mathrm{tr}\left[\mathbf{E}\left(\frac{\partial\mathbf{V}}{\partial\theta_i}\right)\mathbf{E}\left(\frac{\partial\mathbf{V}}{\partial\theta_j}\right)\right].$$

▶ **Exercise 12** Verify the expressions for $\frac{\partial L_1}{\partial\theta_i}$, $\frac{\partial^2 L_1}{\partial\theta_i\partial\theta_j}$, and $\mathrm{E}\left(\frac{\partial^2 L_1}{\partial\theta_i\partial\theta_j}\right)$ given by (16.26)-(16.29).

Solution Writing L_1 as

$$L_1(\boldsymbol{\theta};\,\mathbf{y}) = -\frac{1}{2}\log|\mathbf{V}| - \frac{1}{2}\log|\breve{\mathbf{X}}^T\mathbf{V}^{-1}\breve{\mathbf{X}}| - \frac{1}{2}\mathbf{y}^T\mathbf{E}\mathbf{y}$$

and taking advantage of the expression for $\frac{\partial\mathbf{E}}{\partial\theta_i}$ derived in the solution to Exercise 16.11, we find, using Theorem 2.21.1, that

$$\frac{\partial L_1}{\partial\theta_i} = -\frac{1}{2}\left(\frac{\partial[\log|\mathbf{V}| + \log|\breve{\mathbf{X}}^T\mathbf{V}^{-1}\breve{\mathbf{X}}|]}{\partial\theta_i}\right) - \frac{1}{2}\left(\frac{\partial\mathbf{y}^T\mathbf{E}\mathbf{y}}{\partial\theta_i}\right)$$
$$= -\frac{1}{2}\left[\mathrm{tr}\left(\mathbf{V}^{-1}\frac{\partial\mathbf{V}}{\partial\theta_i}\right) - \mathrm{tr}\{(\breve{\mathbf{X}}^T\mathbf{V}^{-1}\breve{\mathbf{X}})^{-1}\breve{\mathbf{X}}^T\mathbf{V}^{-1}\left(\frac{\partial\mathbf{V}}{\partial\theta_i}\right)\mathbf{V}^{-1}\breve{\mathbf{X}}\} - \mathbf{y}^T\mathbf{E}\left(\frac{\partial\mathbf{V}}{\partial\theta_i}\right)\mathbf{E}\mathbf{y}\right]$$
$$= -\frac{1}{2}\left[\mathrm{tr}\left(\mathbf{V}^{-1}\frac{\partial\mathbf{V}}{\partial\theta_i}\right) - \mathrm{tr}\{\mathbf{V}^{-1}\breve{\mathbf{X}}(\breve{\mathbf{X}}^T\mathbf{V}^{-1}\breve{\mathbf{X}})^{-1}\breve{\mathbf{X}}^T\mathbf{V}^{-1}\left(\frac{\partial\mathbf{V}}{\partial\theta_i}\right)\} - \mathbf{y}^T\mathbf{E}\left(\frac{\partial\mathbf{V}}{\partial\theta_i}\right)\mathbf{E}\mathbf{y}\right]$$
$$= -\frac{1}{2}\mathrm{tr}\left(\mathbf{E}\frac{\partial\mathbf{V}}{\partial\theta_i}\right) + \frac{1}{2}\mathbf{y}^T\mathbf{E}\left(\frac{\partial\mathbf{V}}{\partial\theta_i}\right)\mathbf{E}\mathbf{y}.$$

Using Theorem 2.21.1 again, along with repeated use of the chain rule, we obtain

$$\frac{\partial^2 L_1}{\partial\theta_i\partial\theta_j} = \frac{\partial}{\partial\theta_j}\frac{\partial L_1}{\partial\theta_i}$$
$$= \frac{\partial}{\partial\theta_j}\left[-\frac{1}{2}\mathrm{tr}\left(\mathbf{E}\frac{\partial\mathbf{V}}{\partial\theta_i}\right) + \frac{1}{2}\mathbf{y}^T\mathbf{E}\left(\frac{\partial\mathbf{V}}{\partial\theta_i}\right)\mathbf{E}\mathbf{y}\right]$$
$$= -\frac{1}{2}\mathrm{tr}\left[\left(\frac{\partial\mathbf{E}}{\partial\theta_j}\right)\left(\frac{\partial\mathbf{V}}{\partial\theta_i}\right)\right] - \frac{1}{2}\mathrm{tr}\left[\mathbf{E}\left(\frac{\partial^2\mathbf{V}}{\partial\theta_i\partial\theta_j}\right)\right] + \frac{1}{2}\mathbf{y}^T\left(\frac{\partial\mathbf{E}}{\partial\theta_j}\right)\left(\frac{\partial\mathbf{V}}{\partial\theta_i}\right)\mathbf{E}\mathbf{y}$$
$$+\frac{1}{2}\mathbf{y}^T\mathbf{E}\left(\frac{\partial^2\mathbf{V}}{\partial\theta_i\partial\theta_j}\right)\mathbf{E}\mathbf{y} + \frac{1}{2}\mathbf{y}^T\mathbf{E}\left(\frac{\partial\mathbf{V}}{\partial\theta_i}\right)\left(\frac{\partial\mathbf{E}}{\partial\theta_j}\right)\mathbf{y}$$
$$= \frac{1}{2}\mathrm{tr}\left[\mathbf{E}\left(\frac{\partial\mathbf{V}}{\partial\theta_j}\right)\mathbf{E}\left(\frac{\partial\mathbf{V}}{\partial\theta_i}\right)\right] - \frac{1}{2}\mathrm{tr}\left[\mathbf{E}\left(\frac{\partial^2\mathbf{V}}{\partial\theta_i\partial\theta_j}\right)\right] - \frac{1}{2}\mathbf{y}^T\mathbf{E}\left(\frac{\partial\mathbf{V}}{\partial\theta_j}\right)\mathbf{E}\left(\frac{\partial\mathbf{V}}{\partial\theta_i}\right)\mathbf{E}\mathbf{y}$$

$$+\frac{1}{2}\mathbf{y}^T\mathbf{E}\left(\frac{\partial^2\mathbf{V}}{\partial\theta_i\partial\theta_j}\right)\mathbf{E}\mathbf{y}-\frac{1}{2}\mathbf{y}^T\mathbf{E}\left(\frac{\partial\mathbf{V}}{\partial\theta_i}\right)\mathbf{E}\left(\frac{\partial\mathbf{V}}{\partial\theta_j}\right)\mathbf{E}\mathbf{y}$$
$$=\frac{1}{2}\mathrm{tr}\left[\mathbf{E}\left(\frac{\partial\mathbf{V}}{\partial\theta_j}\right)\mathbf{E}\left(\frac{\partial\mathbf{V}}{\partial\theta_i}\right)\right]-\frac{1}{2}\mathrm{tr}\left[\mathbf{E}\left(\frac{\partial^2\mathbf{V}}{\partial\theta_i\partial\theta_j}\right)\right]+\frac{1}{2}\mathbf{y}^T\mathbf{E}\left(\frac{\partial^2\mathbf{V}}{\partial\theta_i\partial\theta_j}\right)\mathbf{E}\mathbf{y}$$
$$-\mathbf{y}^T\mathbf{E}\left(\frac{\partial\mathbf{V}}{\partial\theta_i}\right)\mathbf{E}\left(\frac{\partial\mathbf{V}}{\partial\theta_j}\right)\mathbf{E}\mathbf{y}.$$

Finally, by Theorems 4.2.4, 11.1.6d, and 2.10.3 we obtain

$$\mathrm{E}\left(\frac{\partial^2 L_1}{\partial\theta_i\partial\theta_j}\right)=\frac{1}{2}\mathrm{tr}\left[\mathbf{E}\left(\frac{\partial\mathbf{V}}{\partial\theta_j}\right)\mathbf{E}\left(\frac{\partial\mathbf{V}}{\partial\theta_i}\right)\right]-\frac{1}{2}\mathrm{tr}\left[\mathbf{E}\left(\frac{\partial^2\mathbf{V}}{\partial\theta_i\partial\theta_j}\right)\right]$$
$$+\frac{1}{2}\left[\boldsymbol{\beta}^T\mathbf{X}^T\mathbf{E}\left(\frac{\partial^2\mathbf{V}}{\partial\theta_i\partial\theta_j}\right)\mathbf{E}\mathbf{X}\boldsymbol{\beta}\right]$$
$$+\frac{1}{2}\mathrm{tr}\left[\mathbf{E}\left(\frac{\partial^2\mathbf{V}}{\partial\theta_i\partial\theta_j}\right)\mathbf{E}\mathbf{V}\right]-\boldsymbol{\beta}^T\mathbf{X}^T\mathbf{E}\left(\frac{\partial\mathbf{V}}{\partial\theta_i}\right)\mathbf{E}\left(\frac{\partial\mathbf{V}}{\partial\theta_j}\right)\mathbf{E}\mathbf{X}\boldsymbol{\beta}$$
$$-\mathrm{tr}\left[\mathbf{E}\left(\frac{\partial\mathbf{V}}{\partial\theta_i}\right)\mathbf{E}\left(\frac{\partial\mathbf{V}}{\partial\theta_j}\right)\mathbf{E}\mathbf{V}\right]$$
$$=-\frac{1}{2}\mathrm{tr}\left[\mathbf{E}\left(\frac{\partial\mathbf{V}}{\partial\theta_i}\right)\mathbf{E}\left(\frac{\partial\mathbf{V}}{\partial\theta_j}\right)\right].$$

Empirical BLUE and BLUP

17

This chapter presents exercises on empirical best linear unbiased estimation and empirical best linear unbiased prediction and provides solutions to those exercises.

▶ **Exercise 1** Show that, when it exists and is unique, the maximum likelihood estimator of $\boldsymbol{\theta}$ under a normal positive definite general mixed linear model is even and translation invariant. Show that the same result also holds for the REML estimator (when it exists and is unique).

Solution According to (16.12),

$$L_0(\boldsymbol{\theta}; \mathbf{y}) = -\frac{1}{2}\log|\mathbf{V}| - \frac{1}{2}\mathbf{y}^T[\mathbf{V}^{-1} - \mathbf{V}^{-1}\mathbf{X}(\mathbf{X}^T\mathbf{V}^{-1}\mathbf{X})^-\mathbf{X}^T\mathbf{V}^{-1}\mathbf{y}.$$

Thus,

$$\begin{aligned} L_0(\boldsymbol{\theta}; -\mathbf{y}) &= -\frac{1}{2}\log|\mathbf{V}| - \frac{1}{2}(-\mathbf{y})^T[\mathbf{V}^{-1} - \mathbf{V}^{-1}\mathbf{X}(\mathbf{X}^T\mathbf{V}^{-1}\mathbf{X})^-\mathbf{X}^T\mathbf{V}^{-1}(-\mathbf{y}) \\ &= -\frac{1}{2}\log|\mathbf{V}| - \frac{1}{2}\mathbf{y}^T[\mathbf{V}^{-1} - \mathbf{V}^{-1}\mathbf{X}(\mathbf{X}^T\mathbf{V}^{-1}\mathbf{X})^-\mathbf{X}^T\mathbf{V}^{-1}\mathbf{y} \\ &= L_0(\boldsymbol{\theta}; \mathbf{y}), \end{aligned}$$

which establishes that maximum likelihood estimator of $\boldsymbol{\theta}$, when it exists and is unique, is even. Furthermore, by Theorem 11.1.6a

$$[\mathbf{V}^{-1} - \mathbf{V}^{-1}\mathbf{X}(\mathbf{X}^T\mathbf{V}^{-1}\mathbf{X})^-\mathbf{X}^T\mathbf{V}^{-1}]\mathbf{X} = \mathbf{V}^{-1}\mathbf{X} - \mathbf{V}^{-1}\mathbf{X} = \mathbf{0},$$

D. L. Zimmerman, *Linear Model Theory*,
https://doi.org/10.1007/978-3-030-52074-8_17

which implies that, for any p-vector $\mathbf{c}$,

$$\begin{aligned} L_0(\boldsymbol{\theta}; \mathbf{y} + \mathbf{X}\mathbf{c}) &= -\frac{1}{2}\log|\mathbf{V}| - \frac{1}{2}(\mathbf{y} + \mathbf{X}\mathbf{c})^T[\mathbf{V}^{-1} - \mathbf{V}^{-1}\mathbf{X}(\mathbf{X}^T\mathbf{V}^{-1}\mathbf{X})^-\mathbf{X}^T\mathbf{V}^{-1}(\mathbf{y} + \mathbf{X}\mathbf{c}) \\ &= -\frac{1}{2}\log|\mathbf{V}| - \frac{1}{2}\mathbf{y}^T[\mathbf{V}^{-1} - \mathbf{V}^{-1}\mathbf{X}(\mathbf{X}^T\mathbf{V}^{-1}\mathbf{X})^-\mathbf{X}^T\mathbf{V}^{-1}\mathbf{y} \\ &= L_0(\boldsymbol{\theta}; \mathbf{y}). \end{aligned}$$

This establishes that the maximum likelihood estimator of $\boldsymbol{\theta}$, when it exists and is unique, is translation invariant. Since the REML log-likelihood function differs from the profile log-likelihood function by a term that is not functionally dependent on $\mathbf{y}$, the REML estimator of $\boldsymbol{\theta}$ (when it exists and is unique) likewise is even and translation invariant.

▶ **Exercise 2** Consider the problem of E-BLUP under the normal positive definite general mixed linear model considered in this chapter. Let $\tau = \mathbf{c}^T\boldsymbol{\beta} + u$ be a predictable linear combination; let $\hat{\boldsymbol{\theta}}$ be an even, translation-invariant estimator of $\boldsymbol{\theta}$; and let $\tilde{\tau}$ and $\tilde{\tilde{\tau}}$ be the BLUP and E-BLUP of τ. Assume that the joint distribution of $\mathbf{e}$ and u is multivariate normal and that the prediction error variance of $\tilde{\tilde{\tau}}$ exists.

(a) Expand $[\tilde{\tilde{\tau}} - \tilde{\tau}(\boldsymbol{\theta})]^2$ in a Taylor series (in $\hat{\boldsymbol{\theta}}$) about $\boldsymbol{\theta}$ to obtain the second-order approximation

$$[\tilde{\tilde{\tau}} - \tilde{\tau}(\boldsymbol{\theta})]^2 \doteq [(\partial\tilde{\tau}(\boldsymbol{\theta})/\partial\boldsymbol{\theta})^T(\hat{\boldsymbol{\theta}} - \boldsymbol{\theta})]^2.$$

(b) Use the result of part (a) to argue that an approximation to the prediction error variance of the E-BLUP is

$$\mathrm{var}(\tilde{\tilde{\tau}} - \tau) \doteq \mathrm{var}(\tilde{\tau} - \tau) + \mathrm{tr}[\mathbf{A}(\boldsymbol{\theta})\mathbf{B}(\boldsymbol{\theta})],$$

where $\mathbf{A}(\boldsymbol{\theta}) = \mathrm{var}[\partial\tilde{\tau}(\boldsymbol{\theta})/\partial\boldsymbol{\theta}]$ and $\mathbf{B}(\boldsymbol{\theta}) = \mathrm{E}[(\hat{\boldsymbol{\theta}} - \boldsymbol{\theta})(\hat{\boldsymbol{\theta}} - \boldsymbol{\theta})^T]$.

Solution

(a)

$$\tilde{\tilde{\tau}} = \tilde{\tau}(\boldsymbol{\theta}) + \left(\frac{\partial\tilde{\tau}(\boldsymbol{\theta})}{\partial\boldsymbol{\theta}}\right)^T(\hat{\boldsymbol{\theta}} - \boldsymbol{\theta}) + O(\|\hat{\boldsymbol{\theta}} - \boldsymbol{\theta}\|^2),$$

implying that

$$[\tilde{\tilde{\tau}} - \tilde{\tau}(\boldsymbol{\theta})]^2 \doteq \left[\left(\frac{\partial\tilde{\tau}(\boldsymbol{\theta})}{\partial\boldsymbol{\theta}}\right)^T(\hat{\boldsymbol{\theta}} - \boldsymbol{\theta})\right]^2.$$

(b) The proof of Theorem 17.1.3 established that $\text{var}(\tilde{\hat{\tau}} - \tau) = \text{var}(\tilde{\tau} - \tau) + \text{var}(\tilde{\hat{\tau}} - \tilde{\tau})$, so it suffices to show that $\text{var}(\tilde{\hat{\tau}} - \tilde{\tau}) \doteq \text{tr}(\mathbf{A}(\boldsymbol{\theta})\mathbf{B}(\boldsymbol{\theta})]$, where $\mathbf{A}(\boldsymbol{\theta}) = \text{var}[\partial\tilde{\tau}(\boldsymbol{\theta})/\partial\boldsymbol{\theta})$ and $\mathbf{B}(\boldsymbol{\theta}) = \text{E}[(\hat{\boldsymbol{\theta}} - \boldsymbol{\theta})(\hat{\boldsymbol{\theta}} - \boldsymbol{\theta})^T]$. Define $\mathbf{a} = (a_i) = \frac{\partial\tilde{\tau}(\boldsymbol{\theta})}{\partial\boldsymbol{\theta}}$ and $\mathbf{b} = (b_i) = \hat{\boldsymbol{\theta}} - \boldsymbol{\theta}$, and observe that

$$\text{E}(\mathbf{a}) = \frac{\partial}{\partial\boldsymbol{\theta}}\text{E}[\tilde{\tau}(\boldsymbol{\theta})] = \frac{\partial}{\partial\boldsymbol{\theta}}(\mathbf{c}^T\boldsymbol{\beta}) = \mathbf{0}$$

(assuming that the expectation and differentiation can be interchanged), implying further that $\text{E}(\mathbf{a}\mathbf{a}^T) = \text{var}(\mathbf{a}) = \mathbf{A}(\boldsymbol{\theta})$. Then,

$$\begin{aligned}
\text{var}[\tilde{\hat{\tau}} - \tilde{\tau}(\boldsymbol{\theta})] &= \text{E}\{[\tilde{\hat{\tau}} - \tilde{\tau}(\boldsymbol{\theta})]^2\} \\
&\doteq \text{E}[(\mathbf{a}^T\mathbf{b})^2] \\
&= \text{E}\left(\sum_{i=1}^m a_i b_i \sum_{j=1}^m a_j b_j\right) \\
&= \sum_{i=1}^m \sum_{j=1}^m \text{E}(a_i a_j)\text{E}(b_i b_j) \\
&= \sum_{i=1}^m \sum_{j=1}^m [\mathbf{A}(\boldsymbol{\theta})]_{ij}[\mathbf{B}(\boldsymbol{\theta})]_{ij} \\
&= \sum_{i=1}^m \sum_{j=1}^m [\mathbf{A}(\boldsymbol{\theta})]_{ij}[\mathbf{B}(\boldsymbol{\theta})]_{ji} \\
&= \text{tr}[\mathbf{A}(\boldsymbol{\theta})\mathbf{B}(\boldsymbol{\theta})],
\end{aligned}$$

where the unbiasedness of the BLUP was used for the first equality, part (b) was used to obtain the approximate equality in the second line, and the independence of $\tilde{\tau}$ and $\hat{\boldsymbol{\theta}} - \boldsymbol{\theta}$ was used to obtain the fourth line.

Printed in the United States
by Baker & Taylor Publisher Services